Verocytotoxigenic *E. coli*

PUBLICATIONS IN FOOD SCIENCE AND NUTRITION

Books

VEROCYTOTOXIGENIC *E. coli*, G. Duffy, P. Garvey and D.A. McDowell
OPEN DATING OF FOODS, T.P. Labuza and L.M. Szybist
NITRITE CURING OF MEAT: *N*-NITROSAMINE PROBLEM, R.B. Pegg and F. Shahidi
DICTIONARY OF FLAVORS, D.A. DeRovira
FOOD SAFETY: THE IMPLICATIONS OF CHANGE, J.J. Sheridan *et al.*
FOOD FOR HEALTH IN THE PACIFIC RIM, J.R. Whitaker *et al.*
DAIRY FOODS SAFETY: 1995–1996, A COMPENDIUM, E.H. Marth
OLIVE OIL, SECOND EDITION, A.K. Kiritsakis
MULTIVARIATE DATA ANALYSIS, G.B. Dijksterhuis
NUTRACEUTICALS: DESIGNER FOODS III, P.A. Lachance
DESCRIPTIVE SENSORY ANALYSIS IN PRACTICE, M.C. Gacula, Jr.
APPETITE FOR LIFE: AN AUTOBIOGRAPHY, S.A. Goldblith
HACCP: MICROBIOLOGICAL SAFETY OF MEAT, J.J. Sheridan *et al.*
OF MICROBES AND MOLECULES: FOOD TECHNOLOGY AT M.I.T., S.A. Goldblith
MEAT PRESERVATION, R.G. Cassens
S.C. PRESCOTT, PIONEER FOOD TECHNOLOGIST, S.A. Goldblith
FOOD CONCEPTS AND PRODUCTS: JUST-IN-TIME DEVELOPMENT, H.R. Moskowitz
MICROWAVE FOODS: NEW PRODUCT DEVELOPMENT, R.V. Decareau
DESIGN AND ANALYSIS OF SENSORY OPTIMIZATION, M.C. Gacula, Jr.
NUTRIENT ADDITIONS TO FOOD, J.C. Bauernfeind and P.A. Lachance
NITRITE-CURED MEAT, R.G. Cassens
POTENTIAL FOR NUTRITIONAL MODULATION OF AGING, D.K. Ingram *et al.*
CONTROLLED/MODIFIED ATMOSPHERE/VACUUM PACKAGING, A.L. Brody
NUTRITIONAL STATUS ASSESSMENT OF THE INDIVIDUAL, G.E. Livingston
QUALITY ASSURANCE OF FOODS, J.E. Stauffer
SCIENCE OF MEAT & MEAT PRODUCTS, 3RD ED., J.F. Price and B.S. Schweigert
ROLE OF CHEMISTRY IN PROCESSED FOODS, O.R. Fennema *et al.*
NEW DIRECTIONS FOR PRODUCT TESTING OF FOODS, H.R. Moskowitz
PRODUCT DEVELOPMENT & DIETARY GUIDELINES, G.E. Livingston, *et al.*
SHELF-LIFE DATING OF FOODS, T.P. Labuza
POSTHARVEST BIOLOGY AND BIOTECHNOLOGY, H.O. Hultin and M. Milner

Journals

JOURNAL OF FOOD LIPIDS, F. Shahidi
JOURNAL OF RAPID METHODS AND AUTOMATION IN MICROBIOLOGY, D.Y.C. Fung and M.C. Goldschmidt
JOURNAL OF MUSCLE FOODS, N.G. Marriott and G.J. Flick, Jr.
JOURNAL OF SENSORY STUDIES, M.C. Gacula, Jr.
FOODSERVICE RESEARCH INTERNATIONAL, C.A. Sawyer
JOURNAL OF FOOD BIOCHEMISTRY, N.F. Haard and B.K. Simpson
JOURNAL OF FOOD PROCESS ENGINEERING, D.R. Heldman and R.P. Singh
JOURNAL OF FOOD PROCESSING AND PRESERVATION, B.G. Swanson
JOURNAL OF FOOD QUALITY, J.J. Powers
JOURNAL OF FOOD SAFETY, T.J. Montville and D.G. Hoover
JOURNAL OF TEXTURE STUDIES, M.C. Bourne, T. van Vliet and V.N.M. Rao

Newsletter

FOOD, NUTRACEUTICALS AND NUTRITION, P.A. Lachance and M.C. Fisher

Verocytotoxigenic *E. coli*

Edited by

Geraldine Duffy, Ph.D

Food Safety Dept., Teagasc, The National Food Centre
Dunsinea, Castleknock, Dublin 15, Ireland

Patricia Garvey, Ph.D

Food Safety Dept., Teagasc, The National Food Centre
Dunsinea, Castleknock, Dublin 15, Ireland

David A. McDowell, Ph.D

Food Microbiology Research Unit
School of Applied Medical Sciences And Sport Studies
Faculty of Science, University of Ulster
Jordanstown, Newtownabbey BT37OQB Northern Ireland

FOOD & NUTRITION PRESS, INC.
TRUMBULL, CONNECTICUT 06611 USA

FOOD & NUTRITION PRESS, INC.

6527 Main Street
Trumbull, Connecticut 06611 USA

Library of Congress Control Number: 2001135366

ISBN: 0-917678-52-4

CONTRIBUTORS

CHRISTOPHER BAYLIS, Campden and Chorleywood Food Research Assoc., Chipping Campden, Gloucestershire GL55 6LD, UK

MARÍA ISABEL BERNÁRDEZ, Laboratorio de Referencia de *E. coli* (LREC), Departamento de Microbiología y Parasitología, Facultad de Veterinaria, Universidad de Santiago de Compostela, Campus de Lugo, 27002 Lugo, Spain

LOTHAR BEUTIN, Division of Emerging Bacterial Pathogens (P13), *Escherichia coli* Reference Lab., Robert Koch Institute, Nordufer 20, D13353 Berlin, Germany

MARTIN BITZAN, Department of Pediatrics, Wake Forest University School of Medicine and Baptist Medical Center, Medical Center Boulevard, Winston-Salem, North Carolina 27157-1081, USA

JESÚS E. BLANCO, Laboratorio de Referencia de E. coli (LREC), Departamento de Microbiología y Parasitología, Facultad de Veterinaria, Universidad de Santiago de Compostela, Campus de Lugo, 27002 Lugo, Spain

JORGE BLANCO, Laboratorio de Referencia de *E. coli* (LREC), Departamento de Microbiología y Parasitología, Facultad de Veterinaria, Universidad de Santiago de Compostela, Campus de Lugo, 27002 Lugo, Spain.

MIGUEL BLANCO, Laboratorio de Referencia de *E. coli* (LREC), Departamento de Microbiología y Parasitología, Facultad de Veterinaria, Universidad de Santiago de Compostela, Campus de Lugo, 27002 Lugo, Spain

ELISABETH BORCH, SIK Swedish Institute for Food and Biotechnology, Ideon, SE 223 70 Lund, Sweden

CATHY A. BROWN, Athens Diagnostic Laboratory, College of Veterinary Medicine, University of Georgia, Athens, Georgia, USA, 30602

ALFREDO CAPRIOLI, Istituto Superiore di Sanità, Laboratorio di Medicina Veterinaria, Viale Regina Elena 299, 00161 Rome, Italy

PETER A. CHAPMAN, Public Health Laboratory, Herries Road, Sheffield S5 7BQ, United Kingdom

BERNARD CHINA, Faculté de Médecine Vétérinaire, Université de Liège, Sart Tilman, Bât B43a, B-4000 Liège, Belgium

JOHN COIA, Scottish *E. coli* O157 Reference Laboratory, Dept. of Clinical Microbiology, Western General Hospital, Edinburgh EH4 2XU, Scotland

M.E. COLEMAN, Food Safety and Inspection Service, Washington, D.C.

ENNE DE BOER, Inspectorate for Health Protection, P.O. Box 202, 7200 AE Zutphen, The Netherlands

MICHAEL P. DOYLE, Center for Food Safety, Department of Food Science and Technology, Georgia Experiment Station, University of Georgia, Griffin, Georgia, USA, 30223

GERALDINE DUFFY, Teagasc, The National Food Centre, Dunsinea, Castleknock, Dublin 15, Ireland

S. DUNDAS, Monklands Hospital, Airdrie, Lanarkshire L6 0JS, United Kingdom

RAYMOND ELLARD, Food Safety Authority of Ireland, Abbey Court, Lower Abbey St., Dublin 1, Ireland.

PATRICIA GARVEY, Teagasc, The National Food Centre, Dunsinea, Castleknock, Dublin 15, Ireland

FRÉDÉRIC GOFFAUX, Faculté de Médecine Vétérinaire, Université de Liège, Sart Tilman, Bât B43a, B-4000 Liège, Belgium

ENRIQUE A. GONZÁLEZ, Laboratorio de Referencia de *E. coli* (LREC), Departamento de Microbiología y Parasitología, Facultad de Veterinaria, Universidad de Santiago de Compostela, Campus de Lugo, 27002 Lugo, Spain

S. GORIETTI, Laboratorio di Epidemiologia e Biostatistica, Istituto Superiore di Sanità, Rome, Italy

CARLTON L. GYLES, Dept. of Pathobiology, Ontario Veterinary College, University of Guelph, Guelph, Ontario, Canada

BARRY G. HARMON, Department of Pathology, College of Veterinary Medicine, University of Georgia, Athens, Georgia, USA, 30602

ANNET HEUVELINK, Inspectorate for Health Protection, P.O. Box 202, 7200 AE Zutphen, The Netherlands

HARMEN HOFSTRA, TNO Food and Nutrition Research Institute, P.O. Box 360, 3700 AJ Zeist, The Netherlands

HELGE KARCH, Institut für Hygiene und Mikrobiologie der Universität Würzburg, Josef-Schneider-Str. 2, D-97080 Würzburg, Germany

JACQUES MAINIL, Chaire de Bactériologie et de Pathologie des Maladies Bactériennes, Faculté de Médecine Vétérinaire, Université de Liège, Sart Tilman, Bât B43a, B-4000 Liège, Belgium

PETER J. MCCLURE, Microbiology Unit, Unilever Research, Colworth Laboratory, Colworth House, Sharnbrook, Bedford MK44 1LQ, United Kingdom

DAVID A. MCDOWELL, Food Microbiology Research Unit, School of Applied Medical Sciences and Sport Studies, Faculty of Science, University of Ulster, Jordanstown, Newtownabbey BT37OQB, Northern Ireland

AZUCENA MORA, Laboratorio de Referencia de *E. coli* (LREC), Departamento de Microbiología y Parasitología, Facultad de Veterinaria, Universidad de Santiago de Compostela, Campus de Lugo, 27002 Lugo, Spain

C.A. NARROD, Economic Research Service, Washington, D.C.

EVA NERBRINK, Cerealia R&D, Malmö, Sweden

TRULS NESBAKKEN, Norwegian Meat Research Centre, PO Box 396 Økern, 0513 Oslo, Norway

HILDE NISSEN, MATFORSK, Norwegian Food Research Institute, Osloveien 1, 1430 Ås, Norway
CONOR P. O'LOUGHLIN, Department of Industrial Microbiology, University College Dublin, Belfield, Dublin 4, Ireland
MARÍA PILAR ALONSO, Laboratorio de Referencia de E. coli (LREC), Departamento de Microbiología y Parasitología, Facultad de Veterinaria, Universidad de Santiago de Compostela, Campus de Lugo, 27002 Lugo, Spain
M. POWELL, Office of Risk Assessment and Cost Benefit Analysis, Washington, D.C.
THOMAS M.S. REID, Dept. of Medical Microbiology, Aberdeen Royal Hospital, Foresterhill, Aberdeen AB25 2ZN, United Kingdom
T. ROBERTS, Economic Research Service, Washington, D.C.
FLEMMING SCHEUTZ, The International *Escherichia* and *Klebsiella* Centre (WHO), Department of Gastrointestinal Infections, Statens Serum Institut, Copenhagen S, Denmark
W.D. SCHLOSSER, Food Safety and Inspection Service, College Station, Texas
HERBERT SCHMIDT, Institut für Hygiene und Mikrobiologie der Universität Würzburg, Josef-Schneider-Str. 2, D-97080 Würzburg, Germany
HEATHER J. SHEELEY, Head of Safety, Centre for Applied Microbiology and Research (CAMR), Porton Down, Salisbury, Wiltshire, SP4 OJG, United Kingdom
JAMES J. SHERIDAN, Head, Dept. of Food Safety, The National Food Centre, Dunsinea, Castleknock, Dublin 15, Ireland
HENRY R. SMITH, Laboratory of Enteric Pathogens, Central Public Health Laboratory, London, England
PHILIPPE STORDEUR, Faculté de Médecine Vétérinaire, Université de Liège, Sart Tilman, Bât B43a, B-4000 Liège, Belgium
FIONA THOMSON-CARTER, Department of Medical Microbiology, Grampian University Hospitals Trust, Foresterhill, Aberdeen AB25 2ZN, Scotland
W.T. ANDREW TODD, Monklands Hospital, Airdrie, Lanarkshire L6 0JS, United Kingdom
MATS TÖRNQUIST, Swedish Animal Health Service, Kävlinge, Sweden
ALBERTO E. TOZZI, Laboratorio di Epidemiologia e Biostatistica, Istituto Superiore di Sanità, Rome, Italy
MARY E. UPTON, Department of Industrial Microbiology, University College Dublin, Belfield, Dublin 4, Ireland
IVAR VÅGSHOLM, National Veterinary Institute, Swedish Institute for Infectious Disease Control, Uppsala, Sweden
YNGVILD WASTESON, The Norwegian School of Veterinary Science, PO Box 8146 Dep., Oslo 0033, Norway
RICHARD C. WHITING, U.S. Food and Drug Admin., Center for Food Safety and Applied Nutrition, Washington, D.C.

TONG ZHAO, Center for Food Safety, Department of Food Science and Technology, Georgia Experiment Station, University of Georgia, Griffin, Georgia, USA, 30223

PREFACE

Verocytotoxin producing *Escherichia coli* (VTEC), and in particular, strains of serogroup O157, have emerged as significant pathogens causing a range of severe and potentially fatal illnesses. The European Union has recognised the threat posed by *E. coli* O157:H7 and the need to devise control strategies based on an understanding of VTEC pathogenicity, transmission, survival and growth. It also acknowledges the importance of informing farmers, veterinarians, food producers and health authorities so that each of these groups can act appropriately to reduce the overall hazards posed by these organisms. To contribute to the development and dissemination of effective control strategies, the European Commission funded a Concerted Action Project "A European study on animal, food, and biomedical aspects of verocytotoxigenic *E. coli* including serotype O157:H7, an emerging pathogen" (CT98-3935) within the Agriculture and Agro-industry Framework IV Research Programme (1998-2001). This book, compiled under the auspices of the above project, integrates contributions from project participants and invited contributors, to provide a comprehensive overview of the current state of research on VTEC. It should be of interest to current workers in this area, and those seeking an effective introduction to research on this important pathogen.

This book, containing contributions from the many and diverse research disciplines currently being brought to bear on VTEC, amply demonstrates the success of the EU project in promoting collaboration among scientists from veterinary, food and biomedical backgrounds from 31 participant groups in 12 European countries. It also includes invited contributions from a wider circle of international research leaders in VTEC research, increasing the benefits to be gained from effective communication of the latest research findings, and the means of their application, to end users working in diverse areas of food safety and public health. The focus provided by the project, and the format and content of this book will enable information on the current state of research and its implications to flow to the widest possible audience, preventing duplication of research efforts, and directing future research in this area. As an effective and widely accessible overview, presenting appropriate dissemination of recommendations for dealing with VTEC in Europe, this book should provide a valuable resource for the many disciplines engaged in combating the public health challenges associated with VTEC.

The nomenclature of verotoxin-producing *E. coli* is a complex issue which is still in a state of flux and there are variations in the nomenclature used in different chapters in the book. A table summarising the nomenclature terms for verocytotoxins is provided in the appendix of this book.

The editors would like to thank all those who have contributed chapters to this text, and/or contributed in other ways to the success of the overall project, encompassing 5 international conferences, workshops and related activities on methodology, survival and growth characteristics, virulence and pathogenicity factors, epidemiology, and measures for the control of VTEC. Conference proceedings from these meetings have been published and are available on request from the project coordinator, or can be downloaded from the project web site http://www.research.teagasc.ie/vteceurope. In addition, a series of technical booklets, likely to be of particular interest to food industry and public health surveillance personnel, are available. The considerable management and coordination activities necessary for the delivery of this project and derived publications, including this book, were provided by Teagasc, The National Food Centre, Dublin, Ireland.

We gratefully acknowledge the generosity of the European Commission Framework IV programme for funding the project (CT98 3935) through which this book has been published.

GERALDINE DUFFY
PATRICIA GARVEY
DAVID McDOWELL

CONTENTS

CHAPTER PAGE

1. EMERGENCE OF VEROCYTOTOXIGENIC *E. COLI*, *Geraldine Duffy, Patricia Garvey and D.A. McDowell* 1

METHODOLOGY

2. DETECTION OF VEROCYTOTOXIN-PRODUCING *ESCHERICHIA COLI* O157 ON THE FARM AND AT THE ABATTOIR, *Peter A. Chapman* 11
3. CLINICAL DETECTION OF VEROCYTOTOXIN-PRODUCING *E. COLI* (VTEC), *Flemming Scheutz, Lothar Beutin and Henry R. Smith* 25
4. PRACTICAL CONSIDERATIONS AND DIFFICULTIES ASSOCIATED WITH THE DETECTION OF VEROCYTOTOXIN-PRODUCING *ESCHERICHIA COLI* (VTEC) IN FOODS, *Christopher Baylis, Annet Heuvelink, Harmen Hofstra and Enne de Boer* 57
5. GENERAL RECOVERY, CHARACTERISATION AND TYPING PROTOCOLS FOR VTEC, *Fiona Thomson-Carter* 91

EPIDEMIOLOGY

6. EPIDEMIOLOGY OF VEROCYTOTOXIGENIC *ESCHERICHIA COLI* (VTEC) IN RUMINANTS, *Jorge Blanco, Miguel Blanco, Jesús E. Blanco, Azucena Mora, María Pilar Alonso, Enrique A. González and María Isabel Bernárdez* 113
7. EPIDEMIOLOGY OF VTEC IN NON-RUMINANT ANIMALS, *Yngvild Wasteson* 149
8. EPIDEMIOLOGY OF HUMAN INFECTIONS BY *ESCHERICHIA COLI* O157 AND OTHER VEROCYTOTOXIN-PRODUCING *E.COLI*, *A.E. Tozzi, S. Gorietti, A. Caprioli* 161
9. FOODS AS VEHICLES OF VTEC INFECTION, *Enne de Boer and Annet Heuvelink* 181
10. A CASE STUDY OF CHEESE ASSOCIATED *E. COLI* O157 OUTBREAKS IN SCOTLAND, *T.M.S. Reid* 201

PATHOGENIC ASPECTS OF VTEC INFECTION

11. PATHOGENIC ASPECTS OF VTEC INFECTIONS IN RUMINANTS, *Frédéric Goffaux, Bernard China, Philippe Stordeur and Jacques Mainil* 213
12. PATHOGENIC ASPECTS OF VTEC INFECTION IN NON-RUMINANT ANIMALS, *Carlton L. Gyles* 227
13. PATHOGENIC ASPECTS OF STEC INFECTIONS IN HUMANS, *Herbert Schmidt, Martin Bitzan and Helge Karch* 241

SURVIVAL AND GROWTH

14. HEALTHY ANIMALS AS CARRIERS OF STEC, *Cathy A. Brown, Barry G. Harmon, Tong Zhao, and Michael P. Doyle* 263
15. SURVIVAL AND GROWTH OF VTEC IN THE ENVIRONMENT, *D.A. McDowell and J.J. Sheridan* 279
16. SURVIVAL AND GROWTH OF VEROCYTOTOXIGENIC *E. COLI* IN FOODS, *Geraldine Duffy and Patricia Garvey* 305
17. PREDICTING THE FATE OF VEROCYTOTOXIGENIC *ESCHERICHIA COLI* IN FOODS, *Peter J. McClure* 323

CONTROL MEASURES FOR VTEC

18. RISK ASSESSMENT IN THE CONTROL OF VTEC, *R.C. Whiting, M.E. Coleman, C.A. Narrod, M. Powell, T. Roberts and W.D. Schlosser* 337
19. FARM MANAGEMENT PRACTICES: A SWEDISH CASE STUDY, *Elisabeth Borch, Eva Nerbrink, Ivar Vågsholm and Mats Törnquist* 353
20. CONTROL OF VTEC IN THE MEAT INDUSTRY, *Hilde Nissen and Truls Nesbakken* 367
21. CONTROL OF VTEC IN NON-MEAT FOOD PRODUCTS, *C.P. O'Loughlin and M.E. Upton* 381
22. CLINICAL MANAGEMENT OF *E. COLI* O157 INFECTION, *W.T.A. Todd, S. Dundas and J. Coia* 393
23. VEROCYTOTOXIGENIC *E. COLI:* SAFE LABORATORY PRACTICES, *Heather J. Sheeley* 421
24. VEROCYTOTOXIGENIC *E. COLI* — LEGAL ASPECTS, *Raymond Ellard* 435

APPENDIX 447
INDEX 453

CHAPTER 1

EMERGENCE OF VEROCYTOTOXIGENIC *E. COLI*

GERALDINE DUFFY
and
PATRICIA GARVEY

Teagasc, The National Food Centre
Dunsinea, Castleknock
Dublin 15, Ireland

AND

DAVID A. MCDOWELL

Food Microbiology Research Unit
School of Applied Medical Sciences and Sport Studies
Faculty of Science
University of Ulster
Jordanstown, Newtownabbey BT37OQB, Northern Ireland

In recent years, *E. coli* O157:H7 has achieved considerable notoriety status, seizing public, government and scientific attention. Such a reputation may be justified, considering the abilities of this organism to survive in many environments, including some widely used preservation systems, its low infective dose, the nature of the populations most susceptible to its attack, and the severity and long term nature of its clinical consequences among such groups. The importance of this organism, and the extent to which it is widely well recognised within the public psyche, has prompted considerable concerns among consumers and legislature alike, leading to demands for effective action at all relevant points of the food chain and beyond. Such concerns have spawned considerable inter-sectoral, interagency, and international collaboration, leading to the acquisition of large amounts of valuable information on the nature of this notorious pathogen and the increasing application of this information in the development of effective means to prevent or ameliorate human infection with *E. coli* O157:H7. Thus recent years have seen considerable progress in relation to methodologies, survival characteristics, pathogenic and virulence traits, control measures and epidemiology of VTEC. Much of this progress is usefully reviewed and set in context within other chapters of this book.

There may, however, be a series of more general lessons to be learned from such studies, and from the wider appreciation of *E. coli* O157:H7 as a model emerging pathogen, rather than as a unique adversary. If such a virulent pathogen can emerge from what is arguably the best known and intensively investigated group of human and animal commensals (14) — the most extensively "domesticated" and investigated "genetic test bed" — it is perhaps time to reconsider our relationships with bacteria.

Throughout history, pandemic bacterial infections have modulated the course of human and animal history and evolution, and despite the brief false dawn presented by antibiotic therapy around the middle of the 20th century, it is now increasingly clear that this pattern is likely to continue. If anything, the declining efficacy of antibiotic therapy is a "side show" to the relentless emergence and re-emergence of a series of infective agents capable of causing debilitating and/or fatal conditions in humans, mammals and other life forms (4).

There are nearly as many bacteria in the human intestine as there are cells in the human body, and it is clear that bacteria instigate and modulate many aspects of human physiology, particularly aspects of gut physiology. It is clear that the generally commensal human gut flora, and its rarer pathogenic derivatives, including *E. coli* O157:H7 has been adapting to, competing within, and modulating the gut and gut flora for millennia. Such extended co-evolution is normally considered to favour commensalism, mutualism and/or nonlethal parasitism, as host death is an unfavourable outcome for the parasite (9). Thus, the emergence of life-threatening pathogenic strains or clones could be viewed as aberrant, or at the very least, not to the longer term advantage of the infecting clone. Examination of the emergence of such organisms as *E. coli* O157:H7 may well provide a number of important insights into wider host-parasite interactions, provide pointers as to the mechanisms and future implications of such evolving relationships, and suggest ways of gaining advantage in our interactions with current and emerging pathogens.

Escherichia coli O157:H7 was first implicated in infectious disease in 1982 (17) and is now recognised as a major cause of food/water borne illness in the developed world. It is a newly evolved serotype of *E. coli* which has become pathogenic through the acquisition of a number of virulence factors. But what were the processes which provided this clone with such a strong set of "trump card" characteristics, capable on occasion of avoiding or negating the human defence system and what can this process tell us about the possible emergence of similar pathogens in the future?

Genetic analysis has shown that *E. coli* O157:H7 is clonally distinct from other verocytotoxin (VT)-producing serotypes but closely related to serotype O55:H7, a non VT-producing clone associated with infantile diarrhoea. Serotype O55:H7 has some pathogenic characteristics, i.e., it has the intimin gene, but most strains do not usually posses the EPEC plasmid. Serotype O55:H7 is reported to have acquired the capability for producing VT and enterohaemolysin via horizontal genetic transfer from other pathogens (23). Acquisition of a new serogroup antigen (O157) led to the emergence of a new and highly virulent pathogen (*E. coli* O157:H7). It has been reported by Bilge *et al.* (2) that the acquisition of the O157 antigen resulted from a lateral transfer of an rbf region containing the *rfbE* gene. Such exchanges, in the microbial equivalent of a molecular "car boot sale" starkly demonstrate the fluidity of exchange of genetic materials within and beyond the procaryotic world, and confirm the inadvisability of viewing groups of bacteria as "species" within the classical meaning of that term. The emergence of *E. coli* O157:H7 is a clear and unfortunate demonstration that bacterial "species" should be viewed in temporal and temporary terms, i.e., as sets of associated genes, gaining and losing individual characteristics in response to, as well as independently of, environmental stimuli. Perhaps a more accurate analogy is of a football team, where "star" players are bought, sold, traded, benched, and/or dropped within overall club activities! Thus we should not be surprised if, in the future, other currently commensal or opportunist organisms put together a "winning" team, and emerge suddenly and unexpectedly into the "premier league" of human pathogens.

E. coli O157:H7 is an unusual pathogen in terms of the severity of disease which it causes. The traditional view of evolution among pathogens was that as they evolve, pathogenicity/virulence decreased so as to ensure survival of the host population. More recent models, however, have disregarded the importance of the relationship between host and pathogen and suggest that the evolution of virulence is dependent on the relationship between the parameters of infection and the transmission process. For example, it has been suggested that the induction of diarrhoea by an enteric pathogen may increase the probability of transmission to new susceptible hosts (9). More recent models, however, paid less attention to the "endgame" of infection, i.e., the abilities of the pathogen to evade or overcome host defence, and places more emphasis on the wider parameters of the cycling of pathogen from the host into the environment, survival and transmission within the environment and access to new host systems. For example, it has been suggested that the induction of diarrhoea by an enteric pathogen may increase the probability of transmission to new susceptible hosts (9). Similarly, increased evolutionary durability and/or ability to modulate metabolism in response to environmental signals allow food/water borne pathogens to successfully survive in the distinctly different environs of the external environment (water, soil, etc.), food production and the host gut. Some of these evolutionary traits can be seen in *E. coli* O157:H7. The initial site of attachment for this pathogen in cattle is in the rumen (3). The extent and results of fermentation in the rumen present an acid environment, applying a selective pressure for the development of acid tolerance in *E. coli* O157:H7 (1, 5). In wider environmental terms, acid rain derived from air pollution with sulphur dioxide and oxides of nitrogen may have lowered the pH of many environments (water, soil, etc.) creating selective pressures which favours the survival of acid tolerant bacteria. Such acid tolerance will, however, also enhance the survival of pathogens in low pH foods, and will increase the numbers of organisms surviving host defences (gastric acid), effectively reducing the infective dose necessary to cause disease (23). The continuing impact of such selective pressures may lead to the emergence of other acid tolerant bacteria with enhanced resistance characteristics and virulence potential.

While most emphasis on the genetic mobility underlying the emergence of *E. coli* O157:H7 has correctly focused on its acquisition of a highly effective set of pathogenic/virulence characteristics, some other characteristics have been lost. One important step on the evolutionary process from *E. coli* O55:H7 to *E. coli* O157:H7 involved loss of the abilities to produce the enzyme ß-glucuronidase (GUD), and to ferment sorbitol. These two phenotypic characteristics have been exploited in the development of selective agars for detection of this serotype.

It is important to recognise *E. coli* O157:H7 as the current manifestation of a set of genes, and that the genetic processes which led to this particular format are continuing. Such plasticity and mobility is clearly demonstrated by the verocytotoxins of VTEC. These toxins, important because of their effects in inhibiting protein synthesis within eucaryotic cells, are already known to occur in a number of forms, i.e., VT1, VT2 and VT2 variants, and other forms, perhaps with significantly different characteristics will continue to emerge in the future. Such developments have significant implications for the future detection, recognition and remediation of verocytotoxins in *E. coli* O157:H7 and related strains. As well as such plasticity within clones, it is important to recognise the impact of horizontal evolution in the mobility of *vt* genes. These toxins are encoded by lambda-like phage and under laboratory conditions they have been transferred to non-toxigenic strains. These VT-encoding phages are potentially capable of disseminating the ability to produce toxin to other *E. coli* strains and indeed to other bacterial species as evidenced by the detection of *vt*2 in *Citrobacter fruendii* strains isolated from diarrhoeal samples (19). Such distribution of pathogenic genes, presents a potent means for the sudden and probably unexpected emergence of newly pathogenic bacteria. Thus the emergence of *E. coli* O157:H7 may be clinically unfortunate, but is not unusual. Its significance is that it is one of the first demonstrations of the wider implications of gene evolution and horizontal gene among bacteria, reinforcing the need for greater understanding of the patterns of development and movement of such materials, to enable effective interventions and therapies.

While VT-encoding phages have been induced *in vitro* from a number of VTEC strains and the induced phages used to infect other *E. coli* (18), the conditions for phage and other virulence factor transmission *in vivo* have yet to be established. In general, phage induction can be triggered by various forms of stress, such as exposure to UV light or chemical compounds such as antibiotics, conditions which induce the SOS response. It is possible that many of the processes that are routinely used to kill vegetative bacterial cells inadvertently promote phage induction and release. One procedure under investigation for its potential role in the dissemination of *vt* genes, is the use of sub-therapeutic levels of antibiotics in animal production (7). With further research into the conditions that trigger the processes of horizontal gene transfer, it may be possible to minimise the emergence of new pathogens through the development of intervention strategies that diminish this risk.

The location of the toxin genes within bacteriophage genomes has proven unfortunate in another respect. In some instances, the expression of the verocytotoxin genes have been shown to be linked to late phage gene expression and thus to the induction of the lytic cycle (22). The administration of antibiotic therapy has sometimes exacerbated patient symptoms through the induction of the phage lytic cycle and the concomitant increase in phage and toxin gene expression (21).

The location of virulence associated factors on mobile genetic elements has implications for the survival and persistence of VTEC strains in the environment. It has been documented that phages in general can survive harsh conditions that are capable of eliminating bacterial populations (10). VT-encoding phage have specifically been shown to be more resistant to exposure to environmental conditions, and to chlorination and pasteurisation, than bacterial cells (12). Bacteriophage are also more efficient vectors for DNA transfer than conjugative plasmids as the process does not require intimate contact between bacterial donor and receptor cells. Thus, DNA important to a population can be preserved until a host for lysogenic conversion is reintroduced in an environmental niche (10).

The emergence of *E. coli* O157 and other VTEC as important agents of disease in the past twenty years has caused a re-evaluation of our view of pathogens. Horizontal transmission of virulence factors has played a crucial role in the evolution of these strains. It has long been known that the primary known virulence factors of *E. coli* O157:H7 are associated with transferable DNA elements, and this was recently reaffirmed with the publication of the entire sequence of the chromosome of an *E. coli* O157:H7 strain (15). The accumulation of virulence factors (phage-encoded *vt* genes, pathogenicity island-encoded intimin gene, plasmid-encoded enterohaemolysin) through their acquisition on mobile genetic elements has facilitated a very rapid form of evolution. The ability to acquire such virulence genes may result from increased mutation rates and enhanced recombination abilities (11). LeClerc *et al.* (8) reported that 1% of O157:H7 strains had spontaneous rates of mutation that were 1000 fold higher those of typical *E. coli*. This ability of *E. coli* O157:H7 to hypermutate may even suggest that the pathogen could acquire new factors that will render it even more virulent and/or persistent.

Comparison of the *E. coli* K12 and *E. coli* O157:H7 genome sequences has also identified numerous other strain specific regions of the *E. coli* O157 genome, encompassing up to a quarter of the genome (15). In these strain specific gene clusters, there are many examples of genes encoding candidate virulence factors and alternative metabolic capacities. Codon usage and base composition analysis, and the identification of remnants of prophages and other mobility elements, again demonstrated extensive genetic exchange, and confirm that the extent of horizontal gene transfer and recombination is far greater than was anticipated. The discovery of these additional DNA segments opens up new avenues of research to investigate the possible roles of these factors in the virulence of *E. coli* O157:H7.

E. coli O157:H7 may well be a striking case of such evolutionary changes, but it is not the sole example. Phylogenetic analysis, using sequence data for seven housekeeping genes and for the genes for the major virulence factors of enterohaemorrhagic *E. coli* (EHEC), has demonstrated that *E. coli* O157:H7 and non-O157 VTEC descended from old lineages which acquired similar virulence factors in parallel (16). These authors theorised that many virulence factors had been gained and lost over time in different lineages of pathogenic *E. coli* and reported evidence of recent acquisition of *vt* genes and the EHEC plasmid, whereas appropriation of the locus of enterocyte effacement (LEE) occurred further in the past. The authors concluded that natural selection favoured an ordered acquisition of genes and the progressive build-up of molecular mechanisms that increased virulence. Thus, the acquisition of a similar collection of virulence factors has permitted a diverse group of *E. coli*, with differing metabolic capacities and environmental tolerances, to produce similar disease when introduced into the human population.

Non-O157 VTEC, in particular serogroups O26, O111, O103 and O145, are increasingly linked to human illness. These serogroups display considerable variation with respect to their complement of virulence factors (20). Variations in gene subtypes, e.g., intimin and *vt* genes, integration sites for the LEE pathogenicity island, and in plasmid-encoded virulence factor complement, have been identified (13, 20). In some non-O157 VTEC, an additional virulence factor termed high pathogenicity island (HPI) has been identified which was probably also disseminated in clonal VTEC subgroups by horizontal gene transfer (6). Thus it has become clear that VTEC represent a heterogeneous group of strains containing a mosaic pattern of virulence factors. Such a conclusion reinforces the view that ongoing and frequent genetic shuffling and sharing is a routine element of microbial life, and that such processes will continue to generate new or significantly enhanced/modified pathogens in the future.

It is clear that, within the VTEC serogroups, and indeed in the wider group of pathogenic *E. coli*, genetic transfer and evolution is still ongoing. Examination of surveillance data has revealed the chronological emergence and decline of particular clonal lineages, as the acquisition or rearrangement of genetic material gives rise to progressively more successful clones. For example, there are several instances of upsurges in particular O157 phage types which may reflect lysogenic conversion by new phages or genetic rearrangement between different prophages resident on the chromosome. Furthermore, there has been a dramatic rise in the number of reported cases of non-O157 VTEC infection in recent years. While this may derive from an increasing awareness of the role of non-O157 strains in disease and the development of new methods for their detection, it may also reflect the ongoing molecular evolution of such strains by horizontal gene transfer.

In summary, therefore, it is clear that there are important lessons which can and should be learned from the emergence of *E. coli* O157:H7 as an human pathogen. While much attention has been focused on the specific problems associated with the current symptoms and prognosis of the infections this organism can cause, particularly in relation to at risk and immunocompromised groups, it is important to set the challenges posed by this organism in proper context. The emergence of this pathogen is not a unique, isolated or unlikely to be repeated occurrence. It is more likely to be the first well recognised and investigated representative in an ongoing series of new or significantly modified pathogens which will continue to impact on human health. The particular severity of *E. coli* O157:H7 infection has galvanised research activity in wider areas of microbial ecology, horizontal gene transfer, and host pathogen interaction and communication, and much of this research will underpin work in the prevention and/or amelioration of infection caused by this particular pathogen. However, in more general terms, many of the advances achieved from such studies should have wider and more strategic application in enabling effective and efficient responses to other, as yet unformed or unidentified pathogens which most surely will continue to emerge to exploit human hosts as one particular aspect of the continuing evolutionary interactions between bacteria and their environment.

REFERENCES

1. ARCHER, D.L. 1996. Preservation microbiology and safety: Evidence that stress enhances virulence and triggers adaptive mutations. Trends Food Sci. and Technol. *7*, 91–95.

2. BILGE, S.S., VARY, J.J.C., DOWELL, S.F. and TARR, P.I. 1996. Role of the *Escherichia coli* O157:H7 O side chain in adherence and analysis of an rfb locus. *Infect.* Immun. *64*, 4795–4801.
3. DOYLE, M.P., ZHAO, T., MENG, T. and ZHAO, S. 1997. *Escherichia coli* O157:H7, *In* Food Microbiology: fundamentals and frontiers, (M.P. Doyle, L.R. Beuchat and T.J. Montville, eds.). ASM Press, Washington DC. pp. 171–191
4. FAUCI, A.S. 1998. New and re-emerging diseases: The importance of biomedical research. Emerg. Infect. Dis. *4*, 374–378.
5. FOSTER, J.W. 1995. Low pH adaptation and the acid tolerance response of *Salmonella typhimurium*. Critical Rev. Microbiol. *21*, 215–237.
6. KARCH, H. *et al.* 1999. A genomic island, termed high-pathogenicity island, is present in certain non-O157 Shiga toxin-producing *Escherichia coli* clonal lineages. Infect. Immun. *67*, 5994–6001.
7. KIMMITT, P.T., HARWOOD, C.R. and BARER, M.R. 2000. Toxin gene expression by shiga toxin-producing *Escherichia coli*: the role of antibiotics and the bacterial SOS response. Emerg. Infect. Dis. *6*, 458–465.
8. LECLERC, J.E., LI, B., PAYNE, W.L. and CEBULA, T.A. 1996. High mutation frequencies among *Escherichia coli* and *Salmonella* pathogens. Science *274*, 1208–1211.
9. LEVIN, B.R. 1996. The evolution and maintenance of virulence in microparasites. Emerg Infect Dis. *2*(2), 93–102. Review.
10. MIAO, A.E. and MILLER, S.I. 1999. Bacteriophages in the evolution of pathogen-host interactions. Proc. Natl. Acad. Sci. *96*, 9452–9454.
11. MOXON, E.R., RAINEY, P.B., NOWAK, M.A. and LENSKI, R.E. 1994. Adaptative evolution of highly mutable loci in pathogenic bacteria. Curr. Biol. *4*, 24–33.
12. MUNIESA, M., LUCENA, F. and JOFRE, J. 1999. Comparative Survival of Free Shiga Toxin 2-Encoding Phages and *Escherichia coli* Strains outside the Gut. Appl. Environ. Microbiol. *65*, 5615–5618.
13. OSWALD, E., MARCHES, O., MORABITO, S. and CAPRIOLI, A. 2001. Insertion and stability of the LEE pathogenicity island in EHEC and EPEC: analysis of tRNA hot spots as a useful tool in epidemiological studies. Conferences proceedings on "Epidemiology of Verocytotoxigenic *E. coli*" organised by an EU Concerted Action on VTEC (CT 98-3935). Malahide, Dublin. February 8-10th, 2001. ISBN 1 84170 147 5.
14. PARK, S., WOROBO, R.W. and DURST, R.A. 1999. *Escherichia coli* O157:H7 as an emerging food borne pathogen: a literature review. Critical Reviews in Food Science and Nutrition *39*, 481–502.
15. PERNA, N. *et al.* 2001. Genome sequence of enterohaemorrhagic *Escherichia coli* O157:H7. Nature *409*, 529–533.
16. REID, S.D., HERBELIN, C.J., BUMBAUGH, A.C., SELANDER, R.K. and WHITTAM, T.S. 2000. Parallel evolution of virulence in pathogenic *Escherichia coli. Nature 406*, 64–67.
17. RILEY, L.W. *et al.* 1983. Hemorrhagic colitis associated with a rare *Escherichia coli* serotype. *N. Engl. J. Med. 308*, 681–685.

18. SCHMIDT, H., KÖHLER, B., UNKMEIR, A., BIELASZEWSKA, M. and KARCH, H. 2000. The Role of Stx-Encoding Bacteriophages in Pathogenicity and Virulence of Shiga Toxin-producing *Escherichia coli* (STEC). Conferences proceedings on "Pathogenicity and Virulence of Verocytotoxigenic *E. coli*" organised by an EU Concerted Action on VTEC (CT 98-3935). Liege, Belgium. November 8–10th 1999.
19. SCHMIDT, H., MONTAG, M., BOCKEMUHL, J., HEESEMANN, J. and KARCH, H. 1993. Shiga like toxin II related cytotoxins in *Citrobacter freundii* strains from human and beef samples. *Infect. Immun. 61*, 534–545.
20. SCHMIDT, H., BIELASZEWSKA, M. and KARCH, H. 2001. Characterisation and typing of non O157 shiga toxin producing *Escherichia coli* by molecular methods. Conferences proceedings on "Epidemiology of Verocytotoxigenic *E. coli*" organised by an EU Concerted Action on VTEC (CT 98-3935). Malahide, Dublin. February 8-10th, 2001.
21. SHIOMI, M., TOGAWA, M., FUJITA, K. and MURAJA, R. 1999. Effects of early fluoroquinolones in hemorrhagic colitis due to *Escherichia coli* O157:H7. Pediatr. Int. *41*(2), 228–32.
22. WAGNER, P.L, NEELY, M.N., ZHANG, X., ACHESON, D.W., WALDOR, M.K. and FRIEDMAN, D.I. 2001. Role for a phage promoter in Shiga toxin 2 expression from a pathogenic *Escherichia coli* strain. J. Bacteriol. *183*(6), 2081–5.
23. WHITTAM, T.S. 1998. Evolution of *Escherichia coli* O157:H7 and other shiga toxin producing *E. coli* strains. *In Escherichia coli* and other shiga toxin producing *E. coli* strains, (J. Kaper and A.D. O'Brien, eds.) American Society for Microbiology. pg 195–209.

CHAPTER 2

DETECTION OF VEROCYTOTOXIN-PRODUCING *ESCHERICHIA COLI* O157 ON THE FARM AND AT THE ABATTOIR

P.A. CHAPMAN

Public Health Laboratory
Herries Road
Sheffield S5 7BQ, United Kingdom

INTRODUCTION

The Need to Investigate Farms and Abattoirs

Verocytotoxin-producing (VT+) *Escherichia coli* O157 cause a range of symptoms from mild non-bloody diarrhoea to haemorrhagic colitis (HC) and haemolytic-uraemic syndrome (HUS). In the first documented outbreak of HC caused by *E. coli* O157 (53), which occurred in the Northwest USA in 1982, there was a strong association between infection and prior consumption of ground beef. Reported outbreaks of *E. coli* O157 infection have often been very severe, with high mortality rates, particularly in the elderly. Such outbreaks, following consumption of undercooked beef, occurred in Ontario, Canada, in September 1985, in which 17 of 55 affected residents died (7) and in Lanarkshire, Scotland, in November and December 1996, which resulted in 20 deaths among the 496 people affected. Many other outbreaks world-wide have implicated foods of animal origin, or food or water contaminated with animal manure, as a source of infection. Because of the potential severity of the infection and the steadily rising incidence of infection in many countries, there is often a need to trace a source of infection in an outbreak, or to perform surveillance of the animal population, in order to elucidate the epidemiology and ecology of this organism and thereby enable appropriate intervention measures to be put in place.

Animals Carrying *E. coli* O157 and Other VTEC

Strains of *E. coli* first isolated from diarrhoeal cattle in Argentina in 1977 were later shown to be VT+ *E. coli* O157 (45); these are probably the first documented isolates of the organism from cattle. *E. coli* O157 have also been isolated from healthy cattle, sampled during investigations of possible sources of human infections or at random. The location, animal population, prevalence of *E. coli* O157 and the isolation methods used are summarised in Tables 1 and 2, respectively.

TABLE 1.
SURVEYS OF CATTLE POSSIBLY IMPLICATED IN OUTBREAKS OF *E.COLI* O157 INFECTION. SMAC, DIRECT CULTURE ON SORBITOL MACCONKEY AGAR; MTSB-SMAC, ENRICHMENT CULTURE IN MODIFIED TRYPTONE SOYA BROTH AND SUBCULTURE TO SMAC; CRSMAC, DIRECT CULTURE ON SMAC SUPPLEMENTED WITH CEFIXIME AND RHAMNOSE; IMS/CTSMAC, ENRICHMENT CULTURE IN BUFFERED PEPTONE WATER FOLLOWED BY IMMUNOMAGNETIC SEPARATION AND CULTURE ON SMAC SUPPLEMENTED WITH CEFIXIME AND TELLURITE.

Year	**Location**	**Type of cattle**	**Place of sampling**	**Method used**	**% positive**	**Reference**
1991	Wisconsin, USA	Dairy	Farm	MTSB/SMAC	0	70
1992	Sheffield, UK	Mixed	Abattoir	CRSMAC	3.9	15
1992	Scotland	Calves	Routine submissions	SMAC	0.4	58
1993	Sheffield, UK	Dairy	Farm	IMS/CTSMAC	9.5	42
1993	Galicia, Spain	Calves	Farm	SMAC	0.5	3

TABLE 2.
SURVEYS OF CATTLE SAMPLED AT RANDOM FOR THE PRESENCE OF *E.COLI* O157. SMAC, MTSB/SMAC AND IMS/CTSMAC, AS TABLE 1.

Year	**Location**	**Type of cattle**	**Place of sampling**	**Method used**	**% positive**	**Reference**
1987	Sheffield, UK	Mixed	Abattoir	SMAC	0.9	17
1990	Berlin, Germany	Beef	Abattoir	SMAC	0.8	43
1991	Wisconsin, USA	Mixed	Farm	mTSB/SMAC	0.4	70
1993	Galicia, Spain	Calves	Farm	SMAC	1.8	3
1993	Washington State,	Dairy	Farm	SMAC	0.3	27
	USA	Beef	Farm	SMAC	0.7	
		Beef	Feedlot	SMAC	0.4	
1996	Veneto, Italy	Beef	Farm	IMS/CTSMAC	3.6	20
1996	Sheffield, UK	Mixed	Abattoir	IMS/CTSMAC	15.7	14
1998	Netherlands	Dairy	Farm	IMS/CTSMAC	0.8-22.4	32
1999	Brazil	Mixed	Farm/abattoir	CTSMAC	1.5	8

Various factors may affect the prevalence of detection of *E. coli* O157 in cattle. The method of isolation used has a major impact. We have shown that immunomagnetic separation (IMS) increases the sensitivity of detection of *E.*

coli O157 in bovine faeces by 10- to 100-fold (18), and it is apparent from Tables 1 and 2 that prevalence studies which have used IMS have usually reported higher prevalence rates. The geographical area also has an effect, with the organism being apparently more prevalent in the cattle population in Northwestern USA and in Sheffield than in many other areas. The season of the study has a marked influence on the results. In studies of both dairy cattle and beef cattle in Sheffield, we have shown carriage rates of the organism to be consistently much higher in the summer (14,42). Young animals also tend to carry the organism more frequently than older animals (42) and carriage may be affected by diet (27), with animals which are fed grain silage tending to carry the organism more frequently.

Various other animals, particularly ruminants, have been shown to be reservoirs or vectors of *E. coli* O157 and the location, animal population, prevalence of *E. coli* O157 and the isolation methods used in these studies are summarised in Table 3.

TABLE 3.
OTHER ANIMALS SHOWN TO BE CARRYING *E. COLI* O157. IMS/CTSMAC, AS TABLE 1; MTSB/CTSMAC, ENRICHMENT CULTURE IN MODIFIED TRYPTONE SOYA BROTH AND SUBCULTURE TO CTSMAC.

Year	Location	Type of cattle	Place of sampling	Method used	% positive	Reference
1996	Sheffield, UK	Sheep	Abattoir	IMS/CTSMAC	2.2	14
1996	Idaho, USA	Sheep	Farm	mTSB/CTSMAC	6.2	38
1996	Cornwall, UK	Dog	Farm	IMS/CTSMAC	-	64
1997	Sheffield, UK	Goats, sheep and pigs	Open Farm	IMS/CTSMAC	60.0	12
1997	Sheffield, UK	Deer	Farm	IMS/CTSMAC	33.0	10
1999	Netherlands	Pigs	Abattoir	IMS/CTSMAC	1.4	33
		Turkeys	Abattoir	IMS/CTSMAC	1.3	

ON FARM SAMPLING

Collection of Samples

Direct transmission of *E. coli* O157 from farm animals to man has been reported on several occasions, either by direct contact with animals (51,52,59) or by contact with animal manure (19). It is important that all appropriate safety procedures (23,29,31) are followed to avoid the risk of infection to sampling personnel.

Collection of Rectal Faeces

Ideally, samples of rectal faeces should be collected from individual animals. For convenience, samples of rectal faeces are most easily collected by means of rectal swabs. We have found that standard swabs used for this purpose usually obtain a faecal sample of about 0.4 to 0.5 g and we have used these effectively in a number of investigations (14-17). However, a slightly larger number of positive animals may be detected if the amount of faeces examined is increased to 1 g or more (54), although such samples are more laborious and costly to collect and examine.

The number of samples that would need to be collected to ensure detection of a positive animal in a herd is influenced by: (1) the herd size or population size to be sampled; (2) the within-herd prevalence of the organism sought; and (3) the statistical confidence limit required for the number of positive samples detected. These factors have been reviewed by Cannon and Roe (6). Table 4 is modified from that of Cannon and Roe (6) and shows the number of animal faecal samples that would be needed to detect *E. coli* O157 in a herd within 95% confidence limits; herd sizes up to 200 head are shown and expected prevalences (based on tables 1-3) in the range of 1% to 20%.

Collection of Faecal Pat, Manure and Manure Slurry Samples

Collection of faecal samples per rectum from farm animals needs the services of a qualified veterinarian. Collection of apparently fresh faecal pat specimens from the farm environment may provide a more convenient and less costly means of obtaining samples. However, obtaining a statistically valid number of samples is much more problematic as due allowance has to be made for the fact that several faecal pats may have been produced by the same animal; opinions differ as to the best approach to adopt to this and statistical advice should be sought in individual cases.

Studies of the survival of *E. coli* O157 in bovine faeces have shown that the organism usually survives for longer periods at lower temperatures and in moist conditions. Under laboratory conditions the organism can survive for 70-100 days at 4-8°C (4,26,67). However, the organism survives for much shorter periods if subject to drying or a higher temperature or if contained in bovine faeces applied to grassland (4). Samples should therefore be taken only from faecal pats which are apparently fresh. We have isolated *E. coli* O157 from manure slurry taken from a dairy farm (16) and Kudva *et al.* (37) have shown that the organism may survive in slurry for up to 6 weeks. To maximise the chance of recovery of the organism, all samples should be apparently fresh when collected and should be refrigerated during transport to the laboratory.

TABLE 4.
NUMBER OF FAECAL SAMPLES THAT NEED TO BE EXAMINED TO ENSURE DETECTION OF A POSITIVE ANIMAL WITHIN A GIVEN HERD SIZE (95% CONFIDENCE LIMITS — MODIFIED FROM CANNON AND ROE (1982)).

Herd size	Number of samples needed to detect a positive animal the within-herd prevalence is					
	20%	**15%**	**10%**	**5%**	**2%**	**1%**
10	8	10	10	10	10	10
20	10	12	16	19	20	20
30	11	14	19	26	30	30
40	12	15	21	31	40	40
50	12	16	22	35	48	50
60	12	16	23	38	55	60
70	13	17	24	40	62	70
80	13	17	24	42	68	79
90	13	17	25	43	73	87
100	13	17	25	45	78	96
120	13	18	26	47	86	111
140	13	18	26	48	92	124
160	13	18	27	49	97	136
180	13	18	27	50	101	146
200	13	18	27	51	105	155

Collection of Water Samples

E. coli O157 may survive for periods up to 12 weeks in water at low temperatures and, for reasons that are unclear, the organism survives less well in untreated surface water than it does in treated drinking water (50,66). Contaminated water troughs may play an important role in maintenance and dissemination of *E. coli* O157 on farms (56), as they can be a source of recurrent exposure to the organism. Multiplication of the organism in water sediments in summer has also been demonstrated on farms in Washington State, USA (28). Ideally, therefore, water samples (minimum 100 ml) collected from water troughs should include some of the sediment and again should be refrigerated during transport to the laboratory.

Collection of Milk Samples

E. coli O157 is readily killed by pasteurisation or other heat treatment (21) and unless a pasteurisation failure is suspected there is little to be gained by examining samples of heat-treated milk. In untreated milk the organism survives for at least 14 days at 5-8°C (41,68) but suffers rapid reductions in numbers after 4 days at 22°C (68). Milk samples should therefore be fresh and should be refrigerated during transport to the laboratory.

During investigations of an outbreak of *E. coli* O157 infection associated with consumption of untreated milk (16,42) we isolated the organism on several occasions from milk taken from individual animals but consistently failed to isolate the organism from milk taken from the bulk storage tank, probably due to the dilution factor involved in bulk storage. Therefore, whilst milk may be included as part of the sampling regime, it should not be relied upon as the sole specimen during investigation of a possible outbreak of milk-borne infection.

Collection of Environmental Samples

Survival of *E. coli* O157 in the environment has been less well studied. Although the organism appears to survive for many weeks on contaminated straw and on common structural surfaces such as wood and breeze block (50) the value of examining such surfaces during on farm investigations remains to be determined. Sampling of the dairy environment was found to be useful in tracing an outbreak of infection linked to pasteurised milk in Scotland (65). In this investigation a pipe which carried milk from the pasteurisation apparatus to the bottling machine and a discarded bottling machine rubber both yielded *E. coli* O157 indistinguishable from the outbreak strain. This investigation emphasised the importance of using sensitive methods as both isolates were obtained only by using IMS.

SAMPLING AT THE ABATTOIR

Collection of Samples

During collection of samples, procedures to prevent infection of personnel and other procedures for safe working in an abattoir should be followed (30). The sampling methods for rectal faeces referred to above apply equally well to collecting samples in the abattoir. In the abattoir rectal swabs offer the added advantage of being much more rapidly obtained and therefore interfering less with the smooth running of the slaughter line.

Sampling of Carcasses

Sampling of carcasses is a problem, as even if the method used is effective in recovering *E. coli* O157, it may damage or otherwise contaminate the carcass being sampled. The numbers of *E. coli*, and presumably *E. coli* O157, on the surface of a carcass decline during the first 24 h of storage (5,69) and it is therefore important that carcasses are sampled and examined as soon as possible.

Various methods for sampling meat surfaces, such as scraping, swabbing, adhesion or rinsing, have been described, and they have been reviewed by Patterson (48). However, for a variety of reasons, these techniques are not always efficient at recovering organisms from the surface of meat. Sponges have become widely used for sampling but care should be exercised as several commercially available types of sponge have antibacterial properties that may interfere with recovery of organisms (39). Poor recovery may also be due to penetration of the organism into the tissue during the slaughter process. Anderson *et al.* (2) used a dye model to show that bacteria may not be removed from the carcass during washing and may be driven deeper into the tissue by this process. Microbiological support for this is provided by evidence that bacterial contamination can occur up to a depth of 6 mm in the tissue of a carcass (22).

According to Ingram and Roberts (36) meat excision is the standard method by which all others should be judged. Our results when sampling carcasses by meat excision and surface swabbing would support this: of nine isolates of *E. coli* O157 obtained from beef carcasses, six were obtained only from excised meat and not by swabbing (15). However, meat excision has its drawbacks in that it is restricted to small areas of the carcass, and is laborious and potentially damaging to the carcass (36). Ideally meat should be excised as a very thin slice from a large area of the carcass; this is difficult with a normal knife but Davidson *et al.* (22) have described an instrument, similar to a surgical dermatome, which facilitates such sampling.

CHOICE OF DETECTION METHOD

Selective Culture

Strains of VT+ *E. coli* O157 do not normally ferment sorbitol, whereas many other serogroups of *E. coli* do, and sorbitol MacConkey agar (SMAC) became widely used for their isolation during the late 1980s (17,40,57). However, reports of its selectivity have varied widely and it became clear that other sorbitol non-fermenting organisms could be more troublesome than early reports suggested (17); particularly troublesome organisms in animal faeces and the farm environment include *Proteus* spp., *Aeromonas* spp., *Morganella* spp., and *Providencia* spp. We modified SMAC to include cefixime and potassium tellurite and the new formulation (CT-SMAC) allowed the growth of all strains of VT+ *E. coli* O157 tested, but completely inhibited the growth of 67% of other strains of *E.coli* (including non-verocytotoxigenic *E. coli* O157) and 97% of strains including *Proteus* spp., *Aeromonas* spp., *Morganella* spp., and *Providencia* spp. (71). A comparison was made of SMAC and CT-SMAC for the isolation of VT+ *E. coli* O157 from swabs of rectal faeces taken from cattle immediately after slaughter (71). Of 391 samples screened on CT-SMAC medium, 26 yielded isolates of VT+ *E. coli* O157 whereas CR-SMAC yielded only nine isolates. However, some more recently reported strains of sorbitol-fermenting VT+ *E. coli* O157 (1) isolated from cases of HUS in Germany (kindly supplied by Professor S Aleksic, Hamburg) are sensitive to tellurite and so do not grow on CT-SMAC.

Another feature of VT+ *E. coli* O157 is that, unlike most other *E. coli*, they do not produce β-glucuronidase. The fluorogenic substrate 4-methylumbelliferyl β-D-glucuronide (MUG) and the chromogenic substrate 5-bromo-4-chloro-3-indoxyl-β-D-glucuronide (BCIG) have therefore been advocated for incorporation into culture media for the detection of VT+ *E. coli* O157 (44,60,62). However, the extremely high cost of these substrates would probably preclude their use in the majority of laboratories.

Enrichment Culture

Whereas direct culture may be a suitable technique for the isolation of VT+ *E. coli* O157 from the faeces of patients with acute illness, where the organism is likely to be the predominant organism in the sample, more sensitive methods are needed for its isolation from animal faeces and the environment, where numbers of the organism may be low. Standard methods for the selective isolation of *E. coli* from environmental samples have used growth at 44°C as a selective factor. However, VT+ *E. coli* O157 have an upper temperature for growth of 41°C in selective *E. coli* broth (49) and grows only poorly at temperatures higher than this in non-selective media (24). Standard methods are therefore ineffective for the isolation of VT+ *E. coli* O157 from such samples, and specific methods are needed.

Doyle and colleagues (25,46) have described a modified tryptone broth (mTSB) supplemented with a phosphate buffer, bile salts and either novobiocin or acriflavine. After enrichment culture of samples in mTSB with shaking at 37°C for 24 h the cultures were diluted in phosphate buffered saline (PBS) from 10^{-5} to 10^{-7} before inoculation onto SMAC medium. The rationale for a prolonged incubation followed by dilution of the culture was not explained.

A modification of the *E. coli* medium used in standard methods for food and water microbiology in the USA has also been described (49). However, such media, containing lactose, are possibly best avoided for the enrichment culture of VT+ *E. coli* O157, as many other organisms that are common in the alimentary tract of animals, in particular *Enterococcus* spp. and *Veillonella* spp, can metabolize lactose to compounds that are inhibitory to the growth and survival of VTEC *E. coli* O157:H7 (34,35).

In an attempt to improve enrichment culture for isolation of VT+ *E. coli* O157 from animal and environmental samples we studied the minimum inhibitory concentrations (MICS) of various antimicrobial agents for over 300 VT+ *E. coli* O157 strains and a wide range of other organisms which we had commonly isolated from such samples. We found a combination of vancomycin, cefsulodin and cefixime, to inhibit gram-positive organisms, *Aeromonas* spp. and *Proteus* spp. respectively, to be an effective supplement to buffered peptone water for the enrichment culture of VT+ *E. coli* O157. Using this medium (BPW-VCC) and incubation at 37°C for 6 h, a period that we had found to be optimal for recovery of the organism from inoculated beef samples, we isolated the organism from naturally contaminated beef carcasses for the first time in 1992 (15). Despite our previous successful use of potassium tellurite as a selective agent in the solid CT-SMAC medium (71), we found it to be ineffective as a selective agent in liquid media. This is probably due to antagonism of tellurite metabolism both by the widely-used phosphate buffer system and by the rapid development of anaerobic conditions in the enrichment culture (63). Our attempts to solve this problem have so far been unsuccessful (9).

Immunomagnetic Separation (IMS)

We have shown that enrichment culture in BPW-VCC, followed by IMS with paramagnetic beads coated with an antibody against *E. coli* O157 (Dynabeads anti-*E. coli* O157, Dynal), followed by culture of the beads on CT-SMAC, was approximately one hundredfold more sensitive than direct culture on CR-SMAC or CT-SMAC in detection of VT+ *E. coli* O157 in inoculated bovine faecal samples (18). This increase in sensitivity was confirmed during subsequent monitoring of a dairy herd when VT+ *E. coli* O157 was isolated from 84 (8.2%) of 1024 rectal swabs taken from cattle over a four month period; 23 (27.4%) of the 84 strains were isolated by both direct culture and IMS, whereas 61 (72.6%) of the 84 strains were isolated by IMS only (18). Using the IMS technique VT+ *E. coli* O157 was isolated from two of 279 forestream milk samples from individual cattle in this dairy herd (16,42); neither strain was isolated by direct subculture of the enrichment medium to CT-SMAC.

Alternatives to Culture Techniques

Enzyme immunoassays and immunoblot techniques for the detection of *E. coli* O157 antigen have been described as methods for the detection of the organism in enrichment cultures of environmental samples (25,47,55,61). Although sensitive, these methods may be time-consuming, technically demanding, expensive and prone to give false positive results that cannot be confirmed by culture (55). We have evaluated two commercially available enzyme immunoassays for the detection of *E. coli* O157 in bovine faecal samples (11,13). Both assays gave a sensitivity of detection equal to that of enrichment culture in BPW-VCC followed by IMS and CT-SMAC. However, both were prone to give false positive results that could not be confirmed by our normal culture method; in some cases these positive results were shown to be due to sorbitol fermenting non-toxigenic *E. coli* O157 which were present in the sample and in others the enrichment cultures appeared to react equally well in other enzyme immunoassays for different antigens.

There have also been many reports in the literature of the use of DNA based methods, including the polymerase chain reaction, for detecting *E. coli* O157 and other VTEC in animal faeces and environmental samples. However, like immunoassays, these methods are time-consuming, technically demanding and expensive and therefore do not lend themselves to large farm or abattoir studies; their use is further hampered by the widespread presence of target virulence genes in many *E. coli* other than serogroup O157.

Despite considerable advances in the development of alternatives to selective culture, the latter remains the method of choice in the vast majority of investigations. It is economical, technically undemanding and, above all, it yields an isolate of *E. coli* O157 which enables comparison with strains isolated from different sources using various typing methods; this is a vitally important part of any investigation but a discussion of the methods used is beyond the scope of this review.

REFERENCES

1. ALEKSIC, S., KARCH, H. and BOCKEMUHL J. 1992. A bio-typing scheme for shiga-like (vero) toxin-producing *Escherichia coli* O157 and a list of cross reactions between O157 and other Gram-negative bacteria. Zentrabl Bakt *276*, 221–230.
2. ANDERSON, M.E., MARSHALL, R.T. and DICKSON, J.S. 1992. Estimating depths of bacterial penetration into post-rigor carcass tissue during washing. J. Food Safety *12*, 191–198.
3. BLANCO, M., BLANCO, J., BLANCO, J.E. and RAMOS, J. 1994. Enterotoxigenic, verotoxigenic, and necrotoxigenic *Escherichia coli* isolated from cattle in Spain. Am. J. Vet. Res. *54*, 1446–1451.
4. BOLTON, D.J., BYRNE, C.M., SHERIDAN, J.J., MCDOWELL, D.A. and BLAIR, I.S. 1999. The survival characteristics of a non-toxigenic strain of *Escherichia coli* O157:H7. J. Appl. Microbiol. *86*, 407–411.
5. CALICIOGLU, M., BUEGE, D.R., INGHAM, S.C. and LUCHANSKY, J.B. 1999. Recovery of *Escherichia coli* biotype I and *Enterococcus* spp. during refrigerated storage of beef carcasses inoculated with a fecal slurry. J. Food Protect. *62*, 944–947.
6. CANNON, R.M. and ROE, R.T. 1982. Livestock disease surveys: a field manual for veterinarians. Australian Government Publishing Service, Canberra.

7. CARTER, A.O. *et al.* 1987. A severe outbreak of *Escherichia coli* O157:H7-associated haemorrhagic colitis in a nursing home. N. Engl. J. Med. *317*, 1496–1500.
8. CERQUEIRA, A.M.F., GUTH, B.E.C., JOAQUIM, R.M. and ANDRADE, J.R.C. 1999. High occurrence of shiga toxin-producing *Escherichia coli* (STEC) in healthy cattle in Rio de Janeiro state, Brazil. Vet. Microbiol. *70*, 111–121.
9. CHAPMAN, P.A. 1994. Isolation, identification and typing of Verocytotoxin-producing *Escherichia coli*. PHLS Microbiology Digest *111*, 13–17.
10. CHAPMAN, P.A. and ACKROYD, H.J. 1997. Farmed deer as a potential source of verocytotoxin-producing *Escherichia coli* O157. Vet. Rec. *141*, 314–315.
11. CHAPMAN, P.A., CERDAN MALO, A.T., SIDDONS, C.A. and HARKIN, M. 1997. Use of commercial enzyme immunoassays and immunomagnetic separation systems for detecting *Escherichia coli* O157 in bovine fecal samples. Appl. Environ. Microbiol. *63*, 2459–2553.
12. CHAPMAN, P.A., CORNELL, J. and GREEN, C. 2000. Infection with Verocytotoxin-producing *Escherichia coli* O157 during a visit to an inner city open farm. Epidemiol. Infect. *125*, 531–6.
13. CHAPMAN, P.A. and SIDDONS, C.A. 1996. Evaluation of a commercial enzyme immunoassay (EHEC-Tek) for detecting *Escherichia coli* O157 in beef and beef products. Food Microbiol. *13*, 175–182.
14. CHAPMAN, P.A., SIDDONS, C.A., CERDAN MALO, A.T. and HARKIN, M.A. 1997. A one-year study of *Escherichia coli* O157 in cattle, pigs, sheep and poultry. Epidemiol. Infection *119*, 245–250.
15. CHAPMAN, P.A., SIDDONS, C.A., WRIGHT, D.J., NORMAN, P., FOX, J. and CRICK, E. 1993. Cattle as a possible source of verocytotoxin-producing *Escherichia coli* O157 infections in man. Epidemiol. Infect *111*, 439–447.
16. CHAPMAN, P.A., WRIGHT, D.J. and HIGGINS, R. 1993. Untreated milk as a source of verotoxigenic *E. coli* O157. Vet. Rec. *133*, 171–172.
17. CHAPMAN, P.A., WRIGHT, D.J. and NORMAN, P. 1989. Verotoxin-producing *Escherichia coli* infections in Sheffield: cattle as a possible source. Epidemiol. Infect. *102*, 439–445.
18. CHAPMAN, P.A., WRIGHT, D.J. and SIDDONS, C.A. 1994. A comparison of immunomagnetic separation and direct culture for the isolation of verocytotoxin-producing *E. coli* O157 from bovine faeces. J. Med. Microbiol. *40*, 424–427.
19. CIESLAK, P.R., BARRETT, T.J. and GRIFFIN, P.M. 1993. *Escherichia coli* O157:H7 infection from a manured garden. Lancet *342*, 367.
20. CONEDERA, G., MARANGON, S., CHAPMAN, P.A., ZUIN, A. and CAPRIOLI, A. 1997. *Escherichia coli* O157 in beef cattle at slaughter in Veneto region, Italy. J. Vet. Med. B. *44*, 301–306.
21. D'AOUST, J.Y., PARK, C.E., SZABO, R.A., TODD, E.C.D., EMMONS, D.B. and MCKELLAR, R.C. 1988. Thermal inactivation of *Campylobacter* species, *Yersinia enterocolitica*, and hemorrhagic *Escherichia coli* O157:H7 in fluid milk. J. Dairy Sci. *71*, 3230–3236.
22. DAVIDSON, C.M., TAYLOR, M. and ZELLERMANN, G.G. 1978. Method for sampling beef carcasses. Appl. Environ. Microbiol. *35*, 811–812.

23. DAWSON, A., GRIFFIN, R., FLEETWOOD, A. and BARRET, N.J. 1995. Farm visits and Zoonoses. Commun. Dis. Rep. Rev. *5*, R81–R86.
24. DOYLE, M.P. and SCHOENI, J.L. 1984. Survival and growth characteristics of *Escherichia coli* associated with hemorrhagic colitis. Appl. Environ. Microbiol. *48*, 855–856.
25. DOYLE, M.P. and SCHOENI, J.L. 1987. Isolation of *Escherichia coli* O157:H7 from retail fresh meats and poultry. Appl. Environ. Microbiol. *53*, 2394–2396.
26. FUKUSHIMA, H., HOSHINA, K. and GOMYODA, M. 1999. Long term survival of shiga toxin-producing *Escherichia coli* O26, O111 and O157 in bovine feces. Appl. Environ. Microbiol. *65*, 5177–5181.
27. HANCOCK, D.D., BESSER, T.E., KINSELL, M.L., TARR, P.I., RICE, D.H. and PAROS, M.G. 1994. The prevalence of *Escherichia coli* O157.H7 in dairy and beef cattle in Washington State. Epidemiol. Infect. *113*, 199–207.
28. HANCOCK D.D., BESSER, T.E. and RICE, D.H. 1998. Ecology of *Escherichia coli* O157:H7 in cattle and impact of management practices. In *Escherichia coli* O157:H7 and Other Shiga Toxin-Producing *E. coli* Strains. (J.B. Kaper and A.D. O'Brien, eds.) ASM, Washington, pp. 85–91.
29. HEALTH AND SAFETY EXECUTIVE. 1989. Zoonoses in agriculture — preventing the spread of disease to livestock handlers. Agriculture Information Sheet 2, HSE, 1989.
30. HEALTH AND SAFETY EXECUTIVE. 1997. Priorities for health and safety in the slaughtering industry. Food Information Sheet No 14, HSE, 1997.
31. HEALTH AND SAFETY EXECUTIVE. 1998. Avoiding ill health at open farms — advice to farmers (with teachers' supplement). Agriculture Information Sheet 23, HSE, 1998.
32. HEUVELINK, A.E. *et al.* 1998. Occurrence of verocytotoxin-producing *Escherichia coli* O157 on Dutch dairy farms. J. Clin. Microbiol. *36*, 3480–3487.
33. HEUVELINK, A.E, ZWARTKRUIS-NAHUIS, J.T.M., VANDENBIGGELAAR, F.L.A.M., VAN LEEUWEN, W.J. and DE BOER, E. 1999. Isolation and characterization of verocytotoxin-producing *Escherichia coli* O157 from slaughter pigs and poultry. Int. J. Food Microbiol. *52*, 67–75.
34. HINTON, A., SPATES, G.E., CORRIER, D.E, HUME, M.E., DELOACH, J.R. and SCANLAN, C.M. 1991. In vitro inhibition of *Salmonella typhimurium* and *Escherichia coli* O157:H7 by bacteria isolated from the cecal contents of adult chickens. J. Food Protect. *54*, 496–501.
35. HINTON, A., CORRIER, D.E and DELOACH, J.R. 1992. In vitro inhibition of *Salmonella typhimurium*, and *Escherichia coli* O157:H7 by an anaerobic Gram-positive coccus from the cecal contents of adult chickens. J. Food Protect. *55*, 162–166.
36. INGRAM, P.M. and ROBERTS, T.A. 1976. The microbiology of the red meat carcass and the slaughterhouse. Roy. Soc. Health. J. *96*, 270–276.
37. KUDVA, I.T., BLANCH, K. and HOVDE, C.J. 1998. Analysis of *Escherichia coli* O157:H7 survival in ovine or bovine manure and manure slurry. Appl. Environ. Microbiol. *64*, 3166–3174.
38. KUDVA, I.T., HATFIELD, P.G. and HOVDE, C.J. 1996. *Escherichia coli* O157:H7 in microbial flora of sheep. J. Clin. Microbiol. *34*, 431–433.

39. LLABRÉS, C.M. and ROSE, B.E. 1989. Antibacterial properties of retail sponges. J. Food Protect. *52*, 49–50.
40. MARCH, S.B. and RATNAM, S. 1986. Sorbitol MacConkey medium for the detection of *Escherichia coli* O157:H7 associated with hemorrhagic colitis. J. Clin. Microbiol. *23*, 869–872.
41. MASSA, S., GOFFREDO, E., ALTIERI, C. and NATOLA, K. 1999. Fate of *Escherichia coli* O157:H7 in unpasteurised milk stored at 8°C. Lett. Appl. Microbiol. *28*, 89–92.
42. MECHIE, S.C., CHAPMAN, P.A. and SIDDONS, C.A. 1997. A fifteen month field study of *Escherichia coli* O157 in a dairy herd. Epidemiol. Infect. *118*, 17–25.
43. MONTENEGRO, M.A. *et al.* 1990. Detection and characterisation of fecal verotoxin-producing *Escherichia coli* from healthy cattle. J. Clin. Microbiol. *28*, 1417–1421.
44. OKREND, A.J.G., ROSE, B.E. and LATTUADA, C.P. 1990. Use of 5-bromo-4-chloro-3-indoxyl-β-D-glucuronide in MacConkey sorbitol agar to aid the isolation of *Escherichia coli* O157:H7 from ground beef. J. Food Protect. *53*, 941–943.
45. ØRSKOV, F., ØRSKOV, I. and VILLAR, J.A. 1987. Cattle as a reservoir of verotoxin-producing *Escherichia coli* O157:H7. Lancet 2, 276.
46. PADHYE, N.V. and DOYLE, M.P. 1991. Rapid procedure for detecting enterohemorrhagic *Escherichia coli* O157:H7 in food. Appl. Environ. Microbiol. *57*, 2693–2698.
47. PADHYE, N.V. and DOYLE, M.P. 1991. Production and characterization of a monoclonal antibody specific for enterohemorrhagic *Escherichia coli* of serotypes O157:H7 and 026:H11. J. Clin. Microbiol. *29*, 99–103.
48. PATTERSON, J.T. 1971. Microbiological assessment of surfaces. J. Food Technol. *6*, 63–73.
49. RAGHUBEER, E.V. and MATCHES, J.R. 1990. Temperature range for growth of *Escherichia coli* O157:H7 and selected coliforms in *E. coli* medium. J. Clin. Microbiol. *28*, 803–805.
50. RANDALL, L.P., WRAY, C. and DAVIES, R.H. 1999. Survival of verocytotoxin-producing *Escherichia coli* O157 under simulated farm conditions. Vet. Rec. *145*, 500–501.
51. RENWICK, S.A. *et al.* 1993. Evidence of direct transmission of *Escherichia coli* O157:H7 between calves and a human. J. Infect. Dis. *168*, 792–793.
52. RICE, D.H, HANCOCK, D.D., VETTER, R.L. and BESSER, T.E. 1996. *Escherichia coli* O157 infection in a human linked to exposure to infected livestock. Vet. Rec. *138*, 311.
53. RILEY, L.W. *et al.* 1983. Haemorrhagic colitis associated with a rare *Escherichia coli* serotype. N. Engl. J. Med. *308*, 681–685.
54. SANDERSON, M.W, GAY, J.M., HANCOCK, D.D., GAY, C.C., FOX, L.K. and BESSER, T.E. 1995. Sensitivity of bacteriologic culture for detection of *Escherichia coli* O157:H7 in bovine faeces. J. Clin. Microbiol. *33*, 2616–2619.
55. SERNOWSKI, L.P. and INGHAM S.C. 1992. Frequency of false presumptive positive results obtained using a commercial ELISA kit to screen retail ground beef for *Escherichia coli* O157:H7. J. Food Protect. *55*, 846.

56. SHERE, J.A., BARTLETT, K.J. and KASPAR, C.W. 1998. Longitudinal studies of *Escherichia coli* O157:H7 dissemination on four dairy farms in Wisconsin. Appl. Environ. Microbiol. *64*, 1390–1393.
57. SMITH, H.R., ROWE, B., GROSS, R.J., FRY, N.K. and SCOTLAND, S.M. 1987. Haemorrhagic colitis and Verocytotoxin-producing *Escherichia coli* in England and Wales. Lancet *1*, 1062–1065.
58. SYNGE, B.A. and HOPKINS, G.F. 1992. Verotoxigenic *Escherichia coli* O157 in Scottish calves. Vet. Rec. *130*, 583.
59. SYNGE, B.A., HOPKINS, G.F., REILLY, W.J. and SHARP, J.C.M. 1993. Possible link between cattle and *E. coli* O157 infection in a human. Vet. Rec. *133*, 507.
60. SZABO, R.A., TODD, E.C. and JEAN, A. 1986. A method to isolate *Escherichia coli* O157:H7 from food. J. Food Protect. *49*, 768–772.
61. SZABO, R.A.E., TODD, E., MACKENZIE, J., PARRINGTON, L. and ARMSTRONG, A. 1990. Increased sensitivity of the rapid hydrophobic grid membrane filter enzyme labelled antibody procedure for *Escherichia coli* O157 detection in foods and bovine feces. Appl. Environ. Microbiol. *56*, 3546–3549.
62. THOMPSON, J.S., HODGE, D.S. and BORCZYK, A.A. 1990. Rapid biochemical test to identify verocytotoxin-positive strains of *Escherichia coli* serotype O157. J. Clin. Microbiol. *28*, 2165–2168.
63. TOMAS, J.M. and KAY, W.W. 1986. Tellurite susceptibility and non–plasmid mediated resistance in *Escherichia coli*. Antimicrob. Agents Chemother. *30*, 127–131.
64. TREVENA, W.B., HOOPER, R.S., WRAY, C., WILLSHAW, G.A., CHEASTY, T. and DOMINGUE, G. 1996. Verocytotoxin-producing *Escherichia coli* O157 associated with companion animals. Vet. Rec. *138*, 400.
65. UPTON, P. and COIA, J.E. 1994. Outbreak of *Escherichia coli* O157 infection associated with pasteurised milk supply. Lancet *344*, 1015.
66. WANG, G. and DOYLE, M.P. 1998. Survival of enterohemorrhagic *Escherichia coli* O157:H7 in water. J. Food Protect. *61*, 662–667.
67. WANG, G., ZHAO, T. and DOYLE, M.P. 1996. Fate of enterohemorrhagic *Escherichia coli* O157:H7 in bovine feces. Appl. Environ. Microbiol. *62*, 2567–2570.
68. WANG, G., ZHAO, T. and DOYLE, M.P. 1997. Survival and growth of *Escherichia coli* O157:H7 in unpasteurized and pasteurized milk. J. Food Protect. *60*, 610–613.
69. WARE, L.M., KAIN, M.L, SOFOS, J.N., BELK, K.E. and SMITH, G.C. 1999. Comparison of sponging and excising as sampling procedures for microbiological analysis of fresh beef-carcass tissue. J. Food Protect. *62*, 1255–1259.
70. WELLS, J.G., SHIPMAND, L.D. and GREENE, K.D. 1991. Isolation of *Escherichia coli* serotype O157:H7 and other shiga-like-toxin-producing *E. coli* from dairy cattle. J. Clin. Microbiol. *29*, 985–989.
71. ZADIK, P.M., CHAPMAN, P.A. and SIDDONS, C.A. 1993. Use of tellurite for the selection of verocytotoxigenic *Escherichia coli* O157. J. Med. Microbiol. *39*, 155–158.

CHAPTER 3

CLINICAL DETECTION OF VEROCYTOTOXIN-PRODUCING *E. COLI* (VTEC)

FLEMMING SCHEUTZ

The International Escherichia and Klebsiella Centre (WHO)
Department of Gastrointestinal Infections
Statens Serum Institut
5 Artillerivej
2300 Copenhagen S, Denmark

LOTHAR BEUTIN

Division of Emerging Bacterial Pathogens (P13)
Escherichia coli Reference Lab
Robert Koch Institute
Nordufer 20
D 13353 Berlin, Germany

AND

HENRY R. SMITH

Laboratory of Enteric Pathogens
Central Public Health Laboratory
61, Colindale Avenue
London, NW9 5HT England

ON THE IMPORTANCE OF DETECTION OF VTEC FROM CLINICAL SPECIMENS

Among the different groups of intestinal pathogenic *Escherichia coli*, strains of *E. coli* O157 and other types of Verocytotoxin-producing *E. coli* (VTEC) are considered today as one of the most important human pathogens in developed countries (1,2). Domestic cattle and other ruminants constitute a natural reservoir for VTEC and healthy animals can excrete these organisms in their faeces. In different European countries, VTEC have been associated with 1-2% of diarrhoeal cases and with 61 to 93% of haemolytic uraemic syndrome (HUS) cases (3). In the USA, 85% of 100,000 estimated VTEC infections/year are thought to be foodborne (4) and VTEC O157 was ranked as the fourth most costly foodborne disease (5). Outbreaks of VTEC with hundreds to several thousand cases have been reported from the UK, North America and Japan

(1,6). Outbreaks have been frequently associated with consumption of contaminated foods and water. A high incidence of VTEC O157 infections and HUS has been reported in certain geographic regions of Argentina, Scotland and North America (1,7), while living in areas with high cattle densities, and consumption of beef products, have been identified as risk factors for contracting VTEC infections (8,9).

In the clinical diagnostic laboratory, there are two major reasons to develop, and to improve already established, VTEC detection methods. Firstly, VTEC infections can cause life-threatening complications such as HUS and neurological damage, particularly in infants and the elderly (1,2). Secondly, VTEC infections in humans are currently underdiagnosed, particularly those infections which are caused by non-O157 VTEC strains (2,4,10). The aim of this chapter is to present an overview of the current knowledge of when and how clinical samples should be investigated for evidence of VTEC infection.

Note: Verocytotoxin (VT) is synonymous with Shiga toxin (Stx), and VTEC with STEC.

SAMPLING CRITERIA FOR THE EXAMINATION FOR VTEC

The diagnosis of VTEC is likely to be part of a general diagnostic procedure of other enteric pathogens. Resources, regional and geographical factors will vary considerably between countries, and within each country.

Diagnostic Categories

The following diagnostic categories for patients with diarrhoea have proved to be useful:

- Acute diarrhoea, acquired in the home country
- Persistent diarrhoea, i.e., diarrhoea for more than two weeks
- Travellers' diarrhoea. Travel to foreign country less than two weeks prior to the onset of diarrhoea
- Diarrhoea in the immunocompromised patient
- Patients with HUS or Thrombotic Thrombocytopenic Purpura (TTP), Guillain-Barré syndrome or other sequelae related to infection of the gastrointestinal tract
- Outbreaks which can be divided into family or general:

Family / person-to-person
General: Institutional
Food-borne / community-wide

The transmission routes for VTEC infection are via food/water, person to person and direct or indirect animal contact.

The incubation period can be 1-14 days, but is more usually 4-5 days. 5-10% of patients develop HUS. The infectious dose has been shown to be very low; i.e., for some VTEC serotypes it may be as low as 10-100 organisms.

Prioritising

Examination for VTEC is recommended in the following high priority groups:

High priority

- Diarrhoea is grossly bloody
- In children and elderly with bloody diarrhoea or a history of bloody diarrhoea
- Clinical suspicion of HUS or TTP
  Note: Significant proportions of HUS cases are preceded by non bloody diarrhoea, some even with no diarrhoea at all
- Diarrhoea in contacts (household, work-related, institutional) of persons with confirmed VTEC infection
- Diarrhoea in persons with contact to farm animals or other ruminants including deer
- Diarrhoea in persons consuming raw animal products (fermented sausages, unpasteurised milk, etc.) or raw vegetables and fruits (suspected of having been in contact with animal products, organic fertilisers, manure, contaminated soil or water)
- Diarrhoea in persons bathing or swimming in water contaminated by faecal matter from ruminants

Other criteria for examination may include:

- Non-bloody stools, i.e., all diarrhoeal stools
  Note: In Britain it is recommended that all diarrhoeal stools are examined for VTEC O157 and selected stools for all VTEC
- The clinicians wish to not overlook VTEC in certain patient categories or clinical circumstances, i.e., immunocompromised patients, or during institutional outbreaks
- Abdominal cramps
- Epidemiological follow up or clarification
- Biopsies or autopsies

Risk Groups

Persons that are at an increased risk of acquiring or transmitting a VTEC infection:

- Young infants and contact persons
- Farmers
- Food-handlers (including butchers and cooks)
- Hospital staff (nurses, etc., with direct contact to infected patients)
- Slaughterhouse personnel
- Microbiologists and their staff

Note: The lack of faecal leukocytes or inflammatory bowel disease does not exclude a VTEC infection in the patient

SAMPLING PROCEDURE AND INFORMATION

Type and Number of Specimens

Clinical material includes faeces, rectal swabs, urine and — in a few cases — blood cultures.

Three consecutive stools should be taken separately as early as possible and preferably during the active period of diarrhoea. However, some patients may shed VTEC for periods longer than the usual first week and specimens should be collected, if necessary up to some weeks after the onset of diarrhoea, if early samples were not taken.

Systematic studies of the importance of the number of stool specimens are lacking. Assuming that the diagnostic yield of VTEC is similar to that of other enteric pathogens, the yield is expected to increase with the number of specimens examined. The diagnostic yield of *Salmonella*, *Campylobacter*, *Shigella* and *Yersinia enterocolitica* can be increased by 15.9% if two instead of one stool specimen is examined and a further 7.3% if a third specimen is examined (Table 1).

The risk of failure in VTEC detection is increased in samples from patients undergoing antibiotic treatment.

Note: Antibiotic treatment of children with VTEC O157 infection increases the risk of HUS (11). This has not been shown to be the case for adults and it must be emphasised that the above statement only refers to the detection of VTEC, not to the prognosis upon VTEC infection.

TABLE 1.
NUMBER OF SPECIMENS WITH *SALMONELLA*, *CAMPYLOBACTER*, *SHIGELLA* OR *YERSINIA ENTEROCOLITICA* FROM PATIENTS THAT HAVE HAD 3 SPECIMENS INVESTIGATED
(Department of Gastrointestinal Infections, SSI, from 1/1-1995 to 31/8-1996).

Number of specimens	Number of positives	Increase in number of positives	Percentage increase in diagnostic yield
1	1634		
2		260	15,9%
3		138	7,3%

Information

It is important to include complete patient information of both clinical and epidemiological character with each specimen:

- Foreign travel
- Bloody/non-bloody diarrhoea
- Abdominal cramps
- Fever
- Other ill persons or contacts
- Suspected source of infection
- Occupation (adults) or institutional presence (all ages).

This information serves to assess the relevance of laboratory procedures, but is also crucial in reporting and surveillance of food-borne diseases.

SAMPLE TRANSPORT

Faecal samples should be stored in air-tight containers and are best tested within two hours of collection. If transport cannot be avoided, the specimen should be transported and stored at 4°C and tested within 48 hours of collection. Systematic studies of the importance of transport time and conditions are lacking. Assuming that the loss of diagnostic yield of VTEC is similar to that of other enteric pathogens, the loss is expected to decrease with time. Odds Ratio for loss of detection is 0.83 for *Campylobacter*, 0.88 for *Salmonella*, 0.85 for *Yersinia enterocolitica* and 0.72 for *Shigella* at SSI if transport time is 2 days rather than up to one day.

Two rectal swabs should be transported in appropriate transport medium such as Stuart, Cary-Blair, Amies, buffered glycerol-saline, etc.

DETECTION

The detection of VT production or *vtx* genes in clinical samples without the subsequent isolation of VTEC is incomplete and should be considered only a presumptive result. In order to diagnose all VTEC strains, it is essential that the diagnostic strategy is directed towards VTEC in general, rather than particular subset(s) of VTEC such as VTEC of serogroup O157.

Definition of VTEC

Strains belonging to the group of Verocytotoxin-producing *Escherichia coli* (VTEC) are characterised by the production of one or more types of Verocytotoxins (VT), such as VT1, VT2 or VT2-related toxins.

Occurrence and Exposure

Non-O157 VTEC strains occur more frequently than VTEC O157 in farm animals such as cattle, sheep, goats and pigs and are more likely to contaminate food. Therefore, humans are more often exposed to strains of non-O157 VTEC than to strains of O157 VTEC, and non-O157 VTEC are likely to be more frequently isolated from diarrhoeic humans than VTEC O157 strains.

Note: In many countries, if the clinical criterion is HUS then VTEC O157 strains are isolated more frequently than non-O157 VTEC.

Diversity

In contrast to VTEC O157 strains, non-O157 VTEC show a high diversity in their genotypes, serotypes and other phenotypical traits and resemble non-toxigenic faecal *Escherichia coli* strains in relation to typical biochemical reactions and growth properties on standard culture media. Due to this diversity, the identification of non-O157 VTEC from faecal samples, food or other sources is complicated.

Methods

The above mentioned type diversity means that there is a requirement for rapid, easy, moderate cost methods for the reliable identification and isolation of VTEC strains, irrespective of their serotype. A number of methods suitable for this purpose have been developed, but there is no internationally recognised standard procedure. The range of methods currently includes biological assays, immunological methods, nucleic acid based assays and culture methods. VTEC detection systems are now commercially available in the form of standardised diagnostic kits. Thus, there is no longer any excuse for not diagnosing VTEC in non-specialised routine diagnostic laboratories.

Note: Budget is often given as an important reason for omitting tests in many laboratories. It is important to stress that the definition of appropriate sampling criteria will, in broader terms, save more money within the health care system than the omission of microbiological analysis for VTEC in clinical specimens (12).

SCREENING

Screening serves to identify presumptive VTEC positive specimens. The screening methods are DNA based, biological, immunological or phenotypical tests, such as O grouping of commonly-occurring VTEC serotypes, enterohaemolysin production of the majority of VTEC types, or the failure to ferment sorbitol or produce β-glucuronidase by most VTEC O157.

Unfortunately, many screening methods are incomplete because they are only directed against a subset of the VTEC strains, e.g., O157. It is highly recommended that methods detecting all VTEC types are implemented. No matter which screening method is being employed, it is of utmost importance that the user analyses and considers its limitations. Subsequently, procedures for referral to more specialised laboratories or reference centres must be described, and easily understood by everybody at the primary screening laboratory.

DNA Based Methods

DNA based methods include PCR and probe techniques. PCR is particularly well suited to screening, whereas probes are better suited for characterisation of individual isolates. For this reason, only PCR methods are described in the screening section, and probes are detailed in "Methods of Isolation".

Amplification of target genes present in stool or primary stool cultures has been found to be problematic compared with samples containing pure culture bacterial isolates. Non-specific inhibitors in stools can cause false-negative PCR amplification results. According to published data, the sensitivity of the PCR assay for detection of VTEC in stool specimens was found to vary between 10 to 10^6 CFU/ml.

Note: This very wide range has to be considered before and during implementation. Such factors as presence of inhibitors, purity of target DNA and PCR product detection methods, can have a considerable influence on sensitivity.

PCR. PCR using primers directed towards the *vtx* genes (13), may be performed on DNA extracted from stool (14-18), stool mixed in a reagent cocktail containing Ampdirect (R) A (Shimadzu) (19), on primary culture from colonies or post enrichment using selective (MacConkey, modified TSB, etc.)

or non-selective (BHI, TSB, etc.) broths (20-22). A growing number of multiplex PCR assays have been described and are listed below.

Note: Laboratories can use primers to detect different VT sequences individually, or use multiplex PCR for VT1, VT2 and VT2 variants.

Sensitivity of detection from stool can be increased by enrichment culture, although there is no generally accepted protocol or enrichment procedure for screening for VTEC. Sensitivity may be influenced by the presence of non-target cells (23).

A magnetic capture-hybridisation PCR (MCH-PCR) (24) for food needs to be validated on clinical specimens.

Semi-automated detection of VTEC O157 and non-O157 VTEC by fluorogenic polymerase chain reaction (TaqMan) (25) and an *eae*O157 TaqMan (26) have been described.

Gene Identification. Fourteen PCR systems targeted at *vtx* genes were tested by Bastian *et al.* (27). Only one system (the Lin all-VTs system (28)), detected all *vtx* genes in all the studied strains. Two other screening systems, the Read all-VTs (29) and the Karch all-VTs (30) detected all *vtxs* except the *vtx*2ev found in strain H.I.8 of O group O128 isolated from infant diarrhoea (31). In some laboratories, in-house primers are used, but this approach cannot be recommended, because of difficulties in interlaboratory comparisons.

A number of target genes other than the *vtxs* have been used, in addition to the direct detection of *vtx* by PCR. The ongoing testing of these PCRs against an array of virulence genes and epidemiological markers will serve to identify candidates for the possible inclusion in multiplex PCR or other sequence-based detection methods which may be used in the screening, detection or characterisation of VTEC. Recent examples of such multiplex PCR procedures have been tested by veterinarians (32-34) and food microbiologists (35), but there is an urgent need to have these validated on clinical specimens.

Primers have been directed against either the *rfb* region encoding the O157 antigen (36,37), or the serotype specific gene *fli*C for the H7 flagella including RFLP of the amplified product (38), or both antigens (39). Other combinations of primers in multiplex PCR protocols include the O157 antigen, *eae* and plasmid pO157 (40), and *vtx*, *eae* and *fli*C (H7) (41). Non-O157 O26 has been targeted (15) and selected commonly occurring non-O157 strains (O111, O113) have been targeted, along with O157 from primary faecal culture (42,43). The O antigen detection of *rfb*O111 and *rfb*O157 by PCR has been combined with specific targeting at *vtxs*, *eae*, *ehx*A (44), and an upstream *eae*O157 sequence together with the VTs has been described (45).

Identification of VT2-producing *E. coli* by PCR in combination with a Digoxigenin (DIG)-labelled DNA probe may facilitate the early detection and epidemiological analysis of *vtx*2 and *vtx*2 variants from individual colonies (46).

A DNA hybridisation-hydrophobic-grid membrane filter method detected VTEC strains isolated from food and animal faecal samples, by using PCR to screen for the presence of VTEC after broth enrichment, and then filtering VTEC-positive cultures through hydrophobic-grid membrane filters (HGMFs) subsequently incubated on MacConkey agar. The filters were probed with a digoxigenin-labeled PCR product, generated by amplification of a conserved Verocytotoxin gene sequence. Replication of the growth on filters allowed probe-positive colonies to be recovered (47).

Biological Assays

Screening for VTEC from stool or mixed cultures of bacteria using cytotoxicity assays can be a very good method for identifying presumptive positive specimens. However, cytotoxicity assays need laboratories specifically equipped to carry out cell culture, and the test procedure is relatively time consuming and expensive.

The Vero Cell Assay (VCA). The classical assay using Vero cells to detect Verocytotoxin was first described by Konowalchuk *et al.* (31) and has been used for screening stool cultures for VT in polymyxin B extracts of colony sweeps (VT/PECS) (48). The specificity of cytotoxicity tests for Verocytotoxins can be affected by other cytotoxins which may be present in stool or stool cultures. Thus, neutralisation assays using antisera directed against VT1, VT2 and VT2-variants are necessary to make cytotoxicity assays specific.

Free Faecal Verocytotoxin (FVT). A small portion of stool may be emulsified in PBS, centrifuged and sterile filtered before testing two-fold dilutions of the sterile supernatant in the VCA to detect free faecal VT (FVT) (48). Sterile filtration of the supernatant can be omitted if appropriate antibiotics such as gentamicin (60 μg/ml) and amfotericin B (0.25 μg/ml) are added to the growth medium used in the VCA test. Subsequent neutralisation tests of positive specimens is required by both methods.

Immunological Detection of VT and VTEC-related Surface Antigens

An ELISA technique, combined with enhanced toxin production during culture with mitomycin C, detected VT1 and VT2 from VTEC O157 in mixed culture in which these components represented 1% and 0.025% respectively, of the mixture. VT2 from O157 VTEC added to faecal samples was detected when O157 comprised less than 0.1% of the coliform population (49).

Monoclonal antibodies against VT1 and VT2 in a sandwich ELISA using bacterial cultures from human diarrhoeal faecal specimens is quite sensitive but not very specific (50).

Receptor based ELISAs using the P1 glycoprotein from hydatid cysts (51) or the Gb3 receptor are both sensitive and specific. Thus, the Gb3 based (R)ELISAs detected VT1 in polymyxin extracts of a mixed culture when the toxin-positive strains represented only 1% of the mixture (52), with 100% specificity (53). A modified Elek test for VT1 showed 91% sensitivity and 100% specificity on cultures, but was not tested for VT2 (54). ELISA is both more sensitive and specific than conventional sorbitol-MacConkey agar (SMAC) culture for the detection of O157 from stool specimens (55).

Commercially produced magnetic polymer beads coated with a polyclonal antibody increased the recovery of O157 VTEC from mixed suspensions, but was offset by a loss of specificity (56). Two monoclonal antibodies against the H7 also reacted with H23 and H24, as did commercial polyclonal antisera (57). This cross-reaction is not known to occur at The International *Escherichia* and *Klebsiella* Centre (WHO) and stresses the need to identify the serotype-specific epitopes of H antigens. A monoclonal antibody against O157 has also been described (58).

In general, the (R)ELISAs are rapid and sensitive for the detection of presumptive VTEC positive specimens, especially where low numbers of the organism are present in faeces, and when the infection is caused by non-O157 VTEC (59).

Phenotypical Screening

Biochemical Reactions and Resistance. No common biochemical denominator for all VTEC has been described. The sorbitol- and β-glucuronidase negative phenotypic properties of the majority of VTEC O157 (60) has facilitated detection in mixed flora on different selective media such as sorbitol-MacConkey (SMAC) agar (61,62) or 4-methylumbelliferyl-β-D-glucuronide (MUG) containing medium (63). Other characteristic phenotypical features of O157 include the failure to ferment rhamnose on agar plates and partial resistance to cefixime and tellurite. Modification of SMAC medium by the addition of rhamnose (64), or cefixime and/or tellurite has been reported to improve detection of VTEC O157 because growth of most other *E. coli* is partially or completely inhibited (65,66). These biochemical features have been used individually or in combinations. Examples include hydrolysis of 4-methyl-umbelliferyl-D-glucuronide (MUG) by β-glucuronidase (63) or SMAC plates containing cefixime and rhamnose (CR-SMAC) (65,66), or both cefixime and tellurite (CT-SMAC) for O157 (66,67). However, all these methods not only fail to detect other VTEC but also fail to detect all VTEC O157 because some O157 strains may also be sorbitol positive (68,69), MUG positive (68,70) or not capable of growing on CT-SMAC (14). Finally, false positive sorbitol-negative non-VTEC O157 EPEC (69) or *E. hermannii* (71) may be identified.

O Grouping and H Testing by Agglutination. Diagnostic antisera may be used for presumptive identification and O grouping, by slide agglutination of live cultures directly from plates. Live slide agglutination has been used to screen for EPEC, EIEC and selected ETEC O groups. This simple serological test also makes it possible to identify specimens of some of the most commonly occurring VTEC strains, and to select isolates for further specific virulence testing (72). VTEC belonging to the EPEC O group O26 have been identified in the UK by this screening method (73). In Italy, 18% of presumptive EPEC strains identified using commercially available EPEC antisera were found to be VTEC belonging to EPEC O groups O26, O111 and O128 (74).

VTEC may be found in the following EPEC O groups: O26 (2,75-84), O55 (2,78,84,85), O86 (86), O111 (2,75,78-81,84,87-92), O114 (2), O119 (2), O125 (2), O126 (2), O128 (2,74,78,84).

A number of commercial diagnostic antisera are available for screening. However, most of these kits do not report the actual occurrence of the most common VTEC O groups. It would be relevant to include at least O103, O91, O121 and O145 in these kits because these O groups occur world-wide (79,93,94) and have been isolated from cases of HUS (2). Three commercial latex agglutination kits for O157, and one for H7, were validated to be both sensitive and specific (95).

Note: It should be stressed that VTEC outside this panel of O groups will not be detected. In Denmark, 50% or more VTEC strains will be overlooked if O grouping is used. Many cross reactions occur: O157 has been reported to cross-react with O:IB of *Yersinia pseudotuberculosis*, O:9 of *Yersinia enterocolitica*, O:30 of *Salmonella* group N (96), *E. hermannii* (71,97), *Citrobacter freundii* (98) and *Citrobacter sedlakii* (99). Furthermore, O157 has been reported to cross-react with 12 O antigens of other *E. coli* O groups (96). Only a few of these (O7 and O116) are also registered by The International *Escherichia* and *Klebsiella* Centre (WHO). This stresses the need for validation of the available antisera, biochemical confirmation as *E. coli* and examination for VT production.

Culture Methods (Enterohaemolysin-Agar). The production of VT is associated with the enterohaemolytic phenotype in 75% to 90% of primary VTEC isolates from patients with diarrhoea or HUS. In contrast, only 4.2% of VT-negative faecal VTEC express the enterohaemolytic phenotype (79). Thus, screening for enterohaemolytic colonies can facilitate rapid detection and isolation of enterohaemolysin-positive VTEC from clinical and other samples.

Note: Enterohaemolysin production varies according to serogroup/serotype.

The enterohaemolytic phenotype can be detected by streaking or plating bacteria on washed sheep blood agar (WSBA) plates followed by incubation at 37°C. In contrast to *E. coli* α-haemolysin, which shows visible lysis of erythrocytes after 1 h of incubation of inoculated blood-agar plates at 37°C, the enterohaemolytic phenotype becomes visible only after prolonged (usually overnight) incubation on WSBA, producing zones of lysis which are smaller and more turbid than those of α-haemolytic *E. coli* (100). In most VTEC strains the genes encoding the Ehly-phenotype EHEC-*hly*A were found to be located on the large virulence plasmid (101,102).

The enterohemolysin test is easy to perform and by plating dilutions of stool cultures small quantities (0.03% of the total coliform counts) of enterohaemolytic bacteria can be detected (103). Standardised WSBA-plates ("Enterohemolysin-Agar") which are needed as reagents for detection are commercially available. Antibiotic supplementation of enterohaemolysin-agar plates was found to increase selectivity (104). The presence of large amounts of α-haemolytic *E. coli* can cause problems for identifying small quantities of enterohaemolytic colonies. Bacterial overgrowth with *Proteus*, *Pseudomonas* and gram-positive bacteria present in stool can cause problems for identification of enterohaemolytic *E. coli* on unsupplemented WSBA (103).

Finally, enterohaemolysin production is quite variable, even within the same O group, e.g., O111. This is a big problem as enterohaemolysin genes may be present but their expression is not detected on WSBA plates (105), indicating their absence in many human and bovine isolates (106).

Comparisons of Screening and Identification Methods. Comparisons between the different screening and identification methods have been described in several studies:

- PCR, VCA & Ridascreen (107)
- PCR, immunomagnetic separation (IMS) & direct culture (14,108)
- PCR & VCA (109)
- Colony hybridisation with poly- and oligonucleotide probes, a bead-enzyme-linked immunosorbent assay & PCR (110)
- Free Vero toxins (FVTs), culture for O157 VTEC followed by colony blots, VCA and Southern blots (111)
- VCA, Premier EHEC, PCR and Enterohemolysin-Agar (112,113)
- Immunomagnetic separation, direct culture CR-SMAC and CT-SMAC (114)

METHODS OF ISOLATION

It is essential that specimens found to be positive by one or more of the above mentioned screening procedures are further examined in an attempt to isolate VTEC for subsequent referral and/or typing. Such typing, together with the clinical data is necessary in epidemiological investigations.

Colony immunoblot and colony DNA hybridisation assays are very valuable for the isolation of VT-positive strains from samples found to be positive by one or more of the screening methods. Every other currently available method requires testing of individual colonies for VT production or VT genes. Frequently, VTEC are present in low numbers and isolation may involve testing large numbers of colonies.

Biological Assays (Cytotoxicity Assays Performed with Vero or HeLa Cells)

The isolation of VTEC from stool, stool cultures or mixed cultures of bacteria by using cytotoxicity assays can be very laborious, because individual bacterial isolates have to be tested for cytotoxicity. Moreover, as noted above, cytotoxicity assays need laboratories specifically equipped for cell culture, and the test procedure is relatively time consuming and expensive. For these reasons, cytotoxicity tests are confined to specialist laboratories, and are not suitable for routine diagnostic laboratories.

Cytotoxicity assays can be performed with clinical samples in the form of pure cultures of bacteria. Cell culture toxicity tests are very sensitive and react to VT1, VT2 and VT2-variants, although they may vary in their responses to different VT-types. Nevertheless, the VCA remains the "Gold Standard" method, because it is highly sensitive. It is, however, less specific than PCR and some ELISAs, because it detects different types of toxins. Specificity is increased if the VCA is combined with neutralisation tests.

Immunological Methods (Colony Immunoblot, Enzyme-immuno Assays, Latex Agglutination Tests)

A variety of immunological methods have been developed for the detection of Verocytotoxins from bacterial isolates. The colony immunoblot allows identification and isolation of VT-producing bacterial colonies among a VT-negative background in ratios of 1:1000 to 1:5000. The sensitivity of the assay can be enhanced by antibiotic treatment of bacteria resulting in increased release of Verocytotoxins. The specificity of the assay depends primarily on the antibody which is chosen for detection of VT (115). The employment of the colony immunoblot is labour intensive and may cause problems when large numbers of clinical samples have to be examined within a short period of time.

Until now, VTEC detection by the colony immunoblot is restricted to specialised laboratories since a standardised test kit for the immunoblot is not commercially available.

The addition of mitomycin C significantly enhances the amount of toxin available for immunological detection. An immunoblot assay examined in a field setting was able to detect VT-producing *E. coli* with a high level of sensitivity and specificity. This method is a rapid and reliable alternative to DNA probing, for the detection of VT-producing organisms in stool samples (116).

DNA Based Methods (Colony Hybridisation and PCR Assays)

VTEC specific DNA-methods are based on the detection of *vt*-specific gene sequences, and the presence of *vt*-genes is generally associated with VT production in *E. coli*. Colony hybridisation assays with gene probes specific for *vtx*1, *vtx*2 and *vtx*2 variants have been widely used in studies performed on *E. coli* from animals, humans and food. Like the colony immunoblot, colony hybridisation has the advantage that identification and isolation of VT-producing bacterial colonies is performed in one step. However, the introduction of the colony hybridisation assay to smaller routine diagnostics operations faces the same problem as the colony immunoblot. Both procedures are difficult to perform in non-specialised diagnostic laboratories, where only relatively few samples have to be tested at a time. Moreover, standardised reagents or test kits are not generally available. However, in larger units such as The Department of Gastrointestinal Infections at Statens Serum Institut (SSI) which receives half of the stool cultures performed in Denmark, the colony hybridisation method can easily be implemented. Since 1997, all stools from patients with grossly bloody diarrhoea, and all stools from patients 4 years of age or less with a history of bloody diarrhoea, have been examined for VTEC. Additionally, stools from patients evaluated for travellers or persistent diarrhoea have been cultured for VTEC. The methods used were a combination of colony hybridisation using pooled probes for *vtx*1, *vtx*2 and *ipa*H (VTEC and EIEC), LT, STh and STp (ETEC), and a single *eae* probe (A/EEC including EPEC), with live slide agglutination of suspect colonies with O antisera against the most common VTEC and EPEC serotypes. This approach resulted in the identification of a total of 117 VTEC infections during 1997-99 (annual incidence of 1 per 100,000).

A variety of *vtx* probes have been described. Polynucleotide DNA probes were the first to be used (see Table 2). They require the use of at least one probe for *vtx*1 and one probe for *vtx*2, unless specific variants are being sought. Oligonucleotide DNA probes are described in numerous references, some of which are listed in Table 2.

TABLE 2.
POLYNUCLEOTIDE AND OLIGONUCLEOTIDE DNA PROBES, AND PCR OLIGOS/PRIMERS BY REFERENCE

VT gene	Reference number		
	Polynucleotide	Oligonucleotide	PCR
VT all			(28) (29) (30) (117)
*vtx*1	(110) (118) (119) (120)	(75) (110) (121) (122) (123) (124) (125)	(13) (16) (126) (127)
*vtx*2	(118) (128) (129)	(75) (110) (122) (123) (124) (125) (130) (131) (132)	(16) (46) (127) (133) (134)
*vtx*2 variants:			
*vtx*2vha		(131)	
*vtx*2vhb		(132)	
*vtx*2e		(110) (122) (130)	(135)
*vtx*2ev			(136)

A panel of different PCR amplification protocols for detection of *vtx*-specific gene sequences in clinical and other samples has been published (27). Two different approaches for PCR-based VTEC diagnostic tests have been described. One type of protocol uses primers specific for target sequences situated in the conserved region of the A-subunit of *vtx*1, *vtx*2 and *vtx*2-variants, in order to detect all genotypes of *vtx* by one reaction (28-30,117) (see Table 2). Other protocols use different primer sets for target sequences which are specific for *vtx*1, *vtx*2, *vtx*2e or *vtx*2ev, respectively (see Table 2). Different primer sets can be used in separate PCR amplifications, or in combination within a multiplex PCR.

Although PCR is a very suitable method for screening large numbers of samples, additional laboratory work is required to obtain an VTEC isolate from a PCR-positive sample, because individual colonies have to be tested for *vtx* genes or for VT-production. Specific PCR tests targeting the *vtx* genes will be more available as commercial kits in the near future, which will allow their application in laboratories performing PCR diagnostic assays.

Once an isolate has been established as gene positive, either by probe or by PCR, the isolate must be tested for VT expression (VCA or serological assay such as ELISA or VTEC-RPLA) and biochemically examined.

COMMERCIAL DIAGNOSTIC KITS FOR VTEC

Commercially obtainable kits are generally easy to perform and can be used for detection without needing specialised laboratories and staff. However, subsequent isolation of a VTEC strain from a positive sample can be cost, time and labour intensive, because individual bacterial isolates have to be tested for VT-production.

A number of different diagnostic kits for the detection of VT are available, and are now widely employed in routine diagnostic laboratories in many countries. Some of the kits can be used for detection of VT in stool, stool cultures and bacterial isolates. VT-ELISAs are not equally sensitive for detection of VT1, VT2 and VT2 variants. The sensitivity of purchased VT-ELISAs in the detection of small amounts of VT has been shown to be lower than the Vero cell cytotoxicity assay. Antibiotic treatment of bacteria can enhance the detection sensitivity. False-positive reactions can occur, particularly when faecal material is used for direct testing. A PCR kit for detection of *E. coli* O157:H7 is already commercially available (Qualicon).

Tables 3-6 list commercially available test kits, but are not comprehensive. Rather, they represent those products that were known to the authors at the time of writing. The products are listed in random order without any indication of preference to relation to performance, quality or price. Presumptive tests which screen for selected O groups or just O157 are listed in Table 3.

Although some VTEC strains can be identified by screening for other virulence markers than VT (such as the *eae*-gene) or for phenotypical traits (such as production of enterohemolysin), production of Verocytotoxin(s) is the only property which is common to all types of VTEC. Therefore, a method which is suitable for identification and isolation of all VTEC strains should be based on the detection of Verocytotoxins or their genes. The detection of VT production or of *vtx*-related genes in clinical and other samples, which is not supported by a subsequent attempt to isolate the VTEC strain, is incomplete and should be considered as a presumptive result. Recovery of an isolate is a prerequisite for the further characterisation of the strain of VTEC, necessary for identification of outbreaks and infection sources, and as part of the wider surveillance of VTEC infections including the detection of newly emerging human pathogenic types. For these reasons, tests used for identification of VTEC in clinical or other samples should either be capable of isolating a culture of the target VTEC strain, or should be used in conjunction with another test which is capable of providing the VTEC isolate in pure culture.

TABLE 3.
SEROLOGICAL DETECTION OF *E. COLI* O157 AND OTHER O GROUPS

Name of commercial assay	Manufacturer	Diagnostic principle	Comments
EPEC/VTEC kit	SSI	Slide agglutination	Selected O groups
VTEC kit	Sifin	Slide agglutination	Selected O groups
	Sanofi Pasteur	Slide agglutination	Selected O groups
E. coli O157-AD and -F, O26- and O111F	Denka Seiken	Latex agglutination	Selected O groups
	Innogenetics (former LD Diagnostics)	Slide agglutination	Selected O-groups
E. coli O157	Becton Dickinson	Tube agglutination	*E. coli* O157
RIM	Remel	Latex agglutination	*E. coli* O157
E. coli O157	Oxoid	Latex agglutination	*E. coli* O157
Prolex	Pro Labs	Latex agglutination	*E. coli* O157
EcolexO157	Orion Diagnostics	Latex agglutination	*E. coli* O157
Foss-EIA	Foss Electric	Immunoblot Automated EIA	*E. coli* O157
Petrifilm EHEC	3M	Immunoblot	*E. coli* O157
VIP	Biocontrol	Immunoblot	*E. coli* O157
Reveal & Alert	Neogen	Immunoblot	*E. coli* O157
NOW	Binax	Immunoblot	*E. coli* O157
EZ COLI	Difco	Tube EIA	*E. coli* O157
EHEC-TEK	Organon-Teknika	ELISA magnetic beads	*E. coli* O157
Assurance	Biocontrol	ELISA	*E. coli* O157
HECO157	3M Canada	ELISA	*E. coli* O157
TECRA	TECRA	ELISA	*E. coli* O157
E. coli O157	LMD Lab.	ELISA	*E. coli* O157
VIDAS	BioMérieux	Automated ELISA	*E. coli* O157
ImmunoCard STAT!	Meridian Diagnostics	Immunodiffusion by lateral flow	*E. coli* O157 and H7
Biocard EHEC Test	Ani Biotech Oy	Immunodiffusion by lateral flow	*E. coli* O157
Microgen *E. coli* O157	Microgen Bioproducts Limited	Latex agglutination	*E. coli* O157

TABLE 4.
IMMUNOMAGNETIC SEPARATION (IMS) KITS

Name of commercial assay	Manufacturer	Principle of assay	Comments
Dynabeads	Dynal	immunomagnetic beads	*E. coli* O157 [a]
Captivate™	IDG	Immunomagnetic beads	*E. coli* O157, O111 and O26
EHEC IMS	Denka Seiken	Immunomagnetic beads	*E. coli* O26, O111 and O157

[a] More O groups are being introduced: O26, O103, O111 and O145

TABLE 5.
SCREENING FOR PHENOTYPICAL MARKERS

Name of commercial assay	Manufacturer	Principle of assay	Comments
SMAC	Oxoid, Difco and others	Sorbitol fermentation	NSF-*E. coli* O157
CT-SMAC	CT-Supplement can be ordered from Dynal	Sorbitol fermentation	NSF-*E. coli* O157
Fluorocult *E. coli* O157:H7 Agar	Merck	β-glucuronidase production and sorbitol fermentation	NSF-*E. coli* O157
PGUA	SSI	β-glucuronidase production	NSF-*E. coli* O157
BBL *E. coli* O157 CROMagar	Becton Dickinson	Chromogenic substrates	*E. coli* O157
Rainbow™ Agar O157	Biolog	Chromogenic detection of ß-galactosidase and ß-glucuronidase	NSF-*E. coli* O157
Enterohemolysin-Agar	Oxoid, Heipha	Detection of the enterohaemolytic phenotype	NSF-*E. coli* O157, major EHEC types, O26, O103, O111, O145 and others
Enterohemolysin-Agar + Antibiotic supplements	Heipha	Detection of the enterohaemolytic phenotype	NSF-*E. coli* O157, major VTEC types, O26, O103, O111, O145 and others

TABLE 6.
EIA/ELISA TESTS TOXIN DETECTION

Name of commercial assay	Manufacturer	Principle of assay	Comments
Premier-EHEC	Meridian Diagnostics	ELISA	Non-discriminative detection of VTs
ProSpecT Shigalike Toxins	Alexon	Solid phase EIA	Non-discriminative detection of VTs
Optimun®-verotoxine I & II			Non-discriminative detection of VTs
Ridascreen®-Verotoxin	r-biopharm	ELISA	Non-discriminative detection of VTs
Verotox-F	Denka Seiken, distributed by Oxoid	RPLA	Detects VT1 and VT2+variants separately
VeroTest™	Antex Biologics Inc.	ELISA	Non-discriminative detection of VTs
VTEC-screen	Denka-Seiken	RPLA	Designed for testing mixed stool cultures *i.e.* colony sweeps for VTEC, non-discriminative detection of VTs
Shiga Toxin (Stx) Colony Immunoblot	Sifin	Double membrane agar immunoblot	Detects single colonies

ENRICHMENT METHODS

Enrichment is necessary during screening of faecal samples for VTEC, because the number of VTEC bacteria in clinical samples can be quite variable and may be low, particularly in cases of HUS. The enrichment procedures developed for VTEC O157 need to be evaluated in relation to their abilities to recover non-O157 VTEC. At present, no specific procedures for non-O157 VTEC can be recommended. Cultivation of stool material in liquid non-selective medium such as Tryptose Soy Broth or on plates may increase the number of VT-producing bacteria and may therefore assist in the detection of VTEC which are present in low numbers or in a physiologically stressed state, in the stool samples. In some laboratories sorbitol fermenting colonies are sub-cultured for VT determination when non-sorbitol-fermenting colonies are absent.

Serodiagnosis

An alternative approach to the detection of O157 or non-O157 VTEC infection is the detection of VTEC-related antibodies in patients' sera.

Serodiagnosis can provide evidence of VTEC infection, when screening and isolation tests for O157 and/or non-O157 VTEC are negative, particularly in cases of HUS or in investigation of outbreaks. With the exception of VTEC O157, little is known about the human immune response to different VTEC-related antigens. However, screening for VT-antibodies is not recommended, as such antibodies are not consistently produced in infected humans following a single infection (137).

The detection of patients' antibodies to the lipopolysaccharide (LPS) from VTEC, using dotted LPS extracts, provides a simple, rapid procedure for serodiagnosis of VTEC belonging to any serogroup (138).

The use of ELISA and immunoblotting has provided valuable procedures for detecting evidence of infection with *E. coli* O157 and possibly other VTEC (139). Passive haemagglutination LPS is also a useful tool for diagnosis of infection with VTEC O157 (140,141).

CONCLUSIONS AND RECOMMENDATIONS

The suitability of diagnostic systems for the detection and isolation of VTEC is determined by technical and economical factors. Ideally, a test system should cover all variants of the VT-family, which means all VTEC strains. Thus, the value of an identification assay is determined by its sensitivity and specificity for different types of Verocytotoxins or *vtx*-associated genes. The relative sensitivity and specificity of a diagnostic test can be calculated as the number of samples reacting positive, or negative respectively, in relation to a "gold standard" chosen as an indicator of "true" positive and "true" negative isolates. This type of evaluation can be performed in specialised laboratories with a large collection of VT-positive and VT-negative bacteria, using purified VT-, or *vtx*-DNA probes as standard reagents. Once the limits of a test system have been defined, its clinical value has to be assessed. In this process, its diagnostic sensitivity and specificity must be evaluated using clinical samples such as human stool or bacterial cultures. This type of evaluation should be performed in clinical studies using larger numbers of VTEC positive and negative samples. Any VTEC assay which is developed for diagnostic purposes, or for use in epidemiological studies, should be easily adaptable by other laboratories which require high reproducibility of the test results. In general, commercially produced and supplied test kits are most likely to guarantee a standardised, widely available, product.

Apart from the above technical aspects, economical factors are crucial for a successful implementation of a VTEC diagnostic system. These can be summarised as general costs, the time which is needed to obtain a result, and the practicability of a test. The labour and material expenses must be seen in relation to the expected benefits. Medical indications focusing the routine application of VTEC diagnostic tests towards selected clinical features such as bloody diarrhoea or HUS, certain risk groups (such as paediatric patients) and particular situations (such as outbreak investigations) might be helpful for limiting general costs. Secondly, a test result must be available in time to allow appropriate action. Rapid test results are crucial to avoid the application of inappropriate therapies to patients, and to prevent further spread of pathogens. Finally, a test system which is intended for broad application must be suitable for routine laboratories and should be feasible without demanding specific laboratory equipment or skills of the laboratory staff.

Although many sensitive and specific methods have been developed for the detection of VTEC from samples containing mixed cultures of bacteria, most of these cannot be easily adopted in routine diagnostic laboratories for two main reasons:

1) Lack of generally available standardised test reagents in sufficient quantities (non-commercialised products, laboratory standards and reagents)
2) Problems with the implementation of a test system (too laborious, time consuming, expensive and need for special equipment and specialised personnel)

However, good general detection of VTEC in the routine diagnostic laboratory is essential for establishing an effective surveillance and control system for VTEC-infections in the human population.

Of the different detection systems, only a few (colony-immunoblot, DNA-colony blot) are useful for detection and isolation of VTEC from mixed cultures in a single step procedure. Commercially available diagnostic kits can be used for screening of VTEC (VT-ELISA), for isolation of putative VTEC (Enterohemolysin-Agar) or for examination of bacterial isolates (VTEC-RPLA) and combinations of different test systems are needed for identification and isolation of VTEC in routine diagnostic laboratories. Screening of clinical samples for VTEC can cause problems when low numbers of VTEC are present and suitable techniques requiring selective enrichment of different VTEC have to be developed and tested. Selective media containing antibiotics or growth factors were developed for enrichment of VTEC O157 but it has to be established whether or not these can be used for enrichment of non-O157 VTEC types.

REFERENCES

1. REILLY, A. 1998. Prevention and control of enterohaemorrhagic *Escherichia coli* (EHEC) infections: Memorandum from a WHO meeting. Bull. WHO *76*, 245–255.

2. WHO. 1999. Zoonotic non-O157 shiga toxin-producing *Escherichia coli* (STEC). Report of a WHO scientific working group meeting. Berlin, Germany. WHO/CSR/APH/98.8:1–30.
3. CAPRIOLI, A. and TOZZI, A.E. 1998. Epidemiology of shiga toxin-producing *Escherichia coli* infections in continental Europe. *In Escherichia coli* O157:H7 and Other Shiga Toxin-Producing *E. coli* Strains, (J.B. Kaper and A.D. O'Brien, eds.). Washington: ASM Press, 38–48.
4. MEAD, P.S. *et al.* 1999. Food-related illness and death in the United States. Emerg. Infect. Dis. *5*, 607–625.
5. MARKS, S. and ROBERTS, T. 1993. *E. coli* O157:H7 ranks as the fourth most costly foodborne disease. Food Review *16*, 1–8.
6. SMITH, H.R., ROWE, B., ADAK, G.K. and REILLY, W.R. 1998. Shiga toxin (verocytotoxin)-producing *Escherichia coli* in the United Kingdom. *In Escherichia coli* O157:H7 and Other Shiga Toxin-Producing *E. coli* Strains, (J.B. Kaper and A.D. O'Brien, eds.). Washington: ASM Press, 49–58.
7. LOPEZ, E.L. *et al.* 1989. Hemolytic uremic syndrome and diarrhoea in Argentine children: The role of Shiga-like toxins. J. Infect. Dis. *160*, 469–475.
8. MICHEL, P., WILSON, J.B., MARTIN, S.W., CLARKE, R.C, MCEWEN, S.A. and GYLES, C.L. 1999. Temporal and geographical distributions of reported cases of *Escherichia coli* O157:H7 infection in Ontario. Epidemiol Infect. *122*, 193–200.
9. SLUTSKER, L., RIES, A.A., MALONEY, K., WELLS, J.G., GREENE, K.D., GRIFFIN, P.M, and the *Escherichia coli* O157:H7 study group. 1998. A nation-wide case-control study of *Escherichia coli* O157:H7 infection in the United States. J. Infect. Dis. *177*, 962–966.
10. ACHESON, D.W.K. and KEUSCH, G.T. 1996. Which Shiga Toxin-Producing Types of *E. coli* Are Important? ASM News *62*, 302–306.
11. WONG, C.S., JELACIC, S., HABEEB, R.L., WATKINS, S.L. and TARR, P.I. 2000. The risk of the hemolytic-uremic syndrome after antibiotic treatment of *Escherichia coli* O157:H7 infections. N. Engl. J. Med. *342*, 1930–1936.
12. TARR, P.I. 1995. *Escherichia coli* O157:H7: clinical, diagnostic, and epidemiological aspects of human infection. Clin. Infect. Dis. *20*, 1–10.
13. POLLARD, D.R., JOHNSON, W.M., LIOR, H., TYLER, S.D. and ROZEE, K.R. 1990. Rapid and Specific Detection of Verotoxin Genes in *Escherichia coli* by the Polymerase Chain Reaction. J. Clin. Microbiol. *28*, 540–545.
14. KARCH, H., JANETZKI-MITTMANN, C., ALEKSIC, S. and DATZ, M. 1996. Isolation of enterohemorrhagic *Escherichia coli* O157 strains from patients with hemolytic-uremic syndrome by using immunomagnetic separation, DNA-based methods, and direct culture. J. Clin. Microbiol. *34*, 516–519.
15. LOUIE, M. *et al.* 1998. Application of multiplex PCR for detection of non-O157 verocytotoxin-producing *Escherichia coli* in bloody stools: Identification of serogroups O26 and O111. J. Clin. Microbiol. *36*, 3375–3377.
16. BRIAN, M.J. *et al.* 1992. Polymerase Chain Reaction for Diagnosis of Enterohemorrhagic *Escherichia coli* Infection and Hemolytic-Uremic Syndrome. J. Clin. Microbiol. *30*, 1801–1806.

17. RAMOTAR, K., WALDHART, B., CHURCH, D., SZUMSKI, R. and LOUIE, T.J. 1995. Direct detection of verotoxin-producing *Escherichia coli* in stool samples by PCR. J. Clin. Microbiol. *33*, 519–524.
18. TAKESHI, K. *et al.* 1997. Direct detection by PCR of *Escherichia coli* O157 and enteropathogens in patients with bloody diarrhoea. Microbiol. Immunol. *41*, 819–822.
19. OKAMOTO, H. *et al.* 1999. Direct amplification of *Escherichia coli* O157 vero toxin genes from human faeces by the polymerase chain reaction. Ann. Clin. Biochem. *36*, 642–648.
20. PIÉRARD, D., STEVENS, D., MORIAU, L., LIOR, H. and LAUWERS, S. 1994. Three years PCR screening for VTEC in human stools in Brussels. *In* Recent advances in verocytotoxin-producing *Escherichia coli* infections, (M.A. Karmali and A.G. Goglio, eds.). The Netherlands: Elsevier Science B.V. 33–36.
21. PIÉRARD, D., STEVENS, D., MORIAU, L., LIOR, H. and LAUWERS, S. 1997. Isolation and virulence factors of verocytotoxin-producing *Escherichia coli* in human stool samples. Clin. Microbiol. Infect. *3*, 531–540.
22. THOMAS, A., JIGGLE, B., SMITH, H.R. and ROWE, B. 1994. The detection of Vero cytotoxin-producing *Escherichia coli* and *Shigella dysenteriae* type 1 in faecal specimens using polymerase chain reaction gene amplification. Lett. Appl. Microbiol. *19*, 406–409.
23. WEAVER, J.W. and ROWE, M.T. 1997. Effect of non-target cells on the sensitivity of the PCR for *Escherichia coli* O157:H7. Lett. Appl. Microbiol. *25*, 109–112.
24. CHEN, J.R., JOHNSON, R. and GRIFFITHS, M. 1998. Detection of verotoxigenic *Escherichia coli* by magnetic capture hybridization PCR. Appl. Environ Microbiol. *64*, 147–152.
25. SHARMA, V.K., DEAN NYSTROM, E.A. and CASEY, T.A. 1999. Semi-automated fluorogenic PCR assays (TaqMan) for rapid detection of *Escherichia coli* O157:H7 and other Shiga toxigenic *E. coli*. Mol. Cell. Probe *13*, 291–302.
26. OBERST, R.D. *et al.* 1998. PCR-Based DNA amplification and presumptive detection of *Escherichia coli* O157:H7 with an internal fluorogenic probe and the 5' nuclease (TaqMan) assay. Appl. Environ. Microbiol. *64*, 3389–3396.
27. BASTIAN, S.N, CARLE, I. and GRIMONT, F. 1998. Comparison of 14 PCR systems for the detection and subtyping of *stx* genes in Shiga-toxin- producing *Escherichia coli*. Res. Microbiol. *149*, 457–472.
28. LIN, Z., KURAZONO, H., YAMASAKI, S. and TAKEDA, Y. 1993. Detection of Various Variant Verotoxin Genes in *Escherichia coli* by Polymerase Chain Reaction. Microbiol. Immunol. *37*, 543–548.
29. READ, S.C, CLARKE, M.A., DE GRANDIS, S.A., HII, J., MCEWEN, S. and GYLES, C.L. 1992. Polymerase Chain Reaction for the Detection of Verotoxigenic *Escherichia coli* Isolated from Animal and Food Sources. Mol. Cell. Probes *6*, 153–161.
30. KARCH, H. and MEYER, T. 1989. Single primer pair for amplifying segments of distinct Shiga-like toxin genes by polymerase chain reaction. J. Clin. Microbiol. *27*, 2751–2757.

31. KONOWALCHUK, J., SPEIRS, J.I. and STAVRIC, S. 1977. Vero response to a cytotoxin of *Escherichia coli*. Infect. Immun. *18*, 775–779.
32. FAGAN, P.K., HORNITZKY, M.A., BETTELHEIM, K.A. and DJORDJEVIC, S.P. 1999. Detection of shiga-like toxin (Stx1 and stx2), intimin (*eae*A), and enterohemorrhagic *Escherichia coli* (EHEC) hemolysin (EHEC *hly*A) genes in animal feces by multiplex PCR. Appl. Environ. Microbiol. *65*, 868–872.
33. HU, Y., ZHANG, Q. and MEITZLER, J.C. 1999. Rapid and sensitive detection of *Escherichia coli* O157:H7 in bovine faeces by a multiplex PCR. J. Appl. Microbiol. *87*, 867–876.
34. PAL, A. *et al.* 1999. Shiga-toxin producing *Escherichia coli* from healthy cattle in a semi-urban community in Calcutta, India. Indian J. Med. Res. *110*, 83–85.
35. CHEN, S. *et al.* 1998. An automated fluorescent PCR method for detection of Shiga toxin-producing *Escherichia coli* in foods. Appl. Environ. Microbiol. *64*, 4210–4216.
36. DESMARCHELIER, P.M., BILGE, S.S., FEGAN, N., MILLS, L., VARY, J.C. and TARR, P.I. 1998. A PCR specific for *Escherichia coli* O157 based on the *rfb* locus encoding O157 lipopolysaccharide. J. Clin. Microbiol. *36*, 1801–1804.
37. MAURER, J.J., SCHMIDT, D., PETROSKO, P., SANCHEZ, S., BOLTON, L. and LEE, M.D. 1999. Development of primers to O-antigen biosynthesis genes for specific detection of *Escherichia coli* O157 by PCR. Appl. Environ. Microbiol. *65*, 2954–2960.
38. FIELDS, P.I, BLOM, K., HUGHES, H.J., HELSEL, L.O., FENG, P. and SWAMINATHAN, B. 1997. Molecular characterization of the gene encoding H antigen in *Escherichia coli* and development of a PCR-restriction fragment length polymorphism test for identification of *E. coli* O157:H7 and O157:NM. J. Clin. Microbiol. *35*, 1066–1070.
39. NAGANO, I., KUNISHIMA, M., ITOH, Y., WU, Z.L. and TAKAHASHI, Y. 1998. Detection of verotoxin-producing *Escherichia coli* O157:H7 by multiplex polymerase chain reaction. Microbiol. Immunol. *42*, 371–376.
40. FRATAMICO, P.M., SACKITEY, S.K., WIEDMANN, M. and DENG, M.Y. 1995. Detection of *Escherichia coli* O157:H7 by multiplex PCR. J. Clin. Microbiol. *33*, 2188–2191.
41. GANNON, V.P.J., D'SOUZA, S., GRAHAM, T., KING, R.K, RAHN, K. and READ, S. 1997. Use of the flagellar H7 gene as a target in multiplex PCR assays and improved specificity in identification of enterohemorrhagic *Escherichia coli* strains. J. Clin. Microbiol. *35*, 656–662.
42. PATON, A.W. and PATON, J.C. 1999. Direct detection of shiga toxigenic *Escherichia coli* strains belonging to serogroups O111, O157, and O113 by multiplex PCR. J. Clin. Microbiol. *37*, 3362–3365.
43. PATON, A.W. *et al.* 1996. Molecular microbiological investigation of an outbreak of hemolytic-uremic syndrome caused by dry fermented sausage contaminated with Shiga-like toxin-producing *Escherichia coli*. J. Clin. Microbiol. *34*, 1622–1627.
44. PATON, A.W. and PATON, J.C. 1998. Detection and characterization of shiga toxigenic *Escherichia coli* by using multiplex PCR assays for *stx*(1), *stx*(2), *eae*A,

enterohemorrhagic *E. coli hlyA*, *rfb*(O111), and *rfb*(O157). J. Clin. Microbiol. *36*, 598–602.
45. MENG, J., ZHAO, S., DOYLE, M.P., MITCHELL, S.E. and KRESOVICH, S. 1997. A multiplex PCR for identifying shiga-like toxin-producing *Escherichia coli* O157:H7. Lett. Appl. Microbiol. *24*, 172–176.
46. JACKSON, M.P. 1992. Identification of Shiga-like toxin type II producing *Escherichia coli* using the polymerase chain reaction and a digoxigenin labelled DNA probe. Mol. Cell. Probes *6*, 209–214.
47. TODD, E.C.D. *et al.* 1999. Application of a DNA hybridization-hydrophobic-grid membrane filter method for detection and isolation of verotoxigenic *Escherichia coli*. Appl. Environ. Microbiol. *65*, 4775–4780.
48. KARMALI, M.A. 1987. Laboratory diagnosis of Verotoxin-producing *Escherichia coli* infections. Clin. Microbiol. Newslett. *9*, 65–70.
49. LAW, D., GANGULI, L.A., DONOHUE-ROLFE, A. and ACHESON, D.W. 1992. Detection by ELISA of low numbers of Shiga-like toxin- producing *Escherichia coli* in mixed cultures after growth in the presence of mitomycin C. J. Med. Microbiol. *36*, 198–202.
50. BALL, H.J., FINLAY, D., ZAFAR, A. and WILSON, T. 1996. The detection of verocytotoxins in bacterial cultures from human diarrhoeal samples with monoclonal antibody-based ELISAs. J. Med. Microbiol. *44*, 273–276.
51. ACHESON, D.W.K., KEUSCH, G.T., LIGHTOWLERS, M. and DONOHUE-ROLFE, A. 1990. Enzyme-linked immunosorbent assay for Shiga toxin and Shiga-like toxin II using P1 glycoprotein from hydatid cysts. J. Infect. Dis. *161*, 134–137.
52. ASHKENAZI, S. and CLEARY, T.G. 1990. A method for detecting Shiga toxin and Shiga-like toxin-I in pure and mixed culture. J. Med. Microbiol. *32*, 255–261.
53. BASTA, M., KARMALI, M. and LINGWOOD, C. 1989. Sensitive receptor-specified enzyme-linked immunosorbent assay for *Escherichia coli* verocytotoxin. J. Clin. Microbiol. *27*, 1617–1622.
54. GERMANI, Y., BEGAUD, E. and DESPERRIER, J.M. 1994. Easy-to- perform modified Elek test to identify Shiga-like toxin-producing diarrhoeogenic *Escherichia coli*. Res. Microbiol. *145*, 333–340.
55. PARK, C.H., VANDEL, N.M. and HIXON, D.L. 1996. Rapid immunoassay for detection of *Escherichia coli* O157 directly from stool specimens. J. Clin. Microbiol. *34*, 988–990.
56. MORTLOCK, S. 1994. Recovery of *Escherichia coli* O157:H7 from mixed suspensions: Evaluation and comparison of pre-coated immunomagnetic beads and direct plating. Brit. J. Biomed. Sci. *51*, 207–214.
57. HE, Y.S., KEEN, J.E., WESTERMAN, R.B., LITTLEDIKE, E.T. and KWANG, J. 1996. Monoclonal antibodies for detection of the H7 antigen of *Escherichia coli*. Appl. Environ. Microbiol. *62*, 3325–3332.
58. PERRY, M.B., BUNDLE, D.R., GIDNEY, M.A.J. and LIOR, H. 1988. Identification of *Escherichia coli* serotype O157 strains by using a monoclonal antibody. J. Clin. Microbiol. *26*, 2391–2394.
59. LAW, D., HAMOUR, A.A., ACHESON, D.W.K., PANIGRAHI, H., GANGULI, L.A. and DENNING, D.W. 1994. Diagnosis of Infections with

Shiga-Like Toxin-Producing *Escherichia coli* by Use of Enzyme- Linked Immunosorbent Assays for Shiga-Like Toxins on Cultured Stool Samples. J. Med. Microbiol. *40*, 241–245.

60. RATNAM, S., MARCH, S.B., AHMED, R., BEZANSON, G.S. and KASATIYA, S. 1988. Characterization of *Escherichia coli* serotype O157:H7. J. Clin. Microbiol. *26*, 2006–2012.
61. MARCH, S.B. and RATNAM, S. 1986. Sorbitol-MacConkey medium for the detection of *Escherichia coli* O157:H7 associated with hemorrhagic colitis. J. Clin. Microbiol. *23*, 869–872.
62. KRISHNAN, C., FITZGERALD, V.A., DAKIN, S.J. and BEHME, R.J. 1987. Laboratory investigation of outbreak of hemorrhagic colitis caused by *Escherichia coli* O157:H7. J. Clin. Microbiol. *25*, 1043–1047.
63. THOMPSON, J.S., HODGE, D.S. and BORCZYK, A.A. 1990. Rapid biochemical test to identify verocytotoxin-positive strains of *Escherichia coli* serotype O157. J. Clin. Microbiol. *28*, 2165–2168.
64. SMITH, H.R. and SCOTLAND, S.M. 1993. Isolation and identification methods for *Escherichia coli* O157 and other Vero cytotoxin producing strains. J. Clin. Pathol. *46*, 10–17.
65. CHAPMAN, P.A., SIDDONS, C.A., ZADIK, P.M. and JEWES, L. 1991. An improved selective medium for the isolation of *Escherichia coli* O157. J. Med. Microbiol. *35*, 107–110.
66. ZADIK, P.M., CHAPMAN, P.A. and SIDDONS, C.A. 1993. Use of Tellurite for the Selection of Verocytotoxigenic *Escherichia coli* O157. J. Med. Microbiol. *39*, 155–158.
67. KLEANTHOUS, H., FRY, N.K., SMITH, H.R., GROSS, R.J. and ROWE, B. 1988. The use of sorbitol-MacConkey agar in conjunction with a specific antiserum for the detection of vero cytotoxin-producing strains of *Escherichia coli* O157. Epidemiol. Infect. *101*, 327–335.
68. GUNZER, F., BÖHM, H., RÜSSMANN, H., BITZAN, M., ALEKSIC, S. and KARCH, H. 1992. Molecular detection of sorbitol-fermenting *Escherichia coli* O157 in patients with hemolytic-uremic syndrome. J. Clin. Microbiol. *30*, 1807–1810.
69. SCOTLAND, S.M., WILLSHAW, G.A., CHEASTY, T. and ROWE, B. 1992. Strains of *Escherichia coli* O157:H8 from human diarrhoea belong to attaching and effacing class of *E. coli*. J. Clin. Pathol. *45*, 1075–1078.
70. HAYES, P.S., BLOM, K., FENG, P., LEWIS, J., STROCKBINE, N.A. and SWAMINATHAN, B. 1995. Isolation and characterization of a beta-D-glucuronidase-producing strain of *Escherichia coli* serotype O157:H7 in the United States. J. Clin. Microbiol. *33*, 3347–3348.
71. LIOR, H. and BORCZYK, A.A. 1987. False positive identification of *Escherichia coli* O157. Lancet *1*, 333.
72. LAW, D. 1993. Detection of Verotoxin-Producing Strains of *Escherichia coli*. J. Infect. *26*, 228–229.
73. MAHER, D.P. and STANLEY, P.J. 1992. Verotoxin-producing *Escherichia coli* infection in hospitalised patients with acute gastro-enteritis. J. Infect. *24*, 220–221.

74. GIAMMANCO, A. *et al.* 1996. Characteristics of *Escherichia coli* strains belonging to enteropathogenic *E. coli* (EPEC) serogroups isolated in Italy from children with diarrhea. J. Clin. Microbiol. *34*, 689–694.
75. BITZAN, M. *et al.* 1991. Clinical and genetic aspects of Shiga-like toxin production in traditional enteropathogenic *Escherichia coli*. Int J. Med. Microbiol. *274*, 496–506.
76. LEVINE, M.M. *et al.* 1987. A DNA probe to identify enterohemorrhagic *Escherichia coli* of O157:H7 and other serotypes that cause hemorrhagic colitis and hemolytic uremic syndrome. J. Infect. Dis. *156*, 175–182.
77. SCOTLAND, S.M., WILLSHAW, G.A., SMITH, H.R. and ROWE, B. 1990. Properties of strains of *Escherichia coli* O26:H11 in relation to their enteropathogenic or enterohemorrhagic classification. J. Infect. Dis. *162*, 1069–1074.
78. TAMURA, K., SAKAZAKI, R., MURASE, M. and KOSAKO, Y. 1996. Serotyping and categorisation of *Escherichia coli* strains isolated between 1958 and 1992 from diarrhoeal diseases in Asia. J. Med. Microbiol. *45*, 353–358.
79. BEUTIN, L., ALEKSIC, S., ZIMMERMANN, S. and GLEIER, K. 1994. Virulence factors and phenotypical traits of verotoxigenic strains of *Escherichia coli* isolated from human patients in Germany. Med. Microbiol. Immunol. *183*, 13–21.
80. BOPP, C.A., GREENE, K.D., DOWNES, F.P., SOWERS, E.G., WELLS, J.G. and WACHSMUTH, I.K. 1993. Unusual verotoxin-producing *Escherichia coli* associated with hemorrhagic colitis. J. Clin. Microbiol. *25*, 1486–1489.
81. CAPRIOLI, A. *et al.* 1992. Hemolytic-Uremic Syndrome and Vero Cytotoxin-Producing *Escherichia coli* Infection in Italy. J. Infect. Dis. *166*, 154–158.
82. KARMALI, M.A., STEELE, B.T., PETRIC, M. and LIM, C. 1983. Sporadic cases of haemolytic-uraemic syndrome associated with faecal cytotoxin and cytotoxin-producing *Escherichia coli* in stools. Lancet *I*, 619–620.
83. RICOTTI, G.C., BUONOMINI, M.I., MERLITTI, A., KARCH, H., LUZZI, I. and CAPRIOLI, A. 1994. A fatal case of hemorrhagic colitis, thrombocytopenia, and renal failure associated with verocytotoxin-producing, non-O157 *Escherichia coli*. Clin. Infect. Dis. *19*, 815–816.
84. SCOTLAND, S.M. *et al.* 1996. Use of gene probes and adhesion tests to characterise *Escherichia coli* belonging to enteropathogenic serogroups isolated in the United Kingdom. J. Med. Microbiol. *44*, 438–443.
85. DORN, C.R., SCOTLAND, S.M., SMITH, H.R., WILLSHAW, G.A. and ROWE, B. 1989. Properties of Vero cytotoxin-producing *Escherichia coli* of human and animal origin belonging to serotypes other than O157:H7. Epidemiol Infect. *103*, 83–95.
86. UENO, N., YOSHIKUNI, K., SHINKAWA, N., ARIMA, T. and NAGATA, K. 1999. A 3-year-old case died of hemolytic uremic syndrome and encephalohathy due to EHEC O86 infection, October 1999 - Kagoshima. IASR. *20*, 271–272.
87. CAPRIOLI, A. *et al.* 1994. Communitywide outbreak of hemolytic-uremic syndrome associated with non-O157 verocytotoxin-producing *Escherichia coli*. J. Infect. Dis. *169*, 208–211.

88. ALLERBERGER, F., ROSSBOTH, D., DIERICH, M.P., ALEKSIC, S., SCHMIDT, H. and KARCH, H. 1996. Prevalence and clinical manifestations of shiga toxin-producing *Escherichia coli* infections in Austrian children. Eur J. Clin. Microbiol. Infect. Dis. *15*, 545–550.

89. CAMERON, S., WALKER, C., BEERS, M., ROSE, N. and ANEAR, E. 1995. Enterohaemorrhagic *Escherichia coli* outbreak in South Australia associated with the consumption of mettwurst. Comm. Dis. Intell. *19*, 70–71.

90. DORN, C.R. *et al.* 1993. Characteristics of Vero Cytotoxin Producing *Escherichia coli* Associated with Intestinal Colonization and Diarrhea in Calves. Vet Microbiol. *36*, 149–159.

91. KUDOH, Y.A. *et al.* 1994. Epidemiological surveys on verocytotoxin- producing *Escherichia coli* infections in Japan. *In* Recent advances in verocytotoxin-producing *Escherichia coli* infections, (M.A. Karmali and A.G. Goglio, eds.). The Netherlands: Elsevier Science B.V. 53–56.

92. CAMPOS, L.C., WHITTAM, T.S., GOMES, T.A.T., ANDRADE, J.R.C. and TRABULSI, L.R. 1994. *Escherichia coli* serogroup O111 includes several clones of diarrheagenic strains with different virulence properties. Infect. Immun. *62*, 3282–3288.

93. BOCKEMÜHL, J., ALEKSIC, S. and KARCH, H. 1992. Serological and biochemical properties of Shiga-like toxin (verocytotoxin)-producing strains of *Escherichia coli*, other than O-group 157, from patients in Germany. Zbl Bakt - Int. J. Med. Microbiol. *276*, 189–195.

94. TAMURA, K., IYODA, S. and WATANABE, H. 2000. Serotyping of non-O157 enterohemorrhagic *Escherichia coli* isolates during 1997-1999 in Japan. IASR. *21*, 94.

95. DATZ, M., JANETZKI-MITTMANN, C., FRANKE, S., GUNZER, F., SCHMIDT, H. and KARCH, H. 1996. Analysis of the enterohemorrhagic *Escherichia coli* O157 DNA region containing lambdoid phage gene p and shiga-like toxin structural genes. Appl. Environ. Microbiol. *62*, 791–797.

96. ALEKSIC, S., KARCH, H. and BOCKEMÜHL, J. 1992. A biotyping scheme for Shiga-like (Vero) toxin-producing *Escherichia coli* O157 and a list of serological cross-reactions between O157 and other gram-negative bacteria. Int J. Med. Microbiol. Virol Parasitol Infect. Dis. *276*, 221–230.

97. RICE, E.W., SOWERS, E.G., JOHNSON, C.H., DUNNIGAN, M.E., STROCKBINE, N.A. and EDBERG, S.C. 1992. Serological cross-reactions between *Escherichia coli* O157 and other species of the genus *Escherichia*. J. Clin. Microbiol. *30*, 1315–1316.

98. BETTELHEIM, K.A., EVANGELIDIS, H., PEARCE, J.L., SOWERS, E. and STROCKBINE, N.A. 1993. Isolation of a *Citrobacter freundii* Strain Which Carries the *Escherichia coli* O157 Antigen. J. Clin. Microbiol. *31*, 760–761.

99. PARK, C.H., MARTIN, E.A. and WHITE, E.L. 1998. Isolation of a non-pathogenic strain of *Citrobacter sedlakii* which expresses *Escherichia coli* O157 antigen. J. Clin. Microbiol. *36*, 1408–1409.

100. BEUTIN, L. *et al.* 1989. Close association of verotoxin (shiga-like toxin) production with enterohemolysin production in strains of *Escherichia coli*. J. Clin. Microbiol. *27*, 2559–2564.
101. SCHMIDT, H., BEUTIN, L. and KARCH, H. 1995. Molecular analysis of the plasmid-encoded hemolysin of *Escherichia coli* O157:H7 strain EDL 933. Infect. Immun. *63*, 1055-1061.
102. SCHMIDT, H., KARCH, H. and BEUTIN, L. 1994. The large-sized plasmids of enterohemorrhagic *Escherichia coli* O157 strains encode hemolysins which are presumably members of the *E. coli* alpha-hemolysin family. FEMS Microbiol. Lett. *117*, 189–196.
103. BEUTIN, L., ZIMMERMAN, S. and GLEIER, K. 1996. Rapid detection and isolation of shiga-like toxin (verocytotoxin)-producing *Escherichia coli* by direct testing of individual enterohemolytic colonies from washed sheep blood agar plates in the VTEC-RPLA assay. J. Clin. Microbiol. *34*, 2812–2814.
104. LEHMACHER, A., MEIER, H., ALEKSIC, S. and BOCKEMÜHL, J. 1998. Detection of hemolysin variants of shiga toxin-producing *Escherichia coli* by PCR and culture on vancomycin-cefixime-cefsulodin blood agar. Appl. Environ. Microbiol. *64*, 2449–2453.
105. SCHMIDT, H. and KARCH, H. 1996. Enterohemolytic phenotypes and genotypes of Shiga toxin-producing *Escherichia coli* O111 strains from patients with diarrhoea and hemolytic-uremic syndrome. J. Clin. Microbiol. *34*, 2364–2367.
106. GYLES, C. *et al.* 1998. Association of enterohemorrhagic *Escherichia coli* hemolysin with serotypes of Shiga-like-toxin-producing *Escherichia coli* of human and bovine origins. Appl. Environ. Microbiol. *64*, 4134–4141.
107. BONARDI, S., FONI, E. and MAGGI, E. 2000. Comparison of vero cell assay, polymerase chain reaction and an enzyme immunoassay for identification of verocytotoxin-producing *Escherichia coli* O157:H7. Microbiologica. *23*, 47–53.
108. CUBBON, M.D., COIA, J.E., HANSON, M.F. and THOMSON-CARTER, F.M. 1996. A comparison of immunomagnetic separation, direct culture and polymerase chain reaction for the detection of verocytotoxin-producing *Escherichia coli* O157 in human faeces. J. Med. Microbiol. *44*, 219–222.
109. RAHN, K. *et al.* 1996. Comparison of Vero cell assay and PCR as indicators of the presence of verocytotoxigenic *Escherichia coli* in bovine and human fecal samples. Appl. Environ. Microbiol. *62*, 4314–4317.
110. YAMASAKI, S. *et al.* 1996. Typing of verotoxins by DNA colony hybridization with poly- and oligonucleotide probes, a bead-enzyme-linked immunosorbent assay, and polymerase chain reaction. Microbiol. Immunol. *40*, 345–352.
111. RAMOTAR, K., HENDERSON, E., SZUMSKI, R. and LOUIE, T.J. 1995. Impact of free verotoxin testing on epidemiology of diarrhoea caused by verotoxin-producing *Escherichia coli*. J. Clin. Microbiol. *33*, 1114–1120.
112. BEUTIN, L., GLEIER, K. and ZIMMERMANN, S. 1999. Comparative evaluation of different diagnostic methods for detection of Shiga-toxin producing strains of *Escherichia coli* (VTEC) in human clinical stool specimens. *In* 1. Methods for Verocytotoxigenic *E. coli*, (G. Duffy, P. Garvey, J. Coia, Y. Wasteson and D.A. McDowell, eds.). Dublin, Ireland: 15–24.

113. BEUTIN, L., HORBACH, I., ZIMMERMANN, S. and GLEIER, K. 1997. Comparative evaluation of different diagnostic methods for the detection of verotoxin (shiga-toxin) producing strains of *Escherichia coli* (VTEC) in human clinical stool specimens. J. Lab Med. *21*, 537–546.

114. CHAPMAN, P.A. and SIDDONS, C.A. 1996. A comparison of immunomagnetic separation and direct culture for the isolation of verocytotoxin-producing *Escherichia coli* O157 from cases of bloody diarrhoea, non-bloody diarrhoea and asymptomatic contacts. J. Med. Microbiol. *44*, 267–271.

115. KARCH, H., STROCKBINE, N.A. and O'BRIEN, A.D. 1986. Growth of *Escherichia coli* in the presence of trimethoprim-sulfamethoxazole facilitates detection of shiga-like toxin producing strains by colony blot assay. FEMS Microbiol. Lett. *35*, 141–145.

116. HULL, A.E., ACHESON, D.W.K., ECHEVERRIA, P., DONOHUE-ROLFE, A. and KEUSCH, G.T. 1993. Mitomycin Immunoblot Colony Assay for Detection of Shiga-Like Toxin-Producing *Escherichia coli* in Faecal Samples — Comparison with DNA Probes. J. Clin. Microbiol. *31*, 1167–1172.

117. PATON, A.W., PATON, J.C., GOLDWATER, P.N. and MANNING, P.A. 1993. Direct Detection of *Escherichia coli* Shiga-Like Toxin Genes in Primary Faecal Cultures by Polymerase Chain Reaction. J. Clin. Microbiol. *31*, 3063–3067.

118. NEWLAND, J.W. and NEILL, R.J. 1988. DNA probes for Shiga-like toxin I and II and for toxin-converting bacteriophages. J. Clin. Microbiol. *26*, 1292–1297.

119. WILLSHAW, G.A., SMITH, H.R., SCOTLAND, S.M, FIELD, A.M and ROWE, B. 1987. Heterogeneity of *Escherichia coli* Phages Encoding Vero Cytotoxins: Comparison of Cloned Sequences Determining VT1 and VT2 and Development of Specific Gene Probes. J. Gen. Microbiol. *133*, 1309–1317.

120. KURAZONO, H., SASAKAWA, C., YOSHIKAWA, M. and TAKEDA, Y. 1987. Cloning of a Vero toxin (VT1, Shiga-like toxin I) gene from a VT1-converting phage isolated from *Escherichia coli* O157:H7. FEMS Microbiol. Lett. *44*, 23–26.

121. DE GRANDIS, S., GINSBERG, J., TOONE, M., CLIMIE, S., FRIESEN, J. and BRUNTON, J. 1987. Nucleotide sequence and promoter mapping of the *Escherichia coli* Shiga-like toxin operon of bacteriophage H-19B. J. Bacteriol. *169*, 4313–4319.

122. BROWN, J.E., SETHABUTR, O., JACKSON, M.P., LOLEKHA, S. and ECHEVERRIA, P. 1989. Hybridization of *Escherichia coli* producing Shiga-like toxin I, Shiga-like toxin II, and a variant of Shiga-like toxin II with synthetic oligonucleotide probes. Infect. Immu. *57*, 2811–2814.

123. THOMAS, A., SMITH, H.R., WILLSHAW, G.A. and ROWE, B. 1991. Non-radioactively labelled polynucleotide oligonucleotide DNA probes for selectively detecting *Escherichia coli* strains producing vero cytotoxins VT1, VT2 and VT2 variant. Mol. Cell. Probes *5*, 129–135.

124. KARCH, H. and MEYER, T. 1989. Evaluation of oligonucleotide probes for identification of shiga-like toxin-producing *Escherichia coli*. J. Clin. Microbiol. *27*, 1180–1186.

125. YOH, M. *et al.* 1997. Evaluation of enzyme-labelled oligonucleotide probes to identify enterohaemorrhagic *Escherichia coli*. Microbiol. Immunol. *41*, 879–882.
126. WOODWARD, M.J., CARROLL, P.J. and WRAY, C. 1992. Detection of entero- and verocyto-toxin genes in *Escherichia coli* from diarrhoeal disease in animals using the polymerase chain reaction. Vet Microbiol. *31*, 251–261.
127. CEBULA, T.A., PAYNE, W.L. and FENG, P. 1995. Simultaneous identification of strains of *Escherichia coli* serotype O157:H7 and their shiga-like toxin type by mismatch amplification mutation assay-multiplex PCR. J. Clin. Microbiol. *33*, 248–250.
128. SCOTLAND, S.M, ROWE, B., SMITH, H.R., WILLSHAW, G.A. and GROSS, R.J. 1988. Verocytotoxin-producing strains of *Escherichia coli* from children with haemolytic-uraemic syndrome and their detection by specific DNA probes. J. Med. Microbiol. *25*, 237–243.
129. YUTSUDO, T. *et al.* 1987. Cloning of a Vero toxin (VT2) gene from a VT2-converting phage isolated from *Escherichia coli* O157:H7. FEMS Microbiol. Lett. *48*, 273–276.
130. JACKSON, M.P., NEILL, R.J., O'BRIEN, A.D., HOLMES, R.K. and NEWLAND, J.W. 1987. Nucleotide sequence analysis and comparison of the structural genes for Shiga-like toxin I and Shiga-like toxin II encoded by bacteriophages from *Escherichia coli* 933. FEMS Microbiol. Lett. *44*, 109–114.
131. HII, J.H. *et al.* 1991. Development of verotoxin 2- and verotoxin 2 variant (VT2v)-specific oligonucleotide probes on the basis of the nucleotide sequence of the B cistron of VT2v from *Escherichia coli* E32511 and B2F1. J. Clin. Microbiol. *29*, 2704–2709.
132. THOMAS, A., SMITH, H.R. and ROWE, B. 1993. Use of Digoxigenin- Labelled Oligonucleotide DNA Probes for VT2 and VT2 Human Variant Genes to Differentiate Vero Cytotoxin-Producing *Escherichia coli* Strains of Serogroup O157. J. Clin. Microbiol. *31*, 1700–1703.
133. MARIANI-KURKDJIAN, P. *et al.* 1993. Identification of a clone of *Escherichia coli* O103:H2 as a potential agent of hemolytic-uremic syndrome in France. J. Clin. Microbiol. *31*, 296–301.
134. TYLER, S.D, JOHNSON, W.M., LIOR, H., WANG, G. and ROZEE, K.R. 1991. Identification of verotoxin type 2 variant B subunit genes in *Escherichia coli* by the polymerase chain reaction and restriction fragment length polymorphism analysis. J. Clin. Microbiol. *29*, 1339–1343.
135. JOHNSON, W.M., POLLARD, D.R., LIOR, H., TYLER, S.D. and ROZEE, K.R. 1990. Differentiation of genes coding for *Escherichia coli* verotoxin 2 and the verotoxin associated with porcine oedema disease (VTe) by the polymerase chain reaction. J. Clin. Microbiol. *28*, 2351–2353.
136. JOHNSON, W.M., TYLER, S.D., WANG, G. and LIOR, H. 1991. Amplification by the polymerase chain reaction of a specific target sequence in the gene coding for *Escherichia coli* verotoxin (VTe variant). FEMS Microbiol. Lett. *68*, 227–230.

137. CHART, H., LAW, D., ROWE, B. and ACHESON, D.W.K. 1993. Patients with Haemolytic Uraemic Syndrome Caused by *Escherichia coli* O157 — Absence of Antibodies to Vero Cytotoxin-1 (VT1) or VT2. J. Clin. Pathol. *46*, 1053–1054.
138. CHART, H. and ROWE, B. 1997. A simple dot immunoassay for detecting antibodies to the lipopolysaccharide of Verocytotoxin-producing *Escherichia coli* in patients with haemolytic uraemic syndrome. J. Microbiol. Meth. *28*, 85–88.
139. CHART, H. and JENKINS, C. 1999. The serodiagnosis of infections caused by Verocytotoxin-producing *Escherichia coli*. J. Appl. Microbiol. *86*, 731–740.
140. YAMADA, S., KAI, A. and KUDOH, Y. 1994. Serodiagnosis by Passive Hemagglutination Test and Verotoxin Enzyme-Linked Immunosorbent Assay of Toxin-Producing *Escherichia coli* Infections in Patients with Hemolytic-Uremic Syndrome. J. Clin. Microbiol. *32*, 955–959.
141. YAMADA, S. *et al.* 1993. Detection of Verocytotoxin from Stool and Serological Testing of Patients with Diarrhoea Caused by *Escherichia coli* O157:H7. Microbiol. Immunol. *37*, 111–118.

SUGGESTED FURTHER READINGS

COIA, J.E. 1998. Clinical, microbiological and epidemiological aspects of *Escherichia coli* O157 infection. FEMS Immunol. Med. Microbiol. *20*, 1–9.

FARMER, J.J. III. 1999. *Enterobacteriaceae*: Introduction and identification. *In* Manual of clinical microbiology, (P.R. Murray, E.J. Baron, M.A. Pfaller, F.C. Tenover and R.H. Yolken, eds.). 7th ed. Washington, D.C. ASM Press, 442–458.

GILLIGAN, P.H., JANDA, J.M., KARMALI, M.A. and MILLER, J.M. 1992. Laboratory diagnosis of bacterial diarrhea. (F.S. Nolte, ed.) American Society for Microbiology. Washington, DC. Cumitech *12A*, 1–28.

GRIFFIN, P.M. 1995. *Escherichia coli* O157:H7 and other enterohemorrhagic *Escherichia coli*. *In* Infections of the gastrointestinal tract, (M.J. Blaser, P.D. Smith, J.I. Ravdin, H.B. Greenberg and R.L. Guerrant, eds.). New York: Raven Press, Ltd, 739–761.

Infections of the Gastrointestinal Tract. (M.J. Blaser, P.D. Smith, J.I. Ravdin, H.B. Greenberg and R.L. Guerrant, eds.) New York: Raven Press, 1995:1–1578.

KARCH, H., BIELASZEWSKA, M., BITZAN, M. and SCHMIDT, H. 1999. Epidemiology and diagnosis of Shiga toxin-producing *Escherichia coli* infections. Diagn Microbiol. Infect. Dis. *34*, 229–243.

NATARO, J.P. and KAPER, J.B. 1998. Diarrheagenic *Escherichia coli*. Clin. Microbiol. Rev. *11*, 142–201.

NATARO, J.P. and MARTINEZ, J. 1998. Diagnosis and investigation of diarrheagenic *Escherichia coli*. *In* Molecular Bacteriology, (N. Woodford and A.P. Johnson, eds.). Totowa: Humana Press Inc, 387–406.

PATON, J.C. and PATON, A.W. 1998. Pathogenesis and diagnosis of Shiga toxin-producing *Escherichia coli* infections. Clin. Microbiol. Rev. *11*, 450–479.

STROCKBINE, N.A., WELLS, J.G., BOPP, C.A. and BARRETT, T.J. 1998. Overview of detection and subtyping methods. *In Escherichia coli* O157:H7 and Other Shiga Toxin-Producing *E. coli* Strains, (J.B. Kaper A.D. and O'Brien, eds.). Washington DC: American Society for Microbiology, 331–356.

CHAPTER 4

PRACTICAL CONSIDERATIONS AND DIFFICULTIES ASSOCIATED WITH THE DETECTION OF VEROCYTOTOXIN-PRODUCING *ESCHERICHIA COLI* (VTEC) IN FOODS

CHRISTOPHER BAYLIS

Campden and Chorleywood Food Research Assoc.
Chipping Campden
Gloucestershire GL55 6LD, UK

ANNET HEUVELINK
and
ENNE DE BOER

Inspectorate for Health Protection
P.O. Box 202
7200 AE Zutphen, The Netherlands

AND

HARMEN HOFSTRA

TNO Food and Nutrition Research Institute
P.O. Box 360
3700 AJ Zeist, The Netherlands

INTRODUCTION

Since they were first described by Konowalchuk in 1977 (75), verocytotoxin-producing *Escherichia coli* (VTEC) have become a major cause of haemorrhagic colitis (HC) and haemolytic uraemic syndrome (HUS) in man. The most commonly reported serotype associated with human disease is *E. coli* O157:H7. However, other non-O157 serogroups, in particular O26 and O111, have also been implicated in cases of infection and there have been reports of sporadic cases of infection and outbreaks (21,51,65,69,93). Although there is evidence of infection with non-O157 VTEC in the UK, the true incidence and clinical significance of non-O157 VTEC is currently unknown (146). Whilst some infections have been associated with the consumption of contaminated food, the lack of appropriate methods for detecting these bacteria has made it difficult to establish potential reservoirs of contamination, the true incidence of infection and the epidemiological importance of these other serotypes.

Past studies have shown that VTEC are present in the intestinal tract of a wide range of domestic and wild animals. Carriage rates in ruminants appear to be especially high and contamination of foods from these animals is thought to be a major source of human infection (26). Farm animals, particularly cattle, are known reservoirs for O157 VTEC, although non-O157 serotypes have been reported to be more commonly isolated from food, particularly meat (111,148). However, there are many more serotypes found in cattle and other animals that are rarely or never isolated from humans, and some strains appear to lack the virulence factors commonly found in strains isolated from patients and strains isolated from animals suffering from infection. This is highlighted in a recent report of a strain of O113:H21 associated with HUS that was found to lack the *E. coli* attaching and effacing gene (*eae*) which encodes intimin (102). Similarly, Starr *et al.* (1998) (125) reported a severe case of HUS due to *E. coli* O5:H$^-$, which lacks the *eae*-gene. A recent one-year prospective study of the prevalence of VTEC in cattle, food and children in France, revealed that the majority of VTEC strains (66%) belonged to serotypes not previously associated with severe disease in humans (108).

At present there are no standard methods for the detection of non-O157 VTEC in foods. Serotype O157:H7, unlike other *E. coli*, is a unique clone that has evolved from EPEC serotype O55:H7 (42), which was the most prominent EPEC serotype in the UK in the 1960's (7). It has certain phenotypic characteristics which distinguish it from other *E. coli* strains. These include the absence of β-glucuronidase activity (32) and the inability of most strains to ferment sorbitol within 24 h (144). These features, together with the O157 antigen itself, distinguish this *E. coli* from other strains and have facilitated its detection and isolation from foods. In contrast, there are over 100 genetically diverse serotypes of VTEC which are, however, biochemically similar to typical non-pathogenic *E. coli*. For these reasons, methods being developed for the detection of the VTEC group are generally molecular-based and targeted at the detection of the toxin genes, the toxins themselves or other genes encoding virulence factors possessed by these bacteria. However, despite their ability to produce verocytotoxins, not all strains will necessarily carry the full complement of virulence factors associated with disease and infection in man (105). Moreover, the exact complement and types of virulence factors required for, and their roles in, pathogenesis remains unclear and requires further investigation.

The diversity of serotypes carrying the toxin genes, together with the many differences in the type and complement of virulence factors carried by different strains, along with differences in the specific nature of the toxins they produce, contribute to a very complex situation. Consequently, the development of methods for non-O157 VTEC in foods is still in its infancy. Furthermore, some methods are merely adaptations of methods developed for the detection of these bacteria in clinical samples. These methods are not always suitable or reliable for the detection of all VTEC present in foods. By comparison, since the recognition of the VTEC O157:H7/H⁻ clone as a foodborne pathogen in 1982 (115) methods for the detection and isolation of this serotype have developed considerably. Consequently, there is now a plethora of test kits, new methods and culture media available for the detection and isolation of this *E. coli* O157:H7 or the O157 serogroup in food (88,99,138). Many of these methods rely on the inability of this serogroup to ferment sorbitol, although sorbitol fermenting strains isolated from human disease have now been isolated, especially in Europe (54).

CULTURAL METHODS FOR *E. COLI* O157 AND OTHER VEROCYTOTOXIN PRODUCING *E. COLI*

In common with other foodborne pathogens, the detection and isolation of VTEC from foods requires a specialist approach. However, the detection of these organisms offers greater practical difficulties for the food microbiologist. Foodborne pathogens including VTEC may not be homogeneously distributed throughout the food. They may be present in low numbers, often in the presence of high numbers of competitor organisms including strains of non-pathogenic *E. coli* and other closely related bacteria. It is therefore essential to increase the numbers of target VTEC to detectable levels while inhibiting the background flora.

Enrichment

Two of the most successful enrichment media used for *E. coli* O157 and other VTEC are tryptone soya broth (TSB) and *E. coli* broth (EC), supplemented with selective agents, with or without modifications to their original formulation. In a change from conventional EC broth, the amount of bile salts in modified EC broth (mEC) is reduced to 1.12 g l^{-1} (from 1.50 g), as proposed by Szabo *et al.* (132). This medium is often made more selective by the addition of 20 mg l^{-1} novobiocin (mEC+n) (93). By comparison, modified TSB (mTSB) differs from conventional TSB broth by the addition (per litre) of 1.5 g bile salts No. 3 and 1.5 g dipotassium phosphate (33). This medium is also made more selective by the addition of 20 mg per litre novobiocin (mTSB+n). More recently a standard method for the detection of *E. coli* O157 from foods and animal feeding stuffs has been validated (122). This method, which uses mTSB+n, will eventually become an International Standard (ISO) method for the detection of *E. coli* O157 in foods. However, at present there are no standard methods available for the detection of non-O157 VTEC in foods, and as a number of VTEC serotypes have not been associated with human disease, interpretation of the results would be difficult.

In a study by Chapman *et al.* (26) buffered peptone water (BPW) supplemented with 8 mg l^{-1} vancomycin, 0.05 mg l^{-1} cefixime and 10 mg l^{-1} cefsulodin (BPW-VCC) was successfully used for the detection of *E. coli* O157 in cattle faeces. Subsequently, this enrichment medium has been used for the detection of *E. coli* O157 in foods (19,25,57). However, it may be unsuitable for the examination of some foods, particularly those likely to contain sub-lethally injured cells, because of the sensitivity of such cells to the selective agents used. In a study by Bolton *et al.* (19) the concentrations of cefixime and cefsulodin used in BPW-VCC were reported to act synergistically against strains of *E. coli* O157, leading to failure to detect these bacteria in artificially inoculated foods which were incubated in BPW-VCC before being plated onto normal selective plating media. Furthermore, detection of these bacteria from naturally contaminated foods was less successful from samples enriched in BPW-VCC, than in BPW containing quarter strength and half strength cefsulodin and cefixime, or mTSB. In an earlier study by Weagant *et al.* (142), inhibition of pure cultures of *E. coli* O157:H7 in BPW-VCC was reported, and cefixime was found to be the most inhibitory factor.

Alternative liquid media also used for the cultivation of *E. coli* O157 from foods include modified TSB with vancomycin, cefsulodin and cefixime (EEB) (140), modified tryptone soy broth (TSB) with acriflavin (mTSB+a) (74), modified BPW with casamino acids (mBPW) (112), lauryl tryptose broth (LST) (48), brilliant green bile lactose broth (BRILA) (60), and TSB with cefixime, tellurite and vancomycin (TSB-CTV). The TSB-CTV was originally used for the isolation of *E. coli* O157 from ovine faeces (76).

The efficacies of enrichment media have been tested in a number of studies. Johnson *et al.* (64) analysed food and environmental samples in an investigation of a food-borne outbreak attributed to the consumption of undercooked hamburgers. They found it more difficult to recover viable *E. coli* O157 from mTSB+a enrichments than from mEC+n enrichments when subsequently detected with commercial immunoassays. It was suggested that mTSB+a, although it enhances the production or availability of O157 antigens, may adversely affect cell viability after extended incubation. Sanderson *et al.* (120) found no difference in the rates of recovery of *E. coli* O157:H7 from spiked bovine faeces between TSB containing cefixime and vancomycin, and TSB containing cefixime, vancomycin and tellurite. The rate of isolation of *E. coli* O157 from inoculated minced beef by the immunomagnetic separation (IMS) technique was higher from samples enriched in mEC+n, than in samples enriched in BPW-VCC (6). Fratamico *et al.* (48) reported that enrichment in mEC+n gave similar results to enrichment in LST, when used for the recovery and isolation of *E. coli* O157:H7 from artificially inoculated raw ground beef by IMS. By comparison, EEB was found to be inhibitory to pure cultures of *E. coli* O157, with cefixime appearing to be the most critical factor in this enrichment broth (61). Heckötter *et al.* (60) suggested the simultaneous use of the enrichment media TSB and brilliant green bile lactose broth (BRILA) for the isolation of *E. coli* O157 from foods. Seo *et al.* (123) found mBPW very effective for enrichment of *E. coli* O157:H7 in ground beef.

There are currently no enrichment media specifically recommended for the recovery and growth of non-O157 VTEC. Consequently research in this area is still ongoing, although media developed for the growth and detection of coliforms and *E. coli* in foods are often regarded as satisfactory. In a recent study Aird *et al.* (1999) (4) compared the growth of a range of VTEC strains including O157:H7, O26, O128, O111 and O145 in media commonly used for the isolation of *E. coli*, coliforms and *E. coli* O157. Their results indicated that media developed for the enrichment of O157 gave sub-optimal results in the enrichment of non-O157 VTEC, and that certain strains of O111 and O26 did not grow well in these media. GN broth (Hajna formulation), was one of the best performing media in this study, and has also been used recently, with the addition of novobiocin (GNHAJNAnov), for the isolation of O157 VTEC from cattle faeces (87).

While the concentration(s) and combination(s) of selective agents used are important, incubation temperature also has an significant effect on the efficiency of enrichment media. Temperature not only affects the growth rate of the target organisms, and the time it will take for them to reach detectable levels, it also has marked effects on the type and level of competitor organisms present in the enriched sample. Because the majority of O157 VTEC strains exhibit poor growth in liquid media incubated at 44°C, lower enrichment incubation temperatures are generally used, i.e. 37°C or 42°C. Doyle and Shoeni (32) reported the generation time of O157 VTEC in trypticase soy broth at 42°C as 0.64h compared with 0.49h at 37°C. In a later study by Bolton *et al.* (19), the combination of mTSB and incubation at 42°C was shown to yield the best growth and recovery of O157 VTEC from beef burgers. The choice of incubation temperature, however, is often dependent on the food sample and the associated background microflora. Typically, foods likely to contain relatively low numbers of competitor organisms can be incubated at 37°C without compromising the recovery of the O157 VTEC, although recovery is inherently dependent on the selectivity of the enrichment medium used. To be effective in enriching O157 VTEC cells in raw meats, incubation conditions during selective enrichment should suppress the growth of competing meat microflora, while at the same time maximising the outgrowth of O157 VTEC. As many other non-sorbitol-fermenting organisms in raw meats, like *Hafnia alvei*, show optimal growth at 37°C, enrichment of O157 VTEC at 41-42°C is preferred. Bolton *et al.* (20) found enrichment in mTSB+n at 42°C more selective than at 37°C. However, temperature control is critical as poor growth of these strains has been observed at temperatures above 42°C (109).

Although enrichment with shaking is a common practice in the selective recovery of O157 VTEC, the advantages over static incubation are not well documented. Incubation at 42°C without shaking effectively suppressed ground beef microflora, while allowing good growth of O157 VTEC (18). For routine analytical purposes it is not practical to shake large numbers of raw meat enrichment cultures. However, when short enrichment periods (< 6 h) are used, pre-warmed enrichment broth and incubation with shaking is recommended, to obtain sufficient growth of the target organisms.

The incubation period required will depend on the expected competing microflora and on the format of the next steps in the isolation procedure. In many cases an incubation period of 6 h is used, followed by immunocapture, and plating on selective media. In methods using direct plating, the enrichment broth is generally incubated for a total time of 18 to 24 h.

Resuscitation

Unlike clinical samples, bacterial cells in some foods may have sustained sub-lethal injury during food processing or by the effects of intrinsic factors such as low pH or specific inhibitory substances, associated with particular foods. Such damaged cells are often more sensitive to inhibitory substances, such as bile salts, and antibiotics, normally included in many culture media to suppress background flora. It is therefore very important to be able to identify these stressed and injured cells, because they may very well retain their pathogenic properties. "Normal" enrichment and plating of injured cells is often insufficient to allow their growth and identification. Indeed, many of the previously described media for the selective enrichment of VTEC (such as modified TSB, supplemented with novobiocin (33) have proved less effective in recovering acid or salt-stressed cells, than unmodified TSB (128). This effect is attributed to increased sensitivity to the selective properties of the media, in particular the bile salts in mTSB and the VCC antibiotics used to supplement both mTSB and BPW. This study also showed that BPW-VCC was also more inhibitory to O157:H7 cultures when incubated at 42°C than to cultures incubated at 37°C. Sorbitol MacConkey agar (SMAC) which is used to isolate presumptive *E. coli* O157 which are the mostly slow or sorbitol non-fermenters (83), proved ineffective to recover freeze-, salt- or heat-injured VTECs (3, 31). Improved recovery of injured VTECs was reported by McCleery and Rowe (86) by introducing a brief resuscitation period (2 h at 25°C) on tryptone soya agar (TSA) before adding an overlay with SMAC agar supplemented with MUG (4-methylumbelliferyl-ß-D-glucuronide). A preferable approach for the recovery of injured VTECs would involve non-selective resuscitation, or pre-enrichment as in the "classical" detection of *Salmonella*. This could be done in situations where the competitive flora population is small or has growth requirements that are very different from VTEC. Suitable pre-enrichment media include BPW or mBPW (112).

One important advance has been the inclusion of a membrane fraction of *Escherichia coli* termed Oxyrase® in culture media. It was reported by Yu and Fung (152) that membrane fractions such as Oxyrase® are capable of reducing oxygen concentration, resulting in a reduction in the redox potential of the medium, providing conditions that appear to greatly improve the recovery and growth of injured cells (152). In another study, a universal pre-enrichment medium supplemented with Oxyrase® successfully and simultaneously recovered heat-injured cells of *E. coli* O157:H7 and *Yersinia enterocolitica* from foods containing other organisms (134). More recently, a novel 24 h enrichment system (S.P.R.I.N.T. O157; Oxoid, Basingstoke, UK) that incorporates Oxyrase® and a controlled timed release of bile salts and selective agents into the growth system has been developed for the recovery and growth of *E. coli* O157. This system provides a recovery period of approximately 5 hours before selective agents are introduced into the growth system by disintegration of a timed release capsule. This process, together with the presence of Oxyrase®, appears to enhance the recovery of stressed cells. In a preliminary study this novel system was shown to be superior to the proposed ISO culture method (which used mTSB + n) for the recovery of stressed cells (129).

Immunomagnetic Separation (IMS)

Immunocapture-based separation and concentration techniques include immunological binding (capture), followed by physical separation of the target organisms from a mixed enrichment culture, resulting in concentration of the target organisms. One application of immunocapture is immuno-magnetic separation (IMS) in which samples are mixed with beads coated with antibodies for the target organisms. The target organisms in the sample bind to the immunomagnetic beads, which are then isolated from the bulk sample suspension and other micro-organisms by concentration within a magnetic field. Recovered beads are then plated on a suitable medium and incubated overnight. This technique is now incorporated into many procedures for the detection of VTEC, particularly O157 VTEC, in foods, clinical and environmental samples, where it is a prerequisite step before cultural isolation on plating media in conventional culture methods or before using an end point detection system such an immunoassay. IMS has been shown to be a sensitive method for isolation of *E. coli* O157 from artificially mixed bacterial cultures, inoculated meat samples, and inoculated as well as naturally contaminated bovine faeces (6,28,48,61, 151).

Several commercial IMS systems are available, including Dynabeads™ anti-*E. coli* O157 (Dynal) and Captivate™ O157 (Lab M). The use of IMS has been focused on the detection of *E. coli* O157. However, growing concern over the association between non-O157 VTEC and human infections, together with the limited specificity of methods for these other serogroups, has led to the provision of beads coated with antibodies to other VTEC, notably O26 and O111. One commercially available brand of beads has been recently used for the isolation of serogroup O26 from ground beef and radish sprouts (56). Another brand of beads has shown promising results for the recovery of O26 and O111 from minced beef, milk and apple cider (63).

Although IMS offers improved specificity, the technique can have practical difficulties and provides the potential for cross contamination of samples. Non-specific binding of other organisms to the magnetic beads resulting in carry over of non-target organisms has also been reported (48,92,97,151). This is a particular problem with foods containing high numbers of background organisms, such as ground beef. Solutions to reduce this problem have been suggested, including addition of protein salmine (Protamine) to the culture-bead sample (97), and washing the beads in saline containing 0.002-0.05% Tween 20 (151) or other detergent (48) before plating. The length of time allowed for immunocapture and the number of washes are additional factors that may also influence the amount of non-specific binding, and the levels of success in capturing and retaining the target organism (6). In the study by Okrend *et al.* (97) the numbers of organisms attaching to beads could be varied by changing the time and temperature of incubation. A later study by Bolton *et al.* (19) reported IMS after 6 h incubation as preferable to IMS after 22 h regardless of the suspension medium. Improved detection within the overall IMS procedure i.e., fewer contaminants on selective plating media were found if a pre IMS enrichment temperature of 42°C was applied (20,57).

Foods with a high lipid content, such as dairy products, pose an additional problem within IMS systems. Under normal conditions the complex formed between the beads and bacteria are concentrated into a pellet at the side of the tube by the application of a magnetic field. In samples with a high lipid content, difficulties can be encountered with the efficient formation of this pellet at the side of the tube. Such difficulties can be overcome by leaving 50 μl of liquid in the tube after the first washing step, allowing the pellet to form more easily during subsequent washing steps (141). Furthermore, it is also important to gently resuspend beads by vortex mixing, (not by shaking) during the washing steps. In addition, if the beads-bacteria complex is to be spread over the surface of a plating medium, a swab streak technique should be used. Alternatively, the enriched culture may be screened using an end-point detection device such as an immunoassay.

The combination of IMS with other immunological test systems, including Petrifilm™ HEC Test and VIDAS™ *E. coli* O157, is reported to be a fast and efficient screening procedure for *E. coli* O157 in foods (53, 139). A method combining a 250-min enrichment, IMS and polymerase chain reaction (PCR) enabled detection of a single cfu of *E. coli* O157:H7 in 8 h (52). A rapid method combining IMS with flow cytometry has been found effective in detecting low numbers of *E. coli* O157:H7 in ground beef, juice and milk (123).

Plating Media for VTEC

The *E. coli* O157:H7/H$^-$ clone differs from the vast majority of *E. coli* strains, in that it does not ferment sorbitol in 24 h and the majority of strains are ß-glucuronidase (GUD) negative. These characteristics are used in a variety of media for the specific isolation of *E. coli* O157. Screening for these bacteria using conventional culture methods is usually performed by examining sorbitol MacConkey (SMAC) agar plates for the presence of colourless, i.e., sorbitol-negative colonies. Since it was described by March and Ratnam (83) for the differentiation of *E. coli* O157:H7 from other *E. coli* in clinical samples, SMAC and modified versions of this medium have been universally adopted as the medium of choice for the isolation of *E. coli* O157. One of the problems associated with SMAC medium is that it can support the growth of the other non-sorbitol fermenting (NSF) organisms. Apart from some strains of *E. coli* which are slow sorbitol fermenters, members of other genera are frequently encountered in clinical specimens and food samples, which can give rise to colonies with similar morphologies to colonies of *E. coli* O157. These include *Aeromonas* spp, *Proteus* spp, *Hafnia alvei* and *Escherichia hermannii*. To improve the selectivity of SMAC and the efficiency of recognition of *E. coli* O157, Chapman *et al.* (27) proposed a modified version of this medium containing cefixime and rhamnose (CR-SMAC). The addition of cefixime (0.05 mg/L) is particularly useful for the inhibition of *Proteus* spp., whereas 0.5% rhamnose enables *E. coli* O157 strains that do not ferment rhamnose to be distinguished from other NSF *E. coli* and *E. hermannii* which do ferment rhamnose. In a later study by Zadik *et al.* (153) cefixime-tellurite SMAC (CT-SMAC) was reported to give improved selectivity and isolation of *E. coli* O157. The addition of 2.5 mg/L potassium tellurite into SMAC inhibits NSF organisms including *Providencia* and *Aeromonas* spp., as well as those non-O157 strains of *E. coli* which have been reported to have an atypically lower resistance to tellurite at this concentration. The improved selectivity of CT-SMAC has been shown to be particularly important when recovering *E. coli* O157 from foods containing high concentrations of competitor organisms such as raw meats (5,20).

The majority of *E. coli* O157:H7/H$^-$ isolates do not show GUD activity. This property is used to distinguish these bacteria, using media containing the fluorogenic compound 4-methylumbelliferyl-ß-D-glucuronide (MUG) (95,132), or the chromogenic compound 5-bromo-4-chloro-3-indolyl-ß-D-glucoronide

(BCIG) (96). Such media include MacConkey sorbitol agar with BCIG (MSA-BCIG) (55,96), Lactose Monensin Glucuronate agar (LMG) and buffered MUG agar (BMA) (140), phenol red sorbitol agar with MUG (PRS-MUG) (93) and SD-39 agar (34) as well as some commercially available chromogenic media. CHROMagar® O157 (CHROMagar), a novel commercial medium containing a chromogenic compound (the exact composition has not been published), has also been reported to be a sensitive and specific medium for VTEC O157 (140). On this agar, VTEC O157 can easily be distinguished as pink colonies (10). Some other pathogenic *E. coli* strains, including many O111 isolates, also give pink colonies on CHROMagar®. Biolog recently introduced a new medium on the market named Rainbow™ Agar O157. This medium contains chromogenic substrates specific for ß-galactosidase and GUD. Strains of VTEC O157 form characteristic grey or steel black colonies, typical non-O157 VTEC strains are blue or purple, and most non-verotoxigenic *E. coli* are violet or red (11). The addition of tellurite and novobiocin has been shown to improve the recovery of VTEC O157 on Rainbow™ Agar O157 (126). In general, colour development on chromogenic media may show fluctuations as a result of variations in the incubation temperature, or the use of agar plates that have been stored for some period of time before use.

One technique that uses MUG or chromogenic substrates is the hydrophobic grid membrane filter (HGMF) technique. Although originally developed for detecting typical *E. coli,* commercial kits such as the ISO-GRID, which employs the HGMF principle and the Petrifilm™ HEC Test, which uses an immuno-blotting technique, have been used for the rapid detection of *E. coli* O157 from foods (35,98,107,117,118,131,136). More recently, the HGMF method has been used together with DNA probe hybridisation to isolate low numbers of VTEC, including non-O157 strains from foods (135).

A study by Wallace and Jones (140), using CHROMagar® O157 in conjunction with the HGMF technique reported slightly more sensitivity for the isolation of *E. coli* O157 from cattle faeces, than IMS and plating alone. This study also compared CR-SMAC, CT-SMAC, LMG, BMA and SD-39 agar. Unlike CHROMagar® O157, all of the other media gave rise to high numbers of false-positive colonies. Whilst other *E. coli* and *Citrobacter* spp. also produced typical colonies on CHROMagar® O157, the number of colonies was lower, and it was noted that streaking onto CT-SMAC and CR-SMAC was a useful discriminatory test in greatly reducing the number of colonies requiring confirmation by latex agglutination.

Another commercially available chromogenic medium is BCM® O157:H7 (Biosynth Inc). In a recent study, this medium was reported to be more effective than the USDA recommended MSA-BCIG medium for the isolation of *E. coli* O157:H7 from both artificially and naturally contaminated raw ground beef samples (113). To improve differentiation of *E. coli* O157 from other *E. coli*,

Kang and Fung (67) screened 35 dyes and developed *E. coli* O157:H7 (EOH) medium which uses indigo carmine. This produces *E. coli* colonies that are yellow with a clear zone, whereas colonies of *E. coli* O157:H7 are easily distinguished by their distinct red colour, and absence of a clear zone.

Whereas most media rely on the lack of sorbitol fermentation and GUD activity, these features cannot be used for differentiation among other VTEC serotypes. One of the most important phenotypical characteristics of the majority of VTEC serotypes is their ability to produce enterohaemolysin (15). Studies by Beutin *et al.* (14) and Bettelheim (8) showed that there is a close association between toxin and enterohaemolysin production by VTEC strains. A medium used to confirm enterohaemolysin production is a variation of sheep blood agar containing washed sheep red cells (8,14). This medium was used recently for the isolation of VTEC from foods (62) although this medium is not generally suitable for direct use with foods because of its poor selectivity. In an earlier study (77), enterohaemolysin agar supplemented with 30 mg l^{-1} vancomycin hydrochloride, 20μg l^{-1} cefixime and 3 mg l^{-1} cefsulodin (BVCC) was used as a diagnostic selective medium for the isolation of O157 and non-O157 VTEC from foods and human specimens. As well as VTEC, the study highlighted interference by haemolytic strains of *Serratia*, a haemolytic *Citrobacter freundii* and growth of non-VTEC *E. coli* on this medium.

Whilst selectivity is also an important feature of plating media, studies have shown that some strains of VTEC O157, and in particular stressed cells, are sensitive to the selective agents such media include. This is particularly true for SMAC-based media that have been shown to perform poorly in the recovery of stressed cells (3,31). Furthermore, some strains have been shown to be sensitive to potassium tellurite and cefixime that are both used as selective supplements in SMAC medium (82). Although such effects are less of a problem in relation to cells that have undergone a period of enrichment/recovery, they can have more serious consequences when enrichment procedures would invalidate attempts to estimate cell numbers, for example in studies of the survival and growth of these organisms in foods or liquid media. Alternative selective media, that have shown better recovery and growth of stressed cells than SMAC and other selective media, include modified eosin methylene blue (MEMB) agar (116), Rainbow™ Agar O157 (133), phenol red sorbitol agar with MUG (PRSA-MUG) and PRSA supplemented with 1% pyruvic acid (PRSA-PA) (3,31). Alternative approaches include resuscitation of stressed cells on a membrane placed on a non selective medium for a period of time before transfer onto SMAC (85), or the provision of a 2 h resuscitation period on non-selective medium before overlaying this medium with SMAC + MUG medium (86).

A number of studies comparing the efficacy of selective media for VTEC have been published. Bennett *et al.* (5) found increased rates of isolation of *E. coli* O157 from inoculated minced beef, when enriched or IMS recovered cells were plated on CT-SMAC rather than SMAC agar. Sanderson *et al.* (120) compared SMAC agar, SMAC agar with cefixime and CT-SMAC agar for the isolation of *E. coli* O157:H7 from bovine faeces, directly inoculated from naturally colonised dairy cattle and from orally inoculated calves. It was concluded that CT-SMAC agar was the most sensitive medium. CR-SMAC agar and CT-SMAC agar were found to have similar levels of sensitivity in the isolation of *E. coli* O157 strains from inoculated bovine faeces, although the organism was easier to obtain in pure culture from CT-SMAC (28). Hemorrhagic coli (HC) agar and SMAC agar did not differ significantly in their ability to recover both stressed and unstressed *E. coli* O157:H7 from ice cream and whole milk, but HC agar was superior to SMAC agar in the recovery of unstressed *E. coli* O157:H7 from Brie cheeses (55). Heuvelink *et al.* (61) compared the selective plating media SMAC, CT-SMAC, and Fluorocult™ *E. coli* O157:H7 agar (Merck) using pure bacterial cultures, and found that CT-SMAC was the most selective, while Fluorocult™ *E. coli* O157:H7 agar, containing sorbitol and MUG, was the least selective.

Though the majority of VTEC O157 strains are sorbitol negative or slow sorbitol fermenters and show no GUD-activity, methods using these discriminative properties do not identify all *E. coli* O157. Feng (40) isolated sorbitol-fermenting GUD-positive VTEC isolates of serotype O157:NM from humans in several European countries and such organisms represent the most frequently isolated verotoxigenic serotypes from HUS patients in Germany (54). Such isolates would not have been identified by using SMAC (70). In addition, the isolation of an atypical O157:H7 strain, which did not ferment sorbitol but did produce an active GUD, has been reported from a patient with bloody diarrhoea (58). The occurrence of such phenotypic variants of *E. coli* O157, and of other verotoxigenic serotypes, necessitates the use of techniques that detect virulence characteristics, such as VT or *vt* genes, enterohaemolysin ($\mathrm{Hly}_{\mathrm{EHEC}}$) or hly_{EHEC} genes, and the *E. coli* attaching-and-effacing (*eae*) gene.

CONFIRMATION OF SUSPECT ISOLATES

Presumptive verotoxigenic *E. coli* must be confirmed biochemically as *E. coli* using conventional IMViC (indole, methyl red, Voges-Proskauer and citrate) tests or commercial identification kits (API, Minitek, Enterotube). *E. coli* can also be identified by inoculation onto Levine's eosin methylene blue agar and observation of colonies with a metallic sheen. *E. hermannii*, also sorbitol- and GUD-negative, can be distinguished from *E. coli* O157 by three additional tests in which *E. hermannii* is positive, i.e., growth in the presence of KCN, fermentation of cellobiose, and production of a yellow pigment (22).

A common approach when screening foods for VTEC O157 is to select suspect colonies and confirm the presence of the O157 antigen. Commercially available latex agglutination assays for the rapid presumptive identification of *E. coli* O157 use latex beads coated with antibodies to the O157 antigen which if present gives rise to an antigen-antibody complex that results in the formation of a visible precipitate. Determination of the H7 antigen is not essential for the presumptive identification of O157 verotoxigenic strains. Some *E. coli* O157 strains possess the H7 flagellin, but are non-motile and negative in H-serology. These strains, however, may produce verocytotoxin and are very similar to O157:H7 (130). Other strains possess H7 antigens but are of serogroups other than O157. These strains include both verotoxigenic strains and non-verotoxigenic *E. coli* strains. However, in the same way that antibodies used in immunoassays cross-react with the lipopolysaccharide (LPS) of other bacteria, some commercial latex agglutination assays can also give false-positive results. In an evaluation of three commercial latex agglutination kits, cross-reactions between O157 antisera and strains of *Citrobacter freundii* and *Salmonella* O group N, have been reported (124). One particular species that commonly gives rise to false positive identifications of O157 is *Escherichia hermannii* (22,104). In addition to its serological cross reactivity with O157, *E. hermannii* is also sorbitol and GUD-negative giving rise to presumptive O157 colonies on SMAC and MUG containing media. Fortunately, it can be easily distinguished from O157 by its ability to ferment rhamnose by plating on CR-SMAC and by additional biochemical tests (22). Studies have shown that strains belonging to serogroup O157 produce a very limited number of biochemical profiles (1). Bettelheim (9) noted similar patterns among strains of O157, and also reported the poor fermentative abilities of other non-O157 VTEC strains, including O111:H- and O113:H21. Such poor fermentative abilities were particularly apparent with the five carbohydrates, sorbitol, sucrose, rhamnose, raffinose and salicin, and in combination with the ability to produce enterohaemolysin, these tests have been proposed as a potential method for identifying suspect VTEC strains pathogenic for humans (EHEC) upon initial isolation. However, it was conceded that this approach would not be 100% for all EHEC strains.

In the case of non-O157 VTEC, which are more difficult to identify upon initial isolation, the close association between toxin production and the presence of enterohaemolysin also makes the enterohaemolysin test a valuable supplementary test in the identification of those colonies to be examined for toxin production. In epidemiological studies, isolates are normally then examined by further biochemical tests and genetic typing methods.

DETECTION OF VEROCYTOTOXINS

The Verocell assay (75) used to confirm the presence of verocytotoxins (VT1, VT2 and VT2 variants) is sensitive and relatively simple to perform. However, despite its sensitivity and specificity, the Vero cell assay is generally too expensive and technically demanding for most routine labs. The Verocell assay is intended to be used to directly screen isolates, but not food samples, for the presence of toxins. Furthermore, this assay needs tissue culture facilities, and cytotoxicity, if observed, is not specific for VT. Consequently, positive samples must be confirmed by other methods such as neutralisation with VT-specific antibodies, which can be difficult to interpret, especially with low level toxin-producers. In addition, some VTEC strains produce a variant form of toxin, which may fail to neutralise antiserum. In one study antigenic differences of a VT2 variant toxin was demonstrated, along with differences in cell line susceptibility to the toxin (68). Whilst this particular VT2 variant was found to be highly toxic against Verocells, it was not toxic against HeLa cells. Consequently, serological screening using antibodies to detect some types of toxin and the use of HeLa cells to demonstrate cytotoxicity in some strains of VTEC may give false negative results if the toxins present are variant types. More recently, commercial assays have been used to detect toxin bound to polyclonal antibodies. These provide a convenient method for screening isolates for toxin production and for confirming VTEC colonies previously isolated by other methods. Protocols have also been developed to enable direct detection of toxins in clinical and food samples.

Toxin assays have the added benefit of highlighting the presence of VTEC, irrespective of the serogroup. However, in the absence of O157 VTEC and VTEC of other serogroups that are more easily isolated culturally or detected with commercial assays, confirmation of presumptive positive samples can be difficult, time consuming and labour-intensive. Furthermore, although many immunoassays will detect VT1, they will not necessarily detect all of the VT2 variants (VT2v) that exist (91,103).

As with detection and isolation methods, enrichment methods play an equally important role in toxin detection from foods. However, enrichment methods which produce the high concentrations of cells and/or target antigen required for successful detection in immunoassays, are not necessarily optimal for toxin production. In a study by Palumbo *et al.* (101), growth and toxin production by some *E. coli* strains was reported to occur from 10°C up to 49°C, although highest titres were recorded at the highest growth temperature. In a later study by Weeratna and Doyle (143), production of VT1 in milk and beef was reported to be greatest at 37°C than 25°C and 30°C despite the development of similar populations of VTEC O157. MacLeod and Gyles (81) noted production of VT2v (also termed shiga like toxin; SLTIIv) at 37°C, although this was attributed to better growth and higher numbers of cells. Interestingly, agitation of the culture at 37°C appeared to increase toxin production. Furthermore, pH of the growth medium may also influence toxin production with pH values of 8 to 8.5 being reported as being optimal (81,143). In a study by Buncic and Avery (23), enhanced production of VT2 at 37°C was reported to occur after cold storage under starvation conditions, and after heat shock, with larger amounts of toxin being produced during the first few hours of post-shock incubation.

In addition to providing optimum incubation conditions, toxin production within cultures can be increased by supplementation of the medium with sub-inhibitory levels of suitable antimicrobial agents, such as mitomycin C (81). Karch *et al.* (71) showed that a mixture of trimethoprim and sulfamethoxazole increased toxin production. As well as increasing toxin-production within the culture, it is also important to increase the release of toxin into the medium to enable subsequent detection. Whereas VT2 is often released into the medium during the late logarithmic growth stage, VT1 is largely cell-associated, present in the periplasmic compartment of the cell, and is produced at its highest levels during mid-logarithmic phase. Pre-treatment of the culture supernatants or samples with polymyxin B, which is thought to increase the release of cell bound toxin across the outer membrane into the medium, has been shown to increase the toxin titre and improve the sensitivity of detection of toxins (72,91). For this reason, a polymyxin extraction procedure is often included in many protocols for commercial toxin assays.

There are two examples of commercially available microtitre plate immunoassays for detecting the presence of toxins directly from food enrichment cultures, the RidaScreen Verotoxin kit (R-Biopharm GMbH, Germany), and the Premier EHEC test kit (Meridian Diagnostics Inc, USA). Although the Premier EHEC test kit has been reported to provide a sensitive method for detecting VTEC in ground beef and milk samples (2), false positive results, caused by cross reactions with *Pseudomonas aeruginosa*, have also been reported (13). In response to these findings, the manufacturer is reported to have made modifications to the kit and sample protocols to improve the specificity of the assay.

A third toxin detection kit, the VTEC reverse passive latex agglutination test (RPLA, Oxoid Ltd, Basingstoke, UK) has the added benefit of distinguishing between VT1 and VT2. However, the kit and extraction procedures are intended for the confirmation of VT production by suspect isolates on solid or in liquid media, and not for the detection of toxins directly from food samples. Beutin *et al.* (16) reported that a combination of the enterohaemolysin test and confirmation of toxin production using the VTEC-RPLA provided a convenient and rapid method for screening plates for suspect VTEC in clinical specimens. However, strains of *E. coli* O157:H7 and O157:H^{-}, that do not produce VT but are nevertheless associated with human disease have been described (121). If further studies confirm the importance of these organisms in human disease, and perhaps show that they produce other virulence factors, then additional problems for the food industry will arise.

The Verotox F™ test (Denka Seiken) is a rapid microplate latex agglutination test for detecting and characterisation of verotoxin, detecting VT1, VT2 and VT2c (73). This test gives results after 48 h and is reported to be consistent with PCR results, obtained after 26 to 28 h (61). Comparison of the Verotox-F assay with the standard Vero cell cytotoxicity assay found the Verotox F™ test to be highly sensitive and specific for the detection and characterisation of VTs in culture filtrates of human *E. coli* isolates. Although the Verotox F™ test seems a good alternative for Vero cell assays and PCR, as mentioned previously, several different antigenic variants or types of toxin exist and immunological methods are not likely to detect all VTs with equal efficiency, while some VTs may not be detected at all.

IMMUNOLOGICAL DETECTION

The O-antigen of the O157 serogroup provides a convenient and specific target for immunological methods, such as enzyme-linked immunosorbent assays (ELISAs), colony immunoblot assays, direct immunofluorescent filter techniques, and immunocapture techniques. Additionally, kits enabling the detection of the four most common non-O157 VTEC isolated from human patients (O111, O26, O103 and O145) (145) are likely to become available in the near future. The majority of commercial assays use antibodies targeted against the O-antigen of serogroup O157, whilst some assays also target both the O antigen and the flagella antigen H7. However, some other bacteria, notably *Citrobacter freundii* and *Escherichia hermannii*, also share the same or similar LPS structures with *E. coli* O157 (12,114) and *E. hermannii*, a sorbitol-negative species of the genus *Escherichia*, has also been reported to be agglutinated by O157 antisera (114), resulting in false-positive reactions. Somatic antibody cross-reactions with bacteria of other genera have also been observed, including *Salmonella* group N species, some *Yersinia enterocolitica*, and *Brucella* spp. (66,98). This means that it is not possible to rely solely on rapid immunological tests to identify *E. coli* O157 without further characterisation tests. Whilst detection of the flagella H7 antigen provides greater specificity (59), not all strains of VTEC O157 associated with human infection express this antigen, or the antigen may only be expressed under certain circumstances.

Also, an immunoblot assay based on detecting cells that produce VT1 and VT2 has been developed and successfully used to detect small numbers of *E. coli* O157:H7 in foods (33). The method involves selective enrichment, filtering through a hydrophobic grid membrane filter (HGMF), incubating each filter on nitro-cellulose paper on selective agar at 37°C for 24 h, preparing an immuno-blot (using antiserum to VT1 and VT2) of each nitro-cellulose paper, selecting from the HGMF those colonies that correspond to immuno-positive sites on blots, and confirming isolates as *E. coli* O157 by biochemical and serological tests. Though very sensitive this method is rather complex and labour intensive, and consequently not recommended for routine use.

Although methods based on the HGMF technique were initially popular, plate based immunoassays and more recently, single step devices based on the immunochromotography principle are now the most popular immunological-based formats for screening foods. These assays are less labour intensive than the HGMF technique, and some stages are easily semi or fully automated. The single step devices, such as the Visual Immunoprecipitate assay (VIP) (37), offer greater convenience, are easy to set up and interpret and have the additional benefit of not requiring any expensive equipment.

The ELISA format is the most widely used immunoassay format for the detection of *E. coli* O157 and VTs. The ELISA test systems differ mainly in the antibodies used, test devices and detection reagents. Recently described systems use a polyclonal antibody as capture antibody for *E. coli* O157 and a monoclonal antibody to *E. coli* O157:H7 as the detection antibody. Test devices include the conventional microtitre plate assay, dipsticks, immunofiltration and capillary migration assays. Many ELISA tests have been commercialised and are used as screening tests for the detection of *E. coli* O157 and VTs (see also section 6). Most of these test systems have been designed to give a visual result, and do not require costly instrumentation.

The EHEC-TEK™ Test System for *E. coli* O157:H7 produced by Organon Teknika uses microelisa wells, coated with polyclonal antibodies to the *E. coli* O157:H7 LPS antigen. The second antibody is a horseradish peroxidase-labelled monoclonal antibody that reacts with two outer membrane proteins expressed by *E. coli* O157:H7 and O26:H11. The enrichment broth used contains acriflavine, which enhances the detection of *E. coli* O157:H7 by the monoclonal antibody (100). The specificity of the EHEC-TEK™ ELISA for detection of *E. coli* O157:H7 in raw meats was improved by a modified test protocol, incorporating immunocapture (66). Bennett *et al.* (6) showed this modified ELISA system to be less sensitive than the standard Tecra™ ELISA (Tecra).

The Petrifilm™ HEC Test is a commercial ELISA kit for the rapid screening of foods for the presence of *E. coli* O157. In this method enrichment cultures are inoculated onto 3M Petrifilm™ *E. coli* count plates. After incubation the colonies are replica-plated on disks. *E. coli* colonies on the disk are identified by an immunoblot method using a polyclonal O157-lipopolysaccharide antibody. As the method uses a polyclonal antibody, false-positive reactions frequently occur and all positive reactions must be confirmed. However, it is understood that this method is no longer commercially available.

The EZ Coli™ Rapid Detection System produced by Difco is an immunological assay formatted in a micropipette tip. The tip is filled with a large number of axially oriented filaments. The filaments in the tip are divided in three sectors. The top element is a positive colour gauge, the middle element is a negative colour gauge, and the bottom element is the test area. Polyclonal antibodies specific for *E. coli* O157 are covalently bound to the test filaments. *E. coli* O157 cells attach to this filament, and a secondary antibody conjugated with alkaline phosphatase is added, followed by the addition of the substrate. When *E. coli* O157 is present in a sample, a grey to purple colour will develop in the test area. This detection system has proved to be a simple and fast screening method with identification of negative and presumptive positive samples within 15 to 18 h (44,150).

A simple enzyme immunoassay system for the detection of *E. coli* O157 has been described based on the use of Polymacron™, an inexpensive hydrophobic polyester fabric, as a high-surface-area immunoadsorbent for the rapid capture and subsequent immunoenzymatic detection of *E. coli* O157 antigens. This system is effective for the screening of large numbers of enrichment cultures of food samples and was commercially available as *E. coli* O157 Rapitest™ produced by Kalyx Biosciences (17). Dipstick assays have also been developed for the identification of *E. coli* O157. These include a sensitive dipstick immunoassay based on a hydrophobic polyvinylidine difluoride-based membrane with a polyclonal antibody to *E. coli* O157 as the capture antibody and a monoclonal antibody to *E. coli* O157:H7 as the detection antibody (74).

Other commercial test systems include the Ampcor™ *E. coli* O157:H7 Kit (Ampcor), Micro-Screen™ for *E. coli* O157:H7 (Neogen), VIP™ Test for EHEC (BioControl Systems) (37), Tecra™ *E. coli* O157 visual immunoassay (Tecra) (45), Vidas™ *E. coli* O157 Assay (bioMérieux) (139), Assurance EHEC™ (BioControl Systems) (38). These commercial test kits reduce analysis time, giving presumptive results after one day, compared to cultural procedures which require two days. The detection limit of these systems varies from 10^4-10^6 cells/ml. As *E. coli* O157 cells in food and faeces are usually present in small numbers, overnight enrichment to obtain the target cells in detectable levels is frequently required.

The performance of a variety of commercial test systems was evaluated by detection of *E. coli* O157 strains inoculated into minced beef (61). The Petrifilm™ HEC Test Kit and the Dynabeads™ anti-*E. coli* O157 proved to be significantly more sensitive kits for the detection of O157 VTEC strains than the remaining systems tested. The minimum numbers of O157 VTEC cells detected by these two sensitive methods were not significantly different. The Petrifilm™ HEC Test Kit and the Dynabeads™ anti-*E. coli* O157 kit were about 100 times more sensitive than the Tecra™ *E. coli* O157 visual immunoassay, which was in turn about 100 times more sensitive than the Ampcor™ *E. coli* O157:H7 Kit and the EHEC-TEK™ test. Although not very sensitive, the EHEC-TEK™ kit proved to be the most selective. Strains of VT negative *E. coli* O157 were not detected with this test system. The other test kits appeared to be as sensitive in the detection of VT-negative *E. coli* O157 strains as in the detection of O157 VTEC strains. Following enrichment of the inoculated samples, all these test systems could detect an initial inoculum of less than one VTEC O157 organism per g of minced beef and no significant differences between the tests were found. The test kits were only slightly more sensitive than the conventional cultural approach, however, they did offer advantages in terms of simplicity, high-volume testing, and reducing the time to achieve a presumptive result.

However, despite the potential time and labour savings these devices offer, many still require cultural isolation and identification to confirm the assay result. Furthermore, irrespective of the end-point detection system used, these tests are still dependent upon there being sufficient numbers of target organisms to guarantee detection or isolation. Many immunoassay-based systems and some molecular techniques such as PCR typically require 10^5-10^6 cfu/ml for a positive result. Therefore, despite advances in technology and improved selectivity and rapidity of many assays, these methods are still heavily dependent on successful enrichment. This reliance on growth therefore continues to be the single most important factor that prevents rapid methods providing results within a working day.

MOLECULAR APPROACHES FOR DETECTING VTEC

In contrast to immunological methods, molecular techniques offer greater specificity and the potential to provide accurate and rapid detection, and confirmation, of VTEC in foods. One of the most common molecular approaches used to confirm the presence of VTEC is the detection of nucleotide sequences related to the toxin genes. In the past, DNA probes have been developed for the detection of genes associated with VT1, VT2 and some of the VT2 variants that are not detected by some of the other methods (68,94,119). Colony blot hybridisation with oligonucleotide probes provides a useful tool in epidemiology studies and for the rapid screening of isolates for the presence of toxin genes. However, the PCR technique is now the most popular molecular method for the specific detection of VTEC directly in food samples.

Specific primers have been developed to enable PCR amplification of toxin gene sequences in bacterial cultures (106) and directly in foods (50,110,149). Compared with immunological methods, the PCR technique enables detection of low as well as moderate and high toxin producers. PCR also offers greater specificity, although, because the Shiga toxin of *Shigella dysenteriae* type 1 and VT1 of VTEC are closely related, many PCR assays will also give positive results with these bacteria. Owing to the diversity of the VT2 family, most PCR primers are specifically designed to be used as a screen to detect only *vt*1 and *vt*2 and not the *vt*2 variants (106), whilst others also enable detection of some of the *vt*2 variants (50).

Other targets for DNA probes and PCR include the haemolysin (*hlyA*) gene (75), the *E. coli* attaching-and-effacing (*eae*) gene (80,90,147), the ß-glucuronidase (*uidA*) gene (41,43) and the 60-MDa plasmid found in O157:H7 and other VTEC (78). Primers to the *rfb* gene cluster that encodes the biosynthesis of the O-antigen of *E. coli* O157 have also been developed to enable specific identification of strains of serogroup O157 and the detection of *E. coli* O157 in milk and cattle faeces (84). Other virulence-associated genes that have been targets for PCR, but not widely used for detection in foods, include the EPEC adherence factor (EAF) (46), sequences associated with other plasmid encoded virulence factors (127) and a variety of *Escherichia coli* type III secreted proteins (*Esps*). Multiplex PCR assays have become popular because they enable the simultaneous detection of more than one PCR target. Multiplex PCR assays have been developed to enable simultaneous detection of genes associated with VT1, VT2 together with *uidA* (24), EHEC *eaeA* and the 60MDa plasmid (47), *eaeA* (89), *eaeA* and *hlyA* (36), EHEC and O157 *eaeA*, and a portion of the *fli C* gene that encodes the flagella H7 antigen (49).

The simultaneous detection of more than one target sequence is particularly useful in screening isolates and confirming particular serotypes such as VTEC O157:H7. The combination of virulence-associated genes is also of importance, because other bacteria, including other members of the Enterobacteriaceae may also possess these or similar gene sequences (79). However, there are some limitations that must be considered in using multiplex PCR directly on food samples. Although pathogenic strains are likely to possess a complement of virulence genes, in addition to the ability to produce toxin, it may not be possible to know whether or not the genes detected by the multiplex PCR were present in a single cell, or dispersed among different cells within the population of a food sample. It could also be costly to carry out routine screening by PCR using more than one pair of primers. So the preferred approach probably involves general screening for VTEC, followed by isolation of VTEC strains from positive samples and subsequent screening of the strains with specific primer sets designed to identify various virulence-associated genes. Furthermore, as with standard PCR, interference and inhibition by the food matrix may affect the performance of the assay.

One problem commonly associated with PCR is the amplification of DNA from dead cells present in food samples. However, in most foods, pathogens would normally be expected to be present in foods at levels below the limit of detection by PCR, hence the continued dependence on enrichment. Uyttendaele *et al.* (137) reported that false PCR-positive results were obtained from ground beef if the number of dead cells exceeded 10^8 cfu/g. Below this level, the number of false-positive results depended on sample preparation, in particular the number of washing steps, and DNA extraction procedures. Another difficulty sometimes experienced with foods is the inability to culture the respective bacteria from PCR-positive samples, although this is also dependent on the selectivity of the medium and the techniques used.

FUTURE DEVELOPMENTS

Improved sensitivity and specificity has been accomplished by the development of methods that combine immunological and molecular methods for detection and isolation. These have included the use of biotin-labelled probes to selectively capture specific DNA segments and streptavidin-coated magnetic beads to aid capture of the target sequence, followed by PCR amplification and detection of the specific sequences. This technique has been used for the detection of *vt*1 and *vt*2 in ground beef and has shown close agreement with a VT-ELISA method (29).

More recently, a study using PCR in combination with an ELISA-based technique has been described (30). This method, termed PCR-microplate capture hybridisation assay, uses two pairs of primers specific for *vt*1 gene and *vt*2 and biotin-labelled probes, which enable immobilisation of the PCR products in the streptavidin coated wells of a microtitre plate. During PCR amplification, the PCR product is labelled with dioxigenin (DIG) and hybridised to a specific capture probe. The bound complex is detected using an anti-DIG peroxidase conjugate and a colorimetric substrate. The method showed high specificity, and sensitivity was 10-fold higher than detection of PCR products by conventional electrophoresis. Furthermore, this approach of replacing electrophoresis with this microtitre plate format, introduced into a commercial PCR system for O157 (PROBELIA™, Bio-Rad S.A. France), offers greater potential for semi-automation, rendering PCR more suitable for the routine screening of foods.

Despite the specificity and rapidity that molecular techniques potentially offer, the food industry, and others routinely screening foods, seem reluctant to adopt these methods in routine testing. Reasons for this may include lack of understanding and knowledge of molecular techniques, the need for technicians trained in different skills, and the high capital cost and ongoing consumables costs. The requirement to only use approved methods, and the need to fully validate new methods before routine use, further restricts their uptake. Some current molecular techniques such as PCR offer greater specificity and rapidity, but can often be labour intensive and do not allow for the high sample throughput essential in a busy routine testing laboratory. Consequently, many of the molecular techniques will continue to be confined to clinical and specialist testing laboratories. However, with further improvement, development and automation, some techniques will become more widely accepted and adopted by food testing laboratories.

Future careful consideration in terms of setting of standards and food safety enforcement will be necessary as methods capable of detecting a wider range of VTEC in foods are being developed. The high carriage rates in cattle and sheep, and the diversity of non-O157 VTEC serotypes in some foods, particularly raw meats, will require carefully considered decisions, for economical and food safety reasons. The implementation of appropriate intervention strategies, based on microbiological risk assessment, will require a greater understanding of VTEC and their clinical significance, and will only become reality when methods for the isolation of these organisms have been improved.

REFERENCES

1. ABBOTT, S.L., HANSON, D.F., FELLAND, T.D., CONNELL, S., SHUM, A.H. and JANDA, J.M. 1994. *Escherichia coli* O157:H7 generates a unique biochemical profile on MicroScan conventional gram-negative identification panels. J. Clin. Microbiol. *32*, 823–824.
2. ACHESON, D.W., LINCICOME, L.L., DE BREUCKER, S. and KEUSCH, G.T. 1996. Detection of shiga-like toxin producing *Escherichia coli* in ground beef and milk by commercial enzyme immunoassay. J. Food Prot. *59*, 344–349.
3. AHMED, N.M. and CONNER, D.E. 1995. Evaluation of various media for recovery of thermally-injured *Escherichia coli* O157:H7. J. Food Prot. *58*, 357–360.
4. AIRD, H., BOLTON, F.J., SMITH, H.R. and WRIGHT, P.A. 1999. Growth of verocytotoxin producing *Escherichia coli* (VTEC) in enrichment media. Second International Symposium of the European Group on Enterohemorrhagic *Escherichia coli*. 1999. April 16-17. Brussels, Belgium: Belgian Society for Infectiology and Clinical Microbiology.
5. BENNETT, A.R., MACPHEE, S. and BETTS, R.P. 1995. Evaluation of methods for the isolation and detection of *Escherichia coli* O157 in minced beef. Lett. Appl. Microbiol. *20*, 375–379.
6. BENNETT, A.R., MACPHEE, S. and BETTS, R.P. 1996. The isolation and detection of *Escherichia coli* O157 by use of immunomagnetic separation and immunoassay procedures. Lett. Appl. Microbiol. *22*, 237–243.
7. BETTELHEIM, K.A. 1968. Investigations into the pathogenicity of *Escherichia coli* in human infections. PhD Thesis, University of London: London.
8. BETTELHEIM, K.A. 1995. Identification of enterohaemorrhagic *Escherichia coli* by means of their production of enterohaemolysin. J. Appl. Bacteriol. *79*, 178–180.
9. BETTELHEIM, K.A. 1997. Biochemical studies on enterohaemorrhagic *Escherichia coli* (EHEC). *In* Coliforms and *E. coli*: Problem or Solution? (D.

Kay and C. Fricker, eds). The Royal Society of Chemistry: Cambridge. p. 243–248.

10. BETTELHEIM, K.A. 1998. Reliability of CHROMagar O157 for the detection of enterohaemorrhagic *Escherichia coli* (EHEC) O157 but not EHEC belonging to other serogroups. J. Appl. Microbiol. *85*, 425–428.

11. BETTELHEIM, K.A. 1998. Studies of *Escherichia coli* cultured on Rainbow™ Agar O157 with particular reference to enterohaemorrhagic *Escherichia coli* (EHEC). Microbiol. Immunol. *42*, 265–269.

12. BETTELHEIM, K.A., EVANGELIDIS, H., PEARCE, J.L., SOWERS, E. and STROCKBINE, N.A. 1993. Isolation of a *Citrobacter freundii* strain which carries the *Escherichia coli* O157 antigen. J. Clin. Microbiol. *31*, 760–761.

13. BEUTIN, L. 1996. *Pseudomonas aeruginosa* can cause false-positive identification of verotoxin (shiga-like toxin) production by a commercial enzyme immune assay system for the detection of shiga-like toxins (SLTs). Infection. *24*, 267–268.

14. BEUTIN, L. *et al.* 1989. Close association of verotoxin (Shiga-like toxin) production with enterohemolysin production in strains of *Escherichia coli*. J. Clin. Microbiol. *27*, 2559–2564.

15. BEUTIN, L., PRADA, J., ZIMMERMANN, S., STEPHAN, R., ORSKOV, I. and ORSKOV, F. 1988. Enterohemolysin, a new type of hemolysin produced by some strains of enteropathogenic *E. coli* (EPEC). Zbl. Bakt. Hyg. *267*, 576–588.

16. BEUTIN, L., ZIMMERMAN, S. and GLEIER, K. 1996. Rapid detection and isolation of Shiga-like toxin (verocytotoxin)-producing *Escherichia coli* by direct testing of individual enterohemolytic colonies from washed sheep blood agar plates in the VTEC-RPLA assay. J. Clin. Microbiol. *34*, 2812–2814.

17. BLAIS, B.W., BOOTH, R.A., PHILIPPE, L., PANDIAN, S. and YAMAZAKI, H. 1997. Polymacron™ Enzyme Immunoassay System for detection of *Escherichia coli* O157 inoculated into foods. J. Food Prot. *60*, 98–101.

18. BLAIS, B.W., BOOTH, R.A., PHILLIPPE, L.M. and YAMAZAKI, H. 1997. Effect of temperature and agitation on enrichment of *Escherichia coli* O157:H7 in ground beef using modified EC broth with novobiocin. Int. J. of Food Microbiol. *36*, 221–225.

19. BOLTON, F.J., CROZIER, L. and WILLIAMSON, J.K. 1995. New technical approaches to *Escherichia coli* O157. PHLS Microbiol. Dig. *12*, 67–73.

20. BOLTON, F.J., CROZIER, L. and WILLIAMSON, J.K. 1996. Isolation of *Escherichia coli* O157 from raw meat products. Lett. Appl. Microbiol. *23*, 317–321.

21. BONNET, R. *et al.* 1998. Non-O157:H7 Stx2-producing *Escherichia coli* strains associated with sporadic cases of hemolytic-uremic syndrome in adults. J. Clin. Microbiol. *36*, 1777–1780.

22. BORCZYK, A.A., LIOR, H. and CIEBIN, B. 1987. False-positive identification of *Escherichia coli* O157 in foods. Int. J. Food Microbiol. *4*, 347–349.

23. BUNCIC, S. and AVERY, S.M. 1998. Effects of cold storage and heat-acid shocks on growth and verotoxin 2 production of *Escherichia coli* O157:H7. Food Microbiol. *15*, 319–328.

24. CEBULA, T.A., PAYNE, W.L. and FENG, P. 1995. Simultaneous identification of strains of *Escherichia coli* serotype O157:H7 and their Shiga-like toxin type by mismatch amplification mutation assay-multiplex PCR. J. Clin. Microbiol. *33*, 248–250.

25. CHAPMAN, P.A. and SIDDONS, C.A. 1996. Evaluation of a commercial enzyme immunoassay (EHEC-Tek) for detecting *Escherichia coli* O157 in beef and beef products. Food Microbiol. *13*, 175–182.

26. CHAPMAN, P.A., SIDDONS, C.A., WRIGHT, D.J., NORMAN, P., FOX, J. and CRICK, E. 1993. Cattle as a possible source of verocytotoxin-producing *Escherichia coli* O157 infections in man. Epidemiol. Infect. *111*, 439–447.

27. CHAPMAN, P.A., SIDDONS, C.A., ZADIK, P.M. and JEWES, L. 1991. An improved selective medium for the isolation of *Escherichia coli* O157. J. Med. Microbiol. *35*, 107-110.

28. CHAPMAN, P.A., WRIGHT, D.J. and SIDDONS, C.A. 1994. A comparison of immunomagnetic separation and direct culture for the isolation of verocytotoxin-producing *Escherichia coli* O157 from bovine faeces. J. of Med. Microbiol. *40*, 424–427.

29. CHEN, J., JOHNSON, R. and GRIFFITHS, M.W. 1998. Detection of verotoxigenic *Escherichia coli* by magnetic capture-hybridization PCR. Appl. Environ. Microbiol. *64*, 147–152.

30. COCOLIN, L., ASTORI, G., MANZANO, M., CANTONI, C. and COMI, G. 2000. Development and evaluation of a PCR-microplate capture hybridization method for direct detection of verotoxigenic *Escherichia coli* strains in artificially contaminated food samples. Int. J. Food Microbiol. *54*, 1–8.

31. CONNER, D.E. and HALL, G.S. 1994. Efficacy of selected media for recovery of *Escherichia coli* O157:H7 from frozen chicken meat containing sodium chloride, sodium lactate or polyphosphate. Food Microbiol. *11*, 337–344.

32. DOYLE, M.P. and SCHOENI, J.L. 1984. Survival and growth characteristics of *Escherichia coli* associated with hemorrhagic colitis. Appl. Environ. Microbiol. *48*, 855–856.

33. DOYLE, M.P. and SCHOENI, J.L. 1987. Isolation of *Escherichia coli* O157:H7 from retail fresh meats and poultry. Appl. Environ. Microbiol. *53*, 2394–2396.

34. ENTIS, P. and LERNER, I. 1997. 24-hour presumptive enumeration of *Escherichia coli* O157:H7 in foods by using the ISO-Grid® method with SD-39 agar. J. Food Prot. *60*, 883–890.

35. ENTIS, P. 1997. Direct 24-hour presumptive enumeration of *Escherichia coli* O157:H7 in foods using hydrophobic grid membrane filter followed by serological confirmation: collaborative study. J. AOAC Int. *81*, 403–418.

36. FAGAN, P., HORNITZKY, M.A., BETTELHEIM, K.A. and DJORDJEVIC, S.P. 1999. Detection of shiga-like toxin (*stx1* and *stx2*), intimin (*eaeA*) and enterohemorrhagic *Escherichia coli* (EHEC) hemolysin (EHEC *hlyA*) genes in animal feces by multiplex PCR. Appl. Environ. Microbiol. *65*, 868–872.

37. FELDSINE, P.T., FALBO-NELSON, M.T., BRUNELLE, S.L. and FORGEY, R.L. 1997. Visual immunoprecipitate assay (VIP) for detection of enterohemorrhagic *Escherichia coli* (EHEC) O157:H7 in selected foods: collaborative

study. J. AOAC Int. *80*, 517–529.

38. FELDSINE, P.T., FORGEY, R.L., FALBO-NELSON, M.T and BRUNELLE, S.L. 1997. Assurance *Escherichia coli* O157:H7: a comparative validation study. J. AOAC Int. *80*, 37–42.

39. FELDSINE, P.T., FORGEY, R.L., FALBO-NELSON, M.T. and BRUNELLE, S.L. 1997. *Escherichia coli* O157:H7 visual immunoprecipitate assay: a comparative validation study. J. AOAC Int. *80*, 43–48.

40. FENG, P. 1995. *Escherichia coli* serotype O157:H7: novel vehicles of infection and emergence of phenotypic variants. Emerg. Infect. Dis. *1*, 47–52.

41. FENG, P. 1993. Identification of *Escherichia coli* serotype O157:H7 by DNA probe specific for an allele of *uid*A gene. Mol. Cell. Probes *7*, 151–154.

42. FENG, P., LAMPEL, K.A., KARCH, H. and WHITTAM, T.S. 1998. Genotypic and phenotypic changes in the emergence of *Escherichia coli* O157:H7. J. Infect. Dis. *177*, 1750–1753.

43. FENG, P., LUM, R. and CHANG, G.W. 1991. Identification of *uid*A gene sequences in ß-D-glucuronidase-negative *Escherichia coli*. Appl. Environ. Microbiol. *57*, 320–323.

44. FIRSTENBERG-EDEN, R. and SULLIVAN, N.M. 1997. EZ Coli Rapid Detection System: a rapid method for the detection of *Escherichia coli* O157 in meat and other foods. J. Food Prot. *60*, 219–225.

45. FLINT, S.H. and HARTLEY, N.J. 1995. Evaluation of the TECRA *Escherichia coli* O157 visual immunoassay for tests on dairy products. Lett. Appl. Microbiol. *21*, 79–82.

46. FRANKE, J. *et al.* 1994. Nucleotide sequence analysis of enteropathogenic *Escherichia coli* (EPEC) adherence factor probe and development of PCR for rapid detection of EPEC harboring virulence plasmids. J. Clin. Microbiol. *32*, 2460–2463.

47. FRATAMICO, P.M., SACKITEY, S.K., WIEDMANN, M. and DENG, M.Y. 1995. Detection of *Escherichia coli* O157:H7 by multiplex PCR. J. Clin. Microbiol. *33*, 2188–2191.

48. FRATAMICO, P.M., SCHULTZ, F.J. and BUCHANAN, R.L. 1992. Rapid isolation of *Escherichia coli* O157:H7 from enrichment cultures of foods using an immunomagnetic separation method. Food Microbiol. *9*, 105–113.

49. GANNON, V.P., D'SOUZA, S., GRAHAM, T., KING, R.K., RAHN, K. and READ, S. 1997. Use of the flagellar H7 gene as a target in multiplex PCR assays and improved specificity in identification of enterohemorrhagic *Escherichia coli* strains. J. Clin. Microbiol. *35*, 656–662.

50. GANNON, V.P., KING, R.K., KIM, J.Y. and THOMAS, E.J. 1992. Rapid and sensitive method for detection of shiga-like toxin-producing *Escherichia coli* in ground beef using the polymerase chain reaction. Appl. Environ. Microbiol. *58*, 3809–3815.

51. GOLDWATER, P.N. and BETTELHEIM, K.A. 1994. The role of enterohaemorrhagic *E. coli* serotypes other than O157:H7 as causes of disease. *In* Recent advances in verocytotoxin-producing *Escherichia coli* infections, (M.A. Karmali and A.G. Goglio, eds.). Elsevier Science: Amsterdam. p. 57–60.

52. GOODING, C.M. and CHOUDARY, P.V. 1998. Detection of *Escherichia coli* O157:H7 in ground beef in eight hours. J. Microbiol. Meths. *34*, 89–98.
53. GRIF, K., DIERICH, M.P. and ALLERBERGER, F. 1998. Dynabeads™ plus 3 M Petrifilm HEC™ versus Vitek Immunodiagnostic Assay System™ for detection of *E. coli* O157 in minced meat. Lett. Appl. Microbiol. *26*, 199–204.
54. GUNZER, F., BÖHM, H., RÜSSMANN, H., BITZAN, M., ALEKSIC, S. and KARCH, H. 1992. Molecular detection of sorbitol-fermenting *Escherichia coli* O157 in patients with hemolytic-uremic syndrome. J. Clin. Microbiol. *30*, 1807–1810.
55. HAMMACK, T.S., FENG, P., AMAGUAÑA, R.M., JUNE, G.A., SHERROD, P.S. and ANDREWS, W.H. 1997. Comparison of sorbitol MacConkey and Hemorrhagic Coli agars for recovery of *Escherichia coli* O157:H7 from brie, ice cream, and whole milk. J. AOAC Int. *80*, 335–340.
56. HARA-KUDO, Y., KONUMA, H., NAKAGAWA, H. and KUMAGAI, S. 2000. *Escherichia coli* O26 detection from foods using an enrichment procedure and an immunomagnetic separation method. Lett. Appl. Microbiol. *30*, 151–154.
57. HARA-KUDO, Y., ONOUE, Y., KONUMA, H., NAKAGAWA, H. and KUMAGAI, S. 1999. Comparison of enrichment procedures for isolation of *Escherichia coli* O157:H7 from ground beef and radish sprouts. Int. J. Food Microbiol. *50*, 211–214.
58. HAYES, P.S., BLOM, K., FENG, P., LEWIS, J., STROCKBINE, N.A. and SWAMINATHAN, B. 1995. Isolation and characterization of a ß-D-glucuronidase-producing strain of *Escherichia coli* serotype O157:H7 in the United States. J. Clin. Microbiol. *33*, 3347–3348.
59. HE, Y., KEEN, J.E., WESTERMAN, R.B., LITTLEDIKE, E.T. and KWANG, J. 1996. Monoclonal antibodies for detection of the H7 antigen of *Escherichia coli*. Appl. Environ. Microbiol. *62*, 3325–3332.
60. HECKÖTTER, S., BÜLTE, M. and LÜCKER, E. 1997. Detection of *Escherichia coli* O157 in foods by immunomagnetic separation (IMS). Archiv fur Lebensmittelhygiene, *48*, 85–87.
61. HEUVELINK, A.E., ZWARTKRUIS-NAHUIS, J.T.M. and DE BOER, E. 1997. Evaluation of media and test kits for the detection and isolation of *Escherichia coli* O157 from minced beef. J. Food Prot. *60*, 817–824.
62. HUDSON, J.A., NICOL, C., CAPILL, J. and BENNETT, J. 2000. Isolation of shiga toxin-producing *Escherichia coli* (STEC) from foods using EHEC agar. Lett. Appl. Microbiol. *30*, 109–113.
63. ILLINGWORTH, S., SEYMOUR, S. and TONKIN, C. 1999. Immunomagnetic separation of *Escherichia coli* O111 and O26 using Captivate Particles. *In* Proceedings of EU Concerted Action Survival and Growth Meeting, Agricultural University of Athens, Greece, 6-8 May, p. 140.
64. JOHNSON, J.L., ROSE, B.E., SHARAR, A.K., RANSOM, G.M., LATTUADA, C.P. and MCNAMARA, A.M. 1995. Methods used for detection and recovery of *Escherichia coli* O157:H7 associated with a food-borne disease outbreak. J. Food Prot. *58*, 597–603.

65. JOHNSON, R.P. *et al.* 1996. Growing concerns and recent outbreaks involving non-O157:H7 serotypes of verotoxigenic *Escherichia coli*. J. Food Prot. *59*, 1112–1122.

66. JOHNSON, R.P., DURHAM, R.J., JOHNSON, S.T., MACDONALD, L.A., JEFFREY, S.R. and BUTMAN, B.T. 1995. Detection of *Escherichia coli* O157:H7 in meat by an enzyme-linked immunosorbent assay, EHEC-Tek. Appl. Environ. Microbiol. *61*, 386–388.

67. KANG, D.H. and FUNG, D.Y. 1999. Development of a medium for differentiation between *Escherichia coli* and *Escherichia coli* O157:H7. J. Food Prot. *62*, 313–317.

68. KARCH, H. and MEYER, T. 1989. Evaluation of oligonucleotide probes for identification of shiga-like-toxin-producing *Escherichia coli*. J. Clin. Microbiol. *27*, 1180–1186.

69. KARCH, H., HUPPERTZ, H.-I., BOCKEMÜHL, J., SCHMIDT, H., SCHWARZKOPF, A. and LISSNER, R. 1997. Shiga toxin-producing *Escherichia coli* infections in Germany. J. Food Prot. *60*, 1454–1457.

70. KARCH, H., JANETZKI-MITTMANN, C., ALEKSIC, S. and DATZ, M. 1996. Isolation of enterohemorrhagic *Escherichia coli* O157 strains from patients with hemolytic-uremic syndrome by using immunomagnetic separation, DNA-based methods, and direct culture. J. Clin. Microbiol. *34*, 516–519.

71. KARCH, H., STOCKBINE, N.A. and O'BRIEN, A.D. 1986. Growth of *Escherichia coli* in the presence of trimethoprim-sulfamethoxazole facilitates detection of shiga-like toxin producing strains by colony blot assay. FEMS Microbiol. Lett. *35*, 141–145.

72. KARMALI, M.A., PETRIC, M., LIM, C., CHEUNG, R. and ARBUS, G. 1985. Sensitive method for detecting low numbers of verotoxin-producing *Escherichia coli* in mixed cultures by use of colony sweeps and polymyxin extraction of verotoxin. J. Clin. Microbiol. *22*, 614–619.

73. KARMALI, M.A., PETRIC, M. and BIELASZEWSKA, M. 1999. Evaluation of a microplate latex agglutination method (Verotox-F assay) for detecting and characterizing verotoxins (Shiga toxins) in *Escherichia coli*. J. Clin. Microbiol. *37*, 396–399.

74. KIM, M.S. and DOYLE, M.P. 1992. Dipstick immunoassay to detect enterohemorrhagic *Escherichia coli* O157:H7 in retail ground beef. Appl. Environ. Microbiol. *58*, 1764–1767.

75. KONOWALCHUK, J., SPIERS, J.I. and STAVRIC, S. 1977. Vero response to a cytotoxin of *Escherichia coli*. Infect. Immun. *18*, 775–779.

76. KUDVA, I.T., HATFIELD, P.G. and HOVDE, C.J. 1995. Effect of diet on the shedding of *Escherichia coli* O157:H7 in a sheep model. Appl. Environ. Microbiol. *61*, 1363–1370.

77. LEHMACHER, A., MEIER, H., ALEKSIC, S. and BOCKEMÜHL, J. 1998. Detection of hemolysin variants of shiga toxin-producing *Escherichia coli* by PCR and culture on vancomycin-cefixime-cefsulodin blood agar. Appl. Environ. Microbiol. *64*, 2449–2453.

78. LEVINE, M.M. *et al.* 1987. A DNA probe to identify enterohemorrhagic *Escherichia coli* of O157:H7 and other serotypes that cause hemorrhagic colitis and hemolytic uremic syndrome. J. Infect. Dis. *156*, 175–182.

79. LINBERG, A.M., LJUNGH, Å., AHRNÉ, S., LÖFDAHL, S. and MOLIN, G. 1998. Enterobacteriaceae found in high numbers in fish, minced meat and pasteurised milk or cream and the presence of toxin encoding genes. Int. J. Food Microbiol. *39*, 11–17.

80. LOUIE, M. *et al.* 1994. Sequence heterogeneity of the *eae* gene and detection of verotoxin-producing *Escherichia coli* using serotype-specific primers. Epidemiol. Infect. *112*, 449–461.

81. MACLEOD, D.L. and GYLES, C.S. 1989. Effects of culture conditions on yield of shiga-like toxin IIv from *Escherichia coli*. Can. J. Microbiol. *35*, 623–629.

82. MACRAE, M., REBATE, T., JOHNSTON, M. and OGDEN, I.D. 1997. The sensitivity of *Escherichia coli* O157 to some antimicrobials by conventional and conductance assays. Lett. Appl. Microbiol. *25*, 135–137.

83. MARCH, S.B. and RATNAM, S. 1986. Sorbitol-MacConkey medium for detection of *Escherichia coli* O157:H7 associated with hemorrhagic colitis. J. Clin. Microbiol. *23*, 869–872.

84. MAURER, J.J., SCHMIDT, D., PETROSKO, P., SANCHEZ, S., BOLTON, L. and LEE, M.D. 1999. Development of primers to O-antigen biosynthesis genes for specific detection of *Escherichia coli* O157 by PCR. Appl. Environ. Microbiol. *65*, 2954–2960.

85. MCCARTHY, J., HOLBROOK, R. and STEPHENS, P.J. 1998. An improved direct plate method for the enumeration of stressed *Escherichia coli* O157:H7 from food. J. Food Prot. *61*, 1093–1097.

86. MCCLEERY, D.R. and ROWE, M.T. 1995. Development of a selective plating technique for the recovery of *Escherichia coli* O157:H7 after heat stress. Lett. Appl. Microbiol. *21*, 252–256.

87. MCDONOUGH, P.L., ROSSITER, C.A., REBHUN, R.B., STEHMAN, S.M., LEIN, D.H. and SHIN, S.J. 2000. Prevalence of *Escherichia coli* O157:H7 from cull dairy cows in New York state and comparison of culture methods used during pre-harvest food safety investigations. J. Clin. Microbiol. *38*, 318–322.

88. MENG, J., DOYLE, M.P., ZHAO, T. and ZHAO, S. 1994. Detection and control of *Escherichia coli* O157:H7 in foods. Trends Food Sc. Technol. *5*, 179–184.

89. MENG, J., ZHAO, S., DOYLE, M.P., MITCHELL, S.E. and KRESOVICH, S. 1997. A multiplex PCR for identifying shiga-like toxin-producing *Escherichia coli* O157:H7. Lett. Appl. Microbiol. *24*, 172–176.

90. MENG, J., ZHAO, S., DOYLE, M.P., MITCHELL, S.E. and KRESOVICH, S. 1996. Polymerase chain reaction for detecting *Escherichia coli* O157:H7. Int. J. Food Microbiol. *32*, 103–113.

91. MILLEY, D.G. and SEKLA, L.H. 1993. An enzyme-linked immunosorbent assay-based isolation procedure for verotoxigenic *Escherichia coli*. Appl. Environ. Microbiol. *59*, 4223–4229.

92. MORTLOCK, S. 1994. Recovery of *Escherichia coli* O157:H7 from mixed suspensions: evaluation and comparison of pre-coated immunomagnetic beads and direct plating. Brit. J. Biomed. Sc. *51*, 207–214.

93. NEILL, M.A. 1997. Overview of verotoxigenic *Escherichia coli*. J. Food Prot. *60*, 1444–1446.

94. NEWLAND, J.W. and NEILL, R.J. 1988. DNA probes for shiga-like toxins I and II and for toxin-converting bacteriophages. J. Clin. Microbiol. *26*, 1292–1297.

95. OKREND, A.J.G., ROSE, B.E. and BENNETT, B. 1990a. A screening method for the isolation of *Escherichia coli* O157:H7 from ground beef. J. Food Prot. *53*, 249–252.

96. OKREND, A.J.G., ROSE, B.E. and LATTUADA, C.P. 1990b. Use of 5-bromo-4-chloro-3-indoxyl-beta-D-glucuronide in MacConkey sorbitol agar to aid in the isolation of *Escherichia coli* O157:H7 from ground beef. J. Food Prot. *53*, 941–943.

97. OKREND, A.J.G., ROSE, B.E. and LATTUADA, C.P. 1992. Isolation of *Escherichia coli* O157:H7 using O157 specific antibody coated magnetic beads. J. Food Prot. *55*, 214–217.

98. OKREND, A.J.G., ROSE, B.E. and MATNER R. 1990c. An improved screening method for the detection and isolation of *Escherichia coli* O157:H7 from meat, incorporating the 3M Petrifilm™ Test Kit-HEC for hemorrhagic *Escherichia coli* O157:H7. J. Food Prot. *53*, 936–940.

99. PADHYE, N.V. and DOYLE, M.P. 1992. *Escherichia coli* O157:H7: epidemiology, pathogenesis, and methods for detection in food. J. Food Prot. *55*, 555–565.

100. PADHYE, N.V. and DOYLE, M.P. 1991. Rapid procedure for detecting enterohemorrhagic *Escherichia coli* O157:H7 in food. Appl. Environ. Microbiol. *57*, 2693–2698.

101. PALUMBO, S.A., CALL, J.E., SCHULTZ, F.J. and WILLIAMS, A.C. 1995. Minimum and maximum temperatures for growth and verotoxin production by hemorrhagic strains of *Escherichia coli*. J. Food Prot. *58*, 352–356.

102. PATON, A.W., WOODROW, M.C., DOYLE, R.M., LANSER, J.A. and PATON, J.C. 1999. Molecular characterization of a shiga toxigenic *Escherichia coli* O113:H21 strain lacking *eae* responsible for a cluster of cases of hemolytic-uremic syndrome. J. Clin. Microbiol. *37*, 3357–3361.

103. PERERA, L.P., MARQUES, L.R.M. and O'BRIEN, A.D. 1988. Isolation and characterisation of monoclonal antibodies to shiga-like toxin II of enterohemorrhagic *Escherichia coli* and use of the monoclonal antibodies in a colony enzyme-linked immunosorbent assay. J. Clin. Microbiol. *26*, 2127–2131.

104. PERRY, M.B. and BUNDLE, D.R. 1990. Antigenic relationships of the lipopolysaccharides of *Escherichia hermannii* strains with those of *Escherichia coli* O157:H7, *Brucella melitensis*, and *Brucella abortus*. Infect. Imm. *58*, 1391–1395.

105. PIERARD, D., VAN DAMME, L., MORIAU, L., STEVENS, D. and LAUWERS, S. 1997. Virulence factors of verocytotoxin-producing *Escherichia coli* isolated from raw meats. Appl. Environ. Microbiol. *63*, 4585–4587.

106. POLLARD, D.R., JOHNSON, W.M., LIOR, H., TYLER, S.D. and ROZEE, K.R. 1990. Rapid and specific detection of verotoxin genes in *Escherichia coli* by the polymerase chain reaction. J. Clin. Microbiol. *28*, 540–545.

107. POWER, C.A. *et al.* 1998. Repeatability of the Petrifilm™ HEC test and agreement with a hydrophobic grid membrane filtration method for the enumeration of *Escherichia coli* O157:H7 on beef carcasses. J. Food Prot. *61*, 402–408.

108. PRADEL, N. *et al.* 2000. Prevalence and characterization of shiga toxin-producing *Escherichia coli* isolated from cattle, food, and children during a one-year prospective study in France. J. Clin. Microbiol. *38*, 1023- 1031.

109. RAGHUBEER, E.V. and MATCHES, J.R. 1990. Temperature range for growth of *Escherichia coli* serotype O157:H7 and selected coliforms in *E. coli* medium. J. Clin. Microbiol. *28*, 803–805.

110. READ, S.C. *et al.* 1992. Polymerase chain reaction for detection of verocytotoxigenic *Escherichia coli* isolated from animal and food sources. Mol. Cell. Probes *6*, 153–161.

111. READ, S.C., GYLES, C.L., CLARKE, R.C., LIOR, H. and MCEWEN, S. 1990. Prevalence of verocytotoxigenic *Escherichia coli* in ground beef, pork and chicken in Ontario. Epidemiol. Infect. *105*, 11–20.

112. RESTAINO, L., CASTILLO, H.J., STEWART, D. and TORTORELLO, M.L. 1996. Antibody-direct epifluorescent filter technique and immunomagnetic separation for 10-h screening and 24-h confirmation of *Escherichia coli* O157:H7 in beef. J. Food Prot. *59*, 1072–1075.

113. RESTAINO, L., FRAMPTON, E.W., TURNER, K.M. and ALLISON, D.R. 1999. A chromogenic plating medium for isolating *Escherichia coli* O157:H7 from beef. Lett. Appl. Microbiol. *29*, 26–30.

114. RICE, E.W., SOWERS, E.G., JOHNSON, C.H., DUNNIGAN, M.E., STROCKBINE, N.A. and EDBERG S.C. 1992. Serological cross-reactions between *Escherichia coli* O157 and other species of the genus *Escherichia*. J. Clin. Microbiol. *30*, 1315–1316.

115. RILEY, L.W. *et al.* 1983. Hemorrhagic colitis associated with a rare *Escherichia coli* serotype. New Engl. J.Med. *308*, 681–685.

116. ROCELLE, M., CLAVERO, S. and BEUCHAT, L.R. 1995. Suitability of selective plating media for recovering heat- or freeze-stressed *Escherichia coli* O157:H7 from tryptic soy broth and ground beef. Appl. Environ. Microbiol. *61*, 3268–3273.

117. SAAD, S.M.I. and FRANCO, B.D.G.M. 1999. Performance of Petrifilm™ Kit-HEC for enumeration of *Escherichia coli* O157:H7 in artificially contaminated ground beef. J. Rapid Meths Autom. in Microbiol. *7*, 173–181.

118. SAGE, J.R. and INGHAM, S.C. 1998. Evaluating survival of *Escherichia coli* O157:H7 in frozen and thawed apple cider: potential use of a hydrophobic grid membrane filter-SD-39 agar method. J. Food Prot. *61*, 490–494.

119. SAMADPOUR, M., ONGERTH, J.E. and LISTON, J. 1994. Development and evaluation of oligonucleotide DNA probes for detection and genogrouping of shiga-like toxin producing *Escherichia coli*. J. Food Prot. *57*, 399–402.

120. SANDERSON, M.W., GAY, J.M., HANCOCK, D.D., GAY, C.C., FOX, L.K. and BESSER, T.E. 1995. Sensitivity of bacteriologic culture for detection of *Escherichia coli* O157:H7 in bovine feces. J. Clin. Microbiol. *33*, 2616–2619.
121. SCHMIDT, H., SCHEEF, J., HUPPERTZ, H.I., FROSCH, M. and KARCH, H. 1999. *Escherichia coli* O157:H7 and O157:H(-) strains that do not produce shiga toxin: phenotypic and genetic characterization of isolates associated with diarrhea and hemolytic-uremic syndrome. J. Clin. Microbiol. *37*, 3491–3496.
122. SCOTTER, S., ALDRIDGE, M. and CAPPS, K. 2000. Validation of a method for the detection of *E. coli* O157:H7 in foods. Food Control, *11*, 85–95.
123. SEO, K.H., BRACKETT, R.E. and FRANK, J.F. 1998. Rapid detection of *Escherichia coli* O157:H7 using immunomagnetic flow cytometry in ground beef, apple juice, and milk. Int. J. Food Microbiol. *44*, 115–123.
124. SOWERS, E.G., WELLS, J.G. and STROCKBINE, N.A. 1996. Evaluation of commercial latex reagents for identification of O157 and H7 antigens of *Escherichia coli*. J. Clin. Microbiol. *34*, 1286–1289.
125. STARR, M. *et al.* 1998. Hemolytic-uremic syndrome following urinary tract infection with enterohemorrhagic *Escherichia coli*: case report and review. Clin. Infect. Dis. *27*, 310–315.
126. STEIN, K.O. and BOCHNER, B.R. 1998. Tellurite and novobiocin improve recovery of *E. coli* O157 on Rainbow® Agar O157. *In* proceedings of the 98th Annual Meeting of the American Society for Microbiology, P-80.
127. STEPHAN, R., RAGETTLI, S. and UNTERMANN, F. 2000. Prevalence and characteristics of verotoxin-producing *Escherichia coli* (VTEC) in stool samples from asymptomatic human carriers working in the meat processing industry in Switzerland. J. Appl. Microbiol. *88*, 335–341.
128. STEPHENS, P.J. and JOYNSON, J.A. 1998. Direct inoculation into media containing bile salts and antibiotics is unsuitable for the detection of acid/salt stressed *Escherichia coli* O157:H7. Lett. Appl. Microbiol. *27*, 147–151.
129. STEPHENS, P.J., LEE-SMITH, J.A.M., PHILLIPS, L.E. and HARRISON, M. 1999. Evaluation of a novel 24 hour timed release enrichment system for the rapid isolation of *Escherichia coli* O157 from foods. *In* American Society for Microbiology 99th General Meeting. 1999. 30 May–3 June. Chicago, Illinois.
130. STROCKBINE, N.A., WELLS, J.G., BOPP, C.A. and BARRETT, T.J. 1998. Overview of detection and subtyping methods. *In Escherichia coli* O157:H7 and other Shiga toxin-producing *E. coli* (J.B. Kaper and A.D. O'Brien, eds.). pp. 331–356. Washington, D.C.: American Society for Microbiology.
131. SZABO, R., TODD, E., MACKENZIE, J., PARRINGTON, L. and ARMSTRONG, A. 1990. Increased sensitivity of the rapid hydrophobic grid membrane filter enzyme-labeled antibody procedure for *Escherichia coli* O157 detection in foods and bovine feces. Appl. Environ. Microbiol. *56*, 3546–3549.
132. SZABO, R.A., TODD, E.C.D. and JEAN, A. 1986. Method to isolate *Escherichia coli* O157:H7 from food. J. Food Prot. *49*, 768–772.
133. TAORMINA, P.J., ROCHELLE, M., CLAVERO, S. and BEUCHAT, L.R. 1998. Comparison of selective agar media and enrichment broths for recovering heat-stressed *Escherichia coli* O157:H7 from ground beef. Food Microbiol. *15*, 631–638.

134. THIPPAREDDI, H., PHEBUS, R.K., FUNG, D.Y.C. and KASTNER, C.L. 1995. Use of universal pre-enrichment medium supplemented with oxyrase for the simultaneous recovery of *Escherichia coli* O157:H7 and *Yersinia enterocolitica.* J. Rapid Meths. Aut. Microbiol. *4,* 37–50.

135. TODD, E.C. *et al.* 1999. Application of a DNA hybridization-hydrophobic-grid membrane filter method for detection and isolation of verotoxigenic *Escherichia coli.* Appl. Environ. Microbiol. *65,* 4775–4780.

136. TODD, E.C. *et al.* 1988. Rapid hydrophobic grid membrane filter-enzyme-labeled antibody procedure for identification and enumeration of *Escherichia coli* O157 in foods. Appl. Environ. Microbiol. *54,* 2536– 2540.

137. UYTTENDAELE, M., VAN BOXSTAEL, S. and DEBEVERE, J. 1999. PCR assay for detection of the *E. coli* O157:H7 *eae*-gene and effect of the sample preparation method on PCR detection of heat-killed *E. coli* O157:H7 in ground beef. Int. J. Food Microbiol. *52,* 85–95.

138. VERNOZY-ROZAND, C. 1997. Detection of *Escherichia coli* O157:H7 and other verocytotoxin-producing *E. coli* (VTEC) in food. J. Appl. Microbiol. *82,* 537–551.

139. VERNOZY-ROZAND, C., MAZUY, C., RAY-GUENIOT, S., BOUTRAND-LOEÏ, S., MEYRAND, A. and RICHARD, Y. 1997. Detection of *Escherichia coli* O157 in French food samples using an immunomagnetic separation method and the VIDAS™ *E. coli* O157. Lett. Appl. Microbiol. *25,* 442–446.

140. WALLACE, J.S. and JONES, K. 1996. The use of selective and differential agars in the isolation of *Escherichia coli* O157 from dairy herds. J. Appl. Bacteriol *81,* 663–668.

141. WASTESON, Y. 1998. Detection of *E. coli* O157 from the environment, food and faeces; immuonomagnetic separation and practical tips. *In* Proceedings of the Methods for Verocytotoxigenic *E. coli* meeting of an EU Concerted Action 24-26 November. Western General Hospital, Edinburgh, Scotland: Teagasc, The National Food Centre, Dublin. p. 64–66.

142. WEAGANT, S.D., BRYANT, J.L. and JINNEMAN, K.G. 1995. An improved rapid technique for isolation of *Escherichia coli* O157:H7 from foods. J. Food Prot. *58,* 7–12.

143. WEERATNA, R.D. and DOYLE, M.P. 1991. Detection and production of verotoxin 1 of *Escherichia coli* O157:H7 in food. Appl. Environ. Microbiol. *57,* 2951–2955.

144. WELLS, J.G. *et al.* 1983. Laboratory investigation of hemorrhagic colitis outbreaks associated with a rare *Escherichia coli* serotype. J. Clin. Microbiol. *18,* 512–520.

145. WORLD HEALTH ORGANISATION SCIENTIFIC WORKING GROUP. 1999. Zoonotic non-O157 shiga toxin-producing *Escherichia coli* (STEC). WHO/CSR/APH/98.8. Department of Communicable Disease Surveillance and Response World Health Organisation Berlin, Germany.

146. WILLSHAW, G.A., SCOTLAND, S.M., SMITH, H.R. and ROWE, B. 1992. Properties of Vero cytotoxin-producing *Escherichia coli* of human origin of O serogroups other than O157. J. Infect. Dis. *166,* 797–802.

147. WILLSHAW, G.A., SCOTLAND, S.M., SMITH, H.R., CHEASTY, T., THOMAS, A. and ROWE, B. 1994. Hybridization of strains of *Escherichia coli* O157 with probes derived from the *eaeA* gene of enteropathogenic *E. coli* and the *eaeA* homolog from a Vero cytotoxin-producing strain of *E. coli* O157. J. Clin. Microbiol. *32*, 897-902.

148. WILLSHAW, G.A., SMITH, H.R., ROBERTS, D., THIRLWELL, J., CHEASTY, T. and ROWE, B. 1993. Examination of raw beef products for the presence of Vero cytotoxin producing *Escherichia coli,* particularly those of serogroup O157. J. Appl. Bacteriol. *75*, 420–426.

149. WITHAM, P.K., YAMASHIRO, C.T., LIVAK, K.J. and BATT, C.A. 1996. A PCR-based assay for the detection of *Escherichia coli* shiga-like toxin genes in ground beef. Appl. Environ. Microbiol. *62*, 1347–1353.

150. WOODY, J.M., STEVENSON, J.A., WILSON, R.A. and KNABEL, S.J. 1998. Comparison of the Difco EZ Coli™ Rapid Detection System and Petrifilm™ Test Kit-HEC for detection of *Escherichia coli* O157:H7 in fresh and frozen ground beef. J. Food Prot. *61*, 110–112.

151. WRIGHT, D.J., CHAPMAN, P.A. and SIDDONS, C.A. 1994. Immunomagnetic separation as a sensitive method for isolating *Escherichia coli* O157 from food samples. Epidemiol. Infect. *113*, 31-39.

152. YU, L.S.L. and FUNG, D.Y.C. 1991. Effect of Oxyrase™ enzyme on *Listeria monocytogenes* and other facultative anaerobes. J. Food Safety *11*, 163–175.

153. ZADIK, P.M., CHAPMAN, P.A. and SIDDONS, C.A. 1993. Use of tellurite for the selection of verocytotoxigenic *Escherichia coli* O157. J. Med. Microbiol. *39*, 155–158.

CHAPTER 5

GENERAL RECOVERY, CHARACTERISATION AND TYPING PROTOCOLS FOR VTEC

FIONA THOMSON-CARTER

Department of Medical Microbiology
Grampian University Hospitals Trust
Foresterhill
Aberdeen AB25 2ZN, Scotland

"A microbiologist would rather use another microbiologist's toothbrush than use their method"

Prof. F.J. Bolton, April 2000

The aim of this chapter is to provide an objective technique-oriented overview of methods for recovery, isolation and typing of VTEC. Constant development and adaptation of these three broad areas have resulted in a plethora of methods, but many recently published techniques have not been widely applied in terms of number of isolates analysed, diversity of isolates analysed and duration of study. Therefore their robustness and longevity at the bench is not yet proven.

Several recent reviews cover established methods for recovery, isolation and typing of VTEC from diverse sources (32,94,105,125). Thus appropriate references will be cited to direct the reader to comprehensive descriptions of specific techniques.

RECOVERY

Enrichment

Because some biological materials contain only small numbers of VTEC, which may be unevenly distributed within the specimen milieu, many techniques have been proposed to preferentially increase their numbers, prior to plating out the "enriched" specimen on selective media (see Immunomagnetic separation section below).

Rates of recovery will also be affected by the quality of the specimen: human faecal specimens collected 2 days after onset of symptoms have an isolation rate of 100% for *E. coli* O157, but the rate decreases to 33% for specimens collected at 7 days (19).

Enrichment may be achieved using a range of liquid and/or solid media of different compositions, and such media have been reviewed extensively (82,94,105,125). Protocols for recovery of stressed and sub-lethally injured VTEC from foods are detailed elsewhere (8).

The proposed International Standard (ISO) method for detection of *E. coli* O157 in foods uses modified trytone soy broth with novobiocin (20 mg/L) as the enrichment medium (ISO TC 43/SC 9, 1998). No similar standard exists for selective enrichment of *E. coli* O157 or other VTEC serotypes from human clinical specimens. Within the immunomagnetic separation procedure, the recommended enrichment medium is buffered peptone water (BPW, 20,22). Supplementation of BPW with antibiotics such as vancomycin, cefixime, cefsulodin, and the concentration(s) at which they should be included to achieve optimum isolation of *E. coli* O157 from biological material remain controversial (115).

Detection of *E. coli* O157 in bovine faecal samples is improved by using broth enrichment compared with direct plating (79,107).

Development of Media

As members of the species *Escherichia coli*, VTEC exhibit the common species biochemical characteristics, with the notable exception of *E. coli* O157 which generally does not ferment sorbitol and does not produce β-glucuronidase (1). These differences have been exploited in development of selective media.

Sorbitol-MacConkey agar (SMAC) was developed in 1986 for the detection of *E. coli* O157 (78). The majority of *E. coli* O157 isolates and certain other enterohaemorrhagic *E. coli* serotypes associated with haemolytic ureamic syndrome are non-sorbitol fermenting (85). Sorbitol-fermenting *E. coli* O157 have, however, been reported on several occasions (13,35,53) and the possibility that isolates exhibiting this phenotype could be overlooked on SMAC should be borne in mind. With the exceptions cited above SMAC is not generally applicable in the isolation of non-O157 VTEC. In addition to its use in direct plating SMAC can also be used in combination with other isolation and characterisation methods, for example, the Petrifilm™ colony blot enzyme immunoassay (17).

Since its first description, several modifications of SMAC agar have been made to improve selectivity and differentiation (125), principally CR-SMAC (cefixime and L-rhamnose added to SMAC, 21) and CT-SMAC (cefixime and tellurite added to SMAC, 131). However, supplementation of SMAC with tellurite will inhibit the growth of some strains of sorbitol-fermenting *E. coli* O157.

Supplementation of CT-SMAC with another carbohydrate, salicin, plus 4-methylumbelliferyl-B-D-galactopyranoside, improved selection of *E. coli* O157 from radish sprouts (37). The utility of this medium for human clinical specimens is not yet proven. Inclusion of the β-glucuronidase substrates 4-methylumbelliferyl-D-glucuronide (MUG) or 5-bromo-4-chloro-3-indoxyl-D-glucuronide (BCIG) in SMAC differentiates the mostly β-glucuronidase negative *E. coli* O157 from other *E. coli* (86). The majority of non-O157 VTEC isolates (94-96%) are β-glucuronidase positive, but more recently, β-glucuronidase positive isolates of *E. coli* O157 have been reported (41,44).

The recognition of the above biochemical characteristics and other phenotypic traits, e.g., β-galactosidase production, has led to the development of chromogenic agars capable of isolating and differentiating several VTEC serotypes, including *E. coli* O157 (94,105). However, the routine use of these media in the diagnostic laboratory would have significant cost implications.

The production of enterohaemolysin by several VTEC serotypes has also been exploited in the development of selective media: enterohaemolysin agar (11) and BVCC (vancomycin-cefixime-cefsulodin blood agar, 68). Detection of the enterohaemolytic phenotype on conventional blood agar has been shown to be insufficient for effective screening VTEC O111 isolates (108).

Selective media may be used in combination with other media and methods to improve detection of different enterohaemorrhagic *E. coli* serotypes: including ELISA toxin testing and chromogenic agar (83); pre-enrichment hydrochloric acid treatment (38).

Immunomagnetic Separation

The development of the immunomagnetic separation (IMS) technique has increased the sensitivity of detection of *E. coli* O157 in biological samples by 10- to 100-fold compared with conventional culture and enrichment methods. This technique is now widely used in the isolation of *E. coli* O157 from human, animal, food, water and environmental samples on CT-SMAC plates (20,22,29, 54,102,103). Further recommendations for optimisation of the technique have been made including reduction of non-specific adsorption (121) and exclusion of antibiotics from the enrichment medium (115).

The sensitivity of the IMS technique has been compared favourably with the sensitivities achieved by other screening methods, such as VIDAS (126); *E. coli* O157 Visual Immunoassay (20) and solid-phase laser cytometry (99). Immunomagnetic separation has been used in combination with a fluorescent-bacteriophage assay to improve detection of *E. coli* O157, with a detection limit of 10^4 cells per ml of broth culture (40).

Currently, no IMS based technique is generally available for non-O157 VTEC although magnetic beads coated with antibodies specific for serotypes O26 and O111 are being assessed (49).

DETECTION

The difficulties encountered in isolation of VTEC from biological samples and the requirement for accurate clinical diagnosis, have led to assessment and development of methods which, although not recovering viable target cells, can provide evidence of their presence (past or continuing) in the tested material.

Direct and Indirect Detection of Verotoxins and VT Genes

The classical method for indirectly demonstrating the presence of VTEC in biological specimens is by observation of the effects of verotoxins on Vero cell monolayers (14,65). Although the necessity for cell culture is a disadvantage, this assay is sensitive and specific (when correctly performed), providing confirmation of the presence of verotoxin(s) and, by extrapolation, VTEC. The Vero cell monolayer assay, in combination with PCR has been found to be a good predictor of the presence of VTEC, in both human and animal faecal specimens (100).

The presence of verotoxins in biological specimens may also be detected using commercially available immunological kits: Verotox-F latex agglutination (58); VTEC-RPLA latex agglutination (12); Premier EHEC ELISA (59,75).

DNA and oligonucleotide probes can detect VT genes in pure cultures of VTEC isolated from human faecal specimens or by direct screening of primary isolation plates by colony hybridisation (91). Screening with probes usefully differentiates any VTEC present from the background flora.

A digoxigenin-labelled PCR probe derived from the VT gene sequence has been applied, in combination with a membrane filter method, to detect and isolate VTEC from food and animal faecal samples (120). In ground beef 100% VTEC strains were recovered and the detection limit was 0.1 cfu per gram. The major advantage of this technique is the capability to detect *and subsequently isolate* low numbers of *vt*-positive colonies from biological samples.

Polymerase Chain Reaction Detection

The increasing capability of many diagnostic laboratories to perform the polymerase chain reaction (PCR), with its manifest advantages of speed and sensitivity, has resulted in the development of diverse methods exploiting this technology. In general terms, the PCR method can be performed directly on biological specimens, or may require prior extraction of target DNA. Particular problems are associated with the application of this method in the direct examination of animal and human faecal specimens. Best results are obtained when specimens are collected as soon as possible after the onset of symptoms. The presence of bilirubin and bile salts in the specimen inhibit PCR amplification, and centrifugation of samples to remove particulate matter and inhibitors is recommended (101). For these and other reasons prior extraction of DNA from faecal specimens, before PCR analysis, is recommended. Three commercially available kits compared favourably with an in-house method in their ability to extract *E. coli* O157 DNA from human faecal specimens for PCR (47). Comparison of methods for preparing template DNA from bovine faecal specimens demonstrated that boiling (alone) gave the most consistent results (114).

Several PCR methods have been devised for detection of *E. coli* O157 in foods (milk, 31; ground beef, 84; meat and milk, 26) and environmental samples. Application of restriction-site-specific PCR to *E. coli* isolated from cattle feedlot samples correctly identified *E. coli* O157 strains and therefore could be used for detection of *E. coli* O157 in environmental samples (64).

Between 18 and 37 EHEC genome equivalents were detected in enriched animal faecal specimens using a multiplex PCR for *stx*1, *stx*2, *eae*A and *hly*A (33). Intestinal sections of humans and animals can also be subjected to PCR analysis, 1-2 colony forming units of *eae*-positive *E. coli* can be detected (48). Eleven virulence genes, including *stx* and *eae* genes, were detected using a minimum of four multiplex reactions in *E. coli* isolated from dogs (90).

The first PCR method for direct detection of toxin genes was capable of detecting fewer than 10 VTEC per ml in crude faecal culture extracts (92). Subsequently sensitivities of 10^2 cfu and 10^7 cfu per 0.1 g stool were reported for VT1- and VT2-producing *E. coli*, respectively (101).

Multiplex PCR reactions for detection of *E. coli* O157 have also been applied to human specimens, initially targeting the *eae*, *slt* I and *slt* II genes, and also the 60 MDa plasmid sequences (36). These experiments were performed on enrichment cultures of human faecal specimens, but could also be extended to food and environmental specimens. Multiplex PCR can also detect the *vt* and *eae* genes of VTEC of serogroups O26 and O111 in stool specimens (72).

The sensitivity of PCR in the detection of VTEC sequences (*vt*1 and *vt*2) has been improved by inclusion of a magnetic capture step (25), allowing 10^0 cfu of VTEC per gram of ground beef to be detected after 15 hours of non-selective enrichment.

PCR of enriched samples is now a component of a number of commercial detection systems. BAX screening for *E. coli* O157:H7 (Qualicon) gave better detection rates in ground beef than conventional methods (52). The Probelia PCR system (Sanofi Diagnostics Pasteur) specifically detects *E. coli* O157:H7 in food samples within 24 hours, using a microplate assay to detect an amplified proprietary sequence unique to *E. coli* O157:H7. To date, neither of these systems has been routinely applied to specimens of animal or human origin.

Immunological Techniques

Immunological detection techniques are generally of two types: latex agglutination and enzyme linked immunosorbent assays (ELISA), based on bacterial characteristics. Monoclonal antibodies to *E. coli* O157 fimbrial antigen (61) and *E. coli* O26 surface proteins (60) can be used in diagnosis of infection. A colony lift immunoassay based on a horseradish peroxidase-labelled antibody has been developed to detect and quantitate *E. coli* O157 in retail meats (50). Several commercial kits are now available which target a range of bacterial components (O157 antigen, Premier *E. coli* O157, 88; Tecra *E. coli* O157 visual immunoassay, 20; O157 lipopolysaccharide and H7 antigen, ImmunoCard STAT!, 75, 113). These methods differ in specificity and sensitivity and any positive results must be confirmed by culture or demonstration of verocytotoxin production.

ELISA detection methods are amenable to automation, leading to the development of VIDAS, a rapid sandwich ELISA (126) and EiaFoss, combined IMS and ELISA (110), both of which claim high sample throughputs. A procedure combining filtration capture with immunoelectrochemical detection of *E. coli* O157 has been reported to detect as few as 50 bacterial cells (15).

Serodiagnosis

In the absence of a culture-positive specimen *E. coli* O157, infection may be diagnosed serologically. Sera can be tested for antibodies (IgM and/or IgG) to O157 lipopolysaccharide using ELISA or immunoblot (24). Faecal specimens may also be tested for IgA (109). Serodiagnosis of non-O157 VTEC is more problematic (23), and cross-reactions, particularly with *Brucella abortus* and *Yersinia enterocolitica*, have been reported (23).

IDENTIFICATION AND CONFIRMATION

Biochemical Tests

Suspect colonies isolated on selective agar are usually identified as *E. coli* by their biochemical characteristics using, e.g., API 20E, Biolog. Latex agglutination kits are available to confirm the identity of *E. coli* O157:H7 (Wellcolex, Oxoid, Pro-Lab) but not other VTEC serotypes. The sensitivities of several latex kits have been evaluated (112). Antisera to 43 O-types and 22 H-types, suitable for use in slide or tube agglutination confirmation tests, are commercially available (Denka Seiken).

Polymerase Chain Reaction Identification

A multiplex PCR assay that targets the β-glucuronidase (*uid*A) gene and the *vt* 1 and *vt* 2 genes can be used to identify bacterial isolates as *E. coli* O157 and to simultaneously determine their *vt* sub-type (18). A commercial kit based on detection of the *uid*A gene is available.

TYPING

Typing and subtyping, i.e., detailed descriptive characterisation and subsequent grouping of pathogenic bacteria are essential if human cases of infection are to be linked within epidemiological investigations, and the sources of infection traced. In research studies, typing data provides information on the individual organism, its clonal type and the gross population.

Historically the first typing techniques focused mainly on the phenotype of VTEC. This approach has, to a certain extent, been superseded by the emergence of genotypic based methods, although phenotypic methods still have some utility within hierarchical typing schemes. It is now possible, using appropriate genotypic techniques, to differentiate individual isolates, even at the nucleotide sequence level, and to construct robust databases with the data accrued.

Leaving aside such issues as the local availability of suitable expertise and resources, the selection of typing techniques to be used in any instance relies on a number of factors. These include, among others, the microbiological questions that have to be answered, the extent of the problem, the urgency with which answers are needed, any legal implications and the robustness required of the data elicited.

Recent extensive studies have analysed several VTEC serotypes from animal, human and environmental sources, including food. Many different typing and subtyping methods were applied to effectively characterise these diverse VTEC (2,45,81,96,97,132).

Phenotypic Methods

Serotyping. Serotyping is the most commonly used method of typing VTEC (57). The procedure for slide and/or tube agglutination itself is relatively simple, and adaptation of the conventional method has been described (10). Access to a complete panel of antisera is required (173 distinct 'O' and 56 distinct 'H' serotypes are recognised and a significant proportion of these are VTEC). In the past, only specialist reference centres had this facility but more recently, some antisera have become commercially available.

Biotyping. Several methods have conventionally been used in biotyping of VTEC. Some are extensions of isolation techniques, exploiting particular biochemical characteristics of the bacteria, e.g., ability to ferment sorbitol and produce β-glucuronidase. Other unusual characteristics, e.g., citrate utilisation may also be used (1).

More recently, differences have been identified in the distribution of the core oligosaccharide type among enterohaemorrhagic *E. coli* serogroups O157, O111, O26 and other VTEC serogroups, which could be exploited in the typing of isolates in the future (4).

Bacteriophage Typing. A phage-typing scheme for *E. coli* O157 has been developed that recognises 82 different types and certain subgroups within types 2 and 49 (62,63). Phage typing alone, however, may not provide the level of discrimination required for epidemiological investigation. Thus, for example, one study noted that 62.5% Scottish human O157 isolates recovered in a one-year period belonged to phage type 21/28 (119). Supplementary methods are therefore required. A development of the classical phage typing scheme (using microtitre plates to improve discrimination and data processing) is currently being validated (Dr. I. Toth, personal communication). No phage typing schemes are available for non-O157 VTEC.

Genotypic Methods

Multilocus Enzyme Electrophoresis. Multilocus enzyme electrophoresis (MLE) has been applied extensively in studies of bacterial population genetics. This method identifies alleles of common housekeeping enzymes, allowing phylogenetic relationships within groups of bacteria to be established. MLE analyses of 1300 isolates of *E. coli* representing 16 serotypes including VTEC O26:H11, O55:H7, O111:H2, O128:H2 and O157:H7 demonstrated that 72% of isolates belonged to 15 major electrophoretic types, with each type corresponding to a widely distributed bacterial clone. Little variation was observed among 369 *E. coli* O157:H7 isolates, with 95% being assigned to one electropherotype (129). This method indicates that the O157:H7 clonal group is only distantly related to other VTECs (128).

Plasmid Analysis. *E. coli* O157 and VTEC of other serotypes possess a 60mDa plasmid. Although not a definitive method of typing, detection of this plasmid may be useful in confirming the identity of isolates. In *E. coli* O157 the plasmid may be detected using a probe (69). Other plasmids e.g. pColD157, may be present (46). The plasmid profiles of VTEC other than serotype O157:H7 are highly variable in genetic arrangement and composition (16).

Restriction Fragment Length Polymorphism (RFLP) Analysis. Isolates of VTEC can be analysed and differentiated using probes of different specificities and target sequences. Such probes can detect: polymorphisms of ribosomal RNA genes (116); insertion sequences (67); phage lambda-related sequences (89); verotoxin genes and their polymorphisms (118); verotoxin-encoding phage-related sequences (106). A factor to be considered in RFLP and PCR (see below) analyses of these genes, however, is the reported loss of VT genes on subcultivation of clinical isolates of *E. coli* (55).

As the genetic heterogeneity of particular loci differs, the level of discrimination obtained will vary. Thus, supplementary analyses may be required. For example, ribosomal RNA genes of *E. coli* O157, demonstrate little heterogeneity and therefore may not usefully differentiate among O157 isolates, although they could enable differentiation of O157 from other VTEC serotypes. An automated procedure has been developed for ribotyping (124). Sufficient variation among the other genomic characteristics exists at the nucleotide sequence level to allow differentiation of groups of closely related isolates of O157 and other serotypes (2).

Pulsed-field Gel Electrophoresis. Restriction fragment length polymorphisms of the VTEC genome can also be analysed by pulsed-field gel electrophoresis (PFGE) for which a limited, standardised protocol has been agreed (98). A more rapid PFGE protocol has been used successfully (39). Macrorestriction patterns of large DNA fragments resolved by PFGE have been used extensively in the investigation of sporadic cases and outbreaks of *E. coli* O157 infection (2,3,6,66). PFGE has also proved invaluable in molecular characterisation and epidemiology of infection involving other VTEC serotypes: O26 (132), O111 (81,93), O118 (130). Isolates with indistinguishable macrorestriction patterns are generally considered to be associated. Currently, however, no specific interpretative rules exist for VTEC PFGE data, only general guidelines for certain bacterial pathogens (117). Minor variation in macrorestriction patterns (1-2 fragment(s) absent/present) has been observed in VTEC, including isolates considered to be epidemiologically linked (2,97). Previously, certain fragment differences have been shown to be insignificant, in that they did not materially affect the grouping of outbreak strains (51). Any interpretation of VTEC PFGE data should recognise reports of variation in the size of the *E. coli* O157 genome (43) and clonal turnover (53). PFGE studies of experimental populations of *E. coli* B have noted short-term chromosome evolution over 2000 generations (9). A total of 11 changes occurred, 8 of which affected chromosome size and 3 of which altered restriction sites in strains cultured at 32°C and/or 42°C. However, no changes were observed in strains cultured at 37°C. The effect of chromosome evolution on VTEC macrorestriction patterns obtained under experimental conditions outlined here may therefore be minimal.

Standardisation of PFGE protocols and electronic transmission of gel images could create a genetic database with which macrorestriction patterns for unknown isolates could be compared for epidemiological purposes, outbreak investigation and definition of clonal relationships (98).

Polymerase Chain Reaction Typing. Since description of the first PCR assay to detect VT genes in *E. coli* (95) many others have been developed for detection, identification and subtyping of VTEC. Multiplex assays in which more than one pair of primers is used to target sequences encoding different traits or serotype-specific polymorphisms of the same gene have been described. In multiplex PCR assays, however, problems may arise from changes in concentrations and ratios of reagents and the presence of more than one pair of primers (80).

VTEC target sequences exhibiting sufficient variation to enable subtyping include: variants of the VT genes (7); other proven and potential virulence traits (*eaeA*, 104, 87; *eae*, *slt*I, *slt*II, 27; *eaeA*, *hlyA*, *rfb* O157, 92); *fliC* genes (34, 127); 16S-23S rDNA spacer regions (42); lambdoid phage gene p and *vt* structural genes (30). Genomic polymorphisms can also be targeted by arbitrarily-primed PCR (AP-PCR, 76), enterobacterial repetitive intergenic consensus PCR (ERIC PCR, 73) and amplified fragment length polymorphism analysis (AFLP, 2, 5).

DNA Sequencing. The availability of automated sequencers and improved sequence data analysis programs have led to assessment of gene sequencing as a subtyping method. However, the 16S rRNA genes of several *E. coli* serotypes, including O157, have been found to provide insufficient variations at sequence level to support useful discrimination of isolates (Dr. P. Carter, personal communication), although the gene encoding the VT2 B-subunit (Dr. P. Carter, personal communication), and the *eae* genes of several VTEC serotypes (71), do exhibit some sequence diversity.

Multilocus sequence typing (MLST), a development of multilocus enzyme electrophoresis, is based on determining the DNA sequence of housekeeping genes (77). Initially proposed for the typing of *Neisseria meningitidis*, this technique may be readily applied to other bacterial genera, although no data for VTEC are currently available. A major advantage of future systematic application of MLST to VTEC would be the creation of a WWW (Worldwide Web) database of robust, portable nucleotide sequence data. Submission of sequence data from unknown isolates for comparison with the database would inform real-time epidemiological investigations, while simultaneously accruing the raw data necessary to draw definitive conclusions on population genetics and evolution.

CONCLUSIONS AND FUTURE DEVELOPMENTS

Sustained efforts to increase our knowledge of VTEC have led to the development and application of many disparate methods for the recovery, detection and typing of these organisms. Not all of these are equally useful, for a number of reasons including significant differences in sensitivity, specificity and applicability. The value of these methods in fulfilling these criteria, however, will only be proven when they have been generally applied in laboratories world-wide, to many VTEC of different serotypes, isolated from diverse sources. In the meantime, microbiologists will continue to rely on their own individually-derived or preferred methods. The recent development of the proposed ISO method for the detection of *E. coli* O157 in foods and similar exercises may help provide a focus to achieve a better degree of overall consensus.

The desired criteria for methods to definitively type VTEC include aspects of general utility: low cost, ease of application in routine laboratories, high levels of reproducibility, between samples and between laboratories. For molecular methods, further desirable characteristics are the enhanced ability to discriminate and differentiate isolates usefully and the potential for creation of electronic databases. Currently, no single method can fulfil all these requirements. Some candidate "one-stop" methods, e.g., AFLP and MLST, are not yet generally available. In the immediate future, therefore, the recently recommended hierarchical, combinatorial approach (2,97) will continue to be applied to achieve maximum discriminate and differentiate among VTEC and *E. coli* O157 isolates.

In the longer term, new technologies, including biosensors (28,123) and high-density DNA microarrays (79,111,122), capable of detecting minute quantities of target sequences in biological specimens, may have a role in the recovery, detection and characterisation of VTEC. Until then, the resourcefulness of microbiologists in rising to meet the challenges presented by VTEC will continue to be tested.

REFERENCES

1. ALEKSIC, S., KARCH, H. and BOCKEMUHL, J. 1992. A biotyping scheme for Shiga-like (Vero) cytotoxin-producing *Escherichia coli* O157 and a list of serological cross-reactions between O157 and other Gram-negative bacteria. Zentral. Bakteriol. *276*, 221–230.
2. ALLISON, L.J., CARTER, P.E. and THOMSON-CARTER, F.M. 2000. Characterization of a recurrent clonal type of *Escherichia coli* O157:H7 causing major outbreaks of infection in Scotland. J. Clin. Microbiol. *38*, 1632–1635.

3. ALLISON, L.J., STIRRAT, A. and THOMSON-CARTER, F.M. 1998. Genetic heterogeneity of *Escherichia coli* O157:H7 in Scotland and its utility in strain subtyping. Eur. J. Clin. Microbiol. Infect. Dis. *17*, 844–848.
4. AMOR, K., HEINRICHS, D.E., FRIRDICH, E., ZIEBELL, K., JOHNSON, R.P. and WHITFIELD, C. 2000. Distribution of core oligosaccharide types in lipopolysaccharides from *Escherichia coli*. Infect. Immun. *68*, 1116–1124.
5. ARNOLD, C., METHERELL, L., WILLSHAW, G., MAGGS, A. and STANLEY, J. 1999. Predictive fluorescent amplified-fragment length polymorphism analysis of *Escherichia coli*: high-resolution typing method with phylogenetic significance. J. Clin. Microbiol. *37*, 1274–1279.
6. BARRETT, T.J. *et al.* 1994. Laboratory investigation of a multistate food-borne outbreak of *Escherichia coli* O157:H7 by using pulsed-field gel electrophoresis and phage typing. J. Clin. Microbiol. *32*, 3013–3017.
7. BASTIAN, S.N., CARLE, I. and GRIMONT, F. 1998. Comparison of 14 PCR systems for the detection and subtyping of *stx* genes in Shiga-toxin-producing *Escherichia coli*. Res. Microbiol. *149*, 457–472.
8. BAYLIS, C. 2000. ibid.
9. BERGTHORSSON, U. and OCHMAN, H. 1999. Chromosomal changes during experimental evolution in laboratory populations of *Escherichia coli*. J. Bacteriol. *181*, 1360–1363.
10. BETTELHEIM, K.A. and THOMPSON, C.J. 1987. New method of serotyping *Escherichia coli*: implementation and verification. J. Clin. Microbiol. *25*, 781–786.
11. BEUTIN, L. *et al.* 1989. Close association of verotoxin (Shiga-like toxin) production with enterohemolysin production in strains of *Escherichia coli*. J. Clin. Microbiol. *27*, 2559–2564.
12. BEUTIN, L., ZIMMERMANN, S. and GLEIER, K. 1996. Rapid detection and isolation of Shiga-like toxin (verocytotoxin)-producing *Escherichia coli* by direct testing of individual enterohemolytic colonies from washed sheep blood agar plates in the VTEC-RPLA assay. J. Clin. Microbiol. *34*, 2812–2814.
13. BIELASZEWSKA, M. *et al.* 1998. Isolation and characterization of sorbitol-fermenting Shiga toxin (Verocytotoxin)-producing *Escherichia coli* O157:H- strains in the Czech Republic. J. Clin. Microbiol. *36*, 2135–2137.
14. BITZAN, M., LUDWIG, K., KLEMT, M., KONIG, H., BUREN, J. and MULLER-WIEFEL, D.E. 1993. The role of *Escherichia coli* O157 infections in the classical (enteropathic) haemolytic uraemic syndrome: results of a Central European, multicentre study. Epidemiol. Infect. *110*, 183–196.
15. BREWSTER, J.D. and MAZENKO, R.S. 1998. Filtration capture and immuno-electrochemical detection for rapid assay of *Escherichia coli* O157:H7. J. Immunol. Methods. *211*, 1–8.
16. BRUNDER, W., SCHIDT, H., FROSCH, M. and KARCH, H. 1999. The large plasmids of Shiga-toxin-producing *Escherichia coli* (STEC) are highly variable genetic elements. Microbiol. *145*, 1005–1014.
17. CALICCHIA, M.L., PARKER, E.L., GAMBREL-LENARZ, S. and MATNER, R.R. 1997. Use of the Petrifilm™ Test Kit-HEC direct blot enzyme immunoassay method without swab pre-enrichment to screen for low levels of *Escherichia coli*

O157:H7 on beef carcasses by surface swabbing. J. Food Protect. *60*, 870–873.

18. CEBULA, T.A., PAYNE, W.L. and FENG, P. 1995. Simultaneous identification of strains of *Escherichia coli* serotype O157:H7 and their Shiga-like toxin type by mismatch amplification mutation assay-multiplex PCR. J. Clin. Microbiol. *33*, 248–250.
19. CHAPMAN, P.A. 1994. Isolation, identification and typing of Vero cytotoxin-producing *Escherichia coli* O157. PHLS Microbiol. Dig. *11*, 13–17.
20. CHAPMAN, P.A., CERDAN MALO, A.T., SIDDONS, C.A. and HARKIN, M. 1997a. Use of commercial enzyme immunoassays and immunomagnetic separation systems for detecting *Escherichia coli* O157 in bovine fecal samples. Appl. Environ. Microbiol. *63*, 2549–2553.
21. CHAPMAN, P.A., SIDDONS, C., ZADIK, P.M. and JEWES, L. 1991. An improved selective medium for the isolation of *Escherichia coli* O157. J. Med. Microbiol. *35*, 107–110.
22. CHAPMAN, P.A., SIDDONS, C.A., CERDAN MALO, A.T. and HARKIN, M.A. 1997b. A 1-year study of *Escherichia coli* O157 in cattle, sheep, pigs and poultry. Epidemiol. Infect. *119*, 245–250.
23. CHART, H. 1993. Serodiagnosis of infections caused by *Escherichia coli* O157:H7 and other VTEC. Serdiagn. Immunother. Infect. Dis. *1*, 8–12.
24. CHART, H., SMITH, H.R., SCOTLAND, S.M., ROWE, B., MILFORD, D.V. and TAYLOR, C.M. 1991. Serological identification of *Escherichia coli* O157:H7 infection in haemolytic uraemic syndrome. Lancet *337(8734)*, 138–140.
25. CHEN, J., JOHNSON, R. and GRIFFITHS, M. 1998. Detection of verotoxigenic *Escherichia coli* by magnetic capture-hybridization PCR. Appl. Environ. Microbiol. *64*, 147–152.
26. CHEN, S. *et al.* 1998. An automated fluorescent PCR method for detection of Shiga toxin-producing *Escherichia coli* in foods. Appl. Environ. Microbiol. *64*, 4210–4216.
27. CHINA, B., PIRSON, V. and MAINIL, J. 1996. Typing of bovine attaching and effacing *Escherichia coli* by multiplex in vitro amplification of virulence-associated genes. Appl. Environ. Microbiol. *62*, 3462–3465.
28. CORNELL, B.A. *et al.* 1997. A biosensor that uses ion channel switches. Nature *387*, 580–583.
29. CUBBON, M.D., COIA, J.E., HANSON, M.F. and THOMSON-CARTER, F.M. 1996. A comparison of immunomagnetic separation, direct culture and polymerase chain reaction for the detection of verocytotoxin-producing *Escherichia coli* O157 in human faeces. J. Med. Microbiol. *44*, 219–222.
30. DATZ, M., JANETZKI-MITTMANN, C., FRANKE, S., GUNZER, F., SCHMIDT, H. and KARCH, H. 1996. Analysis of the enterohemorrhagic *Escherichia coli* O157 DNA region containing lambdoid phage gene p and Shiga like toxin structural genes. Appl. Environ. Microbiol. *62*, 791–797.
31. DESMARCHELIER, P.M., BILGE, S.A., FEGAN, N., MILLS, L., VARY, J.C. JR. and TARR, P.I. 1998. A PCR specific for *Escherichia coli* O157 based on the *rfb* locus encoding O157 lipopolysaccharide. J. Clin. Microbiol. *36*, 1801–1804.

32. EASTON, L. 1997. *Escherichia coli* O157: occurrence, transmission and laboratory detection. Br. J. Biomed. Sci. *54*, 57–64.
33. FAGAN, P.K., HORNITZKY, M.A., BETTELHEIM, K.A. and DJORDJEVIC, S.P. 1999. Detection of Shiga-like toxin (*stx*1 and *stx*2), intimin (*eae*A), and enterohemorrhagic *Escherichia coli* (EHEC) hemolysin (EHEC *hly*A) genes in animal feces by multiplex PCR. Appl. Environ. Microbiol. *65*, 868–872.
34. FIELDS, P.I., BLOM, K., HUGHES, H.J., HELSEL, L.O., FENG, P. and SWAMINATHAN, B. 1997. Molecular characterization of the gene encoding H antigen in *Escherichia coli* and development of a PCR-restriction fragment length polymorphism test for identification of *E. coli* O157:H7 and O157:NM. J. Clin. Microbiol. *35*, 1066–1070.
35. FRATAMICO, P.M., BUCHANAN, R.L. and COOKE, P.H. 1993. Virulence of an Escherichia coli O157:H7 sorbitol-positive mutant. Appl. Environ. Microbiol. *59*, 4245–4252.
36. FRATAMICO, P.M., SACKITEY, S.K., WIEDMANN, M. and DENG, M.Y. 1995. Detection of *Escherichia coli* O157:H7 by multiplex PCR. J. Clin. Microbiol. *33*, 2188–2191.
37. FUJISAWA, T., SATA, S., AIKAWA, K., TAKAHASHI, T., YAMAI, S. and SHIMIDA, S. 2000. Modification of sorbitol MacConkey medium containing cefixime and tellurite for isolation of *Escherichia coli* O157:H7 from radish sprouts. Appl. Environ. Microbiol. *66*, 3117–3118.
38. FUKUSHIMA, H., HOSHINA, K. and GOMYODA, M. 2000. Selective isolation of *eae*-positive strains of Shiga toxin-producing *Escherichia coli*. J. Clin. Microbiol. *38*, 1684–1687.
39. GAUTOM, R.K. 1997. Rapid pulsed-field gel electrophoresis protocol for typing of *Escherichia coli* O157:H7 and other Gram-negative organisms in one day. J. Clin. Microbiol. *35*, 2977–2980.
40. GOODRIDGE, L., CHEN, J. and GRIFFITHS, M. 1999. Development and characterization of a fluorescent-bacteriophage assay for detection of *Escherichia coli* O157:H7. Appl. Environ. Microbiol. *65*, 1397–1404.
41. GUNZER, F., BOHM, H., RUSSEMANN, H., BITZAN, M., ALEKSIC, S. and KARCH, H. 1992. Molecular detection of sorbitol-fermenting *Escherichia coli* O157 in patients with hemolytic uremic syndrome. J. Clin. Microbiol. *30*, 1807–1810.
42. GURTLER, V. and STANISICH, V.A. 1996. New approaches to typing and identification of bacteria using the 16S-23S rDNA spacer region. Microbiol. *142*, 3–16.
43. HARSONO, K.D., KASPAR, C.W. and LUCHANSKY, J.B. 1993. Comparison and genomic sizing of *Escherichia coli* O157:H7 isolates by pulsed-field gel electrophoresis. Appl. Environ. Microbiol. *59*, 3141–3144.
44. HAYES, P.S., BLOM, K., FENG, P., LEWIS, J., STROCKBINE, N.A. and SWAMINATHAN, B. 1995. Isolation and characterization of a B-glucuronidase-producing strain of *Escherichia coli* serotype O157:H7 in the United States. J. Clin. Microbiol. *33*, 3347–3348.

45. HEUVELINK, A.E. *et al.* 1998. Isolation and characterization of verocytotoxin-producing *Escherichia coli* O157 strains from Dutch cattle and sheep. J. Clin. Microbiol. *36*, 878–882.

46. HOFINGER, C., KARCH, H. and SCHMIDT, H. 1998. Structure and function of plasmid pColD157 of enterohemorrhagic *Escherichia coli* O157 and its distribution among strains from patients with diarrhea and hemolytic-uremic syndrome. J. Clin. Microbiol. *36*, 24–29.

47. HOLLAND, J.L., LOUIE, L., SIMOR, A.E. and LOUIE, M. 2000. PCR detection of *Escherichia coli* O157:H7 directly from stools: evaluation of commercial extraction methods for purifying fecal DNA. J. Clin. Microbiol. *38*, 4108–4113.

48. HUBBARD, A.L., HARRISON, D.J., MOYES, C. and MCORIST, S. 1998. Direct detection of *eae*-positive bacteria in human and veterinary colorectal specimens by PCR. J. Clin. Microbiol. *36*, 2326–2330.

49. ILLINGWORTH, D.S., SEYMOUR, S. and TONKIN, C. 1999. Immunomagnetic separation of *E. coli* O111 and O26 using Captivate particles. *In* Survival and growth of verocytotoxigenic *E. coli*. Proceedings of the meeting of CT98-3935 VTEC in Europe (G. Duffy, P. Garvey, J. Coia, Y. Wasteson and D.A. McDowell, eds.) Teagasc, Ireland, pp. 140.

50. INGRAM, D.T., LAMICHHANE, C.M., ROLLINS, D.M., CARR, L.E., MALLINSON, E.T. and JOSEPH, S.W. 1998. Development of a colony lift immunoassay to facilitate rapid detection and quantification of *Escherichia coli* O157:H7 from agar plates and filter monitor membranes. Clin. Diagn. Lab. Immunol. *5*, 567–573.

51. IZUMIYA, H. *et al.* 1997. Molecular typing of enterohemorrhagic *Escherichia coli* O157:H7 isolates in Japan by using pulsed field gel electrophoresis. J. Clin. Microbiol. *35*, 1675–1680.

52. JOHNSON, J.L., BROOKE, C.L. and FRITSCHEL, S.J. 1998. Comparison of the BAX for screening/*E. coli* O157:H7 method with conventional methods for detection of extremely low levels of *Escherichia coli* O157:H7 in ground beef. Appl. Environ. Microbiol. *64*, 4390–4395.

53. KARCH, H., BOHM, H., SCHMIDT, H., GUNZER, F., ALEKSIC, S. and HEESEMANN, J. 1993. Clonal structure and pathogenicity of Shiga-like toxin-producing sorbitol-fermenting *Escherichia coli* O157:H-. *31*, 1200–1205.

54. KARCH, H., JANETZKI-MITTMANN, C., ALEKSIC, S. and DATZ, M. 1996. Isolation of enterohemorrhagic *Escherichia coli* O157 strains from patients with hemolytic-uremic syndrome by using immunomagnetic separation, DNA-based methods, and direct culture. J. Clin. Microbiol. *34*, 516–519.

55. KARCH, H., MEYER, T., RUSSMANN, H. and HEESEMANN, J. 1992. Frequent loss of Shiga-like toxin genes in clinical isolates of *Escherichia coli* upon subcultivation. Infect. Immun. *60*, 3464–3467.

56. KARCH, H., RUSSMANN, H., SCHMIDT, H., SCHWARZKOPF, A. and HEESEMANN, J. 1995. Long-term shedding and clonal turnover of enterohemorrhagic *Escherichia coli* O157 in diarrheal diseases. J. Clin. Microbiol. *33*, 1602–1605.

57. KARMALI, M.A. 1989. Infection by verocytotoxin-producing *Escherichia coli*. Clin. Microbiol. Rev. *2*, 15–38.

58. KARMALI, M.A., PETRIC, M. and BIELASZEWSKA, M. 1999. Evaluation of a microtitre latex agglutination method (Verotox-F assay) for detecting and characterizing verotoxins (Shiga toxins) in *Escherichia coli*. J. Clin. Microbiol. *37*, 396–399.

59. KEHL, K.S., HAVENS, P., BEHNKE, C.E. and ACHESON, D.W.K. 1997. Evaluation of the Premier EHEC assay for detection of Shiga toxin-producing *Escherichia coli*. J. Clin. Microbiol. *35*, 2051–2054.

60. KERR, P. *et al.* 1999. Use of a monoclonal antibody against an *Escherichia coli* O26 surface protein for detection of enteropathogenic and enterohemorrhagic strains. Clin. Diagn. Lab. Immunol. *6*, 610–614.

61. KERR, P., BALL, H., FINLAY, D., POLLOCK, D., WILSON, I. and MACKIE, D. 1998. Diagnostic application of a fimbrial *Escherichia coli* O157 monoclonal antibody. *In* Methods for verocytotoxigenic *E. coli*. Proceedings of meeting of CT98–3935 VTEC in Europe (G. Duffy, P. Garvey, J. Coia, Y. Wasteson and D.A. McDowell, eds.), Teagasc, Ireland, pp. 67–70.

62. KHAKRIA, R., DUCK, D. and LIOR, H. 1990. Extended phage typing scheme for *Escherichia coli* O157:H7. Epidemiol. Infect. *105*, 511–520.

63. KHAKRIA, R., MULVEY, M., AHMED, R., WOODWARD, D. and JOHNSON, W. 1997. Abstr. 3rd International Symposium and Workshop on Shiga Toxin (Verocytotoxin)- Producing *Escherichia coli* Infections, abstr. V124/I.

64. KIMURA, R., MANDREL, R.E., GALLAND, J.C., HYATT, D. and RILEY, L.W. 2000. Restriction-site-specific PCR as a rapid test to detect enterohemorrhagic *Escherichia coli* O157:H7 strains in environmental samples. Appl. Environ. Microbiol. *66*, 2513–2519.

65. KONOWALCHUK, J., SPEIRS, J.I. and STAVRIC, S. 1977. Vero response to a cytotoxin of *Escherichia coli*. Infect. Immun. *18*, 775–779.

66. KRAUSE, U., THOMSON-CARTER, F.M. and PENNINGTON, T.H. 1996. Molecular epidemiology of *Escherichia coli* O157:H7 by pulsed-field gel electrophoresis and comparison with that by bacteriophage typing. J. Clin. Microbiol. *34*, 959–961.

67. LAWRENCE, J.G., OCHMAN, H. and HART, D.L. 1992. The evolution of insertion sequences within enteric bacteria. Genet. *131*, 9–20.

68. LEHMACHER, A., MEIER, H., ALEKSIC, S. and BOCKEMUHL, J. 1998. Detection of hemolysin variants of Shiga toxin-producing *Escherichia coli* by PCR and culture on vancomycin-cefixime-cefsulodin blood agar. Appl. Environ. Microbiol. *64*, 2449–2453.

69. LEVINE, M.M. *et al.* 1987. A DNA probe to identify enterohemorrhagic *Escherichia coli* of O157:H7 and other serotypes that cause hemorrhagic colitis and hemolytic uremic syndrome. J. Infect. Dis. *157*, 175–182.

70. LIPSHUTZ, R.J., FODOR, S.P.A., GINGERAS, T.R. and LOCKHART, D.J. 1999. High density synthetic oligonucleotide arrays. Nature Genet. *21*, 20–24.

71. LOUIE, M. *et al.* 1994. Sequence heterogeneity of the *eae* gene and detection of verotoxin-producing *Escherichia coli* using serotype-specific primers. Epidemiol. Infect. *112*, 449–461.

72. LOUIE, M. *et al.* 1998. Application of multiplex PCR for detection of non-O157 verocytotoxin-producing *Escherichia coli* in bloody stools: identification of serogroups O26 and O111. J. Clin. Microbiol. *36*, 3375–3377.
73. LOUWS, F.J., FULBRIGHT, D.W., STEPHENS, C.T. and DE BRUIJN, F.J. 1994. Specific genomic fingerprints of phytopathogenic *Xanthomonas* and *Pseudomonas* pathovars and strains generated with repetitive sequences and PCR. Appl. Environ. Microbiol. *60*, 2286–2295.
74. MACKENZIE, A. *et al.* 2000. Performance of the ImmunoCard STAT! *E. coli* O157:H7 test for detection of *Escherichia coli* O157:H7 in stools. J. Clin. Microbiol. *38*, 1866–1868.
75. MACKENZIE, A.M.R. *et al.* 1998. Sensitivities and specificities of Premier *E. coli* O157 and Premier EHEC enzyme immunoassays for diagnosis of infection with verotoxin (Shiga-like toxin)-producing *Escherichia coli*. J. Clin. Microbiol. *36*, 1608–1611.
76. MADICO, G., AKOPYANTS, N.S. and BERG, D.E. 1995. Arbitrarily primed PCR DNA fingerprinting of *Escherichia coli* O157:H7 strains by using template from boiled cultures. J. Clin. Microbiol. *33*, 1534–1536.
77. MAIDEN, M.C. *et al.* 1998. Multilocus sequence typing: a portable approach to the identification of clones within populations of pathogenic microorganisms. Proc. Natl. Acad. Sci. *95*, 3140–3145.
78. MARCH, S.B. and RATNAM, S. 1986. Sorbitol-MacConkey medium for detection of *Escherichia coli* O157:H7 associated with hemorrhagic colitis. J. Clin. Microbiol. *23*, 869–872.
79. MCDONOUGH, P.L., ROSSITER, C.A., REBHUN, R.B., STEHMAN, S.M., LEIN, D.H. and SHIN, S.J. 2000. Prevalence of *Escherichia coli* O157:H7 from cull dairy cows in New York State and comparison of culture methods used during preharvest food safety investigations. J. Clin. Microbiol. *38*, 318–322.
80. MENG, J., ZHAO, S., DOYLE, M.P., MITCHELL, S.E. and KRESOVICH, S. 1997. A multiplex PCR for identifying Shiga-like toxin-producing *Escherichia coli* O157:H7. Lett. Appl. Microbiol. *24*, 172–176.
81. MORABITO, S. *et al.* 1999. Molecular characterisation of verocytotoxin-producing *Escherichia coli* of serogroup O111 from different countries. J. Med. Microbiol. *48*, 891–896.
82. NATARO, J.P. and KAPER, J.B. 1998. Diarrheagenic *Escherichia coli*. Clin. Microbiol. Rev. *11*, 142–201.
83. NOVICKI, J., DALY, J.A., MOTTICE, S.L. and CARROLL, K.C. 2000. Comparison of sorbitol MacConkey agar and a two-step method which utilizes enzyme-linked immunosorbent assay toxin testing and a chromogenic agar to detect and isolate enterohemorrhagic *Escherichia coli*. J. Clin. Microbiol. *38*, 547–551.
84. OBERST, R.D. *et al.* 1998. PCR-based DNA amplification and presumptive detection of *Escherichia coli* O157:H7 with an internal fluorogenic probe and the 5' nuclease (TaqMan) assay. Appl. Environ. Microbiol. *64*, 3389–3396.
85. OJEDA, A. *et al.* 1995. Sorbitol-negative phenotype among enterohemorrhagic *Escherichia coli* strains of different serotypes and from different sources. J. Clin. Microbiol. *33*, 2199–2201.

86. OKREND, A.J.G., ROSE, B.C. and LATTUADA, C.P. 1990. Use of 5-bromo-4-chloro-3-indoxyl-D-glucuronide in MacConkey sorbitol agar to aid in the isolation of *Escherichia coli* O157:H7 from ground beef. J. Food Protect. *53*, 941–943.
87. OSWALD, E., SCHMIDT, H., MORABITO, S., KARCH, H., MARCHES, O. and CAPRIOLI, A. 2000. Typing of intimin genes in human and animal enterohemorrhagic and enteropathogenic *Escherichia coli*: characterization of a new intimin variant. Infect. Immun. *68*, 64–71.
88. PARK, C.H., VANDEL, N.M. and HIXON, D.L. 1996. Rapid immunoassay for detection of *Escherichia coli* O157 directly from stool specimens. J. Clin. Microbiol. *34*, 988–990.
89. PAROS, M., TARR, P.I., KIM, H., BESSER, T.E. and HANCOCK, D.D. 1993. A comparison of human and bovine *Escherichia coli* O157:H7 isolates by toxin genotype, plasmid profile and bacteriophage lambda-restriction fragment length polymorphism profile. J. Infect. Dis. *168*, 1300–1303.
90. PASS, M.A., ODEDRA, R. and BATT, R.M. 2000. Multiplex PCRs for identification of *Escherichia coli* virulence genes. J. Clin. Microbiol. *38*, 2001–2004.
91. PATON, A.W. and PATON, J.C. 1998. Detection and characterization of Shiga toxigenic *Escherichia coli* by using multiplex PCR assays for *stx*1, *stx*2, *eaeA*, enterohemorrhagic *E. coli hlyA*, *rfb*O111, and *rfb*O157. J. Clin. Microbiol. *36*, 598–602.
92. PATON, A.W., PATON, J.C., GOLDWATER, P.N. and MANNING, P.A. 1993. Direct detection of *Escherichia coli* Shiga-like toxins genes in primary fecal cultures by polymerase chain reaction. J. Clin. Microbiol. *31*, 3063–3067.
93. PATON, A.W. *et al.* 1996. Molecular microbiological investigation of an outbreak of hemolytic-uremic syndrome caused by dry fermented sausage contaminated with Shiga-like toxin-producing *Escherichia coli*. J. Clin. Microbiol. *34*, 1622–1627.
94. PATON, J.C. and PATON, A.W. 1998. Pathogenesis and diagnosis of Shiga toxin-producing *Escherichia coli* infections. Clin. Microbiol. Rev. *11*, 450–479.
95. POLLARD, D.K., JOHNSON, W.M., LIOR, H., TYLER, S.D. and ROZEE, K.R. 1990. Rapid and specific detection of verotoxin genes in *Escherichia coli* by the polymerase chain reaction. J. Clin. Microbiol. *28*, 540–545.
96. PRADEL, N. *et al.* 2000. Prevalence and characterization of Shiga toxin-producing *Escherichia coli* isolated from cattle, food and children during a one-year prospective study in France. J. Clin. Microbiol. *38*, 1023–1031.
97. PRESTON, M.A., JOHNSON, W., KHAKRIA, R. and BORCZYK, A. 2000. Epidemiologic subtyping of *Escherichia coli* O157 strains isolated in Ontario by phage typing and pulsed-field gel electrophoresis. J. Clin. Microbiol. *38*, 2366–2368.
98. PULSENET 1998. Standardized molecular subtyping of foodborne bacterial pathogens by pulsed-field gel electrophoresis. Centers for Disease Control and Prevention, Atlanta, Ga.
99. PYLE, B.H., BROADAWAY, S.C. and MCFETERS, G.A. 1999. Sensitive detection of *Escherichia coli* O157:H7 in food and water by immunomagnetic separation and solid-phase laser cytometry. Appl. Environ. Microbiol. *65*, 1966–1972.

100. RAHN, K. *et al.* 1996. Comparison of Vero cell assay and PCR as indicators of the presence of verocytotoxigenic *Escherichia coli* in bovine and human faecal samples. Appl. Environ. Microbiol. *62*, 4314–4317.
101. RAMOTAR, K., WALDHART, B., CHURCH, D., SZUMSKI, R. and LOUIE, T.J. 1995. Direct detection of verotoxin-producing *Escherichia coli* in stool samples by PCR. J. Clin. Microbiol. *33*, 519–524.
102. REID, S.D., BETTING, D.J. and WHITTAM, T.S. 1999. Molecular detection and identification of intimin alleles in pathogenic *Escherichia coli* by multiplex PCR. J. Clin. Microbiol. *37*, 2719–2722.
103. REID, T.M.S *et al.* 2000. ibid.
104. REID, T.M.S., CASHMORE, C. and BEVERIDGE, L. 1999. The Central Scotland *E. coli* O157 outbreak 1996: investigation of food and related samples. *In* Methods for verocytotoxigenic *E. coli*. Proceedings of meeting of CT98-3935 VTEC in Europe (G. Duffy, P. Garvey, J. Coia, Y. Wasteson and D.A. McDowell, eds.) Teagasc, Ireland, pp. 1–6.
105. REISSBRODT, R. 1998. Isolation and characterization of enterohemorrhagic *Escherichia coli*. Remedia-Klin. Mikrobiol. *2*, 344–351.
106. RIETRA, P.J.G.M., WILLSHAW, G.A., SMITH, H.R., FIELD, A.M., SCOTLAND, S.M. and ROWE, B. 1989. Comparison of verocytotoxin-encoding phages from *Escherichia coli* of human and bovine origin. J. Gen. Microbiol. *135*, 2307–2318.
107. SANDERSON, M.W., GAY, J.M., HANCOCK, D.D., GAY, C.C., FOX, L.K. and BESSER, T.E. 1995. Sensitivity of bacteriologic culture for detection of *Escherichia coli* O157:H7 in bovine feces. J. Clin. Microbiol. *33*, 2616–2619.
108. SCHMIDT, H. and KARCH, H. 1996. Enterohemolytic phenotypes and genotypes of Shiga toxin-producing *Escherichia coli* O111 strains from patients with diarrhea and hemolytic-uremic syndrome. J. Clin. Microbiol. *34*, 2364–2367.
109. SIDDONS, C.A. and CHAPMAN, P.A. 1993. Detection of serum and faecal antibodies in haemorrhagic colitis caused by *Escherichia coli* O157. J. Med. Microbiol. *39*, 408–415.
110. SMITH, P.J., ILLINGWORTH, D.S. and COATES, D. 1999. Evaluation of the EiaFoss assay for the detection of *Escherichia coli* O157 in foods, faeces and environmental samples. *In* Methods for verocytotoxigenic *E. coli*. Proceedings of the 1st meeting of CT98-3935 (G. Duffy, P. Garvey, J. Coia, Y. Wasteson and D.A. McDowell, eds.) Teagasc, Ireland, pp. 53–63.
111. SOUTHERN, E., MIR, K., SHCHEPINOV, M. 1999. Molecular interactions on microarrays. Nature Genet. *21*, 5–9.
112. SOWERS, E.G., WELLS, J.G. and STROCKBINE, N.A. 1996. Evaluation of commercial latex reagents for identification of O157 and H7 antigens of *Escherichia coli*. J. Clin. Microbiol. *34*, 1286–1289.
113. STAPP, J.R. *et al.* 2000. Comparison of *Escherichia coli* O157:H7 antigen detection in stool and broth cultures to that in sorbitol-MacConkey agar stool cultures. J. Clin. Microbiol. *38*, 3403–3406.

114. STEWART, D.S., TORTORELLO, M.L. and GENDEL, S.M. 1998. Evaluation of DNA preparation techniques for detection of the SLT-1 Gene of *Escherichia coli* O157:H7 in bovine faeces using the polymerase chain reaction. Lett. Appl. Microbiol. *26*, 93–97.

115. SYNGE, B.A. *et al.* 2000. Preliminary results from epidemiological studies in cattle in Scotland. Proceedings of "Zoonotic infections in livestock and the risk to public health", Royal College of Physicians, Edinburgh, pp. 10–17.

116. TARKKA, E., AHMAN, H. and SIITONEN, A. 1994. Ribotyping as an epidemiologic tool for *Escherichia coli*. Epidemiol. Infect. *112*, 264–274.

117. TENOVER, F. *et al.* 1995. Interpreting chromosomal DNA restriction patterns produced by pulsed-field gel electrophoresis: criteria for bacterial strain typing. J. Clin. Microbiol. *33*, 2233–2239.

118. THOMAS, A., SMITH, H.R. and ROWE, B. 1993. Use of digoxigenin-labelled oligonucleotide DNA probes for VT2 and VT2 human variant genes to differentiate Verocytotoxin-producing *Escherichia coli* strains of serogroup O157. J. Clin. Microbiol. *31*, 1700–1703.

119. THOMSON-CARTER, F.M. 2000. Annual report of the Scottish Reference Laboratory for *E. coli* O157 and Campylobacter spp.

120. TODD, E.C.D. *et al.* 1999. Application of a DNA hybridization-hydrophobic-grid membrane filter method for detection and isolation of verotoxigenic *Escherichia coli*. Appl. Environ. Microbiol. *65*, 4775–4780.

121. TOMOYASU, T. 1998. Improvement of the immunomagnetic separation method selective for *Escherichia coli* O157 strains. Appl. Environ. Microbiol. *64*, 376–382.

122. TROESCH, A. *et al.* 1999. *Mycobacterium* species identification and rifampin resistance testing with high density DNA probe arrays. J. Clin. Microbiol. *37*, 49–55.

123. TURNER, A.P.F. 1997. Biosensors: switching channels makes sense. Nature *387*, 555.

124. VAN DER PLAS, J., DIRKSE, M., HARTOG, B. and HOFSTRA, H. 1999. Molecular typing of VTEC. *In* Methods for verocytotoxigenic *E. coli*. Proceedings of meeting of CT98-3935 VTEC in Europe (G. Duffy, P. Garvey, J. Coia, Y. Wasteson and D.A. McDowell, eds.) Teagasc, Ireland, pp. 158–161.

125. VERNOZY-ROZAND, C. 1997. Detection of *Escherichia coli* O157:H7 and other verocytotoxin-producing *E. coli* (VTEC) in food. J. Appl. Microbiol. *82*, 537–551.

126. VERNOZY-ROZAND, C., MAZUY, C., RAY-GUENIOT, S., BOUTRAND-LOEI, S., MEYRAND, A. and RICHARD, Y. 1997. Detection of *Escherichia coli* O157 in French food samples using an immunomagnetic separation method and the VIDAS *E. coli* O157. Lett. Appl. Microbiol. *25*, 442–446.

127. WANG, L., ROTHEMUND, D., CURD, H. and REEVES, P. 2000. Sequence diversity of the *Escherichia coli fliC* genes: implication for a DNA-based typing scheme for *E. coli* O157:H7. J. Clin. Microbiol. *38*, 1786–1790.

128. WHITTAM, T.S., WACHSMUTH, I.K. and WILSON, R.A. 1998. Genetic evidence of clonal descent of *Escherichia coli* O157:H7 associated with hemorrhagic colitis and hemolytic uremic syndrome. J. Infect. Dis. *157*, 1124–1133.

129. WHITTAM, T.S., WOLFE, M.L., WACHSMUTH, I.K., ORSKOV, F., ORSKOV, I. and WILSON, R.A. 1993. Clonal relationships among *Escherichia coli* strains that cause haemorrhagic colitis and infantile diarrhoea. Infect. Immun. *61*, 1619–1629.

130. WIELER, L.H. *et al.* 2000. Enterohemorrhagic *Escherichia coli* (EHEC) strains of serogroup O118 display three distinctive clonal groups of EHEC pathogens. J. Clin. Microbiol. *38*, 2162–2169.

131. ZADIK, P.M., CHAPMAN, P.A. and SIDDONS, C.A. 1993. Use of tellurite for the selection of verocytotoxigenic *Escherichia coli* O157:H7. J. Med. Microbiol. *39*, 155–158.

132. ZHANG, W.-L. *et al.* 2000. Molecular characteristics and epidemiological significance of Shiga toxin-producing *Escherichia coli* O26 strains. J. Clin. Microbiol. *38*, 2134–2140.

CHAPTER 6

EPIDEMIOLOGY OF VEROCYTOTOXIGENIC *ESCHERICHIA COLI* (VTEC) IN RUMINANTS

JORGE BLANCO, MIGUEL BLANCO, JESÚS E. BLANCO, AZUCENA MORA
MARÍA PILAR ALONSO, ENRIQUE A. GONZÁLEZ and
MARÍA ISABEL BERNÁRDEZ

Laboratorio de Referencia de E. coli (LREC)
Departamento de Microbiología y Parasitología
Facultad de Veterinaria
Universidad de Santiago de Compostela
Campus de Lugo, 27002 Lugo, Spain

INTRODUCTION

Verotoxin (VT) producing *Escherichia coli* (VTEC), and particularly strains of serotype O157:H7, have emerged as food poisoning pathogens which can cause a severe and potentially fatal illness. It is a major cause of gastro-enteritis that may be complicated by haemorrhagic colitis (HC) or the haemolytic uraemic syndrome (HUS), the main cause of acute renal failure in children. Since its identification as a pathogen in 1982, VTEC O157:H7 has been the cause of a series of outbreaks especially in the UK, the USA, Canada and Japan (42,55).

Domestic ruminants, especially cattle and sheep, have been implicated as the principal reservoir of VTEC causing human infections. Transmission occurs through consumption of undercooked meat, unpasteurized dairy products, and vegetables or water contaminated by faeces of carriers, because VTEC strains are found as part of the normal intestinal flora of the animals. Wild ruminants are considered to represent other potential infection sources for humans (4,34,54,60).

VTEC elaborate two potent phage-encoded cytotoxins called verotoxins (VT1 and VT2) or Shiga-like toxins (SLT-I and SLT-II). In addition to toxin production, they produce another virulence-associated factor, a protein called intimin, which is responsible for intimate attachment of VTEC to the intestinal epithelial cells, causing attaching and effacing (AE) lesions in the intestinal mucosa. Intimin is encoded by the chromosomal gene *eaeA*, which is part of a pathogenicity island termed the locus for enterocyte effacement (LEE). Another factor that may also affect virulence of VTEC is the enterohaemolysin (enterohaemorrhagic *E. coli* haemolysin, EHEC-Hly) (5,9,51,71).

Although VTEC strains that cause human infections belong to a large number of O:H serotypes (Table 1), most outbreaks and sporadic cases of HC and HUS have been attributed to strains belonging to the enterohaemorrhagic serotype O157:H7. Unlike other *E. coli* strains, VTEC O157:H7 do not ferment sorbitol, and are β-glucuronidase negative. These differences make it easier to identify O157:H7 strains in food products and clinical samples. Human infections caused by VTEC O157:H7 are under nation-wide surveillance in many countries. However, the detection of other non-O157 VTEC infections is often limited to a small number of specialized laboratories because VTEC O157 colonies are more easily detectable on some culture media (CTSMAC medium) (23) than non-O157 VTEC types, which are thus often missed during laboratory analysis of stool and food specimens. As VTEC non-O157 are more prevalent in animals, and as contaminants in foods, humans are probably more frequently exposed to these strains. Infections with some non-O157 VTEC types, such as O26:H-, O26:H11, O91:H-, O103:H2, O111:H-, O113:H21, O118:H16, O128:H2, O145:H-, O145:H28 and O146:H21, are frequently associated with severe illness in humans, but the role of other non-O157 VTEC types in human disease needs further examination. Severe diarrhoea (especially HC) and HUS were closely associated with VTEC types carrying the *eaeA* gene for intimin. In contrast, other serotypes of non-O157 VTEC which are negative for the *eaeA* gene have been rarely implicated in severe human illness and have been more frequently found among asymptomatic carriers, in uncomplicated cases of diarrhoea, or among adult patients. Nevertheless, production of intimin is not essential for pathogenesis, because a number of sporadic cases of HUS are caused by *eae*-negative non-O157 VTEC strains. Furthermore, VTEC O104:H21 and O113:H21 strains lacking *eaeA* gene were responsible for an outbreak and a cluster of three HUS cases in the USA and Australia (11,19,40,42,64,73,81).

ECOLOGY OF VTEC O157:H7 ON RUMINANT FARMS

VTEC O157:H7 is virtually ubiquitous in both dairy and beef cattle. Although early studies appeared to show that the agent was present on a minority of cattle farms, repeated sampling using sensitive screening methods has demonstrated that VTEC O157:H7 is, at least occasionally, present on most farms (75 to 90%) in America and Europe. Epidemiological data support the concept that VTEC O157:H7 infections in cattle are relatively transient events, among individual animal or herds (30,45,46,76). Cattle are temporal incubators for this pathogen, as they are for other non-pathogenic strains of *E. coli*, and as other transient non-pathogenic *E. coli* strains, VTEC O157:H7 can sometimes colonize multiple members of a group simultaneously. Persistence of VTEC O157:H7 in cattle typically lasts for 2 months or less. Long-term carriers have

TABLE 1.
O:H SEROTYPES OF HUMAN VTEC

O1: H1,H2,H7,H20,**H-**	O71:H-	O123:H19,H49
O2:H1,H2,**H5**,**H6**,**H7**,H11,H27,**H29**, H44,H-	O73:H34	O124:H-
O3:H10	O74	O125:H8,H-,**H?**
O4:**H5**,H10,H40,H-	O75:H1,**H5**,H8,**H-**	O126:H2,H8,H11,H20,H21,**H27**,H-
O5:H16,**H-**	O76:H7,H19	O127
O6:H1,**H2**,**H4**,H12,H28,H29,**H31**,H34, H49,**H-**	O77:H4,H7,H18,H41,**H-**	O128:**H2**,**H7**,H8,H10,H12,**H25**,H31, **H45**,**H-**
O7:H4,H8	O78:H-	O129:H-
O8:**H2**,H9,H11,**H14**,H19,**H21**,H25,H-	O79:**H7**,H14	O130:**H11**
O9:H7,H21,**H-**	O80:H-	O131:H4
O11:**H2**,H49,**H-**	O81:H?	O132:H-
O12:H-	O82:H5,H8,H-	O133:H53,H-
O14:**H-**	O83:**H1**,H-	O134:**H25**
O15:**H2**,H8,H27,**H-**	O84:H2,**H-**	O137:**H6**,**H41**
O16:H6,H21,**H-**	O85:**H10**,**H23**,**H-**	O138:H2
O17:H18,H41	O86:H10,H40,**H-**	O141:H2,H8,H-
O18:H7,H12,H15,**H-**,**H?**	O88:H25,H-	O142
O20:H7,**H19**,H-	O89:H-	O143:H-
O21:**H5**,**H?**	O90:H-	O144:H-
O22:**H1**,**H5**,**H8**,H16,H40,H-	O91:H4,**H10**,H14,H15,**H21**,H40,**H-**	O145:H4,H8,H16,**H25**,H26,**H28**,**H-**
O23:H7,H16	O92:H3,H11	O146:**H8**,H14,H21,H28,H-
O25:**H2**,H14,H-	O93:H19	O148:H28
O26:H2,H8,**H11**,H12,H21,H32,**H-**	O95:H-	O150:H8,H10,H-
O27:H30,H-	O98:H8,**H-**	O152:H4
O28:H25,H35,H-	O100:**H25**,H32	O153:H2,H11,H12,H21,**H25**,H30, H33,H-
O30:H2,H21,H23	O101:H9,**H-**	O154:**H4**,**H19/H20**,**H-**
O37:H41	O102:H6	O156:H4,H27,H-
O38:H21	O103:**H2**,H4,H6,H7,H11,H18,H21,H25,**H-**	O157:**H7**,**H-**
O39:H4,H8,H28	O104:**H2**,H7,H16,**H21**,**H-**	O160:**H?**
O40:H2,H8	O105:**H18**,H19,H20	O161:**H-**
O41:H2	O106	O162:H4
O44	O107:H27	O163:**H19**,**H25**,H-
O45:H2,H7,H-	O109:H2	O165:H10,**H19**,H21,**H25**,**H-**
O46:H2,**H31**	O110:H19,H28,H-	O166:H12,H15,H28
O48:**H21**	O111:**H2**,**H7**,**H8**,H11,H21,H30,H34,H40, H49,**H-**,**H?**	O168:**H-**
O49:H10,**H-**	O112:**H2**,H19,H21	O169:H-
O50:**H7**,H-	O113:H2,H4,H5,H7,**H21**,H32,H53,H-	O171:H2,**H-**
O52:H19,H23,H25	O114:H4,H48,**H?**	O172:**H-**,**H?**
O55:**H6**,**H7**,H9,**H10**,H19,**H-**,**H?**	O115:**H10**,H18	O173:**H2**
O60:H-	O116:H4,H19,H-	O174(OX3):**H2**,H8,**H21**,**H-**
O64:H25	O117:**H4**,H7,H19,H28,H-	O175(OX7):H16
O65:H16	O118:H2,**H12**,**H16**,**H30**,H-	OX177:H-
O68:H-	O119:**H5**,**H6**,H25,H-	OX178:H7
O69:H11,**H-**	O120:H19	OX179:H8
O70:H11	O121:H8,H11,**H19**,**H-**	

Serotypes found in VTEC strains isolated from patients with HUS are in bold.
References: (7,11,18,19,55,93, Blanco *et al.* unpublished results).

not been found and shedding of the microorganism has not been associated with any recognised disease. Faecal shedding of VTEC O157:H7 is more common in calves (2 to 24 months of age) than in adults (46). Higher prevalence in calves is consistent with greater susceptibility of this age group to colonisation by transient *E. coli* strains. The environment, particularly water and feed, seems to play an important role in VTEC O157:H7 ecology. Water troughs on farms have been frequently found to contain VTEC O157:H7, and evidence suggests an important role for contaminated water troughs in the transmission of the bacteria. VTEC O157:H7 can survive in water trough sediments for at least 4 months and appears to multiply in such environments. VTEC O157:H7 can contaminate and readily multiply in moist feeds. Thus, certain farm management practices, particularly those related to maintenance and multiplication of the agent in feed and water, may provide practical means to substantially reduce the prevalence of VTEC O157:H7 in cattle farms (45). In a study of numerous environmental sites on dairy farms, evidence of VTEC was found in composite samples of the calf feeders (19%), calf-barn surfaces (18%), cow feeders (15%), cow-barn surfaces (11%) and individual milk filters. VTEC O157:H7 has also been isolated from manure and slurry, and the presence of manure on pastures and vegetable gardens has been linked to human infections. Experimental models of survival of VTEC O157:H7 suggest that contaminated farm environments may remain sources of VTEC O157:H7 and possibly other VTEC for several months. VTEC O157:H7 survived for at least 10-12 weeks in cattle faeces and at least 20 weeks in soil (45). A variety of dietary factors have been proposed as factors contributing to the increased prevalence of VTEC O157:H7. Evidence that supports a role for these factors remains inconclusive and, in some cases, controversial. However, some dietary factors have shown sufficient connections to prevalence to warrant further investigation. Some of these connections indicate a negative relationship to *E. coli* (e.g., cottonseed and clover feeding) and others indicate positive relationships to *E. coli* (e.g., ionophore feeding) (76). Diez-Gonzalez *et al.* (33) have suggested that grain feeding creates an acidic environment in the gut of cattle which leads to the selection of acid-tolerant *E. coli*, including VTEC O157:H7 strains, which are then heavily shed in the faeces. The authors propose that cattle producers could solve the *E. coli* problem by feeding hay to cattle before slaughter. But the majority of other studies available contradict the claims of Diez-Gonzalez *et al.* (33). Epidemiological data of Hancock *et al.* (43) demonstrated no difference between VTEC O157:H7 prevalences in grain-fed and hay-fed cattle. Kudva *et al.* (56) observed that forage feeding resulted in the highest levels of VTEC O157:H7 in a sheep model. The ecology of the VTEC O157:H7 in sheep appears to be similar to that in cattle (45,46,76).

SCREENING METHODS USED FOR DETECTION VTEC O157:H7 IN RUMINANTS

Big differences in prevalence shown in the studies listed in Table 2 are, at least in part, artefacts of the different methods used to detect VTEC O157:H7. Testing for this organism is far from being standardised, and the tests in use vary markedly in their sensitivity. Sensitive methods are needed in the identification of VTEC O157:H7, since this pathogen may be present in bovine faeces in very small numbers. Several selective enrichment media have been described. Among these, buffered peptone water with vancomycin, cefsulodin and cefixime (BPWvcc), tryptone soy broth with cefixime and vancomycin (TSBcv), modified TSB with novobiocin (mTSBn) and modified *E. coli* broth with novobiocin (mECn) seem to be the most appropriate (23,31,61,80). However, Synge (86) found that BPW without antibiotics was significantly better than BPWvcc. Selective sorbitol MacConkey agar with tellurite and cefixime (CTSMAC) is the most commonly used isolation medium for VTEC O157:H7. However, the growth of some O157 strains on CTSMAC agar is inhibited by tellurite-sensitivity. Therefore, the use of a second isolation medium such as SMAC is recommended. Immunomagnetic separation (IMS) following enrichment, with plating of the concentrated target cells onto CTSMAC and SMAC media appears to be the most sensitive and cost-effective method for the isolation of VTEC O157:H7 from bovine faeces (23,31). Enrichment culture in BPWvcc, followed by IMS and culture on CTSMAC, was approximately 100-fold more sensitive than direct culture on CTSMAC for detection of VTEC O157:H7 in inoculated bovine faecal samples (24). This increase in sensitivity was confirmed during subsequent monitoring of a dairy herd, when VTEC O157:H7 was isolated from 84 (8%) of 1024 rectal swabs taken from cattle over a 4-month period; 23 (27%) of the 84 strains were isolated by both direct culture and IMS, whereas 61 (73%) of the 84 strains were isolated by IMS only (23). Heuvelink *et al.* (47) and Blanco *et al.* (unpublished results) found similar results to those of Chapman *et al.* (23,24). In the study of Heuvelink *et al.* (47) the use of IMS resulted in a 7-fold increase in the rate of isolation of VTEC O157:H7 from adult cattle compared with the rate determined using selective enrichment in mECn and plating onto CTSMAC (57 positive cases with IMS *versus* only 9 positive cases without IMS). However, Sanderson *et al.* (80) found that the use of IMS provided relatively small additional increases in sensitivity. They concluded that the culture of 0.1-g fecal samples on cotton-tipped swabs with TSBcv for enrichment followed by plating of 10^{-3} and 10^{-4} dilutions on CTSMAC appears to be a cost-effective and sensitive method for large-scale studies involving hundreds of bovine faecal samples per week.

TABLE 2.
SURVEYS OF PREVALENCE OF VTEC O157:H7 IN CATTLE

Study (Reference)	Country	Sampling Year	Prevalence				Type of Cattle Studied	Screening Method
			Animals		Herds or Farms			
Blanco *et al.* (15,16)	Spain	1993-1994	1/686	0.1%	1/51	2%	Beef and dairy cattle on farms	CTSMAC
Blanco *et al.* (17)	Spain	1995	7/383	2%			Beef cattle at an abattoir	BPWvcc+IMS+CTSMAC
Blanco *et al.* (17)	Spain	1998-1999	55/471	12%	32/145	22%	Feedlot cattle	BPWvcc+IMS+CTSMAC
Bonardi *et al.* (20)	Italy	1997-1998	59/450	13%	28/133	21%	Cattle at an abattoir	mTSBn+IMS+CTSMAC
Montet *et al.* (59)	France	1998	1/300	0.3%			Slaughtered heifers	Automated immunoconcentration
Heuvelink *et al.* (48)	Netherlands	1996	75/1,152	7%	7/10	70%	Dairy cattle on farms	mECn+IMS+CTSMAC
De Zutter *et al.* (32)	Belgium		41/648	6%	34/467	7%	Slaughtered cattle	BPW+IMS+CTSMAC
Møller & Frydendahl (67)	Denmark	1999	4/82	5%				IMS+CTSMAC
Vold *et al.* (89)	Norway	1995	6/1,970	0.3%	2/197	1%	Heifers and milking cows	IMS+CTSMAC
Wateson & Lassen (91)	Norway	1996	19/504	4%	7/105	7%	Imported meat cattle	IMS+CTSMAC
Johnsen *et al.* (52)	Norway	1998	1/680	0.1%	1/367	0.3%	Slaughtered cattle	BPWvcc+IMS+CTSMAC
Eriksson *et al.* (35)	Sweden	1997-1998	88/631	14%				BPW+IMS+CTSMAC
Andlovic & Marinsek (3)	Slovenia	1996-1997	2/250	0.8%			Slaughtered cattle	TSB+SMAC
Cizek *et al.* (27)	Czech Rep.	1997-1998	72/365	20%			Feedlot cattle	mECn+IMS+CTSMAC
McEvoy *et al.* (62)	Ireland	1998-1999	7/250	3%			Cattle at an abattoir	BPWvcc+IMS+CTSMAC
Synge & Hopkins (84)	Scotland	1992-1993	13/5,237	0.2%				SMAC
Synge & Hopkins (85)	Scotland	1993-1995	65/441	15%	13/40	33%	Herds with human incidents	IMS+CTSMAC
Chapman *et al.* (22)	England	1992	78/2,103	4%			Cattle at an abattoir	CRSMAC
Richards *et al.* (78)	England-Wales	1994-1995	54/6,495	0.8%	50/5,037	1%	Submitted for diagnostic	BPWvcc+IMS+CTSMAC
Chapman *et al.* (25)	England	1995-1996	752/4,800	16%			Cattle at an abattoir	BPWvcc+IMS+CTSMAC

Sanderson *et al* (80)	USA		24/351	7%			Dairy cattle	TSBcv+CTSMAC
Wells *et al.* (92)	USA		18/1,266	1%	11/25	44%	Dairy cattle	mTSB+SMAC
Hancock *et al.* (43)	USA	1991-1992	22/5,582	0.4%	9/85	11%	Dairy and beef cattle	TSBv+SMAC
Garber *et al.* (38)	USA	1991-1992	25/6,894	0.4%	19/1,068	2%	Dairy heifers	
Zhao *et al.* (99)	USA	1993	31/965	3%	18/64	28%	Dairy cattle	mTSB+ELISAO157+SMAC
Hancock *et al.* (44)	USA	1994	210/11881	2%	63/100	63%	Feedlot cattle	TSBcv or mECn+CTSMAC
McDonough *et al.* (61)	USA	1995	22/1,668	1%			Cull dairy cows-packing plant	mEC+SMAC
Garber *et al.* (39)	USA	1996	52/4,361	1%	22/91	24%	Dairy cows	mECn+CTSMAC
Laegreid *et al.* (58)	USA	1997	61/878	7%	13/15	87%	Beef calves at weaning	GNBvcc+IMS+CTSMAC
Wilson *et al.* (98)	Canada	1992-1993	12/1,478	0.8%	7/80	9%	Dairy cattle	SMAC
Rahn *et al.* (75)	Canada	1993	2/406	0.5%	2/8	25%	Dairy cattle	MacConkey broth+BHIB+PCR-VTgenes
Jackson *et al.* (50)	Canada	1995	59/95	62%			Farm with human incidents	mECn+IMS+CTSMAC
Van Donkersgoed *et al.* (88)	Canada	1995-1996		8%		20%	Cattle at processing	mTSB+IMS+CTSMAC
Cobbold & Desmarchelier (28)	Australia		11/588	2%			Dairy cattle	mECn+PCR-VTgenes
Miyao *et al.* (65)	Japan	1992-1994	7/387	2%			Slaughtered cattle	Sorbitol-IPA-bile salt agar
Vuddhakul *et al.* (90)	Thailand	1998	1/55	2%			Dairy and beef cattle	mECn+IMS+SMAC

Abbreviations for screening methods used. Selective enrichment with: mECn, modified *E.coli* broth with novobiocine; BPW, buffered peptone water ; BPWvcc, BPW supplemented with vancomycin, cefixime and cefsulodin; GNBvcc, gram negative broth with vancomycin, cefixime and cefsulodin ; TSB, tryptone soya broth; mTSBn, modified TSB supplemented with novobiocin; TSBv, TSB with vancomycin; TSBcv, TSB with cefixime and vancomycin. IMS, immunomagnetic separation of *E.coli* O157. Plating on: SMAC, sorbitol MacConkey agar; CTSMAC, SMAC with cefixime and potassium tellurite; CRSMAC, cefixime rhamnose SMAC. Studies based on methods of very low sensitivity screening for VTEC O157 are not reflected in this table.

VTEC O157:H7 PREVALENCE IN CATTLE

Laboratory factors may not be the only cause of differences among studies (as presented in Table 2). Prevalence may also be affected by sampling origin (abattoir *versus* feedlot), regional and seasonal variations, etc. Despite the limitations of prevalence data, eight generalisations can be made:

- The bovine reservoir of VTEC O157:H7 is now widely recognised.
- VTEC O157:H7 appears to be widespread in cattle and it has a wide geographic distribution in Europe and America.
- VTEC O157:H7 is virtually ubiquitous in both dairy and beef cattle. Although early studies appeared to show that the agent was present on a minority of cattle farms, repeated sampling has demonstrated that VTEC O157:H7 is, at least occasionally, present on most farms. Herd prevalence for VTEC O157:H7 can be quite high, up to 90%.
- Prevalence in individual cattle within herds ranges from 5% to 50%, 10 to 100-fold higher than initial estimates obtained using insensitive methods.
- VTEC O157:H7 is more common in calves than in older cattle.
- Fecal carriage is detected in all seasons, but shedding rates are subjected to seasonal effects with an increase in spring and summer.
- Surveys suggest that fecal shedding is intermittent and variable in cattle. In longitudinal studies, herds which were culture-negative for several months showed periods of positivity of a few weeks with 10-25% of cattle shedding the micro-organism.
- Horizontal transmission of VTEC O157:H7 occurs in groups of animals. Contaminated water and feed seem to serve as vehicles in the spread and persistence of bacteria within herds.

In Spain, between 1993 and 1999, three different studies on VTEC in cattle were carried out. An initial survey in 1993/94 used the conventional methods available at that time, i.e., direct plating on CTSMAC medium and selection of non-sorbitol fermenting colonies for VT testing (Vero cell test and PCR) and serotyping (15,16). In this study VTEC O157:H7 was recovered from 1 (0.6%) of 161 calves and 0 (0%) of 525 cows from 1 (2%) of 51 farms examined.

The two later Spanish studies used IMS (17). During the survey in 1995 VTEC O157:H7 was isolated from 7 (2%) of the 383 slaughtered cattle in an abattoir. In the survey in 1998/99, VTEC O157:H7 was isolated from 55 (12%) of the 471 feedlot calves (4 to 8 months of age). Although, only a mean of 3 animals per herd was sampled, VTEC O157:H7 strains were detected in 32 (22%) of 145 feedlots examined. When we sampled a higher number of animals (9 to 56) VTEC O157:H7 was found in 3 (60%) of 5 feedlots investigated. Individual prevalences in the 3 positive feedlots were: 23% (13 of 56), 22% (2 of 9) and 8% (1 of 13). Farms were visited only on one occasion to collect fecal samples (17, unpublished data). In Italy, Bonardi *et al.* (20) carried out a study to estimate the prevalence of VTEC O157:H7 in cattle at slaughter in the area of northern Italy in which two community outbreaks of VTEC infections had occurred. VTEC O157:H7 strains were isolated from 13% of the total cattle examined and from 21% of the farms. Higher individual isolation rates were observed from feedlot cattle (17%) and dairy cows (16%), although no isolates were obtained from veal calves. The isolation rate was higher during warm weather (18%), falling to an average of 3% during the winter months. In France, Montet *et al.* (69) examined 300 faecal samples collected from 300 heifers at a slaughterhouse during 1998. After enrichment, these samples were tested using an automated immunoconcentration system. Only 1 (0.3%) VTEC O157:H7 strain was isolated. Pradel *et al.* (74) found similar results, i.e., 1 (0.2%) of 471 faecal samples from cattle. However, VTEC O157:H7 prevalence rates in French studies are probably underestimated because low sensitivity screening methods were used.

A number of German studies have also used methods of low sensitivity. Between 1987 and 1996, six separate investigations on VTEC in cattle were carried out in different parts of Germany (11). One survey aimed to specifically select VTEC O157:H7 using only a modified haemorrhagic colitis agar but no VTEC O157:H7 were isolated. VTEC O157:H7 was detected in only one of the studies, where 2 (0.8%) of 259 cattle were found to excrete the agent (68). Since all other studies were negative for VTEC O157:H7, the total frequency of this serotype was calculated as 0.1% of the 1,596 cattle investigated (11).

In The Netherlands, Heuvelink *et al.* (48) tested all the cattle on 10 dairy farms during the period from September through November 1996. Cattle on 7 (70%) of the 10 farms were positive for VTEC O157:H7, with a proportion of cattle infected varying from 0.8% to 22%. On the 7 positive farms, the excretion rate was higher in calves aged 4 to 12 months (21%). In a follow-up study, two VTEC O157:H7 positive farms and two negative farms visited in the prevalence study were revisited five times at intervals of approximately 3 months. Each farm tested positive on at least one occasion during the survey period. The proportion of cattle infected varied from 0% to 61%. Excretion rates peaked in summer and were lower in winter. VTEC O157:H7 was also isolated from faecal samples from horses, ponies, and sheep and from milk filters and stable flies. In almost all cases, different farms had different strain types. The predominance of a single type at each sampling suggests that horizontal transmission is an important factor in dissemination of VTEC O157:H7 within the farm environment. The presence of more than one strain type, both simultaneously and over the time, suggests that there was more than one source of VTEC O157:H7 on the farms. Furthermore, this study demonstrated that the VTEC O157:H7 status of the farm cannot be ascertained from a single visit testing a small number of cattle (48). In another study, Heuvelink *et al.* (47) sampled cattle at the main slaughterhouses in The Netherlands during 1995 and 1996. VTEC O157:H7 strains could be isolated from 57 (11%) of the 540 adult cattle and only 2 (0.5%) of the 397 veal calves. In Belgium, De Zutter *et al.* (32) sampled cattle at slaughterhouses and found VTEC O157:H7 in 6% of the animals. The age distribution of positive animals was: in 1 year old animals 23 (14%) out of 166; in two year old animals, 8 (5%) out of 154; and in animals 3 or more years old, 10 (3%) out of 323. The herd prevalence was 7% (32).

In Denmark, Møller Nielsen and Frydendahl (67) isolated VTEC O157:H7 in 4 (5%) of the 82 cattle samples analysed. In Norway, in 1995, faecal samples collected from 197 cattle herds during the grazing season were screened for VTEC O157:H7 using IMS and PCR (89). Although Vol *et al.* (89) used sensitive methods, only six (0.3%) of the 1,970 animals were positive, and these animals all belonged to one of two herds. In 1996 and 1997, faeces from 1,641 cattle sampled at abattoir were analysed but no VTEC O157:H7 were detected (91). In 1996, fecal samples from 504 cattle (mostly meat animals) imported to Norway from Sweden, Denmark and Finland were analysed for VTEC O157:H7 by IMS. *E. coli* O157:H7 was found in 19 animals (4%), representing 7 (7%) of the 105 analysed herds. However, isolates from only 3 herds were found to be *vt1*+ and/or *vt2*+ by PCR analysis (91). Johnsen *et al.* (52), examined 680 cattle sampled in three slaughterhouses during 1998, using the most sensitive method for detection, but only found 1 animal (0.1%) positive for VTEC O157:H7. In Sweden, Eriksson *et al.* (35) detected VTEC O157:H7 in 88 (14%) of the 631 animals sampled in 6 cattle herds in 1997 and 1998. The highest prevalence within a herd was 66% and the lowest prevalence was 0%. The faecal samples from the 631 animals were tested both as individual samples and as 127 pooled samples (with each pool containing 5 individual samples). Of the 127 pooled samples, 32 included ≥1 positive individual sample for VTEC O157:H7. The pathogen was detected in 29 (90%) of these pooled samples. The results indicate that pooling of cattle faeces is possible, but may lead to a slight loss in sensitivity and a loss of positive findings, particularly when the levels of VTEC O157:H7 are low (35). In Slovenia, Andlovic and Marinsek (3) examined 250 rectal swabs from slaughtered cattle and identified VTEC O157:H7 in 2 (0.8%) animals. In the Czech Republic, between October 1997 and September 1998, 365 samples of cattle rectum contents, 12 samples of untreated slurry, 70 samples of fresh bird droppings and 19 colons of synanthropic rodents on a large-scale feedlot cattle farm were collected and tested for the presence of VTEC O157:H7 by Cizek *et al.* (27). VTEC O157:H7 was demonstrated in 72 (20%) cattle, 6 (50%) untreated slurry samples and in 4 (40%) of the rats. Shedding of VTEC O157:H7 by other rodents and birds was not observed.

In Ireland, McEvoy *et al.* (62) examined 750 samples from a commercial Irish beef abattoir between June 1998 and June 1999. VTEC O157:H7 was detected in 3% of faecal, 0.8% of rumen, and 3% of carcass samples. It was not detected in any sample between early November and late March. In Scotland numerous studies have been done in cattle. In a survey (1992/1993) using direct plating on SMAC only, 13 (0.2%) of 5,237 cattle samples were positive for VTEC O157:H7 (84). Another study, between 1993 and 1995 using IMS, detected VTEC O157:H7 in 65 (15%) of 441 samples from 40 herds investigated, with 13 positive herds (85). Recent studies (87) estimated prevalence of shedding of VTEC O157:H7 in Scottish beef cattle to be 9% at animal level, with 24% of herds infected. Shedding rates were statistically higher in housed animals, and were subject to seasonal effects with an increase in shedding in spring. Farms that restocked with homebred replacements had a lower probability of being identified as a positive herd. Another Scottish study examining factors influencing the shedding of VTEC O157:H7 in beef suckler cows also noted higher shedding in spring and autumn, with strong positive associations with housing and the feeding of home-grown barley (87). Numerous studies have been also been carried out on cattle in England and Wales. Chapman *et al.* (22) investigated a local South Yorkshire abattoir associated with a cluster of cases of human infection with VTEC O157:H7 in 1992 in the Sheffield area. VTEC O157:H7 was isolated from 78 (4%) of 2,103 bovine rectal swabs. The most common phage types found were 2 and 8, the types implicated in the cluster of human cases (22). A Sheffield dairy herd implicated in VTEC O157:H7 infection in humans was studied for 15 months (28 visits were made between May 1993 to July 1994) following the outbreak (63). VTEC O157:H7 was isolated from 153 (4%) of 3,593 bovine rectal swabs. The maximum prevalence at one visit was 14% in lactating cows, 40% in non-lactating cows, 56% in calves and 68% in heifers. Eighty-seven (74%) of the animals which excreted VTEC O157:H7 did so on only one occasion but 23 (32%) of 73 cows and heifers and 7 (16%) of 44 calves which excreted the organism did so on more than one occasion. Between November 1993 and May 1994 there was no evidence of excretion in any group. All isolates of VTEC O157:H7 obtained were of the same phage type, toxin type and plasmid profile. In a third study by Chapman *et al.* (25) in the Sheffield area samples of rectal faeces were collected immediately after slaughter from 400 cattle each month for a 1-year period (1995-1996) and VTEC O157:H7 was isolated from 752 (16%) of such samples. VTEC O157:H7 was found in 13% of the beef cattle and 16% of the dairy cattle. The monthly prevalence of VTEC O157:H7 in cattle varied from 5% to 37% and was highest in spring and late summer. A recent (1999-2000) on-farm study (87) of cattle herds in England and Wales isolated VTEC O157:H7 from 35% dairy, 48% suckler and 53% fattening herds. Overall, 5% of the cattle tested were excreting VTEC O157:H7, although in infected herds the average figure was 10% (6 to 15%). Excretion was significantly more common in cattle under 24 months and in fattening herds (86).

In the majority of studies carried out in the USA and Canada, the most sensitive isolation technique ISM has not been employed and, therefore, prevalence rates in many of these studies have probably been underestimated. According to officials of the US Department of Agriculture (USDA), almost 90% of ground beef sampled since September 1999 has been found to be contaminated with VTEC O157:H7. A USDA feedlot survey from summer 1999 showed VTEC O157:H7 contaminating 50% of feedlot cattle (49). In a study carried out in 1991 and 1992 in USA by Hancock *et al.* (43), VTEC O157:H7 was found in 10 (0.3%) of 3,570 faecal samples from dairy cattle in 5 (8%) of 60 herds, in 10 (0.7%) of 1,412 faecal samples from pastured beef cattle in 4 (16%) of 25 herds and in 2 (0.3%) of 600 feedlot beef cattle. In a study of beef cattle published by Hancock *et al.* (44) in 1997 in which 120 faecal samples were collected from each of 100 feedlots in 13 USA states on a single day, VTEC O157:H7 was found in fecal samples from 63% of the feedlots and, overall, from 2% of 11,881 faecal samples. Prevalence rates between feedlots and between pens within feedlots were highly variable, with highest rates for pens holding cattle recently entering the feedlots (32-53%). Similar prevalences were found in all geographic regions sampled (44). In another US study by Hancock *et al.* (45), VTEC O157:H7 shedding in 14 cattle herds was determined by culture at approximately 1 month intervals, for up to 13 months (May 1993 to July 1994). The overall prevalence was 1% (113/10,832 faecal samples) and 9 of the 14 herds were detected as positive. Herds shown positive by a previous study (2 years before) maintained a higher prevalence of positive cattle (1.9%) than herds which had been negative 2 years before (0.2%). Weaned heifers had a higher prevalence (1.8%) than unweaned calves (0.9%) or adults (0.4%). For all herds, the highest prevalence occurred in summer months. No positive samples were found on 63% of all sampling visits to 7 positive herds. Thus, the typical pattern of VTEC O157:H7 excretion in cattle herds appears to be characterised by short periods of relatively high prevalence of excretion separated by longer periods of reduced or undetectable shedding. The duration of VTEC O157:H7 excretion by individual cattle was <1 month in 35 (63%) of the 56 cattle (45). Laegreid *et al.* (58) using IMS determined the prevalence of VTEC O157:H7 in beef calves at weaning (prior to arrival at the feedlot) sampled in the 5 major beef cattle producing USA states in 1997. Thirteen (87%) of the 15 herds were found to give at least one positive isolation of VTEC O157:H7. Within positive herds, prevalence ranged from 2 to 20%. All herds had a high prevalence of anti-O157 antibodies, with the frequency in individual animals ranging from 63 to 100% within seropositive herds. This study indicates that VTEC O157:H7 infection before weaning, prior to entry into feedlots, is widespread. Furthermore, serologic evidence suggests that most calves (83%) and all herds (100%) have probably been exposed to VTEC O157:H7. However, high rates of seropositivity to the O157 LPS should be interpreted in conjunction with the recognition of the occurrence of non-VTEC of serogroup O157 in cattle, as well as potential less specific cross-reactions with antibodies active against other bacteria (58). In Canada, Wilson *et al.* (98) and Rahn *et al.* (75) detected VTEC O157:H7 in <1% of animals sampled, but this was probably because these authors used non-sensitive screening methods. In contrast, Jackson *et al.* (50) using IMS isolated VTEC O157:H7 from 59 (62%) of 95 cattle in an Ontario dairy farm implicated in a case of human infection. Van Donkersgoed *et al.* (88) examining faecal samples collected from cattle at processing during a 1-year period (1995-1996) in an abattoir in Alberta (Canada), found that the prevalence of VTEC O157:H7 was lower on direct culture (3%) on CTSMAC than on IMS-enrichment culture (8%). They also found that the individual prevalence and percentage of positive herds was higher for yearling cattle than for culled cows. Furthermore, the prevalence of VTEC O157:H7 was significantly higher in the summer than during the other three seasons (88). In surveys carried out in Australia (28), Japan (65) and Thailand (90) only 2% of cattle sampled were positive for VTEC O157:H7. However, the VTEC O157:H7 prevalence rates in these countries as noted in other cases, are probably underestimates due to the low sensitivity of the methods used.

NON-O157 VTEC PREVALENCE IN CATTLE

Detection of all VTEC requires testing for verotoxins (VT1 and VT2) or the genes encoding verotoxins (*vt1* and *vt2*). In most studies, non-O157 VTEC have been detected by Vero cell toxicity test, DNA probes or PCR amplification techniques specific for *vt* genes (Table 3). While testing of specimens such as stools or mixed cultures by the above VT-specific tests can provide evidence of the presence of VTEC, obtaining isolates from VTEC positive samples is laborious. Non-O157 VTEC have no consistent features other than toxin production or toxin genes to differentiate them from other *E. coli*. However, culture on CTSMAC after hydrochloric acid treatment of samples facilitates the isolation of *eaeA*+ VTEC strains belonging to serotype O157:H7 and other serotypes (37). Isolates are obtained by testing numerous individual colonies for verotoxins or *vt* genes. The more colonies are tested, the more sensitive the technique used will be and so, higher rates of prevalence will be found. Thus, when the Vero cell test was used to examine 5 and 10 colonies per animal, prevalence rates in cattle of 21% (15) and 35% (16) respectively, have been found, but when mixed cultures with approximately 100 colonies were assayed by PCR the proportion of positive animals was >75% (unpublished results). Thus, we found a very high proportion of cattle excreting non-O157 VTEC in

TABLE 3.
SURVEYS OF PREVALENCE OF VTEC (O157:H7 AND NON-O157) IN CATTLE. MOST COMMON NON-O157 SEROTYPES.

Study (Reference)	Country	Prevalence				Predominant Non-O157 Serotypes[c]	Screening Method
		VTEC[a]		VTEC O157:H7			
Blanco *et al.* (15,16,17)	Spain	334/ 1,069	31%	8/1,069	0.7%	O113:H21, O22:H8, O26:H11, O77:H41, O171:H2, O20:H19, O105:H18, O4:H4, O8:H2, O116:H21, O156:H-, O91:H21, O113:H4, OX3:H2, O2:H27, O82:H8, O103:H2, OX3:H21, OX3:H-, O2:H29, O128:H-	Vero cell test, CTSMAC
Pradel *et al.* (74)	France	330/471	70%	1/471	0.2%	O113:H21, OX3:H2, OX3:H21, OX178:H19, O113:H4, O6:H10, O171:H2, O46:H38, O172:H21, O91:H10, O22:H8, O22:H16, O109:H-, O74:H42	PCR
Beutin *et al.* (8)	Germany	30/142	21%	0/142	0%	O146:H21, O113:H21, O20:H19, O76:H21, O156:H21	Vero cell test
Montenegro *et al.* (68)	Germany	28/259	11%	2/259	0.8%	O116:H21, O82:H8, O113:H21, O136:H12, O3:H-, O10:H21, O22:H8, O126:H20	DNA hybridization
Richter *et al* (79)	Germany	97/204	48%	0/204	0%	O113:H21, O22:H8, O171:H2, O116:H21, O113:H4, O153:H25	VT-colony immunoblot
Nielsen & Frydendahl (67)	Denmark	40/82	49%	4/82	5%		PCR, IMS+CTSMAC
Andlovic & Marinsek (3)	Slovenia	51/250	20%	2/250	0.8%		Vero cells, TSB+SMAC
Willshaw *et al.* (96)	UK	31/180	17%	0/180	0%	O113:H4, O172:H21, O163:H19, O168:H8, O46:H38	DNA hybridization
Wells *et al.* (92)	USA	45/322[b]	14%	18/1,26 6	1%	O25,O84:H-, O171:H2, O22:H8, O121:H7, O111:H-, OX3:H-, O15:H27, O45:H2, O45:H-, OX3:H21	HeLa cells and DNA probe, mTSB+SMAC

Cray *et al.* (29)	USA	77/1,305	6%	2/1,305	0.2%	O26, O5, O111, O103, O84,	DNA hybridization
Wilson *et al.* (97)	Canada	271/ 1,790	15%	0/1,790	0%	OX3:H-, O8:H-, O153:H-, OX3:H16, OX3:H21, O103:H39, O103:H-, O121:H-	Vero cell test
Rhan *et al.* (75)	Canada	105/406	26%	2/406	0.5%	O26:H11, O153:H25, O156:H-	PCR
Wilson *et al.* (97)	Canada	679/ 1,478	46%	12/ 1,478	0.8%		Vero cell test, PCR, SMAC
Van Donkersgoed *et al.* (88)	Canada		43%		8%		Vero, mTSB+IMS+CTSMA C
Parma *et al* (72)	Argentina	80/244	33%	0/244	0%	O26:H11, OX3:H21, O5:H-, O39:H49, O171:H2, O20:H19, O20:H7, O103:H-, O117:H7, O141:H8	Vero cell test
Cerqueira *et al.* (21)	Brazil	139/197	71%	3/197	2%		PCR
Sidjabat & Bensink (83)	Australia	18/105	17%				PCR
Fagan *et al.* (36)	Australia	56/226	25%				PCR
Cobbold & Desmarcheleir (28)	Australia	98/588	17%	11/588	2%	O26:H11	PCR
Miyao *et al.* (65)	Japan	94/387	24%	7/387	2%	O145:H-, O45:H-, O45:H8, O16:H21, O22:H- , O2:H29, O22:H8, O42:H25, O74:H19, O109:H16, O109:H-, O113:H21, O113:H-, O132:H2, O153:H25	Vero cell test

[a] Prevalence of VTEC (O157:H7 and non-O157 strains).

[b] Prevalence of non-O157 VTEC.

[c] By order of frequency

faeces. In the studies included in Table 3, higher numbers of *E. coli* colonies per animal were tested and 6% to 71% of the cattle were found to excrete VTEC. In these surveys the low prevalence of VTEC O157:H7 in cattle contrasts with the high proportion of cattle excreting non-O157 VTEC. Therefore, non-O157 VTEC are more frequently found in gastrointestinal tract of cattle than VTEC O157:H7.

The results of our studies indicate that VTEC colonisation is widespread among healthy cattle in Spain. Between 1993 and 1995, 1,069 healthy cattle were examined for VTEC colonisation (Table 3). VTEC-positive animals were found in 95% of the farms examined and the estimated proportions of positive cattle in each farm ranged from 0 to 100% (Table 4). The overall prevalence rates of VTEC colonisation were estimated to be 37% in calves and 27% in cows (Table 5). In a study of healthy cattle in Wisconsin and Washington, VTEC strains were found on 80% of dairy cattle farms examined: in 8% of adult cows and 19% of heifers and calves (92). Similar results were reported in a Canadian study in Ontario dairy cattle (97), which estimated prevalence rates of VTEC infection in cows and calves as 9% and 25%, respectively. In another study of healthy cattle in Germany, VTEC strains were recovered from 17% of dairy cows and 9% of bulls (68). High rates of occurrence of VTEC was detected by Pradel *et al.* (74) in France (70%) using PCR to test a loopful of colonies grown on Drigalski agar, and by Cerqueira *et al.* (21) in Brazil (71%), using confluent growth collected from cystine lactose electrolyte deficient agar. In the French study, a total of 330 (70%) of the 471 fecal samples collected from healthy cattle at a slaughterhouse were positive for the *vt* genes, with higher rates in August. The higher rates of occurrence in the Brazilian study was found among dairy cattle (82% *versus* 53% in beef cattle). In contrast with other studies cited above, French and Brazilian studies did not observe differences between prevalence of *vt* genes in relation to the ages of the animals tested.

In Spain, *vt2*+ and *vt1*+ *vt2*+ strains were present in similar proportion in calves and cows. In contrast, *vt1*+ and *eaeA*+ strains were more commonly recovered from calves than from cows (Table 5). VTEC O157:H7 was detected in only 8 (0.7%) of the 1,069 animals investigated. Interestingly, the majority of *eaeA*+ non-O157 VTEC strains, and the 8 VTEC O157:H7 strains were isolated from calves, confirming that young animals are the most important reservoir of highly pathogenic *eaeA*+ VTEC strains. Our results indicate that pathogenic *eaeA*+ VTEC strains are very frequently found among VTEC isolated from calves with diarrhoea. Thus, the *eaeA* gene was significantly more frequent among VTEC strains isolated from diarrhoeic calves (72%) than from healthy calves (26%) or cows (4%) (Table 6).

In a study by Mohammad *et al.* (66) VTEC were significantly associated with diarrhoea in cattle and buffalo calves. Thus, VTEC were isolated from 27% ($P < 0.01$) and 29% ($P < 0.025$) of ill cattle and buffalo calves,

TABLE 4.
PREVALENCE OF VTEC STRAINS IN 19 FARMS (HEALTHY CATTLE) (Spain, 1993)

Farm	Cows		Calves		Serogroups of VTEC strains
1	4/16	25%	3/12	25%	O22 (3)[a] , O116 (1), O171 (4), NT[b] (1)
2	5/7	71%	1/5	20%	O22 (2), NT (5)
3	2/12	17%	5/6	83%	O8 (1), O26 (1), O91 (1), O113 (3), NT (1)
4	9/13	69%	0/1	0%	O4 (1), O8 (1), O22 (5), O113 (1), NT (2)
5	4/15	27%	6/11	54%	O2 (3), O8 (3), O77 (1), O171 (3), NT (1)
6	0/9	0%	0/4	0%	
7	8/21	38%	2/5	40%	O2 (4), O20 (1), O22 (1), O136 (1), O165 (2), NT (4)
8	14/31	45%	0/7	0%	O8 (1), O22 (1), O77 (1), O105 (5), O113 (6), O162 (1), NT (1)
9	1/5	20%	0/5	0%	O91 (1), O113 (1), NT (1)
10	4/16	25%	0/1	0%	O22 (1), O82 (1), O91 (1), NT (1)
11	7/24	29%	1/5	20%	O4 (2), O41 (2), O77 (2), O113 (2), O150 (1), NT (2)
12	6/9	67%	2/2	100%	O90 (1), O103 (2), O113 (1), O132 (1), O136 (1), O146 (1), O163 (1), O171 (1), NT (3)
13	3/13	23%	0/3	0%	O74 (1), O113 (2)
14	1/3	33%	1/2	50%	O77 (1), O163 (2)
15	7/7	100%	3/3	100%	O20 (1), O22 (4), O77 (2), O78 (1), O91 (1), NT (1)
16	3/31	10%	2/6	33%	O105 (4), O113 (2), NT (1)
17	2/7	29%	2/4	50%	O92 (1), O113 (1), NT (2)
18	6/9	67%	2/4	50%	O26 (1), O82 (6), O136 (1)
19	7/20	35%	3/4	75%	O8 (1), O9 (1), O77 (3), O113 (1), NT (4)

[a]Number of animals yielding VTEC strains belonging to indicated serogroups.
[b]Not typeable strains.
Reference: (16).

TABLE 5.
PREVALENCE OF VTEC IN HEALTHY CATTLE IN SPAIN (1993-1995)

Age group	No. of animals	No. of animals with VTEC strains													
		Total		*vt1*		*vt2*		*vt1 vt2*		*eaeA*		non-O157		O157:H7[a]	
Calves	480	176	37%	42	9%	104	22%	40	8%	45	9%	168	35%	8	2%
Cows	589	158	27%	17	3%	99	17%	54	9%	6	1%	158	27%	0	0%

[a]Screening method: Direct plating on CTSMAC.
References: (15-17).

TABLE 6.
VIRULENCE GENES AND ENTEROHAEMORRHAGIC SEROTYPES IN BOVINE VTEC STRAINS ISOLATED FROM DIARRHOEIC AND HEALTHY CATTLE IN SPAIN

Characteristics of VTEC	No. of VTEC strains isolated from					
	Calves				Cows	
	Diarrhoeic (n=32)		Healthy (n=223)		Healthy (n=184)	
vt1	26	81%	54	24%	20	11%
vt2	4	12%	122	55%	108	59%
vt1 vt2	2	6%	47	21%	56	30%
EaeA	23	72%	57	26%	7	4%
Enterohaemorrhagic[a]	14	44%	29	13%	0	0%[b]

[a] Enterohaemorrhagic serotypes most frequently associated with HUS: O26:H11, O103:H2, O111:H- and O157:H7.
[b] Sixty-five (35%) of the 184 strains belonged to non-O157 serotypes less frequently isolated from patients with HUS.
References: (12-17).

respectively, compared with < 6% of healthy control calves. In contrast, our results indicate that VTEC may be components of the normal intestinal flora of healthy calves, because we have isolated VTEC significantly ($P < 0.05$) more frequently from healthy (18%) than from diseased calves (9%) (12). However, both studies did not distinguish between VT1 and VT2 strains. Now, we know the type of verotoxins produced and we observed that VTEC strains isolated from diarrhoeic calves usually synthesise VT1 only (81%), whereas VTEC strains obtained from healthy cattle produce VT2 only, or both VT1 and VT2 verotoxins (Table 6). Similar results were found by Wieler *et al.* (94) in Germany characterising VTEC strains isolated from calves with and without diarrhoea. While the total number of animals positive for VTEC did not vary significantly between healthy and diarrhoeic calves, a significantly higher percentage of VT1-producing *E. coli* was found in diarrhoeic calves, suggesting a pathogenic role in neonatal calf diarrhoea. The majority of VTEC that cause intestinal disease in calves are VT1+ only and possess the *eaeA* gene responsible for the intestinal attaching and effacing lesions, and belong to the following serotypes: O5:H-, O8:H8, O20:H19, O26:H11, O103:H2, O111:H-, O118:H16, O128:H- and O145:H+ (60). Interestingly, 6 of these 9 serotypes were also found among VTEC strains that cause HUS in humans (Table 1). Therefore, some non-O157 VTEC VT1+ *eaeA*+ are enteropathogenic for both humans and ruminants. We found that 44% of VTEC strains isolated from diarrhoeic calves belong to human enterohemorrhagic serotypes (O26:H11, O103:H2, O111:H- and O157:H7) *versus* only 13% and 0% of VTEC strains isolated from healthy calves and cows (Table 6). VTEC O157:H7 strains have only exceptionally been associated with clinical disease in animals and appear not to be pathogenic for cattle and sheep. However, Dean-Nystrom *et al.* (30) have shown that VTEC O157:H7 can be pathogenic for neonatal calves.

VIRULENCE GENES AND SEROTYPES OF VTEC ISOLATED FROM CATTLE

We have characterised 432 bovine non-O157 VTEC strains isolated in Spain. PCR demonstrated that 99 (23%) of 432 bovine non-O157 VTEC strains carried *vt*1 genes, 232 (54%) possessed *vt*2 genes, and 101 (23%) carried both *vt*1 and *vt*2 genes (Table 7). Non-O157 VTEC strains belonged to 63 different O serogroups. However, 73% of the isolates fell within 23 serogroups: O2, O4, O8, O20, O22, O26, O41, O64, O77, O82, O91, O103, O105, O113, O116, O126, O128, O136, O141, O162, O163, O171 and OX3. A correlation was found between the serogroups and type of VT produced. Thus, the majority of strains of serogroups O26, O64, O103, O128 and O136 were $VT1^+$, whereas the majority of strains of serogroups O2, O4, O77, O91, O113, O116, O162, O163 and O171 were $VT2^+$. In contrast, the $VT1^+VT2^+$ strains predominated among strains belonging to serogroups O20, O22, O82, O105 and O126. In a significant number of animals we have isolated VTEC strains belonging to 2 or 3 different serogroups. In some farms, the majority of VTEC strains belonged to a reduced number of serogroups, but in other farms the VTEC strains isolated belonged to a large number of serogroups (Table 4). Enterohemolysin (EHEC-*HlyA*) and *eaeA* virulence genes were detected in 244 (56%) and in 75 (17%) of the 432 bovine non-O157 VTEC strains, respectively. Numerous authors have underlined the strong association between the carriage of *eaeA* gene and the capacity of VTEC strains to cause severe human disease, especially HUS. This important virulence gene was found in significantly fewer bovine non-O157 VTEC (75 of 432, 17%) than human non-O157 VTEC strains (46 of 102, 45%), supporting evidence that they contribute to expression of virulence by human VTEC, and that many bovine non-O157 VTEC are less pathogenic for humans. Similar results were observed in other studies comparing human and bovine VTEC strains (5,53,64). Pradel *et al.* (74) found the *eaeA* gene in only 9 (5%) of the 186 bovine non-O157 VTEC strains isolated in France. Serotypes and intimin types determined in bovine VTEC strains isolated in Spain are indicated in Table 8. Intimin types expressed by non-O157:H7 VTEC strains are different from the $\gamma 1$ intimin of VTEC O157:H7 strains.

TABLE 7.
VIRULENCE GENES OF HUMAN, BOVINE AND OVINE NON-O157 VTEC STRAINS ISOLATED IN SPAIN

Origin	Total	*vt1*		*vt2*		*vt1 vt2*		*eaeA*		EHEC-*Hly*	
Human	102	43	42%	34	34%	25	25%	46	45%	64	63%
Cattle	432	99	23%	232	54%	101	23%	75	17%	244	56%
Sheep	379	213	56%	5	1%	161	42%	18	5%	101	27%

References: (17, 18, 77, Blanco *et al.* unpublished data)

TABLE 8.
DIFFERENT *eaeA* GENE DERIVATIVES OF BOVINE VTEC ISOLATED IN SPAIN

Serotypes	Strains	Intimin types
O26:H11	22	$\beta 1$
O26:H-	3	$\beta 1$
O49:H-	1	δ
O84:H2	1	φ
O103:H2	5	ε
O103:H21	1	ε
O103:H-	2	ε
O111:H-	3	$\gamma 2$
O118:H16	2	$\beta 1$
O118:H-	2	$\beta 1$
O126:H-	1	$\beta 1$
O136:H-	1	ε
O138:H48	1	$\beta 2$
O150:H-	1	φ
O156:H-	4	φ
O156:H25	1	φ
O157:H7	81	$\gamma 1$
O163:H21	1	ε
O165:H21	1	ε
O165:H25	1	ε

(Blanco *et al.* unpublished data.)

Cattle harbour many serotypes of VTEC in their gastrointestinal tract. An analysis of the serotypes found within bovine VTEC isolates indicates a great serological diversity. The bovine VTEC isolates obtained in the studies that we have reviewed could be assigned to 125 different O serogroups and 340 O:H serotypes (Table 9). However, only a limited number of serogroups (O2, O5, O8, O20, O22, O26, O45, O82, O91, O103, O113, O116, O153, O156, O157, O171, O172, O174=OX3) and serotypes (O20:H19, O22:H8, O26:H11, O45:H-, O82:H8, O103:H-, O113:H4,H21, O116:H21, O153:H25, O157:H7, O171:H2, O172:H21, O174(OX3):H21,H2,H-) were frequent and commonly found in cattle populations of different countries. The predominant bovine VTEC serotypes in the majority of surveys realised are O113:H21 in Europe, O26:H11 in America and Australia, and O45:H8,H- and O145:H- in Japan (Table 3). Longitudinal studies on separate bovine populations in Germany and in Sri Lanka revealed that typical bovine VTEC strains (O116 and O153) were predominant among the animals and were isolated over long time periods. In contrast, many other "transient" VTEC serotypes occurred only sporadically in a few animals and for a short time (11).

Numerous VTEC serotypes isolated from cattle correspond to serotypes of human VTEC. Thus, of the 340 bovine VTEC serotypes, 135 (40%) have also been recovered from humans, including 49 (42%) of 118 O:H serotypes associated with HUS (Tables 1 and 9). The 432 bovine non-O157 VTEC strains isolated in Spain belonged to 108 different O:H serotypes; however, 60% belonged to one of 21 serotypes listed in Table 3, including 13 serotypes associated with HUS (O2:H29, O8:H2, O20:H19, O22:H8, O26:H11, O91:H21, O103:H2, O105:H18, O113:H21, O128:H-, O174 (OX3):H2, H21,H-). The five most prevalent non-O157 VTEC serotypes (O113:H21, 33 strains; O22:H8, 25 strains; O26:H11, 23 strains; O171:H2, 20 strains; O77:H41, 21 strains) isolated in Spain were also found in VTEC that cause infections in humans. Similarly, among 20 most frequently isolated serotypes in cattle or cattle products in Canada by Johnson *et al.* (53), 18 have been isolated from humans, and 11 of these are serotypes associated with bloody diarrhoea/haemorrhagic colitis and/or HUS. Pradel *et al.* (74) in France, Beutin *et al.* (8), Montenegro *et al.* (68) and Richter *et al.* (79) in Germany, Willshaw *et al.* (96) in UK, Wells *et al.* (92) in the USA and Parma *et al.* (72) in Argentina have also found that many non-O157 VTEC recovered from cattle belonged to serotypes previously associated with human disease. Comparison between human and bovine isolates generally indicates that bovine and human VTEC of the same serotype have similar known virulence-associate properties (14,53,64,82,95). However, the results obtained by Boerlin *et al.* (19) show that distribution of virulence factors and adherence characteristics differ between human and animal strains of the same serotypes. Thus, the results of Boerlin *et al.* (19) suggest that

TABLE 9.
SEROTYPES OF VTEC ISOLATED FROM CATTLE[a]

O1:H18,**H20**	O69:**H11,H-**	O128:**H31**,H35,**H-**
O2:**H5,H7**,H25,**H27,H29**,H39,H45,H49,**H-**	O70:H8,**H11**,H-	O130:**H11**,H38,H43
O3:H-	O73:H?	O131:H2
O4:H4,H7,H11,H16,H25,**H-**	O74:H19,H28,H29,H42,H52,H-	O132:H2,H18,**H-**
O5:H6,H11,H27,**H-**	O75:**H8**,H19	O133
O6:H3,H8,H10,**H34,H49,H-**	O76:**H19**,H21,H25	O136:H1,H12,H16,H19,H-
O7:**H4**,H6,H40	O77:**H18,H41**	O138:H48
O8:**H2**,H8,**H9**,H16,**H19,H25**,H35,**H-**	O79:H19,H-	O139:H8,H19,H51
O9:H4,**H-**	O80:H31,**H-**	O140:H21,H32
O10:H21,H-	O81:H28,H31	O141:**H2**,H7,**H8,H-**
O11:H8,**H-**	O82:H2,**H8**,H40,**H-**	O145:**H8,H16,H25,H28,H-**
O15:H4,H7,H11,H16,H19,**H27,H-**	O84:**H2**,H8,H28,**H-**	O146:H1,**H8,H21**,H32,**H-**
O16:H2,**H21**	O85:**H-**	O147:H11,H29
O17:**H18**,H21,**H41**	O86:H26,H28	O148:H8
O18:H11	O87:H8,H16,H31,H-	O149:H1
O19	O88:H8,H21,**H25,H-**	O150:**H-**
O20:**H7**,H16,**H19**	O90:H24,**H-**	O152:H-
O21:H21	O91:H7,**H10,H14**,H19,**H21**,H49	O153:H7,H9,**H11,H12**,H19,**H21,H25**,H31, H32,**H-**
O22:**H1**,H2,**H8,H16**,H21,**H40,H-**	O92:H-	O156:H7,H8,H19,H21,H25,H46,**H-**
O23:H15,H-	O96:H19,H-	O157:**H7**,H8,**H-**
O25:H5	O98:H25,**H-**	O159:H28
O26:**H11,H21**,**H-**	O101:H4	O160
O28:H9,H21,H40, **H-**	O102:H21	O162:**H4**,H7,H21,H28
O29:H34	O103:**H2**,H16,**H21**,H39,H42,**H-**	O163:H2,**H19**,H21,**H-**
O32:H7,H16,H-	O104:H11,**H21**	O165:**H21,H25,H-**
O35:H21	O105:**H18**	O166:**H28**,H-
O38:H16,**H21**,H39	O106:H42,H-	O167:H21
O39:H7,**H8**,H21,H40,H48,H49,H-	O109:H16,H-	O168:H8
O40:**H8**,H21,H-	O110:H2	O169:H19
O41:**H2**,H14,H51	O111:**H2,H8**,H11,**H-**	O170:H8
O42:H25	O112:**H2,H19**	O171:**H2**,H25
O43:H2	O113:**H2,H4,H7**,H19,**H21,H-**	O172:H21,**H-**
O44:H25,H28,H-	O115:H8,**H18**	O174 (OX3):**H2**,H7,H16,**H21**,H39,H43, H49,**H-**
O45:**H2**,H8,**H-**	O116:H10,H21,H28,**H-**	O175 (OX7):**H16**
O46:**H2**,H38,H-	O117:**H4,H7,H28,H-**	OX177:**H-**
O49:H12,**H-**	O118:**H16,H30,H-**	OX178:H19
O53:H2	O119:H8,**H25,H-**	OX6:H2
O55:H17	O120:H2,H18,H42,H-	OX8:H-
O60:H19	O121:H7,**H-**	"O"K84:H19
O63:H19	O123:H11,H16	
O64:H34,H-	O125:H19	
O65:**H16**	O126:**H8,H20,H21,H27,H-**	
O68:H14,**H-**	O127:H21,H-	

[a] Serotypes of VTEC strains isolated from faecal samples or gastrointestinal tract of cattle. Serotypes found in VTEC strains isolated from beef, milk and other beef products were not listed.

Serotypes found in human VTEC strains are in bold.

References: (7,8,14,17,29,53,65,68,72,74,75,92,93,96-98, Blanco *et al.* unpublished data).

VTEC isolates from humans form a different population from those found in the bovine reservoir or that they are only a subpopulation of the latter.

VTEC PREVALENCE IN SHEEP, GOATS AND OTHER RUMINANTS

Small ruminants have been subjected to fewer surveys compared with cattle (Table 10). Recently, VTEC O157:H7 has been detected in sheep and goat faeces, or at slaughter showing that small ruminants may also represent a source of contamination for humans. Transmission of VTEC O157:H7 and other VTEC serotypes to humans by raw goat milk or home-made raw-milk cheese has been demonstrated. VTEC O157:H7 has also been isolated from wild and farmed deer in the UK and USA and they have been associated with a human outbreak in one case (60).

Prevalence rates of non-O157 VTEC were higher in sheep (67%) and goats (56%) than in cattle (21%) in Germany. In the flocks, the percentage of positive animals varied greatly, from 7 to 84% in goat herds and from 55 to 95% in sheep flocks (8). VTEC strains were detected in all 4 French goat herds, in between 55 and 95% of individual animals (6). In Australia, VTEC were identified in 40% of goats and in 56 to 68% of sheep examined in two studies (36,83). In the USA, Kudva *et al.* (56,57) detected VTEC strains in 43% of sheep. Four groups have surveyed sheep populations using sensitive methods for the detection of VTEC O157:H7. In the UK, a survey of 1,000 sheep at a slaughterhouse found the agent in the faeces of 22 (2%), a rate which is lower than that in cattle (16%) entering the same facility (25). Similarly, in The Netherlands, VTEC O157:H7 strains could be isolated from 2 (4%) of 52 ewes, 2 (4%) of 49 lambs, 57 (11%) of 540 adult cattle and 2 (0.5%) of 397 veal calves (47). In Norway, although a sensitive method was employed, no VTEC O157:H7 were isolated from 364 sheep (52). In contrast, Kudva *et al.* (56,57) detected VTEC O157:H7 in a high percentage (31%) of sheep sampled in a flock in USA. One flock monitored for the presence of VTEC O157:H7 over a 16-month period, displayed a variation in frequency of occurrence (0% to 31%), with animals being culture positive in summer months but not in spring, autumn, or winter (56,57). Faecal swabs obtained from 1,300 lambs on 93 flocks in Spain indicated the presence of VTEC in 68% of the flocks. VTEC O157:H7 strains were isolated from 5 (0.4%) animals, whereas non-O157 VTEC were isolated from 462 (36%) lambs (Table 10). However, the prevalence rate of VTEC O157:H7 in Spain is probably underestimated because IMS methods were not employed (77).

TABLE 10.
SURVEYS OF PREVALENCE OF VTEC IN SHEEP AND GOATS

Study (Reference)	Country	Animals	Prevalence				Predominant Non-O157 Serotypes[b]	Screening Method
			VTEC[a]		VTEC O157:H7			
Rey *et al.* (77)	Spain	Sheep	467/1,300	36%	5/1,300	0.4%	O91:H-, O128:H-, O146:H21, O6:H10, O5:H-, O117: H-, O128:H2, O146:H8, O156:H-, O166:H28, O136:H20, O104:H7, O112:H-, O110:H-	Vero cell test
Beutin *et al.* (8)	Germany	Sheep Goats	80/120 37/66	67% 56%	0 0		Sheep: O5:H-, O91:H-, O128:H2, O146:H21, O77:H4, O123:H10, O119:H25, O146:H8 Goats: O5:H--	Vero cell test
Beutin *et al.* (10)	Germany	Sheep	6-11/25	24 to 44%	0		O128:H2, O125:H14, O146:H21	Vero cell test
Djordjevic *et al.* (33a)	Australia	Sheep Lambs	663/904 426/719	73% 59%	2/904 0/719	0.2% 0%	O91:H-, O5:H-, O128:H2, O123:H-, O85:H49, O75:H-	PCR
Bastian *et al.* (6)	France	Goats	42 -72/76	55 to 95%				PCR
Fagan *et al.* (36)	Australia	Sheep Goats	82/146 4/10	56% 40%				PCR
Sidjabat & Bensink (83)	Australia	Sheep	69/101	68%				PCR
Kudva *et al.* (56, 57)	USA	Sheep	15/35	43%	0-11/35	0 to 31%	O91:H-, O5:H-	DNA hybridization TSB-ctv+CTSMAC
Johnsen *et al.* (52)	Norway	Sheep			0/364	0%		BPWvcc+IMS+CTSMAC
Heuvelink *et al.* (47)	The Netherlands	Sheep			4/101	4%		mEC+IMS+CTSMAC
Chapman *et al.* (25)	England	Sheep			22/1,000	2%		BPWvcc+IMS+CTSMAC

[a] Prevalence of VTEC (O157:H7 and non-O157 strains).
[b] By order of frequency.

VIRULENCE GENES AND SEROTYPES OF VTEC ISOLATED FROM SHEEP

We have characterised 379 ovine non-O157 VTEC strains isolated in Spain. PCR demonstrated that 213 (56%) of 379 ovine VTEC strains carried *vt1* genes, 5 (1%) possessed *vt2* genes and 161 (42%) both *vt1* and *vt2* genes (Table 7). Non-O157 VTEC strains belonged to 32 different O serogroups. However, 80% (258 of 379) were of one of 12 serogroups: O5 (20 strains), O6 (31 strains), O91 (66 strains), O104 (9 strains), O110 (7 strains), O112 (8 strains), O117 (16 strains), O128 (63 strains), O136 (12 strains), O146 (43 strains), O156 (16 strains), and O166 (13 strains). A correlation was found between the serogroups and the VT type produced. Thus, the majority of strains of serogroups O6, O104, O110, O112, O117, O136 and O156 were VT1+, whereas the majority of strains of serogroups O91 and O128 were VT1+ VT2+. Strains belonging to serogroups O5, O146 and O166 were VT1+ or VT1+VT2+. In a significant number of sheep we have isolated multiple VTEC strains belonging to 2 or 3 different serogroups. Enterohemolysin (EHEC-*HlyA*) and *eaeA* virulence genes were detected in 101 (27%) and in 18 (5%) of the 379 ovine non-O157 VTEC strains (Table 7). The *eaeA*-positive non-O157 VTEC strains belonged to serotypes: O26:H11 (1 strain), O49:H- (1 strain), O52:H12 (1 strain), O156:H- (12 strains), O156:H11 (1 strain), O156:H25 (1 strain), and not-typeable (1 strain).

Sheep harbour many serotypes of VTEC in their gastrointestinal tract. Ovine VTEC strains belonging to a total of 50 O serogroups and 101 O:H serotypes have been isolated in the studies we have reviewed (Table 11). However, only a reduced number of serotypes (O5:H-, O91:H-, O128:H2, O146:H8 and O146:H21) are frequently and commonly found in sheep populations of different countries. The predominant ovine serotype is the same in Germany, Spain, Australia and USA: O91:H-. Numerous ovine VTEC serotypes were associated with disease in humans. Thus, 53 (52%) of the 101 ovine serotypes have been also recovered from humans, including 23 serotypes associated with HUS. Furthermore, 11 (79%) of the 14 most prevalent non-O157 VTEC serotypes in sheep in Spain were also found among VTEC strains that cause human infections, including 5 serotypes associated with HUS. Examination of non-O157 VTEC serotypes from cattle and sheep revealed remarkable differences, which could point to an animal-host serotype specificity for some serotypes. Thus, 6 of the 14 non-O157 VTEC serotypes more frequently detected for ovine strains in Spain (Table 10) are not included within the 323 non-O157 VTEC serotypes detected in cattle at that date (Table 9). More than half (54%) of bovine non-O157 VTEC present only *vt2* genes, whereas only 1% of strains of ovine origin are *vt2*+ (Table 7). Furthermore, the *eaeA* gene was found in significantly fewer ovine non-O157 VTEC (5%)

than in bovine (17%) and human strains (45%) (Table 7). Although a wide variety of VTEC serotypes has been associated with human disease, O157:H7 strains are the predominant cause of disease in many countries. The dominance of these strains is even more remarkable considering the overall numbers of outbreaks of VTEC disease complicated with HUS. Indeed, only a small number of HUS outbreaks due to non-O157 STEC have been reported to date, all of which were caused mainly by VTEC strains belonging to serotypes O26:H11,H- and O111:H2,H- (40). Like VTEC O157:H7, O26 and O111 strains are positive for the *eaeA* gene. As this important virulence gene is present in only a minimal (5%) proportion of ovine non-O157 VTEC strains, it is probable that many of these ovine strains show low virulence for humans. However, the intimin encoded by the *eaeA* gene is not essential for pathogenesis as a minority of cases of HUS has been caused by non-O157 VTEC strains lacking *eaeA*, including an *eaeA*- VTEC of serotype O113:H21 that was responsible for a cluster of three cases of HUS in Australia (73). Furthermore, many ovine non-O157 VTEC strains belong to the same O:H serotypes as human VTEC strains that are responsible for sporadic cases of HUS and other less severe human infections. Although our results, and those of other workers, indicate that non-O157 VTEC strains of human and animal origin which share serotypes are similar in terms of known virulence-associated factors, further studies are necessary to establish if animal and human strains represent the same clones or are only related subpopulations (5,19,53,64,82,95).

TABLE 11.
O:H SEROTYPES OF VTEC ISOLATED FROM SHEEP

O2:H?	O75:**H8**,H40,H55,**H-**	O119:**H25**,**H-**
O4:**H10**,H25,**H-**	O77:**H4**	O120:H-
O5:H28,**H-**	O79:**H14**	O123:H10,H-
O6:**H2**,H10,**H12**,**H28**,**H49**,**H-**	O84:H16,**H-**	O125:H14,H26,**H-**
O8:**H9**,H16,H49	O85:H49	O126:**H8**
O9:**H-**	O87:H2,H16	O128:**H2**,**H8**,**H10**,H16,H21,**H-**
O11	O88:**H-**	O136:H20,H40,H-
O15:**H-**	O90:H21,H24,**H-**	O141:**H-**
O21:H21,H28	O91:H2,**H21**,H28,H29,**H-**	O146:**H8**,**H21**,**H-**
O26:**H11**	O98:H7	O153:**H25**,**H-**
O30:H12	O103:**H2**,H38	O156:H11,H25,**H-**
O35:H7	O104:**H7**	O157:**H7**,H21,**H-**
O42:H?	O106:H2,H20	O163:H11,**H19**,**H-**
O49:**H-**	O110:**H-**	O165:**H-**
O52:H12,H45	O112:**H21**,H-	O166:**H28**,H49,H-
O55:H?,**H-**	O117:**H-**	O174(OX3):**H2**,**H21**
O71:H12,**H-**	O118:**H12**	

Serotypes found in human VTEC are in bold.
References: (8,10,33a,77, Blanco *et al.* unpublished data)

BACTERIOPHAGE TYPING AND VIRULENCE GENES OF BOVINE AND OVINE VTEC O157:H7

A total of 86 (81 bovine and 5 ovine) VTEC O157:H7 strains isolated in Spain were characterised by bacteriophage typing and PCR of virulence genes. Fourteen phage types were detected in bovine strains: 2, 8, 14, 21, 23, 26, 27, 32, 34, 39, 45, 51, 54 and 63. The distribution of strains among phage types was hyperbolic, with five types (2, 8, 14, 21, 54), accounting for 70% of the investigated strains. Phage types 2 and 8 were the most frequently found, both in human (58%) and bovine (47%) VTEC strains isolated in Spain. Ovine VTEC O157:H7 belonged to phage types 14 (2 strains), 54 (1 strain) and not-typeable (2 strains). PCR demonstrated that 57% of bovine O157:H7 strains possessed the gene *vt2* only, 42% *vt1* and *vt2* genes, and 2% the gene *vt1* only. All strains with phage type 2 were *vt2* only, whereas the majority of strains with phage type 8 were *vt1* and *vt2* positive. All bovine and ovine VTEC O157:H7 strains possessed *eaeA* and EHEC-*hlyA* gene (70, unpublished data). Chapman *et al.* (25) phage typed 752 bovine and 22 ovine VTEC O157:H7 strains isolated in England. Bovine isolates belonged to 17 different phage types, with almost half the isolates being phage type 2 (32%) or phage type 4 (18%). Verocytotoxicity testing showed that 76% of bovine isolates produced VT2 only, 23% produced both VT1 and VT2, and 0.5% VT1 only. Ovine isolates belonged to 7 different phage types: 2 (5 strains), 4 (5 strains), 8 (2 strains), 14 (3 strains), 31, 54 and 82 (1 strain of each) and not-typeable (4 strains). Twenty strains were VT1+VT2+ and 2 VT2+. All bovine and ovine VTEC O157:H7 strains were positive for the *eaeA* gene (25). Heuvelink *et al.* (47) compared 61 ruminant VTEC O157:H7 strains with 33 human VTEC O157:H7 isolated from HUS cases: 30 (91%) of the 33 human isolates appeared to be of the same VTEC O157:H7 strain types as isolated from cattle and sheep. In The Netherlands, bovine VTEC O157:H7 strains were predominantly phage types 8 and 14, whereas ovine strains belonged to phage types 2, 34, 54 and not-typeable (1 strain of each) (47). An interesting change in the occurrence of different phage types was observed in Scotland. In previous studies (1992 to 1995) phage types 2, 49 and 54 were the most frequently found among bovine VTEC O157:H7, but nowadays phage type 21/28 has become more common, both among bovine and human strains (2,86). In wider terms, the predominant phage types among human VTEC O157:H7 strains isolated in Europe were 2, 4, 8, 14, 21/28 and 32 (26). All these phage types have been frequently detected in VTEC O157:H7 strains isolated from cattle (Table 12). DNA macrorestriction fragment analysis by pulsed-field gel electrophoresis (PFGE) represents a complementary procedure for the subtyping of VTEC O157:H7 with the same phage types (2) which in the future could improve our understanding by providing better resolution.

TABLE 12.
PHAGE TYPES OF VTEC O157:H7 STRAINS ISOLATED FROM CATTLE

STUDY (Reference)	COUNTRY	YEARS	PREDOMINANTS PHAGE-TYPES[a]
Mora *et al.* (70)	Spain	1993-1999	PT2 (21), PT8 (17), PT14 (4), PT21 (7), PT23 (3), PT34 (3), PT54 (8)
Heuvelink *et al.* (47)	The Netherlands	1995-1996	PT4 (4), PT8 (10), PT10 (3), PT14 (10), PT31 (7), PT34 (4), PT54 (4)
Heuvelink *et al.* (48)	The Netherlands	1996	PT8 (28), PT14 (28), PT 21 (50), PT 34 (28)
Synge (86)	Scotland	1992-1995 1998	PT2 (21), PT28 (7), PT32 (7), PT49 (16), PT54 (25) PT28 (58), PT32 (15), PT54 (3)
Chapman & Siddons (24)	England	1987-1993	PT1 (8), PT2 (35), PT4 (8), PT8 (7), PT14 (5), PT32 (4), PT34 (6), PT 49 (8)
Richards *et al.* (78)	England/Wales	1994-1995	PT2 (22), PT4 (4), PT8 (3), PT14 (4), PT34 (5), PT49 (3), PT51 (4)
Chapman *et al.* (25)	England	1995-1996	PT1 (20), PT2 (237), PT4 (137), PT8 (44), PT14 (45), PT 21 (16), PT31 (6), PT32 (36), PT34 (85), PT39 (13), PT49 (56), PT54 (5)
Grif *et al.* (41)	Austria	1996	PT8 (4), PT32 (3)
Loui *et al.* (59)	Canada	1986-1994	PT23 (7), PT32 (3), PT34 (6), PT67 (3)
Akiba *et al.* (1)	Japan	1996	PT2 (4), PT8 (6), PT21 (16), PT32 (4), PT34 (12), PT54 (16)

[a]Number of isolates in parentheses.

ACKNOWLEDGEMENTS

This study was supported by grants from the European Commission (FAIR programme CT98-3935 and PL98-4093), from the Fondo de Investigación Sanitaria (FIS 98/1158), from the Comisión Interministerial de Ciencia y Tecnología (CICYT)(ALI98-0616), CICYT-FEDER (1FD1997-2181-C02-01) and from the Xunta de Galicia (XUGA 26105B97 and XUGA 26106B97). A. Mora acknowledges the Xunta de Galicia for research fellowship.

REFERENCES

1. AKIBA, M., MASUDA, T., SAMESHIMA, T. and KATSUDA, K. 1999. Molecular typing of *Escherichia coli* O157:H7 (H-) isolates from cattle in Japan. Epidemiol. Infect. *122*, 337–341.
2. ALLISON, L.J., CARTER, P.E. and THOMSON-CARTER, F.M. 2000. Characterization of a recurrent clonal type of *Escherichia coli* O157:H7 causing major outbreaks of infection in Scotland. J. Clin. Microbiol. *38*, 1632–1635.
3. ANDLOVIC, A. and MARINSEK, J. 1997. STEC isolated from humans, cattle and minced meat. IVC news 7. Notiziario dell´Istituto Superiore di Sanità. *10*, 7 (Suppl. 2) pp. 3-4.
4. ARMSTRONG, G.L., HOLLINGSWORTH, J. and MORRIS, J.G. JR. 1996. Emerging foodborne pathogens: *Escherichia coli* O157:H7 as a model of entry of a new pathogen into the food supply of the developed world. Epidemiol. Rev. *18*, 29–51.
5. BARRET, T.J., KAPER, J.B., JERSE A.E. and WACHSMUTH, I.K. 1992. Virulence factors in Shiga-like toxin-producing *Escherichia coli* isolated from humans and cattle. J. Infect. Dis. *165*, 979–980.
6. BASTIAN, S., CARLE, I., GRIMONT, F. and GRIMONT, P.A.D. 1999. Diversity of shiga toxin-producing *E. coli* in herds of dairy cows and goats. Abstracts of the Second International Symposium of the European Study Group on Enterohemorrhagic *Escherichia coli*. Acta Clinica Belgica. *54*, 49.
7. BETTELHEIM, K.A. Serotypes of verotoxin-producing *Escherichia coli* reported in the literature apart from those belonging to serogroup O157. MicroBionet. www.microbionet.com.au/frames/feature/VTEC
8. BEUTIN, L., GEIER, D., STEINRÜCK, H., ZIMMERMANN, S. and SCHEUTZ, F. 1993. Prevalence and some properties of verotoxin (Shiga-like toxin)-producing *Escherichia coli* in seven different species of healthy domestic animals. J. Clin. Microbiol. *31*, 2483–2488.
9. BEUTIN, L., GEIER, D., ZIMMERMANN, S. and KARCH, H. 1995. Virulence markers of Shiga-like toxin-producing *Escherichia coli* strains originating from healthy domestic animals of different species. J. Clin. Microbiol. *33*, 631–635.
10. BEUTIN, L., GEIER, D., ZIMMERMANN, S., ALEKSIC, S., GILLESPIE, H.A. and WHITTAM, T.S. 1997. Epidemiological relatedness and clonal types of natural

populations of *Escherichia coli* strains producing Shiga toxins in separate populations of cattle and sheep. Appl. Environ. Microbiol. *63*, 2175–2180.

11. BEUTIN, L. 1999. *Escherichia coli* O157 and other types of verocytotoxigenic *E. coli* (VTEC) isolated from humans, animals and food in Germany. *In Escherichia coli* O157 in farm animals. (C.S. Stewart and H.J. Flint, eds.) CABI Publishing, Wallingford, UK, pp. 121–145.
12. BLANCO, M., BLANCO, J., BLANCO, J.E. and RAMOS, J. 1993. Enterotoxigenic, verotoxigenic and necrotoxigenic *Escherichia coli* isolated from cattle in Spain. Am. J. Vet. Res. *54*, 1446–1451.
13. BLANCO, M. *et al.* 1994. Genes coding for Shiga-like toxins in bovine verotoxin-producing *Escherichia coli* (VTEC) strains belonging to different O:K:H serotypes. Vet. Microbiol. *42*, 105–110.
14. BLANCO, M. *et al.* 1996. Prevalence and characteristics of human and bovine verotoxigenic *Escherichia coli* strains isolated in Galicia (north-western Spain). Eur. J. Epidemiol. *12*, 13–19.
15. BLANCO, M. *et al.* 1996. Prevalence and characteristics of *Escherichia coli* serotype O157:H7 and other verotoxin-producing *E. coli* in healthy cattle. Epidemiol. Infect. *117*, 251–257.
16. BLANCO, M. *et al.* 1997. Distribution and characterization of faecal verotoxin-producing *Escherichia coli* (VTEC) isolated from healthy cattle. Vet. Microbiol. *54*, 309–319.
17. BLANCO, M., BLANCO, J.E., MORA, A., GONZÁLEZ, E.A. and BLANCO, J. 2000. Serotypes and virulence genes of verocytotoxigenic *E. coli* (VTEC) isolated from cattle in Spain. *In* Verocytotoxigenic *E. coli* in Europe, 3. Pathogenicity and virulence of verocytotoxigenic *E. coli*. (G. Duffy, P. Garvey, J. Coia, Y. Wasteson and D.A. McDowell, eds.) Teagasc, The National Food Centre, Dublin, p. 183.
18. BLANCO, J.E. *et al.* 2000. Serotypes and virulence genes of verocytotoxigenic *E. coli* (VTEC) isolated from human patients in Spain. *In* Verocytotoxigenic *E. coli* in Europe, 3. Pathogenicity and virulence of verocytotoxigenic *E. coli*. (G. Duffy, P. Garvey, J. Coia, Y. Wasteson and D.A. McDowell, eds.) Teagasc, The National Food Centre, Dublin, p. 191.
19. BOERLIN, P., MCEWEN, S.A., BOERLIN-PETZOLD, F., WILSON, J.B., JOHNSON, R.P. and GYLES, C.L. 1999. Associations between virulence factors of shiga toxin-producing *Escherichia coli* and disease in humans. J. Clin. Microbiol. *37*, 497–503.
20. BONARDI, S. *et al.* 1999. Isolation of Verocytotoxin-producing *Escherichia coli* O157:H7 from cattle at slaughter in Italy. Vet. Microbiol. *67*, 203–211.
21. CERQUEIRA, A.M.F., GUTH, B.E.C., JOAQUIM, R.M. and ANDRADE, J.R.C. 1999. High occurrence of shiga toxin-producing *Escherichia coli* (STEC) in healthy cattle in Rio de Janeiro State, Brazil. Vet. Microbiol. *70*, 111–121.
22. CHAPMAN, P.A., SIDDONS, C.A., WRIGHT, D.J., NORMAN, P., FOX, J. and CRICK, E. 1993. Cattle as a possible source of verocytotoxin-producing *Escherichia coli* O157 infections in man. Epidemiol. Infect. *111*, 439–447.

23. CHAPMAN, P.A., WRIGHT, D.J. and SIDDONS, C.A. 1994. A comparison of immunomagnetic separation and direct culture for the isolation of verocytotoxin-producing *Escherichia coli* O157 from bovine faeces. J. Med. Microbiol. *40*, 424–427.
24. CHAPMAN, P.A. and SIDDONS, C.A. 1994. A comparison of strains of *Escherichia coli* O157 from humans and cattle in Sheffield, United Kingdom. J. Infect. Dis. *170*, 251–252.
25. CHAPMAN, P.A., SIDDONS, C.A., CERDAN MALO, A.T. and HARKIN, M.A. 1997. A 1-year study of *Escherichia coli* O157 in cattle, sheep, pigs and poultry. Epidemiol. Infect. *119*, 245–250.
26. CHEASTY, T. *et al.* 1999. A comparison of verocytotoxin-producing *Escherichia coli* O157 phage types isolated in England and Wales with those from 13 other European countries 1997-1998. Abstracts of the Second International Symposium of the European Study Group on Enterohemorrhagic *Escherichia coli*. Acta Clinica Belgica. *54*, 45.
27. CIZEK, A., ALEXA, P. and SMOLA, J. 1999. Shiga toxin-producing *Escherichia coli* O157 in feedlot cattle and rats from a large-scale farm. Abstracts of the Second International Symposium of the European Study Group on Enterohemorrhagic *Escherichia coli*. Acta Clinica Belgica. *54*, 47.
28. COBBOLD, R. and DESMARCHELIER, P. 2000. A longitudinal study of shiga-toxigenic *Escherichia coli* (STEC) prevalence in three Australian dairy herds. Vet. Microbiol. *71*, 125–137.
29. CRAY, W.C. JR., THOMAS, L.A., SCHNEIDER, R.A. and MOON, H.W. 1996. Virulence attributes of *Escherichia coli* isolated from dairy heifer faeces. Vet. Microbiol. *53*, 369–374.
30. DEAN-NYSTROM, E.A., BOSWORTH, B.T., O'BRIEN, A.D. and MOON, H.W. 1999. Bovine infection with *Escherichia coli* O157:H7. *In Escherichia coli* O157 in farm animals. (C.S. Stewart and H.J. Flint, eds.) CABI Publishing, Wallingford, UK, pp. 51–58.
31. DE BOER, E. and HEUVELINK, A.E. 1999. Evaluation of methods for the detection and isolation of *Escherichia coli* O157 from foods and faeces. *In* Verocytotoxigenic *E. coli* in Europe, 1. Methods for verocytotoxigenic *E. coli*. (G. Duffy, P. Garvey, J. Coia, Y. Wasteson and D.A. McDowell, eds.) Teagasc, The National Food Centre, Dublin, pp. 25–35.
32. DE ZUTTER, L., URADZINSKI, J. and PIERARD, D. 1999. Prevalence of enterohemorrhagic *E. coli* O157 in Belgian slaughter cattle. Abstracts of the Second International Symposium of the European Study Group on Enterohemorrhagic *Escherichia coli*. Acta Clinica Belgica. *54*, 48.
33. DIEZ-GONZÁLEZ, F., CALLAWAY, T.R., KIZOULIS, M.G. and RSSELL, JB. 1998. Grain feeding and the dissemination of acid-resistant *Escherichia coli* from cattle. Science *281*, 1666–1668.
33a. DJORDJEVIC, S.P. *et al.* 2001. Virulence properties and serotypes of shiga toxin-producing *Escherichia coli* from healthy Australian slaughter-age sheep. J. Clin. Microbiol. *39*, 2017–2021.

34. DOYLE, M.P., ZHAO, T., MENG, J. and ZHAO, S. 1997. *Escherichia coli* O157:H7. *In* Food Microbiology Fundamentals and Frontiers. (M.D. Doyle, L.R. Beuchat and T.J. Montville, eds.) ASM Press, Washington, pp. 171–191.
35. ERIKSSON, E., ENGVALL, A. and GUNNARSSON, A. 1999. Effects of pooling of samples and cold storage of pre-enrichment broth on sensitivity for detecting *E. coli* O157 in cattle faeces. *In* Verocytotoxigenic *E. coli* in Europe, 1. Methods for verocytotoxigenic *E. coli,* (G. Duffy, P. Garvey, J. Coia, Y. Wasteson and D.A. McDowell, eds.) Teagasc, The National Food Centre, Dublin 36–40.
36. FAGAN, P.K., HORNITZKY, M.A., BETTELHEIM, K.A. and DJORDJEVIC, S.P. 1999. Detection of shiga-like toxin (*stx1* and *stx2*), intimin (*eae*A), and enterohemorrhagic *Escherichia coli* (EHEC) hemolysin (EHEC *hlyA*) genes in animal faeces by multiplex PCR. Appl. Environ. Microbiol. *65*, 868–872.
37. FUKUSHIMA, H., HOSHINA, K. and GOMYODA, M. 2000. Selective isolation of *eae*-positive strains of shiga toxin-producing *Escherichia coli*. J. Clin. Microbiol. *38*, 1684–1687.
38. GARBER, L. *et al.* 1995. Risk factors for fecal shedding of *Escherichia coli* O157:H7 in dairy calves. J. Am. Vet. Med. Assoc. *207*, 46–49.
39. GARBER, L., WELLS, S., SCHROEDER-TUCKER, L. and FERRIS, K. 1999. Factors associated with fecal shedding of verotoxin-producing *Escherichia coli* O157 on dairy farms. J. Food Prot. *62*, 307–312.
40. GOLDWATER, P.N. and BETTELHEIM, K.A. 1998. New perspectives on the role of *Escherichia coli* O157:H7 and other enterohemorrhagic *E. coli* serotypes in human disease. J. Med. Microbiol. *47*, 1039–1045.
41. GRIF, K. *et al.* 1998. Comparative study of five different techniques for epidemiological typing of *Escherichia coli* O157. Diagn. Microbiol. Infect. Dis. *32*, 165–176.
42. GRIFFIN, P.M. and TAUXE, R.V. 1991. The epidemiology of infections caused by *Escherichia coli* O157:H7, other enterohemarrhagic *E. coli*, and the associated hemolytic uremic syndrome. Epidemiol. Rev. *13*, 61–68.
43. HANCOCK, D.D., BESSER, T.E., KINSEL, M.L., TARR, P.I., RICE, D.H. and PAROS, M.G. 1994. The prevalence of *Escherichia coli* O157:H7 in dairy and beef cattle in Washington State. Epidemiol. Infect. *113*, 199–207.
44. HANCOCK, D.D., RICE, D.H., THOMAS, L.A., DARGATZ, D.A. and BESSER, T.E. 1997. Epidemiology of *Escherichia coli* O157 in feedlot cattle. J. Food Prot. *60*, 462–465.
45. HANCOCK, D.D., BESSER, T.E. and RICE D.H. 1998. Ecology of *Escherichia coli* O157:H7 in cattle and impact of management practices. *In Escherichia coli* O157:H7 and other shiga toxin–producing *E. coli* strains. (J.B. Kaper and A.D. O'Brien, eds.) ASM Press, Washington, pp. 85–91.
46. HARMON, B.G. *et al.* 1999. Faecal shedding and rumen proliferation of *Escherichia coli* O157 in calves: an experimental model. *In Escherichia coli* O157 in farm animals. (C.S. Stewart and H.J. Flint, eds.) CABI Publishing, Wallingford, UK, pp. 59–69.

47. HEUVELINK, A.E. *et al.* 1998. Isolation and characterization of verocytotoxin-producing *Escherichia coli* O157 strains from Dutch cattle and Sheep. J. Clin. Microbiol. *36*, 878–882.
48. HEUVELINK, A.E. *et al.* 1998. Occurrence of verocytotoxin-producing *Escherichia coli* O157 on Dutch dairy farms. J. Clin. Microbiol. *36*, 3480–3487.
49. HOYLE, B. 2000. Renewed concerns over *E. coli* O157:H7 in ground beef. ASM News *66*, 331–332.
50. JACKSON, S.G. *et al.* 1998. *Escherichia coli* O157:H7 diarrhoea associated with well water and infected cattle on an Ontario farm. Epidemiol. Infect. *120*, 17–20.
51. JERSE, A.E., YU, J., TALL, B.D. and KARPER, J.B. 1990. A genetic locus of enteropathogenic *Escherichia coli* necessary for the production of attaching and effacing lesions on tissue culture cells. Proc. Natl. Acad. Sci. USA *87*, 7839–7843.
52. JOHNSEN, G., BERGET, O.I., WASTESON, Y. and HERIKSTAD, H. 2000. *E. coli* O157:H7 in faeces from cattle, sheep and pigs in the south west part of Norway during 1998. *In* Verocytotoxigenic *E. coli* in Europe, 3. Pathogenicity and virulence of verocytotoxigenic *E. coli*. (G. Duffy, P. Garvey, J. Coia, Y. Wasteson and D.A. McDowell, eds.) Teagasc, The National Food Centre, Dublin, p. 186.
53. JOHNSON, R.P. *et al.* 1996. Growing concerns and recent outbreaks involving non-O157:H7 serotypes of verotoxigenic *Escherichia coli*. J. Food Prot. *59*, 1112–1122.
54. JOHNSON, R.P. *et al.* 1999. Human infection with verocytotoxigenic *Escherichia coli* associated with exposure to farms and rural environments. *In Escherichia coli* O157 in farm animals. (C.S. Stewart and H.J. Flint, eds.) CABI Publishing, Wallingford, UK, pp. 147–168.
55. KARMALI, M.A. 1989. Infection by verocytotoxin-producing *Escherichia coli*. Clin. Microbiol. Rev. *2*, 15–38.
56. KUDVA, I.T., HATFIELD, P.G. and HOVDE, C.J. 1996. *Escherichia coli* O157:H7 in microbial flora of sheep. J. Clin. Microbiol. *34*, 431–433.
57. KUDVA, I.T., HATFIELD, P.G. and HOVDE, C.J. 1997. Characterization of *Escherichia coli* O157:H7 and other shiga toxin-producing *E. coli* serotypes isolated from sheep. J. Clin. Microbiol. *35*, 892–899.
58. LAEGREID, W.W., ELDER, R.O. and KEEN, J.E. 1999. Prevalence of *Escherichia coli* O157:H7 in range beef calves at weaning. Epidemiol. Infect. *123*, 291–298.
59. LOUIE, M. *et al.* 1999. Molecular typing methods to investigate transmission of *Escherichia coli* O157:H7 from cattle to humans. Epidemiol. Infect. *123*, 17–24.
60. MAINIL, J. 1999. Shiga/verocytotoxins and Shiga/verotoxigenic *Escherichia coli* in animals. Vet. Res. *30*, 235–257.
61. MCDONOUGH, P.L., ROSSITER, C.A., REBHUN, R.B., STEHMAN, S.M., LEIN, D.H. and SHIN, S.J. 2000. Prevalence of *Escherichia coli* O157:H7 from cull dairy cows in New York State and comparison of culture methods used during preharvest food safety investigations. J. Clin. Microbiol. *38*, 318–322.
62. MCEVOY, J.M., DOHERTY, A.M., SHERIDAN, J.J. and MCGUIRE, L. 2000. The incidence of *Escherichia coli* O157:H7 in faeces, rumen contents and on carcasses in a commercial Irish beef abattoir. *In* Verocytotoxigenic *E. coli* in

Europe, 3. Pathogenicity and virulence of verocytotoxigenic *E. coli,* (G. Duffy, P. Garvey, J. Coia, Y. Wasteson and D.A. McDowell, eds.) Teagasc, The National Food Centre, Dublin p. 197.

63. MECHIE S.C., CHAPMAN, P.A. and SIDDONS, C.A. 1997. A fifteen month study of *Escherichia coli* O157:H7 in a dairy herd. Epidemiol. Infect. *118*, 17–25.
64. MENG, J., ZHAO, S. and DOYLE, M.P. 1998. Virulence genes of shiga toxin-producing *Escherichia coli* isolated from food, animals and humans. Inter. J. Food. Microbiol. *45*, 229–235.
65. MIYAO, Y., KATAOKAT, T., NOMOTO, T., KAI A., ITOH, T. and ITOH, K. 1998. Prevalence of verotoxin-producing *Escherichia coli* harbored in the intestine of cattle in Japan. Vet. Microbiol. *61*, 137–143.
66. MOHAMMAD, A., PEIRIS, J.S.M., WIJEWANTA, E.A., MAHALINGAM, S. and GUNASEKARA, G. 1985. Role of verocytotoxigenic *Escherichia coli* in cattle and buffalo calf diarrhoea. FEMS Microbiol. Lett. *26*, 281–283.
67. MφLLER NIELSEN, E. and FRYDENDAHL, K. 2000. Virulence factors and serotypes of verocytotoxigenic *E. coli* isolated from cattle. *In* Verocytotoxigenic *E. coli* in Europe, 3. Pathogenicity and virulence of verocytotoxigenic *E. coli*. (G. Duffy, P. Garvey, J. Coia, Y. Wasteson and D.A. McDowell, eds.) Teagasc, The National Food Centre, Dublin, p. 184.
68. MONTENEGRO, M.A. *et al.* 1990. Detection and characterization of fecal verotoxin-producing *Escherichia coli* from healthy cattle. J. Clin. Microbiol. *28*, 1417–1421.
69. MONTET, M.P. *et al.* 1999. Direct detection of *Escherichia coli* O157:H7 in heifers´ fecal samples with an automated immunoconcentration system. Abstracts of the Second International Symposium of the European Study Group on Entero-hemorrhagic *Escherichia coli*. Acta Clinica Belgica. *54*, 48.
70. MORA, A. *et al.* 2000. Bacteriophage typing and virulence genes of verocytotoxigenic *E. coli* (VTEC) O157:H7 strains isolated in Spain. *In* Verocytotoxigenic *E. coli* in Europe, 3. Pathogenicity and virulence of verocytotoxigenic *E. coli*. (G. Duffy, P. Garvey, J. Coia, Y. Wasteson and D.A. McDowell, eds.) Teagasc, The National Food Centre, Dublin, p. 189.
71. OSWALD, E., SCHMIDT, H., MORABITO, S., KARCH, H., MARCHÈS, O. and CAPRIOLI, A. 2000. Typing of intimin genes in human and animal entero-hemorrhagic and enteropathogenic *Escherichia coli*: characterization of a new intimin variant. Infect. Immun. *68*, 64–71.
72. PARMA, A.E. *et al.* 2000. Virulence genotypes and serotypes of verotoxigenic *Escherichia coli* isolated from cattle and foods in Argentina. Importance in public health. Eur. J. Epidemiol. *16*, 757–762.
73. PATON, A.W., WOODROW, M.C., DOYLE, R.M., LANSER, J.A. and PATON, J.C. 1999. Molecular characterization of a shiga toxigenic *Escherichia coli* O113:H21 strain lacking *eae* responsible for a cluster of cases of hemolytic-uremic syndrome. J. Clin. Microbiol. *37*, 3357–3361.
74. PRADEL, N. *et al.* 2000. Prevalence and characterization of shiga toxin-producing *Escherichia coli* isolated from cattle, food, and children during a one-year prospective study in France. J. Clin. Microbiol. *38*, 1023–1031.

75. RAHN, K. *et al.* 1997. Persistence of *Escherichia coli* O157:H7 in dairy cattle and the dairy farm environment. Epidemiol. Infect. *119*, 251–259.
76. RASMUSSEN, M.A., WICKMAN, T.L., CRAY, W.C. JR and CASEY, T.A. 1999. *Escherichia coli* O157:H7 and the rumen environment. *In Escherichia coli* O157 in farm animals. (C.S. Stewart and H.J. Flint, eds.) CABI Publishing, Wallingford, UK, pp. 39–49.
77. REY, J. *et al.* 2000. Serotypes and virulence genes of verocytotoxigenic *E. coli* (VTEC) isolated from sheep in Spain. *In* Verocytotoxigenic *E. coli* in Europe, 3. Pathogenicity and virulence of verocytotoxigenic *E. coli*. (G. Duffy, P. Garvey, J. Coia, Y. Wasteson and D.A. McDowell, eds.) Teagasc, The National Food Centre, Dublin, p. 185.
78. RICHARDS, M.S., CORKISH, J.D., SAYERS, A.R., MCLAREN, I.M., EVANS, S.J. and WRAY, C. 1998. Studies of the presence of verocytotoxic *Escherichia coli* O157 in bovine faeces submitted for diagnostic purposes in England and Wales and on beef carcasses in abattoirs in the United Kingdom. Epidemiol. Infect. *120*, 187–192.
79. RICHTER, H. *et al.* 1997. Verotoxin-producing *E. coli* (VTEC) in faeces from cattle slaughtered in Germany. Berl. Münch. Tierärztl. Wschr. *110*, 121–127.
80. SANDERSON, M.W., GAY, J.M., HANCOCK, D.D., GAY, C.C., FOX, L.K. and BESSER, T.E. 1995. Sensitivity of bacteriologic culture for detection of *Escherichia coli* O157:H7 in bovine faeces. J. Clin. Microbiol. *33*, 2616–2619.
81. SCHMIDT, H., GEITZ C., TARR, P.I., FROSCH, M. and KARCH, H. 1998. Non-O157:H7 pathogenic shiga toxin-producing *Escherichia coli:* phenotypic and genetic profiling of virulence traits and evidence for clonality. J. Infect. Dis. *179*, 115–123.
82. SCOTLAND, S.M., WILLSHAW, G.A., SMITH, H.R. and ROWE, B. 1990. Properties of strains of *Escherichia coli* O26:H11 in relation to their enteropathogenic or enterohemorrhagic classification. J. Infect. Dis. *162*, 1069–1074.
84. SIDJABAT-TAMBUNAN, H. and BENSINK, J.C. 1997. Verotoxin-producing *Escherichia coli* from the faeces of sheep, calves and pigs. Aust. Vet. J. *75*, 292–293.
84. SYNGE, B.A. and HOPKINS, G.F. 1994. Studies of verocytotoxigenic *E. coli* O157 in cattle in Scotland and association with human cases. *In* Recent Advances in verocytotoxin-producing *Escherichia coli* infections. (M.A. Karmali and A.G. Goglio, eds.) Elsevier Science, New York, pp. 65–68.
85. SYNGE, B.A. and HOPKINS, G.F. 1996. Verocytotoxin producing *E. coli* O157-Zoonotic implications. XIX World Buiatrics Congress proceedings, Edinburgh, July 8-12, BCVA, pp. 633–637.
86. SYNGE, B.A. 1999. Methods used for isolation and typing of *E. coli* isolates from livestock in Scotland. *In* Verocytotoxigenic *E. coli* in Europe, 1. Methods for verocytotoxigenic *E. coli*. (G. Duffy, P. Garvey, J. Coia, Y. Wasteson and D.A. McDowell, eds.) Teagasc, The National Food Centre, Dublin, pp. 100–105.
87. SYNGE, B. and PAIBA, G. 2000. Verocytotoxin-producing *E. coli* O157. Vet. Rec. *147*, 27.

88. VAN DONKERSGOED, J., GRAHAM, T. and GANNON, V. 1999. The prevalence of verotoxins, *Escherichia coli* O157, and *Salmonella* in the faeces and rumen of cattle at processing. Can. Vet. J. *40*, 332–338.
89. VOLD, L., KLUNGSETH JOHANSEN, B., KRUSE, H. and SKJERVE, E. 1998. Occurrence of shigatoxinogenic *Escherichia coli* O157 in Norwegian cattle herds. Epidemiol. Infect. *120*, 21–28.
90. VUDDHAKUL, V. *et al.* 2000. Isolation and characterization of *Escherichia coli* O157 from retail beef and bovine in Thailand. FEMS Microbiol. Lett. *182*, 343–347.
91. WASTESON, Y. and LASSEN, J. 1998. Occurrence of *E. coli* O157:H7/H- in animals, foods and humans. IVC news 11. Notiziario dell´Istituto Superiore di Sanità. 11:9 (Suppl.2) pp. 1–2.
92. WELLS, J.G. *et al.* 1991. Isolation of *Escherichia coli* serotype O157:H7 and other Shiga-like-toxin-producing *Escherichia coli* from dairy cattle. J. Clin. Microbiol. *29*, 985–989.
93. WHO. Zoonotic non-O157 shiga toxin-producing *Escherichia coli* (STEC). Report of a WHO Scientific Working Group Meeting, Berlin 23-26 June 1998.
94. WIELER, L.H., BAUERFEIND, R. and BALJER, G. 1992. Characterization of Shiga-like toxin producing *Escherichia coli* (SLTEC) isolated from calves with and without diarrhoea. Zbl. Bakt. *276*, 243–253.
95. WIELER, L.H. *et al.* 2000. Enterohemorrhagic *Escherichia coli* (EHEC) strains of serogroup O118 display three distinctive clonal groups of EHEC pathogens. J. Clin. Microbiol. *38*, 2162–2169.
96. WILLSHAW, G.A., CHEASTY, T., JIGGLE, B., ROWE, B., GIBBONS, D. and HUTCHINSON, D.N. 1993. Vero cytotoxin-producing *Escherichia coli* in a herd of dairy cattle. Vet. Rec. *132*, 96.
97. WILSON, J.B. *et al.* 1992. Distribution and characteristics of verocytotoxigenic *Escherichia coli* isolated from Ontario dairy cattle. Epidemiol. Infect. *108*, 423–439.
98. WILSON, J.B. *et al.* 1996. Vero cytotoxigenic *Escherichia coli* infection in dairy farm families. J. Infect. Dis. *174*, 1021–1027.
99. ZHAO, T., DOYLE, M.P., SHERE, J. and GARBER, L. 1995. Prevalence of enterohemorrhagic *Escherichia coli* O157:H7 in a survey of dairy herds. Appl. Environ. Microbiol. *61*, 1290–1293.

CHAPTER 7

EPIDEMIOLOGY OF VTEC IN NON-RUMINANT ANIMALS

YNGVILD WASTESON

The Norwegian School of Veterinary Science
PO Box 8146 Dep.
Oslo 0033, Norway

Although cattle and other ruminants are regarded as the main reservoir of VTEC, these bacteria have also been isolated from a number of non-ruminant animal species. This chapter reviews data published on the occurrence of VTEC in non-ruminant animals. There is an evident need for more knowledge of the distribution of VTEC in these reservoirs, and especially of the pathogenic potential of the many non-O157 VTEC in animals and foods. Many of the risk factors identified in relation to human infection with VTEC O157 are probably similar to the risk factors for the non-O157 VTEC, but there may also be some differences. According to a WHO report on zoonotic non-O157 VTEC, a direct animal-to-human transmission of non-O157 VTEC has not yet been demonstrated (5).

VTEC is known to be the causative agent of oedema disease in pigs, but information on VTEC as a disease-causing agent in other animal species is limited. A more detailed review of pathogenicity in non-ruminant animals is presented by Gyles in Chapter 12 of this volume.

Examination of the literature raises questions on the nature of the VTEC carrier state of these animals: Setting aside the oedema disease strains in pigs, are bovines (or ruminants in general) the real animal reservoir of VTEC? Are the other animal species only transient carriers of VTEC transferred directly from bovines, or indirectly, i.e., through bovine manure or through bovine environments? There is, for example, very little data available on the mechanisms of establishment or multiplication of VTEC in the non-ruminant intestine, although it could be presumed that as a non pathogen in such animals, VTEC behaves like any other commensal *E. coli* in any warm-blooded animal species.

vt genes may move through the animal population or environment within vegetative (VTEC) cells, or by the transduction of VT-encoding phages among existing *E. coli* populations. It is not yet clear which of these mechanisms is the more important, because as yet, little is known about the epidemiology of the VT-encoding phages. Free-living VT2-encoding phages capable of infecting *E. coli* O157:H7 have been isolated from sewage in concentrations of 1-10/ml sewage, persist longer than *E. coli* O157:H7 in water environment, and are more resistant to chlorination and heat treatment than their host bacteria (39,40). *vt* genes have been detected in a number of similar lambdoid phages (38,48). It is not clear whether or not VT phages from one VTEC serogroup are able to infect *E. coli* cells belonging to other VTEC serogroups. If the VT phages have a broad host specificity, i.e., they are able to infect a wide range of *E. coli* serogroups and/or serotypes, VT phages of non-human pathogenic VTEC in a number of animal species may constitute a reservoir of VT phages for human pathogenic strains. More research on the mobility and host specificity of VT phages is needed before these questions can be answered.

VTEC IN PIGS

VTEC are identified as the disease-causing agent in oedema disease, and may also be involved in the postweaning diarrhoea syndrome in pigs. The majority of oedema disease strains belong to the O-groups O138, O139 and O141, and are not considered to have a zoonotic potential. A number of other VTEC serogroups/types have occasionally been found in faecal samples from pigs, and a few of these are also regarded as human pathogens (see below) (9,13,49,54).

Oedema Disease

Oedema disease is an enterotoxemia, occurring in recently weaned pigs. The main clinical symptoms are oedema in the forehead and eyelids, and disorders of the nervous system causing staggering gait and ataxia. The disease often appears with an acute to peracute progression, and some animals may even be found dead without any previous clinical signs. Sporadic cases of disease may occur, but it is more usual that most, if not all, pigs in a litter are affected. In outbreak situations, both the morbidity and case fatality rate can be very high, exceeding 30% and 90%, respectively. Pathological characteristics include subcutaneous, mesenteric and cerebral oedemas. The pathogenic *E. coli* strain can usually be recovered in almost pure culture from the intestine.

Following its first report, in Ireland in 1938 (53), oedema disease was reported from many countries in Europe, as pig farming became more intensive due to an increased access to animal feed, and a need for food to supply the growing population after World War II. Today, the disease is globally distributed, although the incidence rates may vary from country-to-country, and from year to year. The pathogenesis of the disease was not revealed until the oedema disease variant of *vt*, the *vt*2e, was cloned and sequenced in 1988 (24,61), and clinical oedema disease was experimentally reproduced by intravenous administration of purified toxin (37).

As the designation of the oedema disease toxin indicates, *vt2e* is a member of the *vt2*-family. The nucleotide sequences for *vt2*- and *vt2e*-A subunits are highly homologous, while genes for the B subunits are less homologous. This difference is also reflected in the amino acid sequences; the B subunit of VT2e has 84% deduced amino acid sequence homology to the B subunit of VT2, with greater differences between the carboxy terminal regions than between the amino terminal regions. The most closely related VT2 isolated from a human *E. coli* strain is the VT2 isolated from *E. coli* O128:B12 strain H.I.8, which is designated VT2ev (23).

VT2e binds specifically to the glycolipid-receptor globotetraosylceramide (G_{b4}), and not G_{b3} as is the case with other VT's. The difference in susceptibility of cell lines to the cytotoxic effects of VT2e and VT2 is consistent with this difference in receptor binding specificity (18).

The VT-encoding genes in organisms such as *E. coli* O157:H7 are carried by prophages integrated within the host chromosome. In contrast, genes for VT2e have for a long time only been reported to be located on the bacterial chromosome or on plasmids (36). However, Muniesa *et al.* (41) have recently isolated and characterised an infectious VT2e-converting bacteriophage from an VT2e-producing *E. coli* ONT:H- isolate originating from a patient with diarrhoea. It may be hypothesised that all VT-encoding genes originally have been associated with phages, but that the phage elements have been deleted or inactivated during evolution of some strains. A study by Williams Smith *et al.* (62), investigating experimental transfer of VT production, did not detect transfer of the ability to produce toxin from VT+ *E. coli* O138 strains to VT- *E. coli* O138 strains, or from VT+ *E. coli* O139 strains to VT- *E. coli* O139 strains, during growth in mixed cultures. However, that study did detect transfer of VT production when a VT+ *E. coli* O141:K85,88 donor strain was mixed with an *E. coli* K12 recipient strain. The donor strain, isolated from a pig with post-weaning diarrhoea, also produced the LT and ST enterotoxins.

The F18 fimbrial factor is present in 30 - >50% of post-weaning diarrhoea or oedema disease causing strains of *E. coli* strains which are negative for other known colonisation factors, such as F4, F5, F6 or F41 (30). Susceptibility of pigs to colonisation depends on the availability of intestinal receptors, which become more numerous with age, and are produced under control of a chromosomal locus (30). The F18 fimbriae have been shown to exist in two variants; F18ab and F18ac. The F18ab fimbriae are more strongly associated with oedema disease strains of the O139 serogroup, but are poorly expressed *in vitro* and *in vivo*. The F18ac variant is more efficiently expressed *in vitro* and *in vivo,* and is more frequently associated with post-weaning diarrhoea strains of serogroups O141 and O157, which produce the heat-stable enterotoxins (STa, STb) (42).

As noted above, the predominant oedema disease VTEC serogroups are O138, O139 and O141, although certain strains may dominate in certain geographical areas. In Denmark, for example, where oedema disease did not occur until 1994, oedema disease isolates of serogroup O139 belonged to one PFGE cluster, while isolates from other countries did not (1). Such findings indicated an association of Danish strains of VTEC with the movement of animals between breeding and finishing facilities. Results from a recently published multilocus enzyme electrophoresis analysis of *E. coli* O139 strains isolated from oedema disease and post-weaning diarrhoea in Hungary and USA, suggested that the genes coding for VT2e and the F18 fimbriae can occur in a variety of chromosomal backgrounds (43).

Human-pathogenic VTEC in Pigs

According to the list of non-O157 VTEC serotypes published by Bettelheim (8), corresponding serotypes isolated from pigs and from humans with diarrhoea, haemorrhagic colitis or HUS, include the serotypes O9:H-, O91:H-, and O101:H-. Comparison of VTEC O101 from human and porcine origins, using rep-PCR, revealed a high degree of genetic relatedness (21). Furthermore, the structural *vt* genes of the *E. coli* O101 strains were identical, or nearly identical, to the published *vt*2e-sequence from *E. coli* O139 strains. The pathogenic role of VT2e produced by the *E. coli* O101 strains in human disease is not yet fully clarified or understood.

On rare occasions, VTEC O157 have been isolated from pigs. Heuvelink *et al.* (28) reported the isolation of one VTEC O157 from 145 faecal samples (0.7%) from Dutch pigs at slaughter. This isolate contained the *vt*2-, *eae*- and *E-hly* - genes, was Vero cell positive, and belonged to the phage type 4. The pig concerned originated from a farm where only pigs were raised. In Norway and Sweden, VTEC *E. coli* O157 have sporadically been isolated from pigs raised at farms that also bred cattle (12,34,58). A survey of Danish slaughter pigs concluded that these animals were not a source of human VTEC infection in Denmark (26). The USA has reported that their swineherds are free of VTEC O157, based on the data obtained from the National Animal Health Monitoring System (4).

In contrast to these reports, a few countries report a relatively high frequency of VTEC O157 in pigs. In Chile, Rios *et al.* (50) found that 69% of pigs slaughtered in Santiago were colonised with VTEC strains of serogroups O157, O111, and O26, with a toxin profile similar to that of human isolates, and concluded that pigs are an important animal reservoir of VTEC infection in humans. This theory was supported by PFGE typing analysis of a restricted number of VTEC O157 isolates, in which most isolates from pigs clustered with HUS-associated VTEC O157 strains (50). In Japan, a 1.4% carriage rate of

VTEC O157 was recently reported in pigs presented for slaughter, which is almost the same carriage rate as has been reported earlier among cattle in Japan (44). This study found that the porcine isolates had all the same biochemical and genetic markers as VTEC O157 isolates from cattle and humans, and some belonged to phage type 21, a phage type also observed in cattle and humans.

VTEC IN HORSES

There is not very much information available on the occurrence of VTEC in horses. In a descriptive study of *E. coli* from diarrhoeic and healthy foals, 3 of a total of 99 *E. coli* strains were positive when probed for *vt* (29). These strains were characterised as; O1:H- (*vt*1), O156:NM (*vt*2 and LT) and OX3:NM (*vt*1), and were all *eae* negative. Even though the three *vt* positive strains were isolated from diarrhoeic foals, the diarrhoea-causing role of these strains was not further investigated. According to the list of non-O157 VTEC serotypes published by Bettelheim (8), VTEC O1:H- have also been isolated from healthy humans, humans with bloody diarrhoea and/or HUS, while VTEC O156:NM and OX3:NM have been isolated from healthy cattle.

A few reports reveal the isolation of VTEC O157 from horses, and in two cases such occurrence in the horse was associated with disease and with the isolation of similar VTEC O157 from human contacts (15,27,57). The horses affected were situated on farms which also raised cattle, although in some cases the cattle were negative for VTEC O157 at the time of sampling.

VTEC IN DOGS AND CATS

Pet animals such as dogs and cats live in a different and closer association with humans than farm animals. Contact with dogs and/or cats has earlier been identified as a risk factor for contracting the zoonotic infections campylobacteriosis and toxoplasmosis (33,35), and one would assume that if dogs and cats constitute reservoirs for VTEC, these bacteria could be transferred to humans. This issue has recently been reviewed by Beutin (11). In surveys of *E. coli* from dogs and cats, where the isolates have been screened by probes for *vt*1 and *vt*2, only canine isolates were positive. For example, among 29 cats and 25 dogs in Germany analysed for VTEC, 4% of dogs, and none of the cats, were positive (22). Beutin reported that *vt*2 and *vt*2c were frequent in VTEC from dogs, but that these genes did not occur in feline isolates (10). However, an earlier study by Abaas *et al.* (2) screening *E. coli* strains isolated from 22 diarrhoeic and 25 healthy cats in Sweden for enterotoxin and vero cell toxin production, found that 40% of strains from healthy cats, and 95% of strains from diarrhoeic cats, produced VT (2). A Canadian study showed that the overall prevalence of enteric VTEC infection in cats was 12.3% (22/179), as analysed by screening for the presence of verocytotoxin activity in faecal cultures (55). In this study, statistical analysis of case-control data showed no significant association between VTEC infection and diarrhoeal illness. However, there may be geographic variations in the clinical importance of feline VTEC infections.

Interestingly, PCR analysis detected *vt* genes in the majority of the *vt*-positive *E. coli* only when the primers used amplified conserved regions of the gene sequences coding for VT1, VT2 and VT2e, but did not detect *vt* genes when the primers used were specific for *vt*1 or *vt*2. This led the authors to suggest that a novel verocytotoxin gene may exist in *E. coli* isolated from cats. Of the eight serotypes which were associated with VTEC strains in this study, four have previously been isolated from humans (O2:H1, O4:H5, O6:NM, O6:H1) and two from cattle (O2:H1, O6:NM), suggesting that interspecies transmission between cats, humans and cattle may occur. The clinical importance of such transmission is, however, questionable, since the feline variant of VT seems to be different from VTs of human or bovine origin. Prior to the study of Smith *et al.* (54), Tanaka *et al.* (56) described a feline *E. coli* isolate producing a new VT, which they designated VTy. It is not known if the Japanese VTy is similar to the Canadian new VT.

It is not clear whether or not there is an association between VTEC and diarrhoea in dogs. VTEC strains from diarrhoeic dogs may also carry genes for heat-stable enterotoxins, although such co-existence has not been reported in cattle and human VTEC. In a study of 45 diarrhoeic and 57 healthy dogs in Canada, VT2-producing *E. coli* were isolated from 22.2% of the diarrhoeic dogs, but not from healthy dogs (25). In contrast, VT1-producing *E. coli* occurred in similar proportions in diarrhoeic and healthy dogs (8.9% and 12.3%, respectively). When VT1-producing *E. coli* occurred in the diarrhoeic dogs, genes for VT2 and/or the heat stable enterotoxins were simultaneously present, either in the same or different *E. coli* isolates.

VTEC O157, phage type 4, have been isolated from dogs in Canada and in the UK in a number of cases (46,57). From these reports, it is clear that asymptomatic dogs may function as vectors of transmission for VTEC O157:H7 to humans and other animals, although it does not appear to be a common route of transmission.

Cutaneous and renal glomerular vasculopathy (CRGV) is a disease occurring in greyhounds, characterised by thrombotic vascular lesions in the skin and kidneys (20). The epidemiology, pathology and therapeutic response of greyhounds with CRGV are identical to those of humans with HUS, and VTEC (O157:H7 included) have been isolated from greyhounds with CRGV, and, less frequently, from healthy greyhounds. CRGV occurs in young racing greyhounds, and has so far only been reported from USA. The disease is often associated with eating undercooked ground beef, especially ground beef made from meat considered unfit for human consumption.

VTEC IN BIRDS

Live poultry are not regarded as a naturally occurring reservoir of VTEC, at least not VTEC O157. Neither Chapman *et al.* (16), nor Heuvelink *et al.* (28), were able to isolate VTEC O157 from 1000 and 501 faecal chicken samples from England and the Netherlands, respectively. However, one VTEC O157 isolate was detected in a turkey flock at a Dutch turkey fattening farm. It has, however, been shown that *E. coli* O157:H7 is able to colonise the ceca of experimentally infected young chickens (7). It has also been demonstrated that such colonised chickens can be long-term shedders of *E. coli* O157:H7, probably due to the abilities of these organisms to adhere to and penetrate the epithelia of the cecum. Transovarial transmission of the organism has not been demonstrated (52).

Recently, Parreira *et al.* showed that 72% (36 of 50) of isolates of *E. coli* from chicken with swollen head syndrome (SHS) produced a cytotoxin that was active on Vero cells (45). The cytotoxic effect on Vero cells was morphologically similar to that produced by VT, and the effect was neutralised by antiserum against VT2. However, DNA probes for *vt*1 and *vt*2 failed to hybridise with total DNA from the cytotoxic isolates, suggesting the existence of a different, though related, VT2 toxin. SHS occurs initially as an acute rhinitis caused by a pneumovirus, often followed by secondary invasion of subcutaneous tissue by *E. coli*. The role of the proposed new VT2 in this pathogenesis is not yet identified, but the marked characteristic oedema is consistent with the action of a toxin which targets vascular endothelium.

In contrast to the situation in poultry, a reservoir of VTEC has been identified in the wild bird population, suggesting that wild birds may act as a potential vector for dissemination of VTEC, including human pathogenic variants, in the environment. Wallace *et al.* (1997) published a study in which wild birds, mainly gulls, from an urban landfill site and an area of intertidal bay sediments in the United Kingdom were analysed for *E. coli* O157 (59). Of these, 0.9% of bacterial isolates from the landfill site, and 2.9% of isolates from the bay sediment area were positive for VTEC O157. All isolates were of phage type 4 and possessed *vt*1 and *vt*2.

In a pattern which shares some characteristics with that described for *E. coli* isolated from cats and from chickens with SHS, a cytotoxin with a cytopathic effect on Vero cells similar to the effect induced by verotoxins, and neutralised by VT2 antiserum, has been produced from faecal samples from pigeons living in the centre of Rome (19). No *E. coli* strains belonging to the human pathogenic serogroups O157, O26, or O111 were detected in this material. These results indicate that pigeons may also represent a natural reservoir of VTEC, but whether or not this reservoir poses a health hazard to humans is still uncertain. This new *vt*2 variant has now been characterised by

Schmidt *et al.*, showing that the gene sequence is most similar to the sequence of the *vt2* gene variant from the *E. coli* O128:B12 strain H.I.8 (51). The *vt2* variant-carrying strains were all *eae* positive, and harboured a large plasmid negative for *E-hly*, *espP*, *katP* and *etpD*.

VTEC IN AQUATIC ORGANISMS

The intestinal flora of fish is highly influenced by the contemporary water quality. Consequently, fish swimming in water contaminated by VTEC would presumably also acquire these organisms along with the rest of the microbial flora. However, it is not known how long they will carry any acquired VTEC. Fish, and most seafoods, are not regarded as typical vectors for the transmission of VTEC to humans. However, shellfish and mussels harvested from polluted areas are well-known sources of ETEC and EPEC. In a case-control study of sporadic infections with O157 and non-O157 VTEC in Belgium, consumption of fish appeared, surprisingly, as a risk factor for infection (47).

VTEC IN MISCELLANEOUS ANIMALS

In cattle farm environments, VTEC O157 have been isolated from a number of other non-ruminant species. Four of 10 rats (*Rattus norvegicus*) sampled at a large-scale feedlot farm in the Czech Republic were shown to carry VTEC O157 (17). House flies (*Musca domestica*) sampled at 2 cattle farms in Japan, were found to carry VTEC O157 at rates of 1.25% in summer and 2.5% in fall, i.e., at rates similar to the VTEC carriage rate for cattle at the same farms (31). In this study the cow manure on which flies were found was regarded as the contaminant source. VTEC O157 isolates with identical PFGE patterns were detected over a time span of 3 months, a time period involving many fly generations. A report along similar lines by Heuvelink *et al.* noted the sporadic detection of VTEC O157 on flies sampled at cattle farms positive for VTEC O157 (27).

VTEC can presumably occur, on a sporadic basis, in wild animals intermittently sharing habitats with the human or domestic animal population, as exemplified by reports involving foxes and a rabbit (6,14). However, VTEC were not detected from a collection of faecal samples from brown bears, polar bears, harp seal pups, or penguins, which form a group of animals which are rarely if ever exposed to an environment contaminated by faeces from humans or domestic animals (60). Similarly VTEC O157 could not be detected in faecal samples from free-ranging or captive mammalian and avian wildlife in Trinidad and Tobago, nor from a diverse panel of zoo animals from Kansas City Zoological Gardens in Missouri (3,32).

REFERENCES

1. AARESTRUP, F.M., JORSAL, S.E., AHRENS, P., JENSEN, N.E. and MEYLING, A. 1997. Molecular characterization of *Escherichia coli* strains isolated from pigs with edema disease. J. Clin. Microbiol. *35*, 20–24.
2. ABAAS, S., FRANKLIN, A., KÜHN, I., ØRSKOV, F. and ØRSKOV, I. 1989. Cytotoxin activity on Vero cells among *Escherichia coli* strains associated with diarrhea in cats. Am. J. Vet. Res. *50*, 1294–1296.
3. ADESIYUN, A.A. 1999. Absence of *Escherichia coli* O157 in a survey of wildlife from Trinidad and Tobago. J. Wildlife Dis. *35*, 115–120.
4. ANON. 1997. U.S. swine herd appears free of *E. coli* O157:H7. Large Anim. Pract. *18*, 7.
5. ANON. 1998. Zoonotic non-O157 Shiga Toxin-producing *Escherichia coli* (STEC). Report from a WHO scientific working group meeting, Berlin, Germany, 23–26 June.
6. AWAD-MASALMEH, M. and HÖRZER, J. 1999. Serotypes and virulence factors of VTEC isolated from poultry, birds, dogs and cats. Abstract of the Second international symposium of the European study group on enterohemorrhagic *Escherichia coli*, Brussels, April 16–17, 1999.
7. BEERY, J.T., DOYLE, M.P. and SCHOENI, J.L. 1985. Colonization of chicken cecae by *Escherichia coli* associated with hemorrhagic colitis. Appl. Environ. Microbiol. *49*, 310–315.
8. BETTELHEIM, K. www.microbionet.com.au/frames/feature/vtec
9. BEUTIN, L., GEIER, D., STEINRÜCK, H., ZIMMERMANN, S. and SCHEUTZ, F. 1993. Prevalence and some properties of verotoxin (shiga-like toxin)-producing *Escherichia coli* in seven different species of healthy domestic animals. J. Clin. Microbiol. *31*, 2483–2488.
10. BEUTIN, L., GEIER, D., ZIMMERMANN, S. and KARCH, H. 1995. Virulence markers of Shiga-like toxin-producing *Escherichia coli* strains originating from healthy domestic animals of different species. J. Clin. Microbiol. *33*, 631–635.
11. BEUTIN, L. 1999. *Escherichia coli* as a pathogen in dogs and cats. Vet. Res. *30*, 285–298.
12. BORCH, E. 1999. Personal communication.
13. CAPRIOLI, A., NIGRELLI, A., GATTI, R. and ZAVANELLA, M. 1993. Isolation of verotoxin-producing *Escherichia coli* from slaughtered pigs. Eur J. Clin. Microbiol. Infect. Dis. *12*, 227–228.
14. CARDETI, G., FINAZZI, G., TAGLIABUE, S., LOSIO, N., CAPRIOLI, A. and PACCIARINI, M.L. 1999. Detection and characterisation of Shiga toxin-producing *E. coli* (STEC) in different samples from various animal species: one year of experience. Poster, CA-meeting in Liège, 8–10 Nov.
15. CHALMERS, R.M. *et al.* 1997. Vero-cytotoxin-producing *Escherichia coli* O157 in a farmer handling horses. Lancet *349*, 1816.
16. CHAPMAN, P.A., SIDDONS, C.A., MALO, A.T.C. and HARKIN, M.A. 1997. A one-year study of *Escherichia coli* O157 in cattle, sheep, pigs and poultry. Epidemiol. Infect. *119*, 245–250.

17. CIZEK, A., ALEXA, P., LITERAK, I., HAMRIK, J., NOVAK, P. and SMOLA, J. 1999. Shiga toxin-producing *Escherichia coli* O157 in feedlot cattle and Norwegian rats from a large-scale farm. J. Lett. Appl. Microbiol. *28*, 435–439.
18. DEGRANDIS, S., LAW, H., BRUNTON, J., GYLES, C. and LINGWOOD, C.A. 1989. Globotetraosylceramide is recognized by the pig edema disease toxin. J. Biol. Chem. *264*, 12520–12525.
19. DELL'OMO, G. *et al.* 1998. Feral pigeons as a source of verocytotoxin-producing *Escherichia coli*. Vet. Rec. *142*, 309–310.
20. FENWICK, B.W. and COWAN, L.A. 1998. Canine model of hemolytic-uremic syndrome. *In Escherichia coli* O157:H7 and other Shiga toxin-producing *E. coli* strains (J.B. Kaper and A.D. O'Brien, eds.) ASM Press, pp. 268–277.
21. FRANKE, S., HARMSEN, D., CAPRIOLI, A., PIERARD, D., WIELER, L.H and KARCH, H. 1995. Clonal relatedness of Shiga-like toxin-producing *Escherichia coli* O101 strains of human and porcine origin. J. Clin. Microbiol. *33*, 3174–3178.
22. GALLIEN, P. *et al.* 1994. Detection of verotoxin-producing *E. coli* in field isolates from domestic and agricultural animals in Sachsen-Anhalt. Berl Munch Tierarztl Wochenschri *107*, 331–334.
23. GANNON, V.P.J., TEERLING, C., MASRI, S.A. and GYLES, C.L. 1990. Molecular cloning and nucleotide sequence of another variant of the *Escherichia coli* Shiga-like toxin II family. J. Gen. Microbiol. *136*, 1125–1135.
24. GYLES, C.L., DEGRANDIS, S.A., MACKENZIE, C. and BRUNTON, J.L. 1988. Cloning and nucleotide sequence analysis of the genes determining verocytotoxin production in a porcine edema disease isolate of *Escherichia coli*. Microb. Pathog. *5*, 419–426.
25. HAMMERMUELLER, J., KRUTH, S., PRESCOTT, J. and GYLES, C. 1995. Detection of toxin genes in *Escherichia coli* isolated from normal dogs and dogs with diarrhea. Can. J. Vet. Res. *59*, 265–270.
26. HANSEN, F. 1999. Personal communication.
27. HEUVELINK, A.E. *et al.* 1998. Occurrence of verocytotoxin-producing *Escherichia coli* O157 on Dutch dairy farms. J. Clin. Microbiol. *36*, 3480–3487.
28. HEUVELINK, A.E., ZWARTKRUIS-NAHUIS, J.T.M., VAN DEN BIGGELAR, F.L.A.M., VAN LEEUWEN, W.J. and DE BOER, E. 1999. Isolation and characterization of verocytotoxin-producing *Escherichia coli* O157 from slaughter pigs and poultry. Int. J. Food Microbiol. *52*, 67–75.
29. HOLLAND, R.E. *et al.* 1996. Characterization of *Escherichia coli* isolated from foals. Vet. Microbiol. *48*, 243–255.
30. IMBERECHTS, H., BERTSCHINGER, H.U., NAGY, B., DEPREZ, P. and POHL, P. 1997. Fimbrial colonisation factors F18ab and F18ac of *Escherichia coli* isolated from pigs with postweaning diarrhea and edema disease. Adv. Exp. Med. Biol. *412*, 175–183.
31. IWASA, M., MAKINO, S-I., ASAKURA, H., KOBORI, H. and MORIMOTO, Y. 1999. Detection of *Escherichia coli* O157:H7 from *Musca domestica* (Diptera: Muscidae) at a cattle farm in Japan. J. Med Entomol. *36*, 108–112.

32. JACOBSON, A.P., GYIMESI, Z.S. and WILL, L.A. 1998. Microbiologic investigation of enterohemorrhagic *Escherichia coli* O157:H7 infection at a zoo. JAVMA *212*, 637.
33. JENUM, P.A. *et al.* 1996. Incidence of *Toxoplasma gondii* infection in 35,940 pregnant women in Norway and pregnancy outcome for infected women. Am. J. Epidemiol. *144*, 405–412.
34. JOHNSEN, G., BERGET, O.I., WASTESON, Y. and HERIKSTAD, H. 1999. *E. coli* O157:H7 in faeces from cattle, sheep and pigs in the south west part of Norway during 1998. Poster, CA-meeting in Liège, 8-10 Nov.
35. KAPPERUD, G. 1994. Campylobacter infection. Epidemiology, risk factors and preventive measures. J. Norw. Med. Assoc. *114*, 795–799.
36. KARMALI, M.A. 1989. Infection by verocytotoxin-producing *Escherichia coli*. Clin. Microbiol. Rev. *2*, 15–38.
37. MACLEOD, D.L., GYLES, C.L. and WILCOCK, B.P. 1991. Reproduction of edema disease of swine with purified Shiga-like toxin II-variant. Vet. Pathol. *28*, 66–73.
38. MIYAMOTO, H. 1999. Sequence analysis of Stx2-converting phage VT2-Sa shows a great divergence in early regulation and replication regions. DNA Research *6*, 235–240.
39. MUNIESA, M. and JOFRE, J. 1998. Abundance in sewage of bacteriophages that infect *Escherichia coli* O157:H7 and that carry the Shiga toxin 2 gene. Appl. Environ. Microbiol. *64*, 2443–2448.
40. MUNIESA, M., LUCENA, F. and JOFRE, J. 1999. Comparative survival of free Shiga toxin 2-encoding phages and *Escherichia coli* strains outside the gut. Appl. Environ. Microbiol. *65*, 5616–5618.
41. MUNIESA, M., RECKTENWALD, J., BIELASZEWSKA, M., KARCH, H. and SCHMIDT, H. 2000. Characterization of a shiga toxin 2e-converting bacteriophage from an *Escherichia coli* strain of human origin. Infect. Immun. *68*, 4850–4855.
42. NAGY, B. *et al.* 1997. Biological relationship between F18ab and F18ac fimbriae of enterotoxigenic and verotoxigenic *Escherichia coli* from weaned piglets with oedema disease or diarrhea. Microb. Pathog. *22*, 1–11.
43. NAGY, B., WILSON, R.A. and WHITTAM, T.S. 1999. Genetic diversity among *Escherichia coli* isolates carrying f18 genes from pigs with porcine postweaning diarrhea and edema disease. J. Clin. Microbiol. *37*, 1642–1645.
44. NAKAZAWA, M., AKIBA, M. and SAMESHIMA, T. 1999. Swine as a potential reservoir of Shiga toxin-producing *Escherichia coli* O157:H7 in Japan. Emerg. Infect. Dis. Nov–Dec.
45. PARREIRA, V.R. and YANO, T. 1998. Cytotoxin produced by *Escherichia coli* isolated from chickens with swollen head syndrome (SHS). Vet. Microbiol. *62*, 111–119.
46. PARRY, S.M. *et al.* 1995. Haemorrhagic colitis in child after visit to farm visitor centre. Lancet *346*, 572.
47. PIERARD D. *et al.* 1999. A case-control study of sporadic infection with O157 and non-O157 verocytotoxin-producing *Escherichia coli*. Epidemiol. Infect. *122*, 359–365.

48. PLUNKETT III, G., ROSE, D.J., DURFEE, T.J. and BLATTNER, F.R. 1999. Sequence of Shiga toxin 2 phage 933W from *Escherichia coli* O157:H7: Shiga toxin as a phage late-gene product. J. Bacteriol. *181*, 1767–1778.
49. READ, S.C. *et al.* 1992. Polymerase chain reaction for detection of verocytotoxigenic *Escherichia coli* isolated from animal and food sources. Mol. Cell. Probes *6* 153–161.
50. RIOS M. *et al.* 1999. Clonal diversity of Chilean isolates of enterohemorrhagic *Escherichia coli* from patients with hemolytic-uremic syndrome, asymptomatic subjects, animal reservoirs, and food products. J. Clin. Microbiol. *37*, 778–781.
51. SCHMIDT, H., SCHEEF, J., MORABITO, S., CAPRIOLOI, A. and KARCH, H. 1999. A new Shiga toxin 2 variant (Stx2f) from *Escherichia coli* isolated from pigeons. CA-meeting in Liège, 8-10 Nov., poster.
52. SCHOENI, J.L. and DOYLE, M.P. 1994. Variable colonization of chickens perorally inoculated with *Escherichia coli* O157:H7 and subsequent contamination of eggs. Appl. Environ. Microbiol. *60*, 2958–2962.
53. SHANKS, P.L. 1938. An unusual condition affecting the digestive organs of the pig. Vet. Rec. *50*, 356–358.
54. SMITH, H.R. *et al.* 1988. Verocytotoxin production and presence of VT genes in *Escherichia coli* strains of animal origin. J. Gen. Microbiol. *134*, 829–834.
55. SMITH, K.A., KRUTH, S., HAMMERMUELLER, J., GYLES, C. and WILSON, J.B. 1998. A case-control study of verocytotoxigenic *Escherichia coli* infection in cats with diarrhea. Can. J. Vet. Res. *62*, 87–92.
56. TANAKA H. *et al.* 1992. Isolation of verocytotoxin–producing *Escherichia coli* from cattle and pets. Kansenshogaku Zasshi. *66*, 448–455.
57. TREVENA, W.B., HOOPER, R.S., WRAY, C., WILLSHAW, G.A., CHEASTY, T. and DOMINGUE, G. 1996. Vero cytotoxin-producing *Escherichia coli* O157 associated with companion animals. Vet. Rec. *138*, 400.
58. VÅGSHOLM, I. 1998. Personal communication.
59. WALLACE, J.S., CHEASTY, T. and JONES, K. 1997. Isolation of Vero cytotoxin-producing *Escherichia coli* O157 from wild birds. J. Appl. Microbiol. *82*, 399–404.
60. WASTESON, Y. *et al.* 1999. Analysis of faecal samples from wild animals for Verocytotoxin producing *Escherichia coli* and *E. coli* O157. Vet. Rec. *144*, 646–647.
61. WEINSTEIN, D.L., JACKSON, M.P., SAMUEL, J.E., HOLMES, R.K. and O'BRIEN, A.D. 1988. Cloning and sequencing of a Shiga-like toxin type II variant from an *Escherichia coli* strain responsible for edema disease of swine. J. Bacteriol. *170*, 4223–4230.
62. WILLIAMS SMITH, H., GREEN, P. and PARSELL, Z. 1983. Vero cell toxins in *Escherichia coli* and related bacteria: Transfer by phage and conjugation and toxic action in laboratory animals, chickens and pigs. J. Gen. Microbiol. *129*, 3121–3137.

CHAPTER 8

EPIDEMIOLOGY OF HUMAN INFECTIONS BY *ESCHERICHIA COLI* O157 AND OTHER VEROCYTOTOXIN-PRODUCING *E.COLI*

A.E. TOZZI
and
S. GORIETTI

Laboratorio di Epidemiologia e Biostatistica
Istituto Superiore di Sanità
Rome, Italy

AND

A. CAPRIOLI

Istituto Superiore di Sanita
Laboratorio di Medicina Veterinaria
Viale Regina Elena, 299
00161 Rome, Italy

In the past ten years, there have been remarkable changes in the epidemiology of human infections caused by *Escherichia coli* O157 and other Verocytotoxin-producing *E. coli* (VTEC).

VTEC O157 was first identified as a human pathogen capable of causing foodborne disease in 1982. During the 1980s, outbreaks of this new pathogen rapidly became a serious public health problem, although apparently restricted to North America and the United Kingdom (UK) (42). Most of these outbreaks were foodborne and occurred in limited settings, such as families, schools, and day care centres. The food vehicles implicated were mostly inadequately cooked hamburgers or other beef products, such that the infection became known to the public as the "hamburger disease". The role of VTEC belonging to serotypes other than O157, in causing serious human illness, was generally considered to be negligible.

In the 1990s, the occurrence of VTEC infection began to be reported throughout the industrialised world, with large outbreaks in continental Europe, Australia, and Japan. Serogroups such as O26, O103, O111, and O145 were recognized as a frequent cause of bloody diarrhoea and haemolytic-uraemic syndrome (HUS) and were associated with outbreaks. Furthermore, in addition to foods of bovine origin such as beef and unpasteurized milk, an increasing number of outbreaks were associated with the consumption of fruits and vegetables, while episodes involving commercial salami, mayonnaise and yoghurt highlighted the tolerance of *E. coli* O157 to acidic pH and its ability to survive the processes of fermentation and drying. In addition, several waterborne outbreaks associated with drinking, or swimming in contaminated water and with other types of environment-related exposures have been reported. Finally, the most striking feature of the epidemiology of VTEC infections during the 1990s was the occurrence of outbreaks in the community at large, over long periods of time, over wide geographic areas, and involving large numbers of persons.

Despite the occurrence of these events, the understanding of this composite and evolving pattern has not been supported by homogeneous information derived from comparable surveillance systems.

This chapter provides: 1) a description of the epidemiology of VTEC infections in various geographic areas; 2) a review of the emerging routes and vehicles of infection that have been involved in outbreaks; and, 3) basic recommendations for establishing surveillance systems for VTEC infections.

GEOGRAPHIC DISTRIBUTION

To date, VTEC infections have been a much greater problem in industrialized countries than in non-industrialized countries, probably because of differences in the production and distribution of food. However, reports of some VTEC-associated outbreaks from Africa and the difficulties in performing specific laboratory diagnosis suggest that the extent of the problem in non-industrialized countries may be underestimated.

The description reported below is based on heterogeneous information derived from surveillance systems, outbreak reports, or *ad hoc* studies, and thus it may be difficult to make straightforward comparisons.

Europe

The United Kingdom (UK) has the highest reported incidence of *E. coli* O157 infections in Europe. Surveillance is based on the reporting of *E. coli* O157 isolates, the number of which increased progressively from 1981 to 1996, together with the number of outbreaks (90). In the period from 1990 to 1996, the national incidence of infection with *E. coli* O157 per 100,000 population ranged from 1.5 in England and Wales to more than 5 in Scotland, although in eastern Scotland, the incidence has been reported to be as high as 20 cases per 100,000 population (90). Lower rates were observed in Northern Ireland. A survey on HUS conducted in the UK between 1985 and 1988 showed that the annual incidence ranged from 0.65 to 0.91 per 100,000 children aged 0-15 years

(62). In the past 20 years numerous outbreaks of *E. coli* O157 infection have occurred in the UK in different settings and with different modes of transmission (90). The foodborne outbreaks have involved beef, cooked meat, raw vegetables, milk, cheese, and yoghurt. In many cases, cross contamination from raw to cooked or ready-to-eat foods was implicated. The largest foodborne outbreaks occurred in Scotland, and included one episode caused by milk contaminated after pasteurization and affecting over 100 people, mainly children (96), and another outbreak of 496 cases and 20 deaths among elderly people. This outbreak was associated with the consumption of cooked meat originating from a single butcher shop, and distributed over a large area (3).

In Ireland, *E. coli* O157 has been isolated during family outbreaks and from sporadic cases, with the incidence of infection ranging from 0.9 to 2.1 per 100,000 population during the period from 1997 to 1999 (47). Creche-related outbreaks by *E. coli* O157 and *E. coli* O26 occurred in 1998 and 1999, respectively (47).

A recent review of the epidemiology of VTEC infection in continental Europe (20) has revealed some peculiar features in this area: 1) VTEC belonging to non-O157 serogroups such as O26, O103, O111 and O145 have been frequently associated with HUS; 2) sorbitol-fermenting strains of *E. coli* O157, which cannot be isolated using media for typical sorbitol-negative *E. coli* O157, caused a large community outbreak of HUS in Germany (5) and have also been described in Finland, Hungary, and the Czech Republic; 3) outbreaks seem to occur less frequently than in the UK, and 4) typical vehicles such as hamburgers or other beef products have never been identified as sources of infection. According to this review, which considered reports of VTEC infection outbreaks, and studies of persons with diarrhoea or HUS, there seems to exist a geographic pattern. The incidence of VTEC infection appears to be highest in central Europe. Specifically, a HUS incidence of between 1 and 2 cases per 100,000 children of 0-15 years of age, has been reported in Belgium (71), Germany, the Netherlands, and Switzerland (20), whereas in France and Austria the HUS incidence has been reported to be 0.7 and 0.4, respectively (20). In Switzerland and Germany VTEC was isolated in 2.0-9.3% of persons with diarrhoea (20).

With regard to northern Europe, in Scandinavian countries, VTEC infections were rare and mostly travel-associated or imported until 1995, when a large community outbreak by *E. coli* O157 occurred in Sweden (101). Since then, the incidence of *E. coli* O157 infection has increased in Sweden and Norway (98) as well as in Finland, which had an incidence of 0.73 cases per 100,000 population in 1997 (52), when an outbreak involving 15 microbiologically confirmed cases was observed (69).

Little information is available on VTEC infection from Eastern Europe. *E. coli* O157 has been isolated in the Czech Republic (20), Hungary (20), and

Poland (93). Non-O157 VTEC have been isolated in Serbia (20), and a high incidence of HUS has been reported in Lithuania (20). Outbreaks associated with O157 and non-O157 VTEC have been described in the Czech Republic (14,20) and Hungary (20).

The Mediterranean basin appears to be an area with a low incidence of VTEC infection. In Italy, in the period from 1988 to 1999, the annual incidence of HUS remained constant (i.e., between 0.2 and 0.3 cases per 100,000 children of 0-15 years of age). In Italy (20) and Spain (15), VTEC infection was diagnosed in 0.8% and 1.9% of persons with diarrhoea, respectively; however, in both countries, *E. coli* O157 has been rarely isolated from stool samples routinely submitted to clinical laboratories (20). *E. coli* O157 has also been isolated in Croatia (57), Slovenia (6), and Turkey (7), though the reported isolation rates were low. Outbreaks by *E. coli* O157 and *E. coli* O111 have been reported in Italy and Spain (20), and a small outbreak by *E. coli* O157 occurred in a child-care centre in Israel (54). In addition, three outbreaks of *E. coli* O157 infection occurred among tourists in the Canary Islands (15,37).

North America

VTEC infections were first described in North America and since their discovery have continued to be an important cause of morbidity on this continent.

In Canada, they have been part of routine notification since 1990, and the annual incidence of VTEC isolations in the period from 1992 to 1995, ranged from 3.0 to 5.3 cases per 100,000 population (91). The HUS incidence in Alberta in 1987-1991 was 8.1 cases per 100,000 infants of less than 12 months (100). The incidence of VTEC infection and HUS in Canada peaked in 1988-1989 and has subsequently declined, possibly as a result of interventions on the food-production chain, and of consumer education campaigns (91). *E. coli* O157 was isolated from 15-18% of stool specimens from patients with bloody diarrhoea in different studies (42) and represented more than 90% of the VTEC isolates reported, since most laboratories do not look for non-O157. However, in a study on patients with diarrhoea, the ratio between the isolates of VTEC O157 and non-O157 was 1.1 (63), and evidence of VTEC infection was found in 3.2% of cases.

Epidemiological investigations in Canada (91) have reported that, in addition to eating improperly cooked ground beef, important risk factors for VTEC infection include contact with cattle, consumption of well water, and exposure to the rural environment. A total of 316 outbreaks of *E. coli* O157 infection occurring in Canada between 1982 and 1995 were reviewed by Spika *et al.* (1998). Most of them (70.9%) occurred in family settings, 12.3% of them in nursing homes, day care centres, and other institutions, and 9.2% were in the community at large. Ground beef was the most frequently identified source of infection, though raw milk, apple juice, and drinking water were implicated. The largest episode involved an estimated 520 persons in six Inuit communities, with 22 cases of HUS and two deaths (66). The source of *E. coli* O157 was not determined, but caribou meat was suspected, and person-to-person transmission may have played an important role in the spread of infection among children.

VTEC represents a major cause of illness also in the United States (US). Infection with *E. coli* O157 alone has been estimated to cause an annual 20,000 cases of diarrhoea and as many as 250 deaths (16). Sporadic cases and outbreaks are more common in warmer months than in colder months and are reported more frequently in northern states than in southern states. In a study conducted in King's County, Washington, the average annual incidence of HUS was 1.7 cases per 100,000 children 0-15 years of age in the period 1981-1986; in Minnesota the rate was 2.0 cases per 100,000 population in the age group 0-18 years in 1988 (42). In a large multicentre study conducted between 1990 and 1992, *E. coli* O157 was isolated from only 0.4% of the stool specimens routinely examined, yet it represented the most common (7.8%) pathogen found in visibly bloody specimens (89). Only a few studies on the presence of non-O157 VTEC in diarrhoeal stools have been conducted (41). They showed ratios of isolation of *E. coli* O157 to non-O157 VTEC ranging from 1.5:1 to 2:1. Griffin (1998) reviewed 139 outbreaks of *E. coli* O157 infection reported to the US Center for Disease Control and Prevention (CDC) between 1982 and 1996. Of over 3,000 persons involved, 6% developed HUS and 0.6% died. Most episodes were foodborne (67%) or waterborne (10%). Many of the cases of person-to-person transmission occurred frequently in child-care centres. Foodborne outbreaks occurred most frequently in the home and in restaurants, but also in schools and camps. The food items involved included ground beef, roast beef, salami, raw milk, deer meat, cantaloupe, alfalfa sprouts, lettuce, apple cider, and apple juice. The largest episode occurred in 1993 and involved several Western states; this episode affected over 700 persons, causing 55 cases of HUS and the death of 4 children (12). It was due to the consumption of hamburgers from a single fast-food restaurant chain. Early detection of cases resulted in a rapid recall of contaminated ground beef and prevented an estimated 800 infections. Only two outbreaks have been ascribed to non-O157 VTEC: a family outbreak of *E. coli* O111:NM (11) with a case of HUS and a milk-associated episode due to *E. coli* O104:H21 (22).

South America

VTEC infection appears to be common in countries located in the temperate climate, whereas little information is available on the tropical regions.

Argentina has probably the highest incidence of HUS in the world, with approximately 250 cases observed each year, for an incidence of 8 cases per 100,000 population in the age group 0-4 years (85). In the area of Buenos Aires, an incidence of 22 cases per 100,000 population was reported in the same age group (56). VTEC has been frequently isolated from persons with HUS and from those with bloody or watery diarrhoea (56). *E. coli* O157 has been reported to be the most common VTEC involved in human disease (85). The very high incidence of VTEC-associated disease may be related to the very large numbers of cattle raised in that country, the high rate of bovine meat consumption (about 60 kg per person per year), and the widespread habit of consuming undercooked meat (56). *E. coli* O157 has been frequently isolated from cattle at slaughter, and also from surface water, suggesting a role for environment-related exposure (85).

A high incidence of HUS (5 cases per 100,000 population per year in children less than 5 years of age) has also been reported in Uruguay. Evidence of VTEC infection has been found in a considerable percentage of cases, but no information on the VTEC serogroups involved was available (56).

VTEC infections are also common in Chile. In 1992-1994, the incidence of HUS in the Santiago metropolitan area ranged between 3 and 4.2 cases per 100,000 children less than 4 years of age (72). *E. coli* O157 was reported to be the most commonly isolated serogroup from persons with HUS or bloody diarrhoea; *E. coli* O157 has also been frequently found in cattle and pigs at slaughter and in meat products (84).

In Brazil, studies performed on children with acute diarrhoea in Sao Paulo (40) and Rio de Janeiro (87) indicate that VTEC infection is not common in urban areas. However, *E. coli* O157 has been isolated from dairy and beef cattle (27).

Asia

Most of the information on Asia comes from Japan, where the first outbreak, due to *E. coli* O145, was reported in 1984 (53). Between 1985 and 1995, numerous outbreaks of *E. coli* O157 and two of *E. coli* O111 were reported, most of them occurring within families or school settings (53,60). In 1996, Japan experienced multiple outbreaks of *E. coli* O157 infection involving over 11,000 cases and 12 deaths (60) in different prefectures. One of these outbreaks was the largest in the world, involving more than 6,000 children in 47 primary schools in Sakai City, Osaka. The outbreak was epidemiologically linked to the consumption of uncooked white radish sprouts (61). Radish sprouts from the same farm were also associated with two episodes occurring in the same period in a factory cafeteria in Kyoto and in a nursing home in Habikino City (61). The PFGE profiles of the *E. coli* O157 strains isolated from patients in Sakai, Kyoto, and Habikino were indistinguishable, thus supporting the hypothesis that the radish sprouts were the source of the *E. coli* O157 (99). Other episodes were reported to be bacteriologically associated with different food items such as fish flake salad, salad with seafood sauce, and potato salad with fish sausages, but the source of *E. coli* O157 remained unknown. In March 1997, three outbreaks involving a total of 96 cases were caused by white radish sprouts contaminated with *E. coli* O157, coming from the same farm (60). During routine surveillance activities, *E. coli* O157 has been frequently isolated from asymptomatic adults (94), suggesting that the role of secondary transmission is not negligible.

With regard to other Asian countries, high rates of VTEC infection were observed in both diarrhoeal and asymptomatic subjects in Thailand (19) and in Iran (10). However, none of the strains described belonged to serotype O157. In Hong Kong, VTEC, including *E. coli* O157, were isolated from 0.3% of patients with diarrhoea (55), and the isolation of *E. coli* O157 from a case of HUS was also reported in South Korea (75).

Oceania

In general, the published data indicate that, in contrast to the northern hemisphere, non-O157 VTEC infections may be more common than VTEC O157 infections in the southern hemisphere.

A study conducted in Australia (86), rarely found VTEC in patients with diarrhoea, and *E. coli* O157 represented only a minority of VTEC, even among those strains isolated from cases of HUS. The incidence of sporadic cases of HUS in 1994-1995 was 0.62 cases per 100,000 children of 0-15 years of age (35). *E. coli* O111 caused a family outbreak in 1988 (43) and a large community outbreak with 22 cases of HUS due to consumption of contaminated fermented sausage in 1995 (68,35). The only outbreak of *E. coli* O157 was reported in 1996 and involved only three children with diarrhoea (58).

In New Zealand, a few cases of VTEC O157 infection have been described in children with HUS (8). In a study on children with diarrhoea, VTEC strains were isolated from 3% of the patients, but none of them belonged to serogroup O157 (18).

Africa

A large waterborne outbreak associated with *E. coli* O157 occurred in South Africa and Swaziland in 1992. It was likely due to contamination of surface waters by cattle dung and carcasses, which were washed into rivers by heavy rains after prolonged drought (48). *E. coli* O157 was also isolated from patients involved in a very severe outbreak of bloody diarrhoea in Cameroon (34) and from two fatal cases of bloody diarrhoea in the Central African Republic (39).

In Cameroon, an outbreak with cases of bloody diarrhoea, HUS and deaths, was ascribed to a non-O157 VTEC of undetermined serotype on the basis of a positive PCR examination of stool samples for VT and *eae* genes and negative cultures for *E. coli* O157 (39). These reports suggest that VTEC infections may not be uncommon in African countries, but remain mostly unrecognized due to the difficulties in accurate laboratory diagnosis.

EMERGING MODES OF TRANSMISSION

The gastrointestinal tract of cattle and other ruminants represents the major reservoir of *E. coli* O157 and other VTEC. Consequently, from the time VTEC was first discovered until the mid 1990s, meat and milk contaminated during slaughtering and milking represented the most commonly implicated sources and food vehicles of infection (42). Apart from the primary contamination of meat and milk, however, VTEC can survive in bovine faeces and manure for a considerable time (97). Thus, the dispersion of untreated manure in the environment can result in many different items being contaminated with VTEC and acting as secondary vehicles of human infections (31). These include: other animal species in contact with ruminants, such as horses, dogs, and birds; fruits and vegetables fertilized with manure from ruminants or contaminated during harvesting or processing; drinking-water supplies; and surface waters. In fact, in recent years, an increasing number of outbreaks have been associated with foods of non-animal origin or with non-foodborne routes of transmission. Such less common modes of transmission can be difficult to recognize or even to suspect.

Fruits and Vegetables

The spectrum of fruits and vegetables that have been implicated as vehicles of infection in outbreaks is increasing (Table 1). Contaminated sprouts have frequently caused outbreaks of salmonellosis and represent an emerging source of *E. coli* O157 (59). The recent demonstration of viable *E. coli* O157 in the inner tissues of sprouts grown from experimentally contaminated seeds (49) is a matter of particular concern, in that it indicates that washing and even disinfecting the surface may not be sufficient to assure the safety of this ready-to-eat food.

Water

Waterborne outbreaks due to the contamination of drinking or recreational waters are favoured by the low infectious dose of *E. coli* O157 and by its ability to survive in water, especially at cold temperatures. The contamination of large drinking water supplies in urban settings may be due to cross contamination

TABLE 1.
OUTBREAKS OF *E. COLI* O157 ASSOCIATED WITH CONSUMPTION OF FRUITS AND VEGETABLES (YEARS 1985-1999)

Location	Reference	Month, Year	Setting	No. of suspected cases	No. of hospitalized individuals	No. of HUS cases	No. of deaths	Likely source or mode of transmission
UK	54	July, 1985	Community	24	11	0	1	Handling raw potatoes
US (Massachusetts)	13	October-November, 1991	Community	23	6	4	0	Fresh-pressed apple cider
US (Maine)	29	October, 1992	Small farm	4	?	1	1	Manured vegetables
US (Montana)	1	July, 1995	Community	> 40	13	1	0	Leaf lettuce
US (Illinois and Connecticut)	44	May - June, 1996	Community	54	21	3	0	Mesclun lettuce
Japan (Osaka, Sakai City)	60; 61	July, 1996	Primary schools	7,966	606	106	3	White radish sprouts
Japan (Osaka, Habikino City)	60; 61	July, 1996	Nursing home	98	15	?	0	White radish sprout salad
Japan (Kyoto)	61; 99	July, 1996	Factory	> 47	?	3	1	White radish sprouts
Northeast US (Connecticut)	24; 45	October, 1996	Community	14	10	5	0	Unpasteurized commercial apple cider
Western US and British Columbia (Canada)	30	October, 1996	Community	70	25	14	1	Unpasteurized commercial apple juice
Central Japan	60	March, 1997	Community	96	53	?	1	White radish sprouts
US (Michigan and Virginia)	25	June - July, 1997	Community	108	36	2	0	Alfalfa sprouts
US (Indiana)	73	May, 1998	Restaurant	27	?	0	0	Cole slaw
US (Ohio)	76	June - July, 1999	Restaurants	30	?	0	0	Cole slaw
US (Oklahoma)	80	October, 1999	Community	7	4	1	0	Unpasteurized apple cider

between the water and the sewage pipelines. Large numbers of persons can be involved, as reported in episodes that occurred in the US (92), in Japan (4), in Scotland (50), and in Canada (81). Minor water supplies, such as springs or wells, can also be contaminated by animals. Outbreaks associated with contaminated wells have been described in Scotland (36,65) and in the Canary Islands (37,15). The only waterborne outbreak by non-O157 VTEC reported to date occurred in Spain and was due to *E. coli* O111 (15).

Two outbreaks of *E. coli* O157 infection have been associated with swimming in contaminated pools (17,46). They involved a limited number of persons and were probably due to direct contamination of the water by an infected child with diarrhoea.

Outbreaks Associated with Outdoor Activities

An increasing number of outbreaks of *E. coli* O157 infection have occurred among persons involved in outdoor activities (Table 2). It is noteworthy that five such episodes occurred in the period from August to October 1999. Numerous outbreaks have been associated with swimming in contaminated lakes. Water can be contaminated by faeces washed out into the lake from the nearby environment, either from infected swimmers or from animals. Lake-water-associated outbreaks have been reported in both Europe and the US. The recent episode related to exposure to estuarine water in the UK (70) is a matter of concern, since it suggests that *E. coli* O157 can also survive in salt water. Other outbreaks have occurred among persons attending open-air events, such as fairs, music festivals, farm visits or parties. Different sources of infection have been implicated, including well water, exposure to mud contaminated with cattle faeces, and direct contact with animals on display in the fair or in the farm. Three outbreaks (79,82,88) and several sporadic cases (28,67,95) have also been related to direct contact with animals (cattle, dogs, horses, etc.) on farms.

RECOMMENDATIONS FOR THE SURVEILLANCE OF VTEC INFECTION IN HUMANS

Any effective strategy for prevention of infectious diseases must be based on accurate surveillance. As in the case of other zoonotic diseases, a surveillance system for VTEC infections would greatly benefit from the close cooperation of clinicians, epidemiologists, microbiologists, and veterinarians. In particular, surveillance data and analytical epidemiological studies should always be used for addressing microbiological surveys and controls, to avoid the waste of resources (59).

TABLE 2.
OUTBREAKS OF *E. COLI* O157 ASSOCIATED WITH OUTDOOR ACTIVITIES (YEARS 1991-2000)

Location	Reference	Month, Year	Setting	No. of suspected cases	No. of hospitalized individuals	No. of HUS cases	No. of deaths	Likely source or mode of transmission
US (Oregon)	51	Summer, 1991	Lakeside park	> 21	7	3	0	Swimming
The Netherlands	33	June, 1993	Lakeside	4	4	4	0	Swimming
England	88	Summer, 1994	Farm visit	7	?	4	0	Direct animal contact
US (New York State)	2	July, 1994	Lakeside park	12	5	2	0	Swimming
US (Illinois)	23	July, 1995	Lakeside park	12	3	3	0	Swimming
US (Georgia)	38	July, 1996	Trailer park	> 10	4	1	0	Swimming pool
Finland	69	July, 1997	Lakeside	14	7	0	0	Swimming
England	32	June, 1997	Music festival	8	?	1	0	Contaminated mud
US (Georgia)	74	June, 1998	Water park	> 7	?	5	0	Playing in pools
England	70	August, 1999	Estuarine beach	14	3	3	1	Swimming
US (Washington)	77	August, 1999	Lakeside park	32	7	3	0	Swimming
US (Illinois)	78	September, 1999	Private party	329	22	0	0	Beef and contaminated environment
US (New York State)	26	September, 1999	Fair visit	921	65	11	2	Drinking well water
Canada (Ontario)	79	October, 1999	Fair visit	125	?	?	0	Direct animal contact
US (Washington)	82	May, 2000	Petting farm visit	5	3	1	0	Direct animal contact

Surveillance Systems

Although active surveillance systems are more sensitive than passive reporting systems, they are costly and difficult to implement, and the structure of a surveillance system depends mainly on the availability of resources. In any case, surveillance systems should ensure a timely exchange of information between all levels. The data obtained should be used in the prompt recognition and management of outbreaks and for implementing specific strategies to control the spread of the infection.

HUS represents the most characteristic clinical manifestation of VTEC infection in children. Since children with HUS are treated in a limited number of specialized centres for paediatric nephrology, HUS is considered a robust index of the overall frequency of VTEC infection. Therefore, a basic surveillance system may consist of sentinel centres for paediatric nephrology focussing on paediatric HUS cases. However, children with HUS represent only a small proportion of the overall number of cases of VTEC infection (9). Therefore, using HUS as case definition in surveillance systems, though having a high specificity, may result in a very low overall sensitivity of the system itself.

A wider approach could consist of extending the target population to include adults and persons with bloody diarrhoea, which is also strongly associated with this infection. The inclusion of all individuals with diarrhoea would notably increase the sensitivity of the surveillance system, though the proportion of VTEC-positive cases and the cost-effectiveness of the system would be very low (20).

Since the case definition of VTEC infection is based on microbiological diagnosis, the diagnostic capacity of the laboratories participating in a surveillance system represents a critical point. Most laboratories only look for *E. coli* O157 by screening stools on sorbitol-MacConkey (SMAC) medium. However, several non-O157 serogroups and sorbitol-fermenting O157 strains have been associated with severe human illness, especially in some areas, such as continental Europe (20). The epidemiology of non-O157 VTEC infections is still poorly understood in many countries, including those with well-established laboratory-based surveillance for *E. coli* O157 (41,91,90). All of these infections would be missed by the simple stool culture on SMAC medium. Effective identification requires assays that detect VT production or the presence of VT genes. Thus, for surveillance purposes, at least one reference laboratory for a defined geographical area should have the capacity to perform VT assays.

Recognition and Response to Outbreaks

In the absence of a routine surveillance system, the recognition of an outbreak could be delayed until the number of cases increases enough to be fairly evident. Outbreaks may be spread over large geographic areas and may not be immediately evident, especially when the surveillance system detects only HUS cases. In any case, a surveillance system should continuously process and analyse incoming data to detect unexpectedly high numbers of cases. Public health authorities should be ready to respond immediately to any signal coming from the analysis of surveillance data or from health professionals in the field.

The methods for conducting outbreak investigations have been well described elsewhere (83). Any suspected outbreak first needs to be confirmed on the basis of existing data. When the sensitivity of a surveillance system is enhanced by improving the capacity of laboratories to detect VTEC infections, pseudo-outbreaks can be detected and must be properly recognized as such (21). A team specialised in outbreak investigations with epidemiological and microbiological skills should always be available (21,59). When an outbreak is confirmed, an accurate field investigation, possibly accompanied by an analytical epidemiological study, could be the key to promptly identifying risk factors for the acquisition of the infection and in narrowing down the spectrum of potential vehicles of infection to be submitted for microbiological analysis.

Microbiological confirmation of the sources of infection often requires sophisticated microbiological methods, such as DNA fingerprinting subtyping. The identification of the sources and vehicles of VTEC posed significant challenges (59). As reported above, in recent years, many unusual vehicles of infection have been identified due to outbreak investigations. Their low infectious dose, and ability to survive in acidic food and in the environment, make VTEC a group of versatile pathogens, which can contaminate many items. Finally, the possibility that multiple sources of infection may be involved in the same outbreak should always be considered.

Many recent outbreaks (Table 2) have indicated that environmental investigations can be very important. Although proper microbiological sampling of the environment surrounding the outbreak scenario can be difficult, attempts to consider potential sources of infection contaminated with ruminant manure such as surface and well water, soil, and mud, should always be made.

For all of these reasons, microbiological sampling should follow a thorough epidemiological procedure that also considers unusual means of infection. Although outbreak investigations are frequently conducted when the outbreak is over, the information obtained can be crucial in recognizing new means of transmission and in preventing future similar outbreaks.

Control Measures

If a food item is identified as a source of infection, it can be withdrawn from the market or recalled (12,21). Given the distribution of certain foods over large geographical areas, traceback investigations are not always feasible.

However, they are very useful when the food item has a long shelf life or when a novel vehicle of transmission is implicated (21). Proper control measures should be implemented if an environmental source of infection is identified. Another strategy for controlling or preventing VTEC infections is public education. Person-to-person transmission accounts for a significant proportion of VTEC infections, and poor personal hygiene is recognised as a risk factor for many infectious enteric diseases.

Consumer education represents a critical measure for reducing the burden of foodborne infection. Tailored recommendations for the proper management of food and food items can effectively reduce the potential for foodborne transmission and for cross-contamination between raw food and ready-to-eat food. There is evidence that changes in consumer practices may play a role in reducing the number of *E. coli* O157 cases (91). Nevertheless, such prevention strategies need to be adapted to the local epidemiology of VTEC infection, and their impact should be verified.

REFERENCES

1. ACKERS M.L. *et al.* 1998. An outbreak of *Escherichia coli* O157:H7 infections associated with leaf lettuce consumption. J. Infect. Dis. *177*, 1588–1593.
2. ACKMAN, D., MARKS, S., MACK, P., CALDWELL, M., ROOT, T. and BIRKHEAD, G. 1997. Swimming-associated haemorrhagic colitis due to *Escherichia coli* O157:H7 infection: evidence of prolonged contamination of fresh water lake. Epidemiol. Infect. *119*, 1–8.
3. AHMED, S. and DONAGHY, M. 1998. An outbreak of *Escherichia coli* O157:H7 in central Scotland, p. 59–65. *In Escherichia coli* O157:H7 *and other Shiga toxin-producing E. coli strains*. (J.B. Kaper and A.D. O'Brien, eds.), American Society for Microbiology, Washington, D.C.
4. AKASHI, S. *et al.* 1994. A severe outbreak of haemorrhagic colitis and haemolytic uraemic syndrome associated with *Escherichia coli* O157:H7 in Japan. Eur. J. Pediatr. *153*, 650–655.
5. AMMON, A., PETERSEN, L.R. and KARCH, H. 1999. A large outbreak of hemolytic uremic syndrome caused by an unusual sorbitol-fermenting strain of *Escherichia coli* O157:H-. J. Infect. Dis. *179*, 1274–1277.
6. ANDLOVIC, A. 1997. STEC isolated from humans, cattle and minced meat. EVC News 7. Not. Ist. Superiore di Sanità *10* (Suppl. 2), 3.
7. ANG-KÜCÜKER, M. *et al.* 1999. First isolation of *E. coli* O157. IVC News 13. Not. Ist. Superiore di Sanità *12*(Suppl. 1), 1.
8. ANON. 1995. ESR Surveillance and control notes. Update on *Escherichia coli* O157:H7 infection in New Zealand. New Zealand Public Health Report 2, 76.
9. ARMSTRONG, G.L., HOLLINGSWORTH, J. and MORRIS JR., J.G. 1996. Emerging foodborne pathogens: *Escherichia coli* O157:H7 as a model of entry of a new pathogen into the food supply of the developed world. Epidemiol. Rev. *18*, 29–51.

10. ASLANI, M.M., BADAMI, N., MAHMOODI, M. and BOUZARI, S. 1998. Verotoxin-producing *Escherichia coli* (VTEC) infection in randomly selected population of Ilam Province (Iran). Scan. J. Infect. Dis. *30*, 473–476.
11. BANATVALA, N. *et al.* 1996. Shiga-like toxin-producing *Escherichia coli* O111 and associated haemolytic-uremic syndrome: a family outbreaks. Pediatr. Infect. Dis. J. *15*, 1008–1011.
12. BELL, B.P. *et al.* 1994. A multistate outbreak of *Escherichia coli* O157:H7-associated bloody diarrhea and hemolytic uremic syndrome from hamburgers: the Washington experience. JAMA *272*, 1349–1353.
13. BESSER, R.E. *et al.* 1993. An outbreak of diarrhea and hemolytic uremic syndrome from *Escherichia coli* O157:H7 in fresh-pressed apple cider. JAMA *269*, 2217–2220.
14. BIELASZEWSKA, M. *et al.* 1997. Human *Escherichia coli* O157:H7 infection associated with the consumption of unpasteurized goat's milk. Epidemiol Infect *119*, 299–305.
15. BLANCO ALVAREZ, J., BLANCO ALVAREZ, M., BLANCO ALVAREZ, J.E. and MORA GUTIERREZ, A. 1998. VTEC infections in cattle and humans. IVC News 11. Not. Ist. Superiore di Sanità *11*(Suppl.2), 2–4.
16. BOYCE, T.G., SWERDLOW, D.L. and GRIFFIN, P.M. 1995. *Escherichia coli* O157:H7 and the hemolytic-uremic syndrome. N. Engl. J. Med. *333*, 364–368.
17. BREWSTER, D.H., BROWN, M.I., ROBERTSON, D., HOUGHTON, G.L., BIMSON, J. and SHARP, J.C.M. 1994. An outbreak of *Escherichia coli* O157 associated with a children's paddling pool. Epidemiol. Infect. *112*, 441–447.
18. BROOKS, H.J.L., BETTELHEIM, K.A., TODD, B. and HOLDAWAY, M.D. 1997. Non-O157 Vero cytotoxin producing *Escherichia coli:* aetiological agents of diarrhoea in children in Dunedin, New Zealand. Comp. Immun. Microbiol. Infect. Dis. *20*, 163–170.
19. BROWN, E. *et al.* 1989. Determination by DNA hybridization of Shiga-like-toxin producing *Escherichia coli* in children with diarrhea in Thailand. J. Clin. Microbiol. *27*, 291-294.
20 CAPRIOLI, A. and TOZZI, A.E. 1998. Epidemiology of shiga toxin-producing *Escherichia coli* infections in continental Europe. *In Escherichia coli* O157:H7 *and other Shiga toxin-producing E. coli strains.* (J.B. Kaper and A.D. O'Brien, eds.) p. 38–48. American Society for Microbiology, Washington, D.C.
21. CENTERS FOR DISEASE CONTROL AND PREVENTION. 1994. Enhanced detection of *Escherichia coli* O157:H7 infections - New Jersey, July 1994. Morbid. Mortal. Weekly Rep. *44*, 417–418.
22. CENTERS FOR DISEASE CONTROL AND PREVENTION. 1995. Outbreak of acute gastroenteritis attributable to *Escherichia coli* setotype O104:H21 — Helena, Montana, 1994. MMWR Morb. Mortal. Wkly Rep. *44*, 501–503.
23. CENTERS FOR DISEASE CONTROL AND PREVENTION. 1996. Lake-associated outbreak of *Escherichia coli* O157:H7 — Illinois, 1995. Morbid. Mortal. Weekly Rep. *45*, 437–439.

24. CENTERS FOR DISEASE CONTROL AND PREVENTION. 1997. Outbreaks of *Escherichia coli* O157:H7 infection and cryptosporidiosis associated with drinking unpasteurized apple cider — Connecticut and New York, October 1996. Morbid. Mortal. Weekly Rep. *46*, 4–8.
25. CENTERS FOR DISEASE CONTROL AND PREVENTION. 1997. Outbreaks of *Escherichia coli* O157:H7 infection associated with eating alfalfa sprouts — Michigan and Virginia, June-July 1997. Morbid. Mortal. Weekly Rep. *46*, 741–744.
26. CENTERS FOR DISEASE CONTROL AND PREVENTION. 1999. Outbreak of *Escherichia coli* O157:H7 and *Campylobacter* among attendees of the Washington County Fair — New York, 1999. Morbid. Mortal. Weekly Rep. *48*, 803–804.
27. CERQUEIRA, A.M.F., GUTH, B.E.C., JOAQUIM, R.M. and ANDRADE, J.R.C.. 1999. High occurrence of Shiga toxin-producing *Escherichia coli (STEC)* in healthy cattle in Rio de Janeiro State, Brazil. Vet. Microbiol. *70*, 111–121.
28. CHALMERS, R.M. *et al.* 1997. Vero-cytotoxin-producing *Escherichia coli* O157 in a farmer handling horses. Lancet *349*, 1816.
29. CIESLAK, P.R. *et al.* 1993. *Escherichia coli* O157:H7 infection from a manured garden. Lancet *342*, 367.
30. CODY, S.H. *et al.* 1999. An outbreak of *Escherichia coli* O157:H7 infection from unpasteurized commercial apple juice. Ann. Intern. Med. *130*, 202–209.
31. COIA, J.E, SHARP, J.C., CAMPBELL, D.M., CURNOW, J. and RAMSAY, C.N. 1998. Environmental risk factors for sporadic *Escherichia coli* O157 infection in Scotland: results of a descriptive epidemiology study. J. Infect. *36*, 317–321.
32. CRAMPIN, M., WILLSHAW, G. and HANCOCK, R. 1999. Outbreak of *Escherichia coli* O157 infection associated with a Music Festival. Eur. J. Clin. Microbiol. Infect. Dis. *18*, 286–288.
33. CRANSBERG, K. *et al.* 1996. Four cases of hemolytic uremic syndrome — source contaminated swimming water? Clin. Nephrol. *46*, 45–49.
34. CUNIN, P. *et al.* 1999. An epidemic of bloody diarrhea: *Escherichia coli* O157 emerging in Cameroon? Emerg. Infect. Dis. *5*, 285–290.
35. DESMARCHELIER, P.M. 1997. Enterohemorrhagic *Escherichia coli* — the Australian perspective. J. Food Prot. *60*, 1447–1450.
36. DEV, V.J., MAIN, M. and GOULD, I. 1991. Waterborne outbreak of *Escherichia coli* O157. Lancet *337*, 1412.
37. FISHER, I. 1997. Outbreak of *E. coli* O157 infection associated with travel to the Canary Islands. EVC News 7. Not. Ist. Superiore di Sanità *10* (Suppl. 2), 1–2.
38. FRIEDMAN, M.S, ROELS, T., KOEHLER, J.E., FELDMAN, L., BIBB, W.F. and BLAKE, P. 1999. *Escherichia coli* O157:H7 outbreaks associated with an improperly chlorinated swimming pool. Clin. Infect. Dis. *29*, 298–303.
39. GERMANI, Y., SORO, B., VOHITO, M., MOREL, O. and MORVAN, J. 1997. Enterohaemorrhagic *Escherichia coli* in Central African Republic. Lancet *349*, 1670.
40. GIRALDI, R., GUTH, B.E.C. and TRABULSI, L.R. 1990. Shiga-like toxin production among *Escherichia coli* strains and other bacteria isolated from diarrhea in Sao Paulo, Brazil. J. Clin. Microbiol. *28*, 1460–1462.

41. GRIFFIN, P.M. 1998. Epidemiology of Shiga-toxin producing *Escherichia coli* infections in humans in the United States, *In Escherichia coli* O157:H7 and other Shiga toxin-producing *E. coli* strains. (J.B. Kaper and A.D. O'Brien, eds.) p. 15–22. American Society for Microbiology, Washington, D.C.
42. GRIFFIN, P.M. and TAUXE, R.V. 1991. The epidemiology of infections caused by *Escherichia coli* O157:H7, other enterohemorrhagic *E.coli*, and the associated hemolytic uremic syndrome. Epidemiol. Rev. *13*, 60–98.
43. GUNZBURG, S., GRACEY, M., FORBES, D., HEWITT, I. and BETTELHEIM, K.A. 1988. Haemolytic-uremic syndrome and verocytotoxigenic Escherichia coli. Med. J. Aust. *149*, 54–55.
44. HILBORN, E.D. *et al.* 1999. A Multistate outbreak of *Escherichia coli* O157:H7 infections associated with consumption of mesclun lettuce. Arch. Intern. Med. *159*, 1758–1764.
45. HILBORN, E.D. *et al.* 2000. An outbreak of *Escherichia coli* O157:H7 infections and haemolytic uraemic syndrome associated with consumption of unpasteurized apple cider. Epidemiol. Infect. *124*, 31–36.
46. HILDEBRAND, J.M., MAGUIRE, H.C., HOLLIMAN, R.E. and KANGESU, E. 1996. An outbreak of *Escherichia coli* O157 infection linked to paddling pools. Commun. Dis. Rep. Rev. *6*, R33–R36.
47. IGOE, D. 2000. Personal communication.
48. ISAÄCSON, M., CANTER, P.H., EFFLER, P., ARNTZEN, L., BOMANS, P. and HEENAN, R. 1993. Haemorrhagic colitis epidemic in Africa. Lancet *341*, 961.
49. ITOH, Y. *et al.* 1998. Enterohemorrhagic *Escherichia coli* O157:H7 present in radish sprouts. Appl. Environ. Microbiol. *64*, 1532–1535.
50. JONES, I.G. and ROWORTH, M. 1996. An outbreak of *Escherichia coli* O157 and *Campylobacteriosis* associated with contamination of a drinking water supply. Public Health *110*, 277–282.
51. KEENE, W.E. *et al.* 1994. A swimming-associated outbreak of hemorrhagic colitis caused by *Escherichia coli* O157:H7 and *Shigella sonnei*. N. Engl. J. Med. *331*, 579–584.
52. KESKIMÄKI, M., SAARI, M., HEISKANEN, T. and SIITONEN, A. 1998. A Shiga toxin-producing *Escherichia coli* in Finland from 1990 through 1997: prevalence and characteristics of isolates. J. Clin. Microbiol. *36*, 3641–3646.
53. KUDOH, Y. *et al.* 1994. Epidemiological surveys on Verocytotoxin-producing *Escherichia coli* infections in Japan, *In* Recent Advances in Verocytotoxin-Producing *Escherichia coli* infections. (M.A. Karmali and A.G. Goglio, eds.) p. 53–60, Elsevier Science B.V., Amsterdam.
54. LERMAN, Y., COHEN, D., GLUCK, A., OHAD, E. and SECHTER, I. 1992. A cluster of cases of *Escherichia coli* O157 infection in a day-care center in a communal settlement (Kibbutz) in Israel. J. Clin. Microbiol. *30*, 520–521.
55. LEUNG, P., YAM, W.C., NG, W.W.S., PEIRIS, M. and WONG, K.W. 1999. Isolation and characterization of verocytotoxigenic *Escherichia coli* isolated from animal and human sources in Hong Kong. Abstr. Acta Clinica Belgica *54*-1, 38.
56. LÓPEZ, E.L., CONTRINI, M.M. and DE ROSA, M.F. 1998. Epidemiology of shiga toxin-producing *Escherichia coli* in South America, *In Escherichia coli*

O157:H and other Shiga toxin-producing *E. coli* strains. (J.B. Kaper and A.D. O'Brien, eds.) p. 30–37. American Society for Microbiology, Washington, D.C.
57. MATICA, B. 1999. First isolation of *E. coli* O157. IVC News 12. Not. Ist. Superiore di Sanità *12*(Suppl. 1), 1–2.
58. MCCALL, B. *et al.* 1996. An outbreak *Escherichia coli* infection on the Gold Coast. Comm. Dis. Intell. *20*, 236–239.
59. MERMIN, J.H. and GRIFFIN, P.M. 1999. Invited commentary: Public Health in crisis: outbreaks of *Escherichia coli* O157:H7 infections in Japan. Am. J. Epidemiol. *150*, 797–803.
60. MICHINO, H., *et al.* 1998. Recent outbreaks of infections caused by *Escherichia coli* O157:H7 in Japan, *In Escherichia coli* O157:H7 and other Shiga toxin-producing *E. coli* strains. (J.B. Kaper and A.D. O'Brien, eds.) p. 73–81. American Society for Microbiology, Washington, D.C.
61. MICHINO, H. *et al.* 1999. Massive outbreak of *Escherichia coli* O157:H7 infection in schoolchildren in Sakai City, Japan, associated with consumption of white radish sprouts. Am. J. Epidemiol. *150*, 787–796.
62. MILFORD, D.V., TAYLOR, C.M., GUTTRIDGE, B., HALL, S.M., ROWE, B. and KLEANTHOUS, H. 1990. Haemolytic uraemic syndromes in the British Isles 1985-8: association with Verocytotoxin producing *Escherichia coli*. Part 1: clinical and epidemiological aspects. Arch. Dis. Child. *65*, 716–721.
63. MILLEY, D., COLBY, D. and FAST, M. 1995. Verotoxigenic *E. coli* (VTEC) serotypes associated with diarrhea in Manitoba, abstr. C2. Canadian Association for Clinical and Microbiology and Infectious Diseases Annual Meeting, Winnipeg, Manitoba, Canada.
64. MORGAN, G.M. *et al.* 1988. First recognized community outbreak of haemorrhagic colitis due to verotoxin-producing *Escherichia coli* O157:H7 in the UK. Epidemiol. Infect. *101*, 83–89.
65. OATES, K. 1999. Eurosurveillance Weekly, 1999 Aug 12, *3*. Escherichia coli O157 cluster in Scotland. (http://www.eurosurv.org/update/).
66. ORR, P. *et al.* 1994. An outbreak of diarrhea due to verotoxin-producing *Escherichia coli* in the Canadian Northwest Territories. Scand. J. Infect. Dis. *26*, 675–684.
67. PARRY, S.M. *et al.* 1995. Haemorrhagic colitis in child after visit to farm visitor centre. Lancet *346*, 572.
68. PATON, A.W. *et al.* 1996. Molecular microbiological investigation of an outbreak of hemolytic-uremic syndrome caused by dry fermented sausage contaminated with Shiga-like toxin-producing *Escherichia coli*. J. Clin. Microbiol. *34*, 1622–1627.
69. PAUNIO, M. *et al.* 1999. Swimming-associated outbreak of *Escherichia coli* O157:H7. Epidemiol. Infect. *122*, 1–5.
70. PHLS COMMUNICABLE DISEASE SURVEILLANCE CENTRE. 1999. VTEC O157 outbreak linked to beach holidays. Commun. Dis. Rep. Weekly *9*(37), 327–330 (http://www.phls.co.uk/publications/cdr.htm).
71. PIÉRARD, D. *et al.* 1999. Hemolytic uremic syndrome in Belgium: incidence and association with verocytotoxin-producing *Escherichia coli* infection. Clin. Microbiol. Infect. *5*, 16–22.

72. PRADO, V., MARTINEZ, J., ARELLANO, C. and LEVINE, M. 1997. Temporal variation of genotypes and serotypes of enterohemorrhagic *E. coli* isolated from Chilean children with asymptomatic infections or hemolytic uremic syndrome. Rev Méd Chile *125*, 291–297.
73. ProMED post, 1998 Jun 26. *E. coli* foodborne outbreak - USA (Indiana) (http://www.promedmail.org).
74. ProMED post, 1998 Jun 26. *E.coli*, hemolytic uremic syndrome - USA (Georgia) (http://www.promedmail.org)
75. ProMED post, 1998 Nov 9. *E. coli* (EHEC) - South Korea (Seoul) (http://www.promedmail.org).
76. ProMED post, 1999 Aug 15. *E. coli* infections, source identified - USA (Ohio) (http://www.promedmail.org).
77. ProMED post, 1999 Sep 15. *E. coli*, toxigenic, waterborne - USA (Washington) (03) (http://www.promedmail.org).
78. ProMED post, 1999 Sep 23. *E.coli,* toxigenic - USA (Illinois) (04) (http://www.promedmail.org).
79. ProMED post, 1999 Oct 16. *E.coli,* VTEC, cattle, petting zoo — Canada (Ontario) (03) (http://www.promedmail.org).
80. ProMED post, 1999 Oct 19. *E.coli*, toxigenic, apple cider - USA (Oklahoma) (http://www.promedmail.org).
81. ProMED post, 2000 Jun 7. *E.coli, EHEC* - Canada (Ontario) (06) (http://www.promedmail.org).
82. ProMED post, 2000 Jun 9. *E. coli* O157, petting zoo — USA (Washington) (http://www.promedmail.org).
83. REINGOLD, A.L. 1998. Outbreak investigations — a perspective. Emerg. Infect. Dis. *4*, 21–27.
84. RIOS, M. *et al.* 1999. Clonal diversity of Chilean isolates of enterohemorrhagic *Escherichia coli* from patients with hemolytic-uremic syndrome, asymptomatic subjects, animal reservoirs, and food products. J. Clin. Microbiol. *37*, 778–781.
85. RIVAS, M. 2000. Personal communication.
86. ROBINS-BROWNE, R.M., ELLIOTT, E. and DESMARCHELIER, P. 1998. Shiga toxin-producing *Escherichia coli* in Australia, *In Escherichia coli* O157:H7 and other Shiga toxin-producing *E. coli* strains. (J.B. Kaper and A.D. O'Brien, eds.) p. 66–72, American Society for Microbiology, Washington, D.C.
87. ROSA, A.C., MARIANO, A.T., PEREIRA, A.M., TIBANA, A., GOMES, T.A. and ANDRADE, J.R. 1998. Enteropathogenicity markers in *Escherichia coli* isolated from infants with acute diarrhoea and healthy controls in Rio de Janeiro, Brazil. J. Med. Microbiol. *47*, 781–790.
88. SHUKLA, R., SLACK, R., GEORGE, A., CHEASTY, T., ROWE, B. and SCUTTER, J. 1995. *Escherichia coli* O157 infection associated with a farm visitor centre. Commun. Dis. Rep. Rev. *5*, R86–90.
89. SLUTSKER, L. *et al.* 1997. *Escherichia coli* O157:H7 diarrhea in the United States: clinical and epidemiologic features. Ann. Intern. Med. *126*, 505–513.
90. SMITH, H.R., ROWE, B., ADAK, G.K. and REILLY, W.J. 1998. Shiga toxin (verocytotoxin)-producing *Escherichia coli* in the United Kingdom, *In Escherichia*

coli O157:H7 and other Shiga toxin-producing *E. coli* strains. (J.B. Kaper and A.D. O'Brien, eds.) p. 49–58, American Society for Microbiology, Washington, D.C.

91. SPIKA, J.S., KHAKHRIA, R., MICHEL, P., MILLEY, D., WILSON, J. and WATERS, J. 1998. Shiga toxin-producing *Escherichia coli* infections in Canada, *In Escherichia coli* O157:H7 and other Shiga toxin-producing *E. coli* strains. (J.B. Kaper and A.D. O'Brien, eds.) p. 23–29, American Society for Microbiology, Washington, D.C.
92. SWERDLOW, D.L. *et al.* 1992. A waterborne outbreak in Missouri of *Escherichia coli* O157:H7 associated with bloody diarrhea and death. Ann. Intern. Med. *117*, 812–819.
93. SZYCH, J. 2000. Personal communication.
94. TERAJIMA, J., IZUMIYA, H., WADA, A., TAMURA, K. and WATANABE, H. 1999. Shiga toxin-producing *Escherichia coli* O157:H7 in Japan. Emerg. Infect. Dis. *5*, 301–302.
95. TREVENA, W.B., WILLSHAW, G.A., CHEASTY, T., WRAY, C. and GALLAGHER, J. 1996. Vero cytotoxin-producing *E. coli* O157 infection associated with farms. Lancet *347*, 60–61.
96. UPTON, P. and COIA, J.E. 1994. Outbreak of *Escherichia coli* O157:H7 infection associated with pasteurised milk supply. Lancet *344*, 1015.
97. WANG, G., ZHAO, T. and DOYLE, M.P. 1996. Fate of enterohemorrhagic *Escherichia coli* O157:H7 in bovine feces. Appl. Environ Microbiol. *62*, 2567–2570.
98. WASTESON, Y. and LASSEN, J. 1998. Occurrence of *E. coli* O157:H7/H- in animals, foods and humans. IVC News 11. Not. Ist. Superiore di Sanità *11*(Suppl. 2), 1–2.
99. WATANABE, Y. *et al.* 1999. Factory outbreak of *Escherichia coli* O157:H7 infection in Japan. Emerg. Infect. Dis. *5*, 424–428.
100. WATERS, J.R., SHARP, J.C.M. and DEV, V.J. 1994. Infection caused by *Escherichia coli* O157:H7 in Alberta, Canada, and in Scotland: a five-year review, 1987-1991. Clin. Infect. Dis. *19*, 834–843.
101. ZIESE, T., ANDERSON, Y., DE JONG, B., LOFDHAL, S. and RAMBERG, M. 1996. Outbreak of *Escherichia coli* O157 in Sweden. Eurosurveillance *1*, 2–3.

CHAPTER 9

FOODS AS VEHICLES OF VTEC INFECTION

ENNE DE BOER
and
ANNET E. HEUVELINK

Inspectorate for Health Protection
P.O. Box 202
7200 AE Zutphen
The Netherlands

FOODS INVOLVED IN VTEC INFECTIONS

Verocytotoxin (VT)-producing *Escherichia coli* (VTEC) were first isolated as human pathogens in 1982, when *E. coli* strains of a previously rarely isolated serotype, O157:H7, were implicated in two outbreaks of haemorrhage colitis (HC) (93). Illness was associated with consumption of undercooked hamburgers at a fast-food restaurant chain. Since then, strains of *E. coli* O157:H7 and the non-motile variant *E. coli* O157:NM have emerged as among the most significant foodborne pathogens in developed countries, causing severe symptoms including the haemolytic-uraemic syndrome (HUS). Moreover, more than 100 different VTEC serotypes other than O157:H7 have now been recognised to cause human illness (103). Some VTEC serotypes found in food and animals, however, have never, or only rarely, been associated with severe human disease.

Although person-to-person contact is thought to play an important role in the transmission of these organisms, the primary mode of VTEC infections is foodborne transmission, and a variety of foods have been identified as vehicles of VTEC-associated illnesses. The true incidence of foodborne VTEC infections is difficult to estimate for the following main reasons: 1) In many clinical laboratories not all stool specimens are routinely screened for these organisms; 2) Many persons, presenting only mild symptoms, may not seek medical attention; 3) The source of infection is usually not investigated in sporadic cases; 4) Implicated foods are seldom available for microbiological examination; 5) Food histories are usually obtained several days/weeks after exposure; 6) Isolation of VTEC from incriminated foods and environmental specimens is difficult.

Foods of Bovine Origin

Consumption of raw or undercooked foods of bovine origin, especially undercooked ground beef and unpasteurised cow's milk, has been the most common means of transmitting VTEC organisms in sporadic cases and outbreaks of VTEC infection.

Beef Products. From epidemiological studies it could be concluded that the majority of outbreaks of VTEC O157 infections have resulted from transmission of the organism through the consumption of beef, most commonly undercooked ground beef. In studies on risk factors, sporadic infections with VTEC O157 have been associated with consumption of hamburgers (73,82). Other beef products associated with VTEC O157 infections include roast beef (95) and precooked meat patties (20).

Meat becomes contaminated at slaughter, where faecal contamination of the carcasses may occur. Subsequent grinding may introduce the pathogen into the interior of the meat, where VTEC may survive during uneven or inadequate cooking, i.e., if the product does not achieve the required internal temperature (>68°C).

Table 1 shows published reports of outbreaks associated with ground beef. The largest outbreak of VTEC O157 infection in the United States occurred between November 1992 and February 1993 in the western states and included more than 700 infections and 4 deaths. The outbreak was associated with eating hamburgers at restaurants of one fast-food chain (19,33,107). Investigation determined that the ground beef patties epidemiologically linked with cases were produced during specific hours at a meat processing plant (107). Cultures of recalled ground beef patty samples yielded VTEC O157 strains indistinguishable from those isolated from patients. This outbreak resulted in a large increase in studies on the sources, transmission and survival of VTEC O157, and had a dramatic impact on consumer attitudes towards the safety of foods.

Recently large quantities of a commercial brand of frozen ground beef patties and burgers were recalled because of association with VTEC O157 infection (7), demonstrating the important economical losses which may result from the contamination of foods.

Milk and Dairy Products. Raw and improperly pasteurised cow's milk have been implicated in a number of outbreaks of VTEC O157 infection (Table 2). A milkborne outbreak caused by VTEC serotype O104:H21 has been also described (2). These outbreaks were mostly linked with the consumption of milk or milk products from local farms.

TABLE 1.
OUTBREAKS OF VTEC O157 ASSOCIATED WITH GROUND BEEF

Year	Country	Setting	Number of cases	Reference
1982	USA	Restaurant	47	93
1984	USA	Institution	34	96
1986	USA	Restaurant	37	80
1988	USA	Community	61	95
1988	USA	Institution	51	83
1993	USA	Restaurant	>700	19,33
1993	Wales	Community	8	113
1993	USA	Community	20	94
1994	Scotland	Community	22	38
1995	USA	Restaurant	10	5
1997	USA	Retail shops	15	7

TABLE 2.
OUTBREAKS OF VTEC O157 ASSOCIATED WITH MILK AND DAIRY PRODUCTS

Year	Product	Country	Number of cases	Reference
1986	Raw milk	Canada	43	25
1993	Raw milk	USA	14	62
1994	Pasteurised milk	Scotland	>100	108
1995	Goat's milk	Czech Republic	4	22
1996	Raw milk	England	3	6
1996	Pasteurised milk	England	9	34
1998	Unpasteurised cream	England	7	10
1999	Pasteurised milk	England	60	11
1999	Unpasteurised cheese	England	3	13

Foods of Non-bovine Origin

Besides foods of bovine origin many other food products have been mentioned in relation to VTEC infections.

Meat and Meat Products (Non-beef). Several outbreaks of VTEC O157 infection have been associated with small manufacturing butcher-retailers who handle both raw and cooked meats within the shop and who prepare cooked meats for retail customers without taking sufficient hygienic measures to prevent cross-contamination. Improper handling and inadequate cooking of any meat can increase the risk of infection.

A case control study showed infection (14 cases) with *E. coli* O157 to be associated with consumption of meats from a particular shop. Environmental investigations at the shop showed that cooked meats were stored close to raw beef, raw meats were left on preparation surfaces, and the same staff served both raw and cooked meats (102). An outbreak of VTEC O157 infection affecting 26 people was associated with having eaten cold meats in sandwiches supplied by one wholesaler (72). Turkey-roll sandwiches were linked with a party outbreak of HC and HUS caused by VTEC O157 (97). Seven cases of VTEC O157 infection were associated with meat from one butchers shop. VTEC O157 isolates from stools of the cases, from samples of raw meat and from equipment and surfaces in the butcher shop were indistinguishable by pulsed field gel electrophoresis, demonstrating cross contamination of the environment (12). A large VTEC O157 outbreak in 1996 in Scotland was associated with cooked meat products and resulted in 490 cases of infection and 18 deaths (27). The Pennington Group report (84), published after that outbreak, recommends that hazard analysis procedures should be documented, staff trained in food hygiene, and raw meat and unwrapped cooked meat products physically separated, and wherever possible handled by separate staff.

Few outbreaks associated with game have been reported. An outbreak of VTEC O157 infection in humans caused by contaminated deer meat was reported in Japan (18). Home-made jerky from deer meat was implicated as the source of transmission in a community outbreak of VTEC O157 infection in the United States including 6 confirmed and 5 presumptive cases. The drying conditions applied were shown to be insufficient to eliminate all VTEC organisms. As wild deer may be contaminated with VTEC (18,106), venison consumption, contact with deer, and contamination with deer faeces should be considered potential sources of VTEC O157. Game should be handled with the same caution recommended for other meats (63).

Dry fermented meats were also involved in reported outbreaks of VTEC infection. In the western United States an outbreak was linked with consumption of contaminated dry-cured meats. The outbreak involved 23 cases, including two children who developed HUS. *E. coli* O157:H7 was found on pre-sliced product from delicatessen counters (3). In early 1995, 23 children were hospitalised with HUS in South Australia (4). Most required dialysis; one 4-year-old child died. Investigation of the outbreak identified *E. coli* O111:NM as the principal cause and a locally produced semi-dry sausage ("mettwurst") as the source. A sorbitol-positive strain of *E. coli* O157:H- caused HUS in 28 children in Germany. A case-control study showed a correlation between consumption of two sausages, mortadella (cooked) and teewurst (fermented, containing beef), and illness (16).

Raw meat used for the manufacturing of fermented meat products may be contaminated with pathogenic micro-organisms, including VTEC. To suppress the growth of these pathogens in fermented meat products, sufficient reductions in pH and a_w must be achieved during the early stages of fermentation and drying. The addition of starter cultures and the application of the Hazard Analysis Critical Control Point (HACCP) concept are essential instruments for the production of a safe product. Largely as the result of the VTEC O157 outbreak in the western United States, processors of dry and semidry fermented sausage have been required by the United States Department of Agriculture and the Food Safety and Inspection Service to validate that processing of these products results in at least a 5 $\log_{10}$ reduction in counts of VTEC O157 (90).

Vegetables and Fruit Juices. In the past few years, fruits and vegetables have accounted for a growing number of recognised outbreaks. Contamination of fresh produce may be due to contact with faeces from domestic or wild animals at some stage during cultivation or handling. Contamination in the kitchen may be due to cross-contamination from meat products. Examples of vegetables that have been implicated in foodborne outbreaks of VTEC O157 infection include fresh potatoes (78) and lettuce (15,75). In the outbreak related to potatoes, a strong association was shown with preparing raw vegetables, especially potatoes. The handling of these contaminated foods followed by hand-to-mouth transmission was suspected as the route of infection (78).

Outbreaks of *E. coli* O157:H7 infection in the United States were associated with eating alfalfa sprouts grown from one seed lot (8). Because alfalfa seeds are a raw agricultural commodity, they can become contaminated with animal faeces that may harbour pathogens such as *Salmonella* or VTEC O157 during growth, harvest, processing, storage, shipping, or sprouting. The number of organisms present on seeds may increase during the sprouting process.

Radish sprouts in mass-prepared school lunches appeared to be the most likely source of the largest outbreak of VTEC infection yet reported, which occurred in Sakai, Japan, in 1996 and involved over 6,000 cases of HC and more than 100 cases of HUS (77). Several other outbreaks and sporadic cases reported in the same period in Japan might be closely related to the large outbreak in Sakai (105,111). In experimentally contaminated radish seeds, *E.*

coli O157:H7 proliferated 10^3- to 10^5-fold at an early stage of plant growth, including germination. The bacteria were located in edible parts of radish sprouts, not only on the outer surfaces but also in the inner tissues and stomata of cotyledons of radish sprouts (51,59). It was also demonstrated that the edible parts of radish sprouts could be contaminated when the roots were immersed in a bacterial suspension (51). In addition to seed decontamination, prevention of future sprout-related outbreaks will depend on identification of critical control points to reduce the likelihood of contamination during seed production and distribution. Consumers should thoroughly rinse sprouts before eating them.

Although low pH-fruit juices generally will not allow the survival and growth of Enterobacteriaceae, VTEC O157 may persist in these products because of its higher acid-tolerance. In the United States, outbreaks of VTEC O157 infection associated with unpasteurised commercial apple juice and cider have been reported (9,21,35,100). Although the exact mechanisms of contamination for these outbreaks were not clearly determined, manure was suspected to have contaminated the apples. As cattle, sheep, and deer grazing in orchards can carry VTEC O157, the manure of these animals can also contain this pathogen. Recommendations to reduce the risk of producing contaminated juice, include: 1) Preventing the introduction of animal manure into orchards; 2) Avoiding use of apples that have fallen to the ground; 3) Washing and brushing apples before pressing; 4) Using a preservative such as sodium benzoate; 5) Routine pasteurisation (9). The Food and Drug Administration proposed new regulations requiring unpasteurised juices to carry a "warning label" and requiring application of HACCP principles to fruit and vegetable juice processing (35).

As cantaloupe and watermelon are occasionally involved in foodborne illnesses, including VTEC O157 infection, it is important to prevent surface contamination of these fruits in the field, during transport and marketing, and in restaurants. The external surfaces should be thoroughly washed before cutting with sanitised utensils (40).

Other Foods. A number of other ready-to-eat foods may occasionally get contaminated with VTEC O157 by contact with contaminated raw meats, especially during preparation of meals in restaurants and homes.

Mayonnaise and mayonnaise-based dressings and sauces were the most likely foods to have been contaminated, possibly by cross-contamination with meat products, in a series of outbreaks of VTEC O157 infection in the United States (88,112). Survival studies indicated that mayonnaise could harbour VTEC, when stored at refrigeration temperatures (112) and that its inherent acidity is insufficient to ensure the inhibition of growth of *E. coli* O157 in prepared food (88).

Salads, held over for several days, were identified as presenting a high risk during food analysis in an outbreak attributable to VT 1-producing *E. coli* O118:H2 at a school in Japan. The outbreak strain was isolated from a dipper suggesting poor hygienic conditions (52). Sandwiches were probably the source in an outbreak of VTEC O157 infection in a nursing home (30).

Prevalence of VTEC in Foods

There is a need for better information on the current prevalence of VTEC in foods. This information is important to identify the potential contribution of particular foods to the development of VTEC infections, and to allow documentation of risk assessment methods. It is also important within quantification of the effects of strategies for reduction of contamination levels and the provision of accurate advice to consumers, industries and governments. Several prevalence studies have been reported (Table 3-6). However, the potential to compare published prevalence data is limited because of the use of a wide variety of screening methods, and the different areas and settings in which the samples were taken.

VTEC O157

Most studies on the prevalence of VTEC in foods concern VTEC O157, because techniques used are mainly focused on the detection of this serotype.

Beef Products. The most important animal species in terms of human infection is cattle. Reported estimates of the average prevalence of VTEC O157 in cattle range from 0 to almost 10% (54). Beef may become contaminated if intestinal contents contact the meat during slaughter or processing. Thus, if present, VTEC will remain on the external surface and not penetrate into the interior of beef muscle. Heating of the surface to temperatures sufficient to produce a "cooked colour" will eliminate the organisms. Consequently, whole cuts of meat are usually not associated with outbreaks.

However, bacteria present on the surface are distributed throughout the meat by grinding. Since ground beef may include meat from many carcasses, a small number of infected animals can contaminate a large supply of ground beef. Contaminated beef may transfer VTEC O157 to meat grinding equipment, which may later contaminate other lots of raw meat (44). Minced beef products, therefore, pose a greater hazard than intact joints of meat.

Several prevalence studies have reported generally low isolation frequencies of VTEC O157 from beef products, indicating that this pathogen is an occasional contaminant in such products (Table 3).

Reported numbers of the organism in foods are low. The median most probable number of *E. coli* O157 in recalled ground beef patties was 1.5 per gram (range, <0.3-15) or 67.5 organisms per patty (range, <13.5-675) (107). In another study of naturally contaminated samples of meats associated with human cases, numbers of VTEC O157 ranging from <0.3 to 2300 cfu/g were found, confirming that the infective dose for this organism is low (24).

TABLE 3.
PREVALENCE OF VTEC O157 IN RAW BEEF PRODUCTS

Product	Country	Number positive / number analysed (%)	Reference
Ground beef	USA	6/164 (3.7)	42
Tenderloin beef	Malaysia	9/25 (36.0)	87
Ground beef	USA	3/107 (2.8)	81
Ground beef	Ireland	1/586 (0.2)	110
Ground beef	Norway	0/1319 (0)	67
Ground beef	Canada	4/165 (2.4)	1
Ground beef	Egypt	3/50 (6.0)	14
Ground beef	The Netherlands	6/571 (1.1)	57
Beefburgers	UK	3/1015 (0.3)	70
Beef	USA	2/1668 (0.1)	49
Ground beef	Spain	3/58 (5.2)	23
Ground beef	USA	1/76 (0.3)	64
Ground beef	Italy	9/564 (1.6)	36
Ground beef	USA	0/563 (0)	45
Beef products	UK	25/1631 (1.5)	32

Meat and Meat Products (Non-beef). Although cattle are the main source of VTEC O157, this organism can also be found in the faecal flora of many other domestic animals, including sheep (54), pigs (58), turkeys (58) and horses (55) and consequently the meat of these animals may get contaminated with this organism. Table 4 shows the results of prevalence studies of VTEC O157 in meats other than beef.

VTEC O157 has occasionally been isolated from samples of fermented meat, such as salami (43) and dry fermented pork sausage (57).

Milk and dairy products. Raw milk is often contaminated with enteric organisms during collection and therefore consumption may result in a direct risk for those consumers choosing to drink raw milk. Effective pasteurisation eliminates all pathogens from milk, including VTEC O157 (37). However, post-pasteurisation contamination (108) may occur.

TABLE 4.
PREVALENCE OF VTEC O157 IN MEAT AND MEAT PRODUCTS (NON-BEEF)

Product	Country	Number positive / number analysed (%)	Reference
Raw pork	USA	4/264 (1.5)	42
Raw poultry	USA	4/263 (1.5)	42
Raw lamb	USA	4/205 (2.0)	42
Raw poultry	USA	0/3977 (0)	49
Raw chicken	Egypt	2/50 (4.0)	14
Raw lamb	Egypt	1/25 (4.0)	14
Raw minced pork/beef	The Netherlands	2/402 (0.5)	57
Raw minced pork	The Netherlands	1/76 (1.3)	57
Raw pork products	The Netherlands	1/393 (0.3)	57
Meat products	The Netherlands	1/328 (0.3)	57
Raw beef products	The Netherlands	0/223 (0)	57
Raw poultry	The Netherlands	0/819 (0)	57
Raw lamb	The Netherlands	0/46 (0)	57
Raw game	The Netherlands	0/83 (0)	57
Raw sausage	UK	0/1197 (0)	70
Raw meat products	UK	0/118 (0)	70
Cooked meats	UK	0/2192 (0)	70
Raw meats	UK	1/183 (0.5)	69
Raw lamb	UK	26/431 (5.9)	32
Salami	Italy	0/70 (0)	36

VTEC O157 is occasionally isolated from raw milk (Table 5), but usually not from cheeses (26,66,68).

Failure to detect VTEC O157 in milk samples taken from bulk tanks may be due to dilution of contaminated samples from individual cattle to a level of contamination undetectable by most available assays. During two investigations of dairy herds associated with VTEC O157 infection in humans, the organisms could not be isolated from milk from bulk tanks, but were isolated from milk samples from individual cattle (74,114).

Other foods. Very few data are available on the prevalence of VTEC O157 in foods other than meats and dairy products.

Fruits and vegetables may get contaminated from the environment by flies (60,61), bird droppings (109) or by faeces of domestic and feral animals (92). Consumption of food produced and processed in rural areas may be a risk factor for VTEC infections (76).

TABLE 5.
PREVALENCE OF VTEC O157 IN RAW MILK

Country	Number positive / number analysed (%)	Reference
USA	11/115 (9.6)	81
USA	0/603 (0)	50
USA	0/42 (0)	17
Egypt	3/50 (6.0)	14
The Netherlands	0/1011 (0)	53
Germany	1/273 (0.4)	65
Germany	0/245 (0)	68
UK	0/586 (0)	79
Switzerland	0/93 (0)	29
Australia	0/147 (0)	41
Italy	0/227 (0)	36
Italy	0/100 (0)	71

All types of food may occasionally become contaminated with VTEC O157, by cross-contamination with beef or other meats, contaminated surfaces and kitchen utensils during food preparation in the kitchen (91). Home barbecuing may also contribute to the incidence of such infections.

VTEC non-O157

Techniques used to identify VTEC non-O157 are not widely available in most clinical laboratories, consequently infections caused by these pathogens are most likely to be unrecognised.

VTEC not belonging to serotype O157 but to a wide variety of other serotypes, have been isolated from many types of foods (Table 6). Potential virulence factors, other than the ability to produce VT were often not detected in isolated strains, suggesting that many of these VTEC isolates are not pathogenic (86).

Since several countries have reported an increasing number of outbreaks and sporadic cases related to VTEC non-O157 (48), more research is required on the distribution of VT genes among *E. coli* serotypes, and the pathogenicity of those VTEC serotypes isolated from foods, animals and the environment which have not yet been associated with human disease.

TABLE 6.
PREVALENCE OF VTEC NON-O157 IN FOODS

Product	Country	Number positive/number analysed (%)	Reference
Raw beef	Canada	82/225 (36.4)	89
Raw beef	Brazil	20/105 (19.0)	31
Raw beef	Germany	4/567 (0.7)	28
Raw beef	USA	14/60 (23.3)	98
Raw beef	Thailand	8/93 (8.6)	104
Raw beef	Switzerland	1/30 (3.3)	47
Raw beef sausage	Germany	14/168 (8.3)	46
Raw milk	Germany	22/180 (12.2)	46
Raw milk	Canada	15/1720 (0.9)	101
Raw milk cheese	Germany	1/55 (1.8)	28
Raw lamb	Germany	2/78 (2.6)	28
Raw lamb	USA	10/21 (47.6)	98
Raw pork	USA	9/51 (17.6)	98
Raw pork	Thailand	1/111 (0.9)	104
Raw pork	Canada	24/235 (10.6)	89
Raw pork sausage	UK	46/184 (25.0)	99
Raw chicken	UK	0/71 (0)	99
Raw chicken	USA	4/33 (12.1)	98
Raw chicken	Thailand	1/107 (0.9)	104
Raw deer meat	Germany	4/53 (7.5)	106
Raw minced meat	Germany	17/138 (12.3)	46
Raw meats	The Netherlands	29/180 (16.1)	56
Raw meats	Belgium	67/2440 (2.8)	86
Raw fish	USA	6/62 (9.7)	98
Raw vegetables	Thailand	0/130 (0)	104

CONCLUSIONS

Sources of infection are not identified in a substantial proportion of cases of illness due to VTEC. Continuing epidemiological studies should be conducted to assist in more frequent establishment of the sources of VTEC infections, including identification of the causative food(s) in common-source outbreaks.

VTEC infection can be very serious, or fatal, and consequently such illness must be prevented. Most VTEC O157 infections in humans are foodborne and contamination of beef with VTEC O157 during slaughter is the principal route by which this pathogen enters the human food supply. Thus, reduction of VTEC O157 contamination of beef products will reduce the risk of infection to consumers. Such reductions of the potential risks of consuming contaminated beef products has to be achieved by a combination of good hygiene practices by both producers, and consumers, of beef products. The former requires a combination of actions throughout the whole meat production chain, while the latter is a matter of education and providing adequate consumer information on proper handling of foods. In broader terms, total prevention of VTEC O157 colonisation in cattle at the farm level would be the most effective way to reduce contamination of beef products. Other means of limiting risks include processing improvements, at the slaughterhouse level and in the treatment of end-products. For example, regulations in many countries require that all visible faeces be removed from the surface of beef carcasses during slaughter, a policy often referred to as "zero tolerance". Combinations of decontamination treatments can achieve an effective reduction of bacteria on beef carcasses (85).

As the present modes of production and distribution of raw meat are inadequate, those preparing foods should realise that raw products, especially those of bovine origin, may be contaminated with VTEC O157 and other pathogens.

Because very small numbers of VTEC O157 are sufficient to cause infection, it is essential that all cells of this pathogen in foods are killed, and processes that merely reduce their numbers may be inadequate in terms of protecting consumers. Similarly, the very low infective dose of this pathogen means that even minor incidents of cross-contamination may lead to illness. Thus all transfer from raw to ready-to-eat foods must be prevented.

The most important control measures to ensure safety of milk and milk products are proper pasteurisation, and prevention of post-pasteurization contamination. The potential presence of VTEC and other pathogens in raw milk supports the wider argument for banning the sale of raw milk.

Sensitive recovery methods, using specialised culture media, immunological test systems or nucleic acid based tests are available for the detection of VTEC O157. These methods should be used in continuing surveillance programmes aiming to establish the prevalence of this organism in foods and study its routes of transmission to man (39). Coliform counts may be useful in screening batches for the presence of VTEC O157 and may be useful in a HACCP programme for ground beef (107).

Although no governmental microbiological food safety standards for VTEC O157 currently exist in most countries, "zero tolerance" for VTEC in processed foods must be enforced (107).

REFERENCES

1. ANON. 1990. Verotoxin-producing *Escherichia coli* in ground beef in Manitoba. CMAJ *143*, 519–521.
2. ANON. 1995. Outbreak of acute gastroenteritis attributable to *Escherichia coli* serotype O104:H21 — Helena, Montana, 1994. MMWR Morb. Mortal. Wkly. Rep. *44*, 501–503.
3. ANON. 1995. *Escherichia coli* O157:H7 outbreak linked to commercially distributed dry-cured salami — Washington and California, 1994. MMWR Morb. Mortal. Wkly. Rep. *44*, 157–160.
4. ANON. 1995. Community outbreak of hemolytic uremic syndrome attributable to *Escherichia coli* O111:NM — South Australia 1995. MMWR Morb. Mortal. Wkly. Rep. *44*, 550–1, 557–8.
5. ANON. 1996. Outbreak of *Escherichia coli* O157:H7 infection — Georgia and Tennessee, June 1995. MMWR Morb. Mortal. Wkly. Rep. *45*, 249–251.
6. ANON. 1996. VTEC O157 infection in West Yorkshire associated with the consumption of raw milk. Commun. Dis. Rep. CDR Wkly. *6*, 181.
7. ANON. 1997. *Escherichia coli* O157:H7 infections associated with eating a nationally distributed commercial brand of frozen ground beef patties and burgers — Colorado, 1997. MMWR Morb. Mortal. Wkly. Rep. *46*, 777–778.
8. ANON. 1997. From the Centers for Disease Control and Prevention. Outbreaks of *Escherichia coli* O157:H7 infection associated with eating alfalfa sprouts — Michigan and Virginia, June–July 1997. MMWR Morb. Mortal. Wkly. Rep. *46*, 741–744.
9. ANON. 1997. Outbreaks of *Escherichia coli* O157:H7 infection and cryptosporidiosis associated with drinking unpasteurized apple cider — Connecticut and New York, October 1996. MMWR Morb. Mortal. Wkly. Rep. *46*, 4–8.
10. ANON. 1998. VTEC O157 infection and unpasteurised cream — update. Commun. Dis. Rep. CDR Wkly. *8*, 389, 392.
11. ANON. 1999. VTEC O157 phage type 21/28 infection in North Cumbria: update. Commun. Dis. Rep. CDR Wkly. *9*, 105.
12. ANON. 1999. Outbreak of VTEC O157 infection in East Sussex. Commun. Dis. Rep. CDR Wkly. *9*, 219, 222.
13. ANON. 1999. *Escherichia coli* O157 associated with eating unpasteurised cheese — update. Commun. Dis. Rep. CDR Wkly. *9*, 131, 134.
14. ABDUL RAOUF, U.M., AMMAR, M.S. and BEUCHAT, L.R. 1996. Isolation of *Escherichia coli* O157:H7 from some Egyptian foods. Int. J. Food Microbiol.. *29*, 423–426.
15. ACKERS, M.L. *et al.* 1998. An outbreak of *Escherichia coli* O157:H7 infections associated with leaf lettuce consumption. J. Infect. Dis. *177*, 1588–1593.
16. AMMON, A., PETERSEN, L.R. and KARCH, H. 1999. A large outbreak of hemolytic uremic syndrome caused by an unusual sorbitol-fermenting strain of *Escherichia coli* O157:H-. J. Infect. Dis. *179*, 1274–1277.

17. ANSAY, S.E. and KASPAR, C.W. 1997. Survey of retail cheeses, dairy processing environments and raw milk for *Escherichia coli* O157:H7. Lett. Appl. Microbiol. *25*, 131–134.
18. ASAKURA, H. *et al.* 1998. Detection and genetical characterization of Shiga toxin-producing *Escherichia coli* from wild deer. Microbiol. Immunol. *42*, 815–822.
19. BELL, B.P. *et al.* 1994. A multistate outbreak of *Escherichia coli* O157:H7-associated bloody diarrhea and hemolytic uremic syndrome from hamburgers. The Washington experience. JAMA. *272*, 1349–1353.
20. BELONGIA, E.A. *et al.* 1991. An outbreak of *Escherichia coli* O157:H7 colitis associated with consumption of precooked meat patties. J. Infect. Dis. *164*, 338–343.
21. BESSER, R.E. *et al.* 1993. An outbreak of diarrhea and hemolytic uremic syndrome from *Escherichia coli* O157:H7 in fresh-pressed apple cider [see comments]. JAMA. *269*, 2217–2220.
22. BIELASZEWSKA, M. *et al.* 1997. Human *Escherichia coli* O157:H7 infection associated with the consumption of unpasteurized goat's milk. Epidemiol. Infect. *119*, 299–305.
23. BLANCO, J.E. *et al.* 1996. Detection of enterohaemorrhagic *Escherichia coli* O157:H7 in minced beef using immunomagnetic separation. Microbiologia. *12*, 385–394.
24. BOLTON, F.J., CROZIER, L. and WILLIAMSON, J.K. 1996. Isolation of *Escherichia coli* O157 from raw meat products. Lett. Appl. Microbiol. *23*, 317–321.
25. BORCZYK, A.A., KARMALI, M.A., LIOR, H. and DUNCAN, L.M. 1987. Bovine reservoir for verotoxin-producing *Escherichia coli* O157: H7 [letter]. Lancet *1*, 98.
26. BOWEN, D.A. and HENNING, D.R. 1994. Coliform bacteria and *Staphylococcus aureus* in retail natural cheeses. J. Food Prot. *57*, 253–255.
27. BRADBURY, J. 1997. Report on Scottish *E. coli* outbreak released. Lancet *349*, 1073.
28. BULTE, M., HECKOTTER, S. and SCHWENK, P. 1996. Enterohemorrhagic *E. coli* strains (EHEC)-topicality in the Federal Republic of Germany. 2. Incidence of VTEC strains in foods of animal origin. Fleischwirtschaft. *76*, 88–91.
29. BURNENS, A.P., FREY, A., LIOR, H. and NICOLET, J. 1995. Prevalence and clinical significance of verocytotoxin-producing *Escherichia coli* (VTEC) isolated from cattle in herds with and without calf diarrhoea. J. Vet. Med. *B42*, 311–318.
30. CARTER, A.O. *et al.* 1987. A severe outbreak of *Escherichia coli* O157:H7-associated hemorrhagic colitis in a nursing home. N. Engl. J. Med. *317*, 1496–1500.
31. CERQUEIRA, A.M.F., TIBANA, A. and GUTH, B.E.C. 1997. High occurrence of Shiga-like toxin-producing strains among diarrheagenic *Escherichia coli* isolates from raw beef products in Rio de Janeiro City, Brazil. J. Food Prot. *60*, 177–180.

32. CHAPMAN, P.A., SIDDONS, C.A., MALO, A.T. and HARKIN, M.A. 1996. Lamb products as a potential source of *E. coli* O157 [letter]. Vet. Rec. *139*, 427–428.
33. CIESLAK, P.R. *et al.* 1997. Hamburger-associated *Escherichia coli* O157:H7 infection in Las Vegas: a hidden epidemic. Am. J. Public Health. *87*, 176–180.
34. CLARK, A., MORTON, S., WRIGHT, P., CORKISH, J., BOLTON, F.J. and RUSSELL, J. 1997. A community outbreak of Vero cytotoxin producing *Escherichia coli* O157 infection linked to a small farm dairy. Commun. Dis. Rep. CDR Rev. 7, R206–11.
35. CODY, S.H. *et al.* 1999. An outbreak of *Escherichia coli* O157:H7 infection from unpasteurized commercial apple juice. Ann. Intern. Med. *130*, 202–209.
36. COLOMBO, S., PACCIARINI, M.L. and FUSI, P. 1998. Isolation of a new phenotypic variant of *E coli* O157:H7 from food. Vet. Rec. *142*, 144–145.
37. D'AOUST, J.Y., PARK, C.E., SZABO, R.A., TODD, E.C., EMMONS, D.B. and MCKELLAR, R.C. 1988. Thermal inactivation of *Campylobacter* species, *Yersinia enterocolitica*, and hemorrhagic *Escherichia coli* O157:H7 in fluid milk. J. Dairy Sci. *71*, 3230–3236.
38. DAVIS, B.S. and BROGAN, R.T. 1995. A widespread community outbreak of *E. coli* O157 infection in Scotland. Public Health *109*, 381–388.
39. DE BOER, E. and HEUVELINK, A.E. 2000. Methods for the detection and isolation of Shiga toxin-producing *Escherichia coli*. J. Appl. Microbiol; Symp. Supplement *88*, 1335–1435.
40. DEL ROSARIO, B.A. and BEUCHAT, L.R. 1995. Survival and growth of enterohemorrhagic *Escherichia coli* O157:H7 in cantaloupe and watermelon. J. Food Prot. *58*, 105–107.
41. DESMARCHELIER, P.M., BILGE, S.S., FEGAN, N., MILLS, L., VARY JR., J.C. and TARR, P.I. 1998. A PCR specific for *Escherichia coli* O157 based on the rfb locus encoding O157 lipopolysaccharide. J. Clin. Microbiol. *36*, 1801–1804.
42. DOYLE, M.P. and SCHOENI, J.L. 1987. Isolation of *Escherichia coli* O157:H7 from retail fresh meats and poultry. Appl. Environ. Microbiol. *53*, 2394–2396.
43. DUFFY, G. *et al.* 1999. Differences in thermotolerance of various *Escherichia coli* O157:H7 strains in a salami matrix. Food Microbiol. *16*, 83–91.
44. FARRELL, B.L., RONNER, A.B. and WONG, A.C. 1998. Attachment of *Escherichia coli* O157:H7 in ground beef to meat grinders and survival after sanitation with chlorine and peroxyacetic acid. J. Food Prot. *61*, 817–822.
45. FOOD SAFETY AND INSPECTION SERVICE (FSIS). 1996. National federal plant raw ground beef microbial survey August 1993-March 1994. Washington, D.C., United States: Department of Agriculture.
46. GALLIEN, P. *et al.* 1998. Detection of Shiga toxin-producing *Escherichia coli* (STEC) in foods and characterization of isolates. Bundesgesundhbl. *41*, 26–30.
47. GILGEN, M., HUBNER, P., HOFELEIN, C., LUTHY, J. and CANDRIAN, U. 1998. PCR-based detection of verotoxin-producing *Escherichia coli* (VTEC) in ground beef. Res. Microbiol. *149*, 145–154.

48. GOLDWATER, P.N. and BETTELHEIM, K.A. 1998. New perspectives on the role of *Escherichia coli* O157:H7 and other enterohaemorrhagic *E. coli* serotypes in human disease. J. Med. Microbiol. *47*, 1039–1045.
49. GRIFFIN, P.M. and TAUXE, R.V. 1991. The epidemiology of infections caused by *Escherichia coli* O157: H7, other enterohemorrhagic *E. coli*, and the associated hemolytic uremic syndrome. Epidemiol Rev. *13*, 60–98.
50. HANCOCK, D.D., BESSER, T.E., KINSEL, M.L., TARR, P.I., RICE, D.H. and PAROS, M.G. 1994. The prevalence of *Escherichia coli* O157.H7 in dairy and beef cattle in Washington State. Epidemiol. Infect. *113*, 199–207.
51. HARA KUDO, Y. *et al.* 1997. Potential hazard of radish sprouts as a vehicle of *Escherichia coli* O157:H7. J. Food Prot. *60*, 1125–1127.
52. HASHIMOTO, H. *et al.* 1999. Epidemic of gastrointestinal tract infection including hemorrhagic colitis attributable to Shiga toxin 1-producing *Escherichia coli* O118:H2 at a junior high school in Japan. Pediatrics *103*, E2.
53. HEUVELINK, A.E., BLEUMINK, B., VAN DEN BIGGELAAR, F.L., TE GIFFEL, M.C., BEUMER, R.R. and DE BOER, E. 1998. Occurrence and survival of verocytotoxin-producing *Escherichia coli* O157 in raw cow's milk in The Netherlands. J. Food Prot. *61*, 1597–1601.
54. HEUVELINK, A.E. *et al.* 1998. Isolation and characterization of verocytotoxin-producing *Escherichia coli* O157 strains from Dutch cattle and sheep. J. Clin. Microbiol. *36*, 878–882.
55. HEUVELINK, A.E. *et al.* 1998. Occurrence of verocytotoxin-producing *Escherichia coli* O157 on Dutch dairy farms. J. Clin. Microbiol. *36*, 3480–3487.
56. HEUVELINK, A.E., WERNARS, K. and DE BOER, E. 1996. Occurrence of *Escherichia coli* O157 and other verocytotoxin-producing *E. coli* in retail raw meats in the Netherlands. J. Food Prot. *59*, 1267–1272.
57. HEUVELINK, A.E., ZWARTKRUIS NAHUIS, J.T.M., BEUMER, R.R. and DE BOER, E. 1999. Occurrence and survival of verocytotoxin-producing *Escherichia coli* O157 in meats obtained from retail outlets in the Netherlands. J. Food Prot. *62*, 1115–1122.
58. HEUVELINK, A.E., ZWARTKRUIS NAHUIS, J.T.M., VAN DEN BIGGELAAR, F.L.A.M., VAN LEEUWEN, W.J. and DE BOER, E. 1999. Isolation and characterization of verocytotoxin-producing *Escherichia coli* O157 from slaughter pigs and poultry. Int. J. Food Microbiol. *52*, 67–75.
59. ITOH, Y. *et al.* 1998. Enterohemorrhagic *Escherichia coli* O157:H7 present in radish sprouts. Appl. Environ. Microbiol. *64*, 1532–1535.
60. IWASA, M., MAKINO, S., ASAKURA, H., KOBORI, H. and MORIMOTO, Y. 1999. Detection of *Escherichia coli* O157:H7 from Musca domestica (Diptera: Muscidae) at a cattle farm in Japan. J. Med. Entomol. *36*, 108–112.
61. JANISIEWICZ, W.J., CONWAY, W.S., BROWN, M.W., SAPERS, G.M., FRATAMICO, P. and BUCHANAN, R.L. 1999. Fate of *Escherichia coli* O157:H7 on fresh-cut apple tissue and its potential for transmission by fruit flies. Appl. Environ. Microbiol. *65*, 1–5.

62. KEENE, W.E. *et al.* 1997. A prolonged outbreak of *Escherichia coli* O157:H7 infections caused by commercially distributed raw milk. J. Infect. Dis. *176*, 815–818.
63. KEENE, W.E. *et al.* 1997. An outbreak of *Escherichia coli* O157:H7 infections traced to jerky made from deer meat. JAMA. *277*, 1229–1231.
64. KIM, M.S. and DOYLE, M.P. 1992. Dipstick immunoassay to detect enterohemorrhagic *Escherichia coli* O157:H7 in retail ground beef. Appl. Environ. Microbiol. *58*, 1764–1767.
65. KLIE, H., TIMM, M., RICHTER, H., GALLIEN, P., PERLBERG, K.W. and STEINRUCK, H. 1997. [Detection and occurrence of verotoxin-forming and/or shigatoxin producing *Escherichia coli* (VTEC and/or STEC) in milk] Nachweis und Vorkommen von Verotoxin- bzw. Shigatoxin-bildenden *Escherichia coli* (VTEC bzw. STEC) in Milch. Berl Munch Tierarztl Wochenschr. *110*, 337–341.
66. KNAPPSTEIN, K., HAHN, G. and HEESCHEN, W. 1996. Untersuchungen zum vorkommen von verotoxin-bildenden *Escherichia coli* (VTEC) in weichkase. Archiv Lebensmittelhyg. *47*, 59–61.
67. KRUSE, H. and AAS, N.K. 1997. Occurrence of *E. coli* O157 in Norwegian minced meat. *In* Anonymous World Congress on Food Hygiene, The Hague. Wageningen, The Netherlands: Wageningen Pers, 188.
68. KUNTZE, U., BECKER, H., MARTLBAUER, E., BAUMANN, C. and BUROW, H. 1996. Nachweis von verotoxinbildenden *E. coli* stammen in rohmilch und rohmilchkase. Archiv Lebensmittelhyg. *47*, 141–144.
69. LITTLE, C., GILLESPIE, I., DE LOUVOIS, J. and MITCHELL R. 1999. Microbiological investigation of halal butchery products and butchers' premises. Commun. Dis. Public Health 2, 114–118.
70. LITTLE, C.L. and DE LOUVOIS, J. 1998. The microbiological examination of butchery products and butchers' premises in the United Kingdom. J. Appl. Microbiol. *85*, 177–186.
71. MASSA, S., GOFFREDO, E., ALTIERI, C. and NATOLA, K. 1999. Fate of *Escherichia coli* O157:H7 in unpasteurized milk stored at 8 degrees C. Lett. Appl. Microbiol. *28*, 89–92.
72. MCDONNELL, R.J. *et al.* 1997. An outbreak of Verocytotoxin producing *Escherichia coli* O157 infection associated with takeaway sandwiches. Commun. Dis. Rep. CDR Rev. *7*, R201–R205.
73. MEAD, P.S. *et al.* 1997. Risk factors for sporadic infection with *Escherichia coli* O157:H7. Arch. Intern. Med. *157*, 204–208.
74. MECHIE, S.C., CHAPMAN, P.A. and SIDDONS, C.A. 1997. A fifteen month study of *Escherichia coli* O157:H7 in a dairy herd. Epidemiol. Infect. *118*, 17–25.
75. MERMIN, J. *et al.* 1997. A multistate outbreak of *Escherichia coli* O157:H7 infections associated with eating mesclun mix lettuce, abstract V74/I. *In* Anonymous 3rd International symposium and workshop on Shiga toxin (verotoxin)-producing *Escherichia coli* infections. Melville, N.Y., United States: Lois Joy Galler foundation for Hemolytic-Uremic Syndrome Inc. 9.
76. MICHEL, P., WILSON, J.B., MARTIN, S.W., CLARKE, R.C., MCEWEN, S.A. and GYLES, C.L. 1999. Temporal and geographical distributions of reported

cases of *Escherichia coli* O157:H7 infection in Ontario. Epidemiol. Infect. *122*, 193–200.

77. MICHINO, H. *et al.* 1998. Recent outbreaks of infections caused by *Escherichia coli* O157:H7 in Japan. *In Escherichia coli* O157:H7 and other Shiga toxin-producing *E. coli* strains (J.B. Kaper and A.D. O'Brien, eds.) Washington, D.C., United States: ASM Press, 73–81.
78. MORGAN, G.M. *et al.* 1988. First recognized community outbreak of haemorrhagic colitis due to verotoxin-producing *Escherichia coli* O157.H7 in the UK. Epidemiol Infect. *101*, 83–91.
79. NEAVES, R., DEACON, J. and BELL, C. 1994. A survey of the incidence of *E. coli* O157 in the UK Dairy Industry. Int. Dairy J. *4*, 679–696.
80. OSTROFF, S.M. *et al.* 1990. A statewide outbreak of *Escherichia coli* 0157:H7 infections in Washington State. Am J. Epidemiol. *132*, 239–247.
81. PADHYE, N.V. and DOYLE, M.P. 1991. Rapid procedure for detecting enterohemorrhagic *Escherichia coli* O157:H7 in food. Appl. Environ. Microbiol. *57*, 2693–2698.
82. PARRY, S.M., SALMON, R.L., WILLSHAW, G.A. and CHEASTY, T. 1998. Risk factors for and prevention of sporadic infections with vero cytotoxin (shiga toxin) producing *Escherichia coli* O157 [comments]. Lancet *351*, 1019–1022.
83. PAVIA, A.T. *et al.* Hemolytic-uremic syndrome during an outbreak of *Escherichia coli* O157:H7 infections in institutions for mentally retarded persons: clinical and epidemiologic observations. J. Pediatr. *116*, 544–551.
84. PENNINGTON GROUP. 1997. Report on the circumstances leading to the 1996 outbreak of infection with *E. coli* O157 in Central Scotland, the implications for food safety and the lessons to be learned. 1997; Edinburgh, United Kingdom: Stationary Office.
85. PHEBUS, R.K. *et al.* 1997. Comparison of steam pasteurization and other methods for reduction of pathogens on surfaces of freshly slaughtered beef. J. Food Prot. *60*, 476–484.
86. PIERARD, D., VAN DAMME, L., MORIAU, L., STEVENS, D. and LAUWERS, S. 1997. Virulence factors of verocytotoxin-producing *Escherichia coli* isolated from raw meats. Appl. Environ. Microbiol. *63*, 4585–4587.
87. RADU, S. *et al.* 1998. Detection of *Escherichia coli* O157:H7 in the beef marketed in Malaysia. Appl. Environ. Microbiol. *64*, 1153–1156.
88. RAGHUBEER, E.V., KE, J.S., CAMPBELL, M.L. and MEYER, R.S. 1995. Fate of *Escherichia coli* O157:H7 and other coliforms in commercial mayonnaise and refrigerated salad dressing. J. Food Prot. *58*, 13–18.
89. READ, S.C., GYLES, C.L., CLARKE, R.C., LIOR, H. and MCEWEN, S. 1990. Prevalence of verocytotoxigenic *Escherichia coli* in ground beef, pork, and chicken in southwestern Ontario. Epidemiol. Infect. *105*, 11–20.
90. REED, C.A. 1995. Challenge study — *Escherichia coli* O157:H7 in fermented sausage. April 28, 1995 Letter to plant managers. 1995; Washington, D.C., United States: U.S. Department of Agriculture, Food Safety and Inspection Service.

91. REIDA, P. *et al.* 1994. An outbreak due to enterohaemorrhagic *Escherichia coli* O157:H7 in a children day care centre characterized by person-to-person transmission and environmental contamination. Int. J. Med Microbiol. Virol Parasitol Infect. Dis. *281*, 534–543.
92. RICE, D.H., HANCOCK, D.D. and BESSER, T.E. 1995. Verotoxigenic *E coli* O157 colonisation of wild deer and range cattle [letter]. Vet. Rec. *137*, 524.
93. RILEY, L.W. *et al.* 1983. Hemorrhagic colitis associated with a rare *Escherichia coli* serotype. N. Engl. J. Med. *308*, 681–685.
94. ROBERTS, C.L. *et al.* 1995. The role of heightened surveillance in an outbreak of *Escherichia coli* O157.H7. Epidemiol Infect. *115*, 447–454.
95. RODRIGUE, D.C. *et al.* 1995. A university outbreak of *Escherichia coli* O157:H7 infections associated with roast beef and an unusually benign clinical course. J. Infect. Dis. *172*, 1122–1125.
96. RYAN, C.A. *et al.* 1986. *Escherichia coli* O157:H7 diarrhea in a nursing home: clinical, epidemiological, and pathological findings. J. Infect. Dis. *154*, 631–638.
97. SALMON, R.L. *et al.* 1989. A christening party outbreak of haemorrhagic colitis and haemolytic uraemic syndrome associated with *Escherichia coli* O 157.H7. Epidemiol. Infect. *103*, 249–254.
98. SAMADPOUR, M. *et al.* 1994. Occurrence of Shiga-like toxin-producing *Escherichia coli* in retail fresh seafood, beef, lamb, pork, and poultry from grocery stores in Seattle, Washington. Appl. Environ. Microbiol. *60*, 1038–1040.
99. SMITH, H.R., CHEASTY, T., ROBERTS, D., THOMAS, A. and ROWE, B. 1991. Examination of retail chickens and sausages in Britain for vero cytotoxin-producing *Escherichia coli*. Appl. Environ. Microbiol. *57*, 2091–2093.
100. STEELE, B.T., MURPHY, N., ARBUS, G.S. and RANCE, C.P. 1982. An outbreak of hemolytic uremic syndrome associated with ingestion of fresh apple juice. J. Pediatr. *101*, 963–965.
101. STEELE, M.L. *et al.* 1997. Survey of Ontario bulk tank raw milk for foodborne pathogens. J. Food Prot. *60*, 1341–1346.
102. STEVENSON, J. and HANSON, S. 1996. Outbreak of *Escherichia coli* O157 phage type 2 infection associated with eating precooked meats. Commun. Dis. Rep. CDR Rev. *6*, R116–R118.
103. STROCKBINE, N.A., WELLS, J.G., BOPP, C.A. and BARRETT, T.J. 1998. Overview of detection and subtyping methods. *In Escherichia coli* O157:H7 and other Shiga toxin-producing *E. coli* strains. (J.B. Kaper and A.D. O'Brien, eds.) Washington, D.C., United States: ASM Press, 331–356.
104. SUTHIENKUL, O., BROWN, J.E., SERIWATANA, J., TIENTHONGDEE, S., SASTRAVAHA, S. and ECHEVERRIA, P. 1990. Shiga-like-toxin-producing *Escherichia coli* in retail meats and cattle in Thailand. Appl. Environ. Microbiol. *56*, 1135–1139.
105. TAKEDA, Y. 1999. Enterohemorrhagic *Escherichia coli* infection in Japan. Pediatr Int. *41*, 198–201.
106. THOMS, B. [Detection of verotoxin-producing *Escherichia coli* in deer's meat] Nachweis von verotoxinbildenden *Escherichia coli* in rehfleisch. Archiv Lebensmittelhyg. *50*, 52–54.

107. TUTTLE, J. *et al.* 1999. Lessons from a large outbreak of *Escherichia coli* O157:H7 infections: insights into the infectious dose and method of widespread contamination of hamburger patties. Epidemiol. Infect. *122*, 185–192.
108. UPTON, P. and COIA, J.E. 1994. Outbreak of *Escherichia coli* O157 infection associated with pasteurised milk supply [letter]. Lancet. *344*, 1015.
109. WALLACE, J.S., CHEASTY, T. and JONES, K. 1997. Isolation of verocytotoxin-producing *Escherichia coli* O157 from wild birds. J. Appl. Microbiol. *82*, 399–404.

110. WALSH, L., DOOGE, D. and HILL, C. 1997. Screening for *Escherichia coli* O157:H7 in Irish ground beef using two commercial detection systems. Irish Vet. J. *50*, 111–115.
111. WATANABE, Y. *et al.* 1999. Factory outbreak of *Escherichia coli* O157:H7 infection in Japan. Emerg. Infect. Dis. *5*, 424–428.
112. WEAGANT, S.D., BRYANT, J.L. and BARK, D.H. 1994. Survival of *Escherichia coli* O157:H7 in mayonnaise and mayonnaise-based sauces at room and refrigerated temperatures. J. Food Prot. *57*, 629–631.
113. WILLSHAW, G.A., THIRLWELL, J., JONES, A.P., PARRY, S., SALMON, R.L. and HICKEY, M. 1994. Vero cytotoxin-producing *Escherichia coli* O157 in beefburgers linked to an outbreak of diarrhoea, haemorrhagic colitis and haemolytic uraemic syndrome in Britain. Lett. Appl. Microbiol. *19*, 304–307.
114. WRIGHT, D.J., CHAPMAN, P.A. and SIDDONS, C.A. 1994. Immunomagnetic separation as a sensitive method for isolating *Escherichia coli* O157 from food samples. Epidemiol. Infect. *113*, 31–39.

CHAPTER 10

A CASE STUDY OF CHEESE ASSOCIATED *E. COLI* O157 OUTBREAKS IN SCOTLAND

T.M.S. REID

Dept. of Medical Microbiology
Aberdeen Royal Hospital
Foresterhill
Aberdeen AB25 2ZN, United Kingdom

As we enter the 21st century, *E. coli* O157 is firmly established as a cause of outbreaks of foodborne infection (20). While the range of foods implicated grows ever wider, unpasteurised milk and dairy products remain well-recognised vehicles of significant human infection with associated morbidity and mortality (4,9). In Scotland, the retail sale of raw milk was prohibited in 1983 preventing, at a stroke, major outbreaks of milk borne infection. However, such legislation did not eliminate the risk associated with inadequate pasteurisation or post pasteurisation contamination, which was responsible for the Red House Dairy *E. coli* O157 outbreak in West Lothian in 1996 (22). Equally, the legislation did not prohibit the sale of dairy products such as cheese made from unpasteurised milk, thus the current Dairy Products (Hygiene) (Scotland) Regulations 1995 permit the granting of derogations for such raw milk based products with traditional characteristics. In the last 5 years, 3 outbreaks of *E. coli* O157 infection in the Grampian region of Scotland associated with the consumption of unpasteurised cheese have served to highlight the risks involved with such products. Review of these outbreaks allows some insight into the multi-disciplinary problems associated with their investigation.

OUTBREAK 1

On October 31, 1994 *E. coli* O157 was isolated from the stool specimen of a 55 year-old female who had a 24 hour history of diarrhea. During investigation, it emerged that she owned a registered, well established dairy/food business which produced cheese from unpasteurised cows milk supplied by her herd of ten cows, augmented, when necessary, from a neighbouring farm.

When she developed symptoms she removed herself from cheese production activities. Four days later, an 83 year-old retired farmer was admitted to the hospital with bloody diarrhoea. He gave a history of eating cheese from the source farm, as did a 91 year old long stay patient in a local hospital and his son, who had purchased the cheese from a local butcher (Fig. 1). By November 10 the husband, daughter-in-law and grandson of the index case (cheesemaker) had become confirmed cases, as had a nurse working in a local psychiatric hospital who recounted that, as was normal practice on patient outings, cheese had been purchased for the patients from the farm gate. She had sampled a small piece of cheese, subsequently developed bloody diarrhoea, and was hospitalised.

On November 11 a young child and a 64 year-old man were identified as positive cases by the microbiology laboratory. Enquiries revealed that both families had purchased cheese from a travelling fish van. Over the next three days three further family contacts of the early cases, together with a 51 year old lady who had purchased cheese from the fish van, and a 63 year old man who had acquired his cheese from the butcher, became notified cases.

Subsequently, on November 15 a two year-old boy who had been admitted to the hospital with abdominal pain and diarrhoea was proven positive. His father had purchased cheese from a workmate. A day later an elderly couple (male 87, female 75) joined the list of cases. They had eaten cheese at the home of a relative who had obtained it from a brother who had in turn obtained it from a workmate. After extensive investigation by Environmental Health Officers, it emerged that 10 cheeses had been purchased by a bricklayer on a local building site, and that he had subsequently sold them on to colleagues.

Ultimately, 22 cases of *E. coli* O157 infection were confirmed, one of which, the 2 year old girl, developed haemolytic uraemic syndrome. Eight of the cases were asymptomatic. All the human isolates proved to be *E. coli* O157 Phage type 21/28 VT1 Neg VT2 Pos. The elucidation of this somewhat complex and convoluted distribution of cheese from the farm source was made more difficult by virtue of the fact that the cheeses carried no labels in respect to origin, date of production or batch numbers, nor was there any indication that the product had been made from unpasteurised milk.

The cheese was produced without using a starter culture. The raw milk from the morning milking was kept at ambient temperature for 10-12 hours, and combined with milk from the second milking of the day. The merged sample was then heated to 30°C, rennet was added, the curds harvested, and pressed into rounds. It was eventually established that approximately 56 rounds of cheese were produced per week which supplied five retail outlets plus the farm gate sales. It is worth noting that because of demand cheeses were, on occasion, only allowed to mature for two weeks prior to sale.

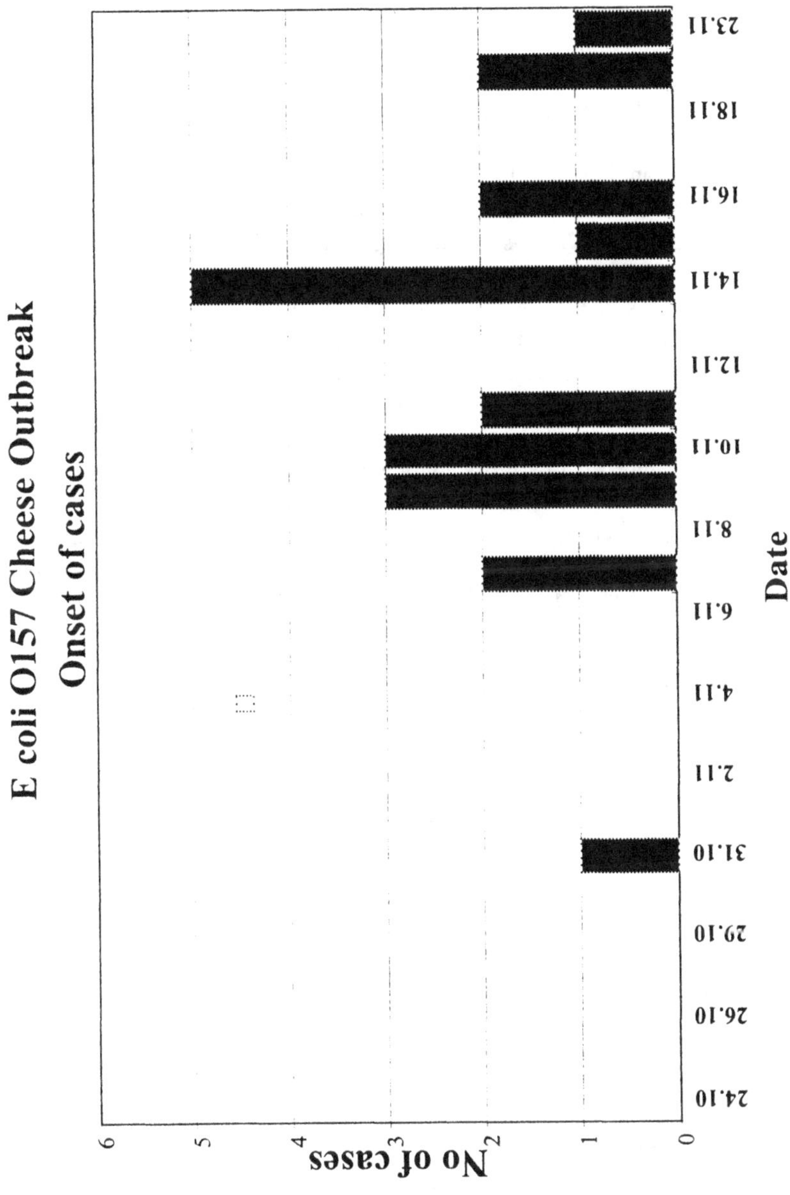

FIG. 1. *E. COLI* O157 CHEESE OUTBREAK: ONSET OF CASES

The cheese making business had been in operation for some 28 years and had maintained very strict hygiene standards throughout. The operators had a very positive working relationship with officers from their local Environmental Health Department. When the problem arose, the cheesemaker voluntarily agreed to cease production and withdraw cheese from sale pending the results from the investigation. On the first visit following notification of the outbreak, Environmental Health Officers obtained samples of cheese, milk and water from the farm, as well as swabs of surfaces and of utensils used in the cheese making process.

Notwithstanding the problems associated with the lack of labeling, numerous cheeses and pieces of cheese were returned and examined for the presence of *E. coli* O157. Using the immunomagnetic separation technique, *E. coli* O157 was isolated from the centre of one cheese, and from 2 other pieces of different cheeses. Of these three isolates, two were Phage Type 21/28 VT1 Neg VT2 Pos and gave pulsed field gel electrophoresis patterns which were indistinguishable from the human isolates. The remaining isolate was untypeable and negative for verotoxins 1 and 2. Detection of the organism within the core of a cheese round suggested contamination had occurred during the manufacturing process, rather than by cross contamination of the formed cheese at a later stage.

Results of environmental swabbing confirmed the overall impression of good hygiene standards. Veterinary Officers had also taken multiple samples of faeces, milk and whey from individual cows in both the herds which supplied unpasteurised milk to the cheese maker. All such raw milk and whey samples proved to be of acceptable quality, and, despite the use of immunomagnetic separation techniques, none of the faeces or milk samples were found to contain *E. coli* O157.

OUTBREAK 2

In early December 1998 the microbiology laboratory reported that *E. coli* O157 had been isolated from the stool specimens of 5 patients resident in the same area of Grampian Region. Immediate Public Health and Environmental Health investigations quickly revealed another complex web of connections which again implicated a locally produced unpasteurised milk cheese, sold at a farm shop. Case 1 was a 58 year-old female who had developed symptoms of abdominal pain and diarrhoea on December 12. Ironically, she had not consumed cheese but had consumed farmhouse butter purchased from a man in a local pub. Subsequent enquiry revealed that the butter was made with milk from the same source. Her husband later proved to be stool positive, but he remained asymptomatic.

A 64 year old female (Case 2) who developed symptoms of bloody diarrhoea and abdominal pain on December 12 was admitted to the Infection Unit where she developed haemolytic uraemic syndrome requiring renal dialysis. Despite active treatment she died in the hospital. She had consumed cheese purchased from the farm shop. Later her daughter was shown to be stool positive although she remained asymptomatic. Case 3 was the 3 year-old grandson of the cheese maker. He developed diarrhoea and abdominal pain and although admitted to hospital as a precautionary measure was subsequently discharged as not suffering any ill effects related to this outbreak. He had not eaten any of the cheese but had drunk the unpasteurised milk. His 7 year-old sister was later shown to be stool positive, but asymptomatic. Case 4 was a 72 year-old female who developed symptoms of diarrhoea and abdominal pain on December 17. She had consumed cheese on two occasions when attending functions at her local British Legion club. The cheese was traced back to the farm shop. Case 5 was a 63 year-old man who lived in the City of Aberdeen but had visited the farm shop and purchased cheeses to take home. He developed symptoms of diarrhoea and abdominal pain on December 18.

Having established the link with the cheese, Environmental Health Officers advised the cheese maker to cease production and sales pending the results of laboratory testing on the cheese. A full inspection of the milking, cheese making and storage facilities was carried out and samples of cheese, water and raw milk, and environmental swabs from various locations, were taken for examination. The overall condition of the premises and in particular the milk room and the cheese room was noted to be good and easily cleaned.

The cheese making operation involved nine milking cows which were machine-milked morning and evening. Their milk was delivered by pipeline to stainless steel heaters in the milk house, treated with commercial starter and rennet, heated and left overnight. The resultant curds were removed in the morning into plastic rings, and placed in a hydraulic press for 24 hours. The formed cheeses were then taken to the sales cabin where they were stored and sold as farm gate sales. Approximately nine cheeses per day were produced. Fortunately, records were kept of each days production and sales, making it possible to trace nearly all the suspect cheeses.

Investigations established that the cheese maker had recently modified her standard procedure of dealing with the "morning" and "evening" milk separately. Her modified practice was to add rennet and starter, allow the mixture to stand overnight, and then combine with the (next) morning milk, rennet and starter. This procedure meant that whey remained in contact with cheese for approximately 22 hours with consequent opportunity for bacterial growth. This change in procedure was deemed unsatisfactory and certainly not in keeping with the Specialist Cheese Makers Code of Best Practice (19).

Investigations also established that cheese was only being stored (to mature) for an average of 5-6 days, a period which is in fact too short to allow adequate maturing of the cheese, a process which includes significant reduction in product pH, and consequential decreases in the total bacterial count of the product. The cheeses were mostly sold whole and once again, carried no labels regarding origin, production date or batch numbers. Some cheeses were, however, supplied cut and packaged to the local farm shops.

E. coli O157 was not isolated from the water in the cheese making premises, from any of the environmental swabs or from milk samples taken from the cows. However, cheeses obtained from the farm shop and cheeses in the possession of patients were found to contain *E. coli* O157 Phage Type 21/28 VT1 Neg VT2 Pos. This strain gave pulsed field gel electrophoresis patterns which were indistinguishable from the clinical isolates.

Veterinary officers obtained milk samples from each of the cows providing milk to the cheese production process. *E. coli* O157 could not be recovered from these samples by immunomagnetic separation, nor from a sample of water from a well in the farmyard, which was considered to have been subject to periodic contamination with surface water.

OUTBREAK 3

In May 1999, a class of seven year-old children in a local primary school were doing a project on cheese making. One of the children prevailed on her grandmother, who had been accustomed to making goats milk cheese, solely for family use, or for use by the residents of her guest house, to make a goats milk cheese. Consequently, an unpasteurised goats milk cheese was produced on May 23, taken to the school the next day and stored in a refrigerator overnight. The teacher and her husband each consumed a portion of the cheese the following day and the remainder was returned to the school on 26 May for consumption by 25 of the 30 members of the class. The children and the teacher developed symptoms of abdominal pain and diarrhoea between May 28-30. Ultimately, 10 children were admitted to the hospital and one developed Haemolytic Uraemic Syndrome requiring peritoneal dialysis for 4 days. Twenty-four of the 25 children and all 3 of the adults who ate the cheese were subsequently shown to be culture positive for *E. coli* O157 Phage Type 21/28 VT1 Neg VT2 Pos. Investigation confirmed that the 5 children who did not eat the cheese had negative stool cultures. Interestingly, the one child who consumed the cheese but remained stool negative was the granddaughter of the cheesemaker who was regularly exposed to the goat and its milk. Six of the children were asymptomatic. All household contacts of positive cases were tested. There were nine additional culture positives (4 adults and 5 children) — secondary family contacts including one child minder who was symptomatic. The grandmother who made the cheese and the man who milked the goat were both negative.

The goat had been milked by an experienced worker who reported that antibacterial gel had been applied to his hands and the goat's teats before milking. The milk was strained through a milk sieve, boiled for one minute, supplemented with rennet, cooled, strained into a bowl and covered with cling film.

Unfortunately, all the goats milk cheese supplied to the school had been consumed. However, samples of the goat's milk were examined. Samples tested showed a coliform count of 160/ml — a level which complies with the Dairy Products (Hygiene) Regulations Scotland 1995. No *E. coli* O157 was found in the milk or in a sample of the private water supply to the smallholding. In addition to the goat there were 9 ewes, 7 lambs, a goose, a duck and 3 hens on the small holding. Faecal samples were taken from all of these and examined for *E. coli* O157 using immunomagnetic separation techniques. The goat, 8 sheep, 6 lambs, the goose and the duck were culture positive for *E. coli* O157 Phage Type 21/28 VT1 Neg VT2 Pos with pulsed field gel electrophoresis patterns which were indistinguishable from the clinical (human) isolates.

An outbreak control committee was speedily convened to co-ordinate the investigation and management of the outbreak. Specific measures instituted to control the outbreak included enhanced cleaning of school toilets, positive reinforcement by teaching staff and letters to parents of the need for thorough hand washing by children, family contacts and parents. Given the high rate of positive results amongst the children, many of whom were asymptomatic, it was agreed to minimise secondary person to person spread by closing the class until sufficient children tested negative on 2 samples taken at least 48 hours apart.

OTHER *E. COLI* OUTBREAKS/INCIDENTS INVOLVING CHEESE

In 1971 (i.e., before *E. coli* O157 was recognised as such) an outbreak of fever, cramps and bloody diarrhoea was linked to soft ripened Camembert cheese imported to the USA from France (14). Invasive *E. coli* O124:B17 was isolated from the stool samples of those infected and also from the incriminated cheeses, which contained 10^5 - 10^7 *E. coli* O124/gram. Investigation of the production plant in France showed that a water filtration system for cleaning water had been malfunctioning. The organism was isolated from a curdling tank and from ripening cheese.

In April 1998 a 12 year-old boy developed haemolytic uraemic syndrome following infection with a VT2 positive strain of *E. coli* O157 from eating cheese made from unpasteurised cow's milk (3). Samples of the same cheese from the shop and from similar cheeses held by a distributor yielded a strain of

E. coli O157 with the same pulsed field gel electrophoresis patterns as the clinical isolates. The manufacturer voluntarily withdrew the cheese from sale. It is not clear if the cheese was contaminated during the production process, or at some later stage.

Three cases of infection with *E. coli* O157 Phage Type 21/28 VT1 Neg VT2 Pos VT2c were reported in Teesdale, Cleveland in patients who had consumed cheese made from unpasteurised milk (Cotherstone Cheese) (5). All the human isolates, an isolate from a core sample of cheese in one of the patient's refrigerators and another core sample of cheese produced three days later in the suspect premises, were indistinguishable by molecular sub-typing using pulsed field gel electrophoresis. Environmental swabs and samples from the dairy herd and slurry proved negative. The cheese maker voluntarily withdrew the cheese from sale.

All these outbreaks underscore the hazards associated with unpasteurised milk and dairy products (15). The risk of infection with milk contaminated with *Salmonella*, *Campylobacter*, verocytotoxin producing *E. coli*, or other pathogens remains undiminished. Equally the case for pasteurisation of milk and milk based products is incontrovertible (7). However there remains a consumer demand for cheese made from unpasteurised milk based on the belief that pasteurisation impairs the organoleptic qualities and the development of full flavour. Derogations for such milk-based products with traditional characteristics are in themselves a potential danger to health.

Raw milk used in cheeses with traditional characteristics does not have to meet the normal prescribed standard in the Dairy Products Regulations. Nevertheless, as a responsible organisation, the Specialist Cheesemakers Association has recommended stricter microbiological criteria for raw milk used for cheesemaking, and takes pains to educate members regarding both the inherent microbiological hazards, and the means of minimising risk as part of adequate hazards analysis (HACCP). As their Code of Best Practice states 'good quality milk is essential for good quality cheese' (19).

Review of the above three outbreaks identifies several common problems, which can best be reduced or eliminated by following a number of guidelines. The best quality cheese is produced from milk taken straight from the milking parlour to the cheesemaking facility without cooling or holding. Failing this, the milk must be rapidly chilled, stored at or below 4°C, and used within 36 hours. Storage, especially above these temperatures for prolonged periods, coupled with the subsequent addition/pooling of later milkings is unacceptable practice.

Enterohaemorrhagic *E. coli* can enter milk via faecal contamination. This organism can survive, but not grow in milk at 5°C, and can begin to multiply in raw cows milk held at 7°C and 15°C for 24 hours. Its growth rate in milk can be reduced by interference by the background microflora (12, 13, 23).

E. coli can survive and even multiply during the manufacture of cheddar cheese (pH 4.95-5.2) (17), with the organisms becoming more prevalent in the curd rather than the whey. If initially present in cheese at concentrations of 1 cfu/ml, *E. coli* O157 populations decline gradually during maturation at 2°C, becoming undetectable after 158 days. Conditions in mature hard cheese (low water activity and pH) are less favorable for the survival and growth of *E. coli* O157 than those conditions that prevail in young, soft and semi-soft (high moisture content) cheeses, although there is little doubt that the organism can survive in many dairy products. Of soft cheeses experimentally inoculated with *E. coli* O157, 93.2% were still found to contain the organism 60 days later(21). Thus, while the use of unpasteurised milk in the manufacture of farmhouse cheese is suggested to preserve unique and desirable characteristics (heat sensitive flavours, aromas and taste) there is no guarantee that the inherent processes of ripening and pH reduction will be sufficient to reduce the numbers of *E. coli* O157 present to a safe level, especially as the infective dose can be less than 10 organisms.

Unpasteurised goats milk used in cheesemaking can be particularly problematical, and in recent years this material has been responsible for outbreaks of *Salmonella paratyphi* B infection (10,11) and presumptive verocytotoxigenic *E. coli* infection (1,8,9) in France, and of *Brucella melitensis* infection in Malta (2). Goats have been noted to carry *E. coli* O157 without any apparent ill effect and have been implicated in cases of human infection associated with a farm visitor centre (18).

In light of the cross contamination which was responsible for the Central Scotland outbreak in 1996 (16) it is debatable whether butcher shops should be handling, cutting and selling cheeses as was the case in Outbreak 1, without strict separation from raw meat and meticulous attention to hygiene. Similarly, handling and cutting of cheese rounds into pieces at the farm shop must be subject to strict hygiene to prevent contamination.

It is essential that cheesemakers label their product clearly, and unequivocally as 'UNPASTEURISED'. Members of the public who demand the fuller flavour of a raw milk cheese must be given this information, in order to make an informed decision on whether or not to purchase such a 'farm cheese'. Such labeling should amount to a health warning as, in the absence of adequate pasteurization, a safe product cannot be guaranteed. The label should also include basic information as to source, batch number and production date to permit traceability in the event of problems. It is not acceptable to use additional human cases as a means of tracing suspect batches, after primary evidence of contamination is obtained. Batch traceability is exceedingly important, because contamination of a batch of cheeses may be totally random and sporadic. Experience in food microbiology laboratories has shown that different rounds, and indeed different wedges from individual rounds in a known positive cheese, may give different results. Given the low infective dose for humans, the food microbiologist is frequently faced with the task of finding a very small number of randomly distributed and often 'stressed' pathogenic *E. coli* O157 organisms, which are greatly outnumbered by background flora. In such circumstances the use of an effective enrichment procedure, followed by a sensitive detection method such as immunomagnetic separation, are essential if meaningful results are to be obtained (6).

E. coli O157 was not detected in any milk samples examined as part of the investigation of the above incidents, which may indicate that random, intermittent excretion or contamination has occurred. Nevertheless, all these outbreaks could have been prevented by pasteurisation and/or adherence to best cheese-making practice, including strict hygiene, to prevent contamination during processing and maturation. Currently, it must be emphasized that effective and consistent application of HACCP is necessary, to ensure a microbiologically safe product.

REFERENCES

1. ANON. 1994. Two clusters of haemolytic uraemic syndrome in France. Communicable Disease Report Weekly *4*, 29.
2. ANON. 1995. Brucellosis associated with unpasteurised milk products abroad. Communicable Diseases Report *5*, 151.
3. ANON. 1998a. VTEC O157 infection and unpasteurised cream — update. Communicable Diseases Report *8*, 167.
4. ANON. 1998b. *Escherichia* O157 in Somerset. Communicable Diseases Report *8* 377, 389–392.
5. ANON. 1999. Outbreak of verocytotoxin producing *Escherichia coli* O157 infection in North Cumbria. Communicable Diseases Report *9*, 95/8.
6. BOLTON, E.J., CROZIER, L. and WILLIAMSON, J.K. 1995. New technical approaches to *Escherichia coli* O157 Optimisation of methods for the isolation of *Escherichia coli* O157 from beefburgers. PHLS Microbiology Digest *12*, 67–70.
7. CAMPBELL, D.M., COWDEN, J.M., MORRIS, G., REILLY, W.J. and O'BRIEN, S.J. 1996. All milk products should be heat treated. British Medical Journal *312*, 1099.
8. CAZENAVE, C., DESENCLOS, J.C., MAILLOT, E., BENOIT, S., DESCHENES, G. and BARON, S. 1993. Enclosion de syndrome hèmolytique et urémique (SHU) dans une communautè rurale du Cher. Bulletin Epidemiologique Hebdomadaire.
9. CHAPMAN, P.A., WRIGHT, D.J. and HIGGINS, R. 1993. Untreated milk as a source of verocytotoxigenic *Escherichia coli* O157. Veterinary Record *133*, 171–172.

10. DESENCLOS, J.C. *et al.* 1996. Large outbreak of *Salmonella enterica* serotype *paratyphi* B infection caused by a goats milk cheese, France, 1993: a case finding and epidemiological study. British Medical Journal *312*, 91–94.
11. GRIMONT, P.A.D. and BOUVET, P. 1991. Les salmonelloses et les shigelloses en 1990 en France. Bulletin Epidèmiologique Hebdomadaire *25*, 102.
12. HEUVELINK, A.E., BLUEMINK, B., VAN DEN BIGGELAAR, F.L.A.M., TE GIFFEL, M.C., BEUMER, R.R. and DE BOER, E. 1998. Occurrence and survival of verocytotoxin producing *Escherichia coli* O157 in raw cow's milk in the Netherlands. J. Food Protection *61*, 1597–1601.
13. HEUVELINK, A.E., DE BOER, E. and BEUMER, R.R. 1999. Survival of verocytotoxin producing *Escherichia coli* O157 in raw cows milk and raw meat products. *In* Survival and Growth of Verocytotoxigenic *E.coli*. Proceedings of a meeting organised by an EU Concerted Action on VTEC (CT98-3935) in Athens, May 6-8.
14. MARIER, R., WELLS, J.G., SWANSON, R.C., CALLAHAN, W. and MEHLMAN, I.J. 1973. An outbreak of enteropathogenic *Escherichia coli* foodborne disease traced to imported French cheese. Lancet *2*, 1376–1378.
15. RAMPLING, A 1996. Raw milk cheeses and salmonella. British Medical Journal *312*, 67–68.
16. REID, T.M.S., CASHMORE, C. and BEVERIDGE, L. 1999. The Central Scotland *E. coli* O157 outbreak 1996, investigation of food related samples. *In* Concerted Action CT98-3935 Verocytotoxigenic *E. coli* in Europe. 1 *Methods for Verocytotoxigenic E. coli.* p. 1–6.
17. REITSMA, C.J. and HENNING, D.R. 1996. Survival of enterohaemorrhagic *Escherichia coli* O157:H7 during the manufacture and curing of cheddar cheese. J. Food Protection *59*, 460–464.
18. SHUKLA, R., SLACK, R., GEORGE, A., CHEASTY, T., ROWE, B. and SCUTTER, J. 1995. *Escherichia coli* O157 infection associated with a farm visitor centre. CDR Review *5*, R86–R90.
19. THE SPECIALIST CHEESEMAKERS ASSOCIATION. 1997. Specialist Cheesemakers' Code of Best Practice. London: Dept. of Health. p 72.
20. SUBCOMMITTEE OF PHLS ADVISORY COMMITTEE ON GASTROINTESTINAL INFECTIONS 2000. Guidelines for the control of infection with verocytotoxin producing *Escherichia coli* (VTEC). Communicable Disease and Public Health *3*, 14–23.
21. SZABO, R. *et al.* 1990. Increased sensitivity of the rapid hydrophobic grid membrane filter enzyme — labeled antibody procedure for *Escherichia coli* O157 detection in foods and bovine faeces. Applied and Environmental Microbiology *56*, 379–381.
22. UPTON, P. and COIA, J.E. 1994. Outbreak of *Escherichia coli* O157 infection associated with pasteurised milk supply. Lancet *344*, 1015.
23. WANG, G., ZHAO, T. and DOYLE, M.P. 1997. Survival and growth of *Escherichia coli* O157:H7 in unpasteurised and pasteurised milk. J. Food Protection *60*, 610–613.

NOTE

The Food Standards Agency, Scotland (2001) has just published a 'Report on small-scale cheese production in Scotland' with recommendations on measures to reduce the incidence of food poisoning outbreaks associated with specialist cheeses.

CHAPTER 11

PATHOGENIC ASPECTS OF VTEC INFECTIONS IN RUMINANTS

FRÉDÉRIC GOFFAUX, BERNARD CHINA, PHILIPPE STORDEUR and JACQUES MAINIL

Faculté de Médecine Vétérinaire
Université de Liège
Sart Tilman, Bât B43a
B-4000 Liège, Belgium

INTRODUCTION

Verotoxigenic *Escherichia coli* (VTEC) interact with ruminants in two ways: (1) intestinal diseases in newborn to four-month old calves and possibly other young ruminants; and (2) carriage by healthy animals, young animals and adults.

This chapter will focus on the clinical aspects of VTEC infections in domestic and wild ruminants. The problem of healthy carriage, of beef and dairy foods as vehicles and of VTEC food poisoning are the subjects of Chapters 9, 10, and 14.

NOMENCLATURE

The terms "verotoxins (VTs)" or "Shiga-like toxins (SLTs)" on one hand and "verotoxigenic *E. coli* (VTEC)" or "Shiga-like toxin-producing *E. coli* (SLTEC)" on the other hand were used for several years by Canadian/British workers and American workers, respectively. More recently the proposal to use the names "Shiga toxins (Stx)" for all these cytotoxins was made, but did not receive unanimous approval and VT/Stx must be considered as synonyms (26). The nomenclature VT and VTEC will be used throughout this chapter.

Another confusing arca concerns the nomenclature of the various VTEC. The term "enterohaemorrhagic *E. coli*" or EHEC, refers to strains associated with haemorrhagic colitis (HC) and its sequelae in humans, and to those isolated from animals, mainly calves with enteritis and bloody diarrhea, which, in addition to the production of VT, are also able to induce the "Attaching-Effacing" (AE) lesions, although bloody diarrhoea is not systematically observed. Until the description of other virulence-associated factors, the name

VTEC will continue to be used for strains isolated from humans and from ruminants, in the presence or absence of clinical signs (19,26), which are not examined for, or are found to be negative for, the production of AE lesions.

HISTORICAL BACKGROUND OF VTEC IN RUMINANTS

Although Konowalchuk *et al.* (18) reported the production of a verotoxin by human and porcine *E. coli* in 1977, the first reports of production of VT by *E. coli* isolated from diarrhoeic calves were published in 1980 (17,41). The association between diarrhoea in calves and the presence of VTEC in their faeces was not confirmed before 1985 (13,25,34). During the following years, several reports were published which provided additional information about the general and specific properties and characteristics of these bovine VTEC (19). However, the history of VTEC in cattle is certainly older than 1980, as retrospective studies have reported the existence of VTEC in collections of bovine strains from the 1960s (1,20,30).

SEROTYPES AND PATHOTYPES OF VTEC IN CALVES

Most VTEC associated with diarrhoea in young calves belong to the following serotypes (19): O5:H-, O8:H8, O20:H19, O26:H11, O103:H2, O111:H-, O111:H8, O111:H11, O118:H16, and O145:H+, some of which are also recovered from humans (Chapter 8). Other serotypes (Chapter 6) are detected much less frequently and many strains are either non-typeable with the available immunsera, or are rough variants. The O157 strains have only exceptionally been associated with clinical diseases.

The bovine VTEC can be subdivided into EHEC and VTEC (see definitions above). The former belong mainly to the O5, O26, O103, O111, O118, and O145 serogroups and are associated with clinical diarrhoea and enteritis. Most of the latter belong to the O8 and O20 serogroups.

CLINICAL SIGNS AND LESIONS OF EHEC AND VTEC INFECTIONS IN CALVES

EHEC are associated with diarrhoea in one to eight week-old calves with a peak between 4 and 5 weeks of age, although newborn and older calves can also be affected (6). The diarrhoea observed is mucoid, sometimes haemorrhagic and differs from the aqueous diarrhoea caused by ETEC. Calf fatality is rare (death rate does not exceed a few 2-3%), but diarrhoea is recurrent even after treatment, leading to dehydration, weakening and a depressed growth rate. Necropsy lesions of enteritis are mainly localized in the large intestine with some extension to the small intestine in severe cases. Localized and diffuse haemorrhages can also be observed. In severe cases, additional diarrheagenic infectious agents are detected, e.g, cryptosporidia, rotavirus, coronavirus, coccidia, and ETEC in very young calves.

Natural and experimental infections in young calves with EHEC belonging to various serotypes (O5:NM; O26:H11; O111:NM; O118:H16, and O157:H7) are characterized by enteritis, mainly of the large intestine, but in severe cases the small intestine is also inflamed. The microscopic lesions consist of Attaching/Effacing lesions (see below) and of neutrophil infiltration. Macroscopic lesions, when present, consist of congestion and localized or diffuse haemorrhages. Mesenteric lymph nodes are frequently enlarged (19,28,35).

The situation in relation to VTEC is equivocal, as some authors regard them as non-pathogenic because they are frequently isolated from healthy as well as from diarrhoeic calves. On the other hand, some studies have isolated VTEC from cases of diarrhoea. Bovine VTEC may thus represent a heterogeneous population of bacteria (31), only some of which are pathogens because they possess additional still uncharacterized virulence-associated properties.

Systemic complications, such as HUS in humans or edema disease in piglets, have never been observed in calves. The most probable reason for this is the absence of specific receptors for the VTs on endothelial cells of small vessels (19,26).

VIRULENCE PROPERTIES OF EHEC AND VTEC

The Verotoxins (VT)

The majority of bovine EHEC associated with diarrhoea in young calves are positive for VT1, and only a few are positive for VT2 or for both toxins. In contrast, bovine EHEC from healthy cattle and VTEC from diarrhoeic or healthy calves and adults can be positive for VT1 and/or VT2. The VT2 is either the prototype toxin or a variant, most frequently Stx2c, but there are some untyped variants. As in human VTEC, localization of the *stx1*, *stx2*, and *stx2c* operons on phages has been reported or suspected (19).

Although the role of VTs in humans and in piglets is well defined through their activity on endothelial cells of small vessels (19,26), the situation is less clear in calves, as HUS or edema disease-like syndromes have never been reported. The observation that EHEC associated with diarrhoea are predominantly positive for VT1 is also unexplained. A role for VTs in the haemorrhagic aspect of the enteritis, but not in the occurrence of the diarrhea, as in humans

(26), is possible. A comparison of the wild-type and isogenic mutant strains in a calf infectious model would confirm that hypothesis. Other properties must thus be expressed by the bovine EHEC and VTEC to explain the diarrhea.

The Attaching/Effacing (AE) Phenotype

Bovine EHEC are capable of causing AE lesions and this phenotype is considered to represent the primary cause of the diarrhoea (13,15,31). AE lesions are also produced by *E. coli* negative for the production of VTs (known as enteropathogenic *E. coli* or EPEC). The AE lesion is characterized and defined by the destruction (Effacement) of the enterocyte microvilli, leading to intimate contact (Attachment) between the bacteria and the enterocyte cytoplasmic membrane. The accumulation of actin and other cytoskeletal components leads to the formation of a pedestal-like structure underneath the bacteria and sometimes a phagocytosis-like appearance (26) (Fig. 1).

A number of AE negative VTEC strains have been associated with human disease. The alternative properties responsible for the occurrence of diarrhoea in these cases have not yet been characterized but some candidates have been identified as discussed in this chapter.

Production of AE lesions is a multistep event characterized by (1) initial adherence to the microvilli of the enterocytes; (2) transduction of a signal into the enterocytes; (3) cellular response with cytoskeleton rearrangement and effacement of the microvilli of the enterocytes; (4) intimate attachment to the enterocyte cytoplasmic membrane; and (5) entry of the bacteria into the eukaryotic cells (26).

Initial Attachment. Although different adherence properties of bovine EHEC to appropriate cells of bovine origin in culture (38,40) have been described, bovine EHEC and VTEC test negative with gene probes derived from genes mediating localized, diffuse and aggregative adherence of human *E. coli* to cell cultures (21). Candidate adhesins have, however, been recently described, i.e., a fimbriae-like structure (42) and a DNA sequence homologous to operons coding for Afa adhesins (11). But until these preliminary data receive confirmation, the initial adhesin of bovine EHEC and VTEC, if any, awaits full identification.

Signal Transduction, Cytoskeleton Rearrangement, Effacement of the Microvilli and Intimate Attachment. The genetic determinants for the next three steps are grouped together on the chromosome forming one pathogenicity island (Pai III) called LEE, for Locus of Enterocyte Effacement. The LEE is not present in human commensal *E. coli*, in other human pathogenic *E. coli*, or in K12 *E. coli*, but is found in all human EHEC and EPEC strains causing AE lesions (9,15,23,24). The size of the LEE in human EHEC and EPEC varies between 35 to 43 kb and several insertion sites have been identified, e.g., *selC* and *pheU* genes. The LEE of bovine EHEC (and EPEC) also vary in size (35 to 48 kb) and have different insertion sites (*selC* gene and others still unidentified) (10,15).

The LEE of human EPEC and EHEC is composed of three functional regions. The left part contains several genes (*esc* and *sep*) coding for a type III secretion system. The right part contains the *esp* genes, coding for proteins responsible for the induction of the AE lesion formation after transduction into the epithelial cell through the type III secretion system. The middle part contains the *eae* gene coding for a 94-97 kDa outer membrane protein, intimin, which mediates, at least in part, the intimate adherence of the bacteria to the enterocyte cytoplasmic membrane, and for the *tir* gene coding for the intimin receptor which is translocated into the enterocyte cytoplasmic membrane by the type III secretion system (4,9,15,24). The LEE of two bovine EHEC studied have a very similar organization (Goffaux, unpublished data). Expression of all these different genes was necessary for the production of AE lesions *in vitro* and *in vivo*, although this had not previously been demonstrated for bovine EHEC and EPEC (15,24,26).

The study of the internal composition of the LEE is also useful with respect to the typing of the EHEC from cattle, foodstuffs, and humans. Seven variants of the *eae* gene ($\alpha1$, $\alpha2$, β, $\gamma1$, $\gamma2$, δ, ε) and three variants of the *tir*, *espA*, *espB*, and *espD* genes (α,β,γ) have been described in EHEC and EPEC from various origins (5,27). Comparison of these genes in human and bovine EHEC (5, China, Goffaux, and Janssen, unpublished data) shows that (1) not all variants, nor variant combinations, have been observed either in bovine or human EHEC (to date, only four variant combinations have been detected: *eaeβ,tirβ,espβ; eaeγ2,tirα,espα; eaeγ1,tirγ,espγ* and *eaeε,tirβ,espβ);* (2) bovine and human EHEC of the same serogroup can share the same variant profile (e.g., *eaeβ,tirβ,espβ* amongst O26 EHEC; *eaeε,tirβ,espβ* amongst O103 EHEC; *eaeγ2,tirα,espα* amongst O111 EHEC; and *eaeγ1,tirγ,espγ* amongst O157 EHEC); (3) the *eaeβ* variant is predominant amongst EHEC isolated from diarrheic calves (75%) but not amongst EHEC isolated from healthy bovines at slaughter.

The cellular response to the translocation of Esp and Tir proteins is basically similar in human EHEC and EPEC, although some differences have been observed in the biochemical cascade leading to the cytoskeleton rearrangement, microvilli effacement and pedestal formation (26). Currently, no data are available on the cellular response following infection with bovine EHEC or EPEC.

Invasive Properties. Invasion of the eukaryotic cells by at least some bovine and human EPEC and EHEC has been observed *in vitro* on cell cultures (8,26) and in *in vivo* animal models (10), in calves (35) and in humans (33). EHEC can actually be isolated from the mesenteric lymph nodes (35). The significance of those observations is still unknown.

The "EHEC Virulence Plasmid"

Human O157:H7 EHEC harbour a ca. 60 MDa plasmid called the "EHEC virulence plasmid", pEHEC or pO157 (15,26). This plasmid is present in O26:H11 EHEC and in most other serotypes of EHEC. The pEHEC carries

(1a)

genes encoding for an enterohaemolysin, i.e., a haemolysin active only on washed red blood cells (*ehx* genes), a type II secretion system (*etp* genes), a catalase-peroxydase (*katP* gene), and a secreted serine protease (*espP* and *pssA* genes) (22,26,36).

(1b)

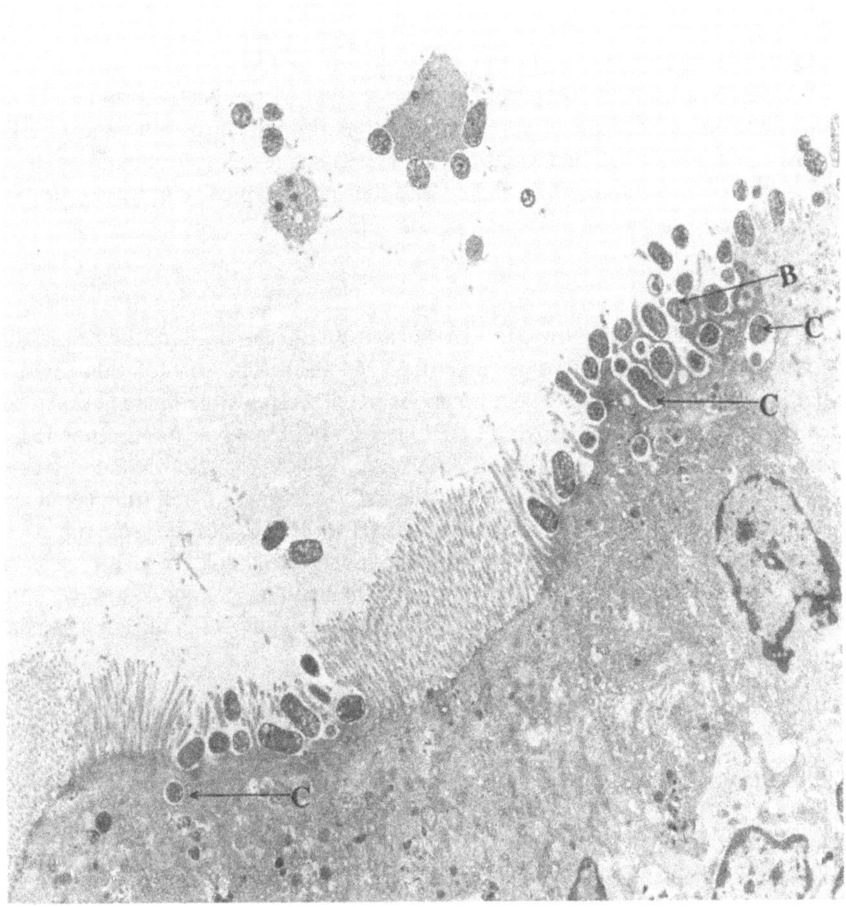

FIG. 1. TRANSMISSION ELECTRON MICROSCOPY OF RABBIT LIGATED INTESTINAL LOOPS INOCULATED WITH A BOVINE EHEC (a) AND A BOVINE EPEC (b) RESPECTIVELY, SHOWING THE EMBEDMENT OF THE *E. COLI* WITHIN THE ENTEROCYTE MICROVILLI, THE EFFACEMENT OF THE MICROVILLI, THE INTIMATE ATTACHMENT OF THE BACTERIA TO THE ENTEROCYTE CYTOPLASMIC MEMBRANE (A), THE PRESENCE OF PEDESTAL STRUCTURES (B) AND OF PHAGOCYTOSIS IMAGES (C) (AUTHOR'S COLLECTION)

The majority of bovine EHEC and VTEC produce the enterohaemolysin phenotype on washed sheep blood cell agar plates, and/or test positive with the pCVD419 DNA probe which corresponds to the *ehx* genes. The EHEC enterohaemolysin phenotype or genotype is associated with the production of VT1 alone or in combination with VT2, in at least 75% of the strains tested, but less often with the production of VT2 alone (50% of the strains tested). The production of enterohaemolysin by bovine EHEC and VTEC is also mediated by a plasmid (38,39, Goffaux and Janssen, unpublished data). It is not known, whether or not this plasmid carries the other genetic determinants present on the pO157 plasmid of human EHEC.

The role of these properties in the survival of the bacteria and/or in the development of disease is purely speculative. Initially a role for the pEHEC plasmid in the initial attachment of EHEC was suggested (16), but no evidence was ever obtained to confirm this role, either phenotypically *in vitro* or *in vivo* (28,37) or by genetic sequencing studies (22,36).

Others

Human O157 and non-O157 EHEC and VTEC can produce the heat-stable toxin of enteroaggregative *E. coli* (EAST1) (32). The role of this toxin in pathogenesis is speculative, but it may be in part responsible for the occurrence of non-bloody diarrhea. Bovine EHEC and VTEC have not, to our knowledge, been tested for the production of EAST1.

Bovine EHEC and VTEC are generally resistant to tellurite (Pohl and Mainil, unpublished data), as are human O157 EHEC and *Shigella* spp. (43). The relevance to the pathogenesis of this resistance is unknown. It is possible that this resistance might play some role in the pathology of these strains and/or in their ecology and survival in the cattle rumen and gut and in the environment.

PATHOGENESIS AND OCCURRENCE OF DIARRHOEA

Natural infection in young calves with EHEC most probably occurs by the oral route. The EHEC mainly colonise the large intestine where they cause the formation of the AE lesions, and sometimes cause haemorrhagic enteritis. However, the actual cause of diarrhoea is still unknown. At one stage the AE lesions were thought to be responsible for the diarrhea, but in volunteers, diarrhoea occurs too rapidly (15,27) to fit with this suggestion. Events which could account for the occurrence of diarrhoea include (1) the production of an enterotoxin, like the EAST1, (2) the activation of the major chloride channel in the enterocyte microvilli as a consequence of the activation of protein kinase C, or (3) the opening of the tight junctions between enterocytes as a consequence of activation of myosin light chain kinase (14). As already mentioned, the haemorrhagic aspect of the diarrhoea is most probably the consequence of the activity of the VTs on the endothelial cells of the vessels of the intestinal wall (26).

SOURCES OF INFECTION AND PUBLIC HEALTH HAZARD

The epidemiology of EHEC and VTEC infection in young calves has not been fully established, but it seems reasonable to speculate that they are carried by healthy individuals, calves and adults, and that the clinical disease is triggered by management problems such as overcrowding, transport or the presence of other pathogens. Diseased calves and healthy cattle can be infected with more than one strain based on serotype, pathotype, and Pulse Field Gel Electrophoresis (PFGE) profile (19).

The main problem is, however, the public health concern about the transmission of bovine EHEC and VTEC to humans (Chapters 8 and 13). To answer that question a comparison of human and bovine strains must be carried out using powerful epidemiological tools (Chapter 5).

Based on serotyping, three categories of EHEC strains could be previously distinguished: those mainly associated with diseases in calves (O5, O118); those which are pathogenic for humans and calves (O26, O103, O111); and those carried by healthy cattle but associated with disease in humans (O157) (19,26). However, the first two categories are becoming less and less clearly defined as the range of serogroups and serotypes isolated from both cattle and humans continues to expand. More precise molecular studies of the bacteria and of their virulence-associated factors and properties (e.g., *stx* genes and phages, LEE, EHEC virulence plasmids) must be carried out to fully compare these bacterial strains and group them according to their pathogenic specificity and public health hazard potential (Chapter 6).

VTEC INFECTIONS IN OTHER DOMESTIC RUMINANTS AND IN WILD RUMINANTS

Historically, small ruminants have been the subject of fewer surveys than cattle. Recently, O157 EHEC were detected in sheep and goats in feces and at slaughter, indicating that such smaller ruminants may also represent a source of contamination for humans. Sheep positive for VTEC of other serotypes have also been detected, with a wide variation from flock to flock (19). In goat herds, the percentage of positive animals also varied greatly, from 7% to 95% (19). Positive strains display various VT profiles and many hybridize with the pCVD419 DNA probe for the *ehx* genes. Transmission of O157 and other EHEC serotypes to humans by raw goats milk or home-made raw-milk cheese has been documented (2,3).

O157 EHEC have also been isolated from wild and farmed deer in the UK and USA and have, in one case, been associated with a human outbreak. VTEC of various serogroups have also been detected in diarrheic and non-diarrheic buffalo calves in Sri Lanka (25) and several fecal samples from various wild ruminant species in nature reserves in Belgium were found positive with using PCR for the *stx1* and *stx2* genes (29).

In more general terms, the association between VTEC infections and diseases in young small ruminants and in wild ruminants has not, to our knowledge, received much attention.

FUTURE OF THE RESEARCH ON VTEC INFECTIONS IN RUMINANTS

Much has been learned about VTEC infections and carriage in ruminants, especially domestic cattle, during the last twenty years, but there are still areas in which our knowledge is inadequate. While there are many areas to be addressed within future research on VT, VTEC and EHEC, there are two particularly important questions:

- what is the ecology of EHEC carriage, especially of the O157 serogroup (Chapter 14)?
- what is the vaccination potential against VTEC and EHEC (Chapter 19)?

The potential for vaccination in cattle is twofold: firstly, while EHEC are known to cause diarrhoea in young calves, the role of VT1 is still speculative as no clinical systemic complications such as HUS have been described. Vaccination against EHEC would thus target the AE phenotype determinants, mainly the intimin, more than the VTs (12), although the primary colonization factor(s) would also be important. Secondly, there is also much excitement and speculation about preventing the carriage of the O157:H7 EHEC by vaccination (7,12).

ACKNOWLEDGMENTS

The authors wish to thank Prof. Albert Kaeckenbeeck, honorary Professor, Dr. Harley W. Moon (Iowa State University, Ames, Iowa), and Dr. Pierre Pohl (Centre d'Etude et de Recherches Vétérinaires et Agrochimiques, Bruxelles,

Belgique), who recently retired, for their help at various stages in their research work and understanding of VTEC infections in cattle.

The research work on bovine Verotoxigenic *Escherichia coli* in the authors' laboratory has been supported financially by the "Institut pour l'encouragement de la Recherche dans l'Industrie et l'Agriculture" (research contract #5631 — 1995-1996) and by the "Ministère des Classes Moyennes et de l'Agriculture" (research contracts #5740 and 5869 — 1997-2000). The authors also thank Ms. Vinciane Pirson and Mr. Etienne Jacquemin for their technical help.

REFERENCES

1. BEUTIN, L. and MÜLLER, W. 1998. Cattle and verotoxigenic *Escherichia coli* (VTEC), an old relationship? Vet. Rec. *142*, 283–284.
2. BIELASZEWSKA, M. *et al.* 1997. Human *Escherichia coli* O157:H7 infection associated with the consumption of unpasteurized goat's milk. Epidemiol. Infect. *119*, 299–305.
3. CASENAVE, C. *et al.* 1993. Eclosion du syndrome hémolytique et urémique (SHU) dans une commune du Cher. Bulletin Epidémiologique Hebdomadaire *48*, 222–224.
4. CHINA, B. and GOFFAUX, F. 1999. Secretion of virulence factors by *Escherichia coli*. Vet. Res. *30*, 181-202.
5. CHINA, B., GOFFAUX, F., PIRSON, V. and MAINIL, J. 1999. Comparison of *eae*, *tir*, *espA*, and *espB* genes of bovine and human attaching and effacing *Escherichia coli* by multiplex polymerase chain reaction. FEMS Microbiol. Letters *178*, 177–182.
6. CHINA, B., PIRSON, V. and MAINIL, J. 1998. Prevalence and molecular typing of attaching and effacing *Escherichia coli* among calf populations in Belgium. Vet. Microbiol. *57*, 1–11.
7. DEAN-NYSTRÖM, E.A., BOSWORTH, B.T., MOON, H.W. and O'BRIEN, A.D. 1998. Bovine infection with Shiga toxin-producing *Escherichia coli in Escherichia coli* O157:H7 and other Shiga toxin-producing *E. coli* strains, (J.B. Kaper and A.D. O'Brien, eds.), pp. 261–267. ASM Press, Washington D.C.
8. DONNENBERG, M.S. and KAPER, J.B. 1992. Enteropathogenic *Escherichia coli*. Infect. Immun. *60*, 3953–3961.
9. DOZOIS, C. and CURTISS III, R. 1999. Pathogenic diversity of *Escherichia coli* and the emergence of "exotic" islands in the gene stream. Vet. Res. *30*, 181–202.
10. GOFFAUX F. *et al.* 1997. Bovine attaching and effacing *Escherichia coli* possess a pathogenesis island related to the LEE of the human enteropathogenic *Escherichia coli* strain E2348/69. FEMS Microbiol. Lett. *154*, 415–421.
11. EECKHOUT, I., DE FILETTE, M., PAESEN, H., IMBERECHTS, H. and D'HOOGHE, I. 1999. Evidence for *afa*-like sequences in a verotoxin-producing *Escherichia coli* of serotype O103 of bovine origin (Poster). Proceedings of the 3rd Meeting of the European Concerted Action CT98-3935 on Verotoxigenic *Escherichia coli* in Europe: Pathogenicity and Virulence. Abstract 4.2.

12. GYLES, C.L. 1998. Vaccines and Shiga toxin-producing *Escherichia coli* in animals, *In Escherichia coli* O157:H7 and other Shiga toxin-producing *E. coli* strains, (J.B. Kaper and A.D. O'Brien, eds.) pp. 434–444. ASM Press, Washington D.C.
13. HALL G.A. *et al.* 1985. Dysentery caused by *Escherichia coli* (S102-9) in calves: natural and experimental disease. Vet. Pathol. *22*, 156–163.
14. KAPER, J.B. 1998. EPEC delivers the good, Trends Microbiol. *6*, 169–172.
15. KAPER, J.B., ELLIOTT, S., SPERANDIOI, V., PERNA, N.T., MAYHEW, G.F. and BLATTNER, F.R. 1998. Attaching- and effacing intestinal histopathology and the locus of enterocyte effacement, *In Escherichia coli* O157:H7 and other Shiga toxin-producing *E. coli* strains, (J.B. Kaper and A.D. O'Brien, eds.) pp. 163–182, ASM Press, Washington D.C.
16. KARCH, H., SCHMIDT, H. and BRUNDER, W. 1998. Plasmid-encoded determinants of *Escherichia coli* O157:H7, *In Escherichia coli* O157:H7 and other Shiga toxin-producing *E. coli* strains, (J.B. Kaper and A.D. O'Brien, eds.) pp. 183–194. ASM Press, Washington D.C.
17. KASHIWAZAKI, M., OGAWA, T., ISAYAMA, Y., AKAIKE, Y., TAMURA, K. and SAKAZAKI, R. 1980. Detection of Vero cytotoxic strains of *Escherichia coli* isolated from diseased animals. Natl Inst. Anim. Health Q. (Yatabe) *20*, 116–117.
18. KONOWALCHUK, J., SPEIRS, J.I. and STAVRIC, S. 1977. Vero response to a cytotoxin of *Escherichia coli*. Infect. Immun. *18*, 775–779.
19. MAINIL, J. 1999. Shiga/Verocytotoxins and Shiga/Verotoxigenic *Escherichia coli* in animals. Vet. Res. *30*, 235–258.
20. MAINIL, J.G. *et al.* 1987. Shiga-like toxin production and attaching effacing activity of *Escherichia coli* associated with calf diarrhea. Am. J. Vet. Res. *48*, 743–748.
21. MAINIL, J.G., JACQUEMIN, E., KAECKENBEECK, A. and POHL, P. 1993. Association between the effacing (*eae*) gene and the Shiga-like toxin-encoding genes in *Escherichia coli* isolates from cattle. Am. J. Vet. Res. *54*, 1064–1068.
22. MAKINO, K. *et al.* 1998. Complete nucleotide sequence of 93-kb and 3.3 kb plasmids of an enterohaemorrhagic *Escherichia coli* O157:H7 derived from Sakai outbreak. DNA Res. *28*, 1–9.
23. MC DANIEL, T.K., JARVIS, G., DONNENBERG, M.S. and KAPER, J.B. 1995. A genetic locus of enterocyte effacement conserved among diverse enterobacterial pathogens. Proc. Natl Acad. Sci. USA *23*, 1664–1668.
24. MILON, A., OSWALD, E. and DE RYCKE, J. 1999. Rabbit EPEC: a model for the study of enteropathogenic *Escherichia coli*. Vet. Res. *30*, 203–220.
25. MOHAMMAD, A., PEIRIS, J.S.M., WIJEWANTA, E.A., MAHALINGAM, S. and GUNASEKARA G. 1985. Role of verocytotoxigenic *Escherichia coli* in cattle and buffalo calf diarrhea. FEMS Microbiol. Lett. *26*, 281–283.
26. NATARO, J.P. and KAPER, J.B. 1998. Diarrheagenic *Escherichia coli*. Clin. Microbiol. Rev. *11*, 142–201.

27. OSWALD, E., SCHMIDT, H., MORABITO, S., KARCH, H., MARCHES, O. and CAPRIOLI, A. 2000. Typing of intimin genes in human and animal enterohemorrhagic and enteropathogenic *Escherichia coli*: characterization of a new intimin variant. Infect. Immun. *68*, 64–71.
28. PATON, J.C. and PATON, A.W. 1998. Pathogenesis and diagnosis of Shiga toxin-producing *Escherichia coli* infections. Clin. Microbiol. Rev. *11*, 450–479.
29. PIRSON, V., CHINA, B., GOFFAUX, F. and MAINIL, J. 1999. Prevalence and properties of verotoxigenic *Escherichia coli* (VTEC) in the faeces of wild ruminants in Belgium (Poster). Proceedings of the 3rd Meeting of the European Concerted Action CT98-3935 on Verotoxigenic *Escherichia coli* in Europe: Pathogenicity and Virulence. Abstract 2.7.
30. POHL, P., DAUBE, G., LINTERMANS, P., KAECKENBEECK, A. and MAINIL, J.G. 1991. Description de 70 souches d'*Escherichia coli* d'origine bovine possédant les gènes des vérotoxines. Ann. Méd. Vét. *135*, 267–272.
31. POHL, P. *et al.* 1997. Différences entre les pathotypes et les sérogroupes des *Escherichia coli* vérotoxinogènes isolées de veaux sains et celles isolées de veaux souffrant de diarrhée. Ann. Méd. Vét. *141*, 155–159.
32. SAVARINO, S.J. *et al.* 1996. Enteroaggregative *Escherichia coli* heat-stable enterotoxin is not restricted to enteroaggregative *Escherichia coli*. J. Infect. Dis. *173*, 1019–1022.
33. SCALETSKY, I.C.A., PEDROSO, M.Z. and FAGUNDES-NETO, U. 1996. Attaching and effacing enteropathogenic *Escherichia coli* O18ab invades epithelial cells and cause persistent diarrhea. Infect. Immun. *64*, 4876–4881.
34. SHERWOOD, D., SNODGRASS, D.R. and O'BRIEN, A.D. 1985. Shiga-like toxin production from *Escherichia coli* associated with calf diarrhea. Vet. Rec. *116*, 217.
35. STORDEUR, P., CHINA, B., CHARLIER, G., ROELS, S. and MAINIL, J. 2000. Clinical signs, reproduction of attaching/effacing lesions, and enterocyte invasion after oral inoculation of an O118 enterohaemorrhagic *Escherichia coli* in neonatal calves. Microb. Infect. *2*, 17–24.
36. TOBE, T., HAYASHI, T., HAN, C.G., SCHOLLNIK, G.K., OHTSUBO, E. and SASAKAWA, C. 1999. Complete DNA sequence and structural analysis of the enteropathogenic *Escherichia coli* adherence factor plasmid. Infect. Immun. *67*, 5455–5462.
37. TZIPORI, S. *et al.* 1987. Role of a 60-megadalton plasmid and shiga-like toxins in the pathogenesis of infection caused by enterohaemorrhagic *Escherichia coli* O157:H7 in gnotobiotic piglets. Infect. Immun. *55*, 3117–3125.
38. WIELER, L.H. 1997. Bestimmung von Virulenzfaktoren boviner Shiga-Toxin-bildender *Escherichia coli*-(STEC-) Stämme als Bewertungsgrundlage ihrer klinischen Bedeutung für Rind und Mensch, "Habilitationthesis", Universität Giessen (Germany), Fachbereich Veterinärmedizin.
39. WIELER, L.H., BAUERFEIND, R. and BALJER, G. 1992. Characterization of Shiga-like toxin-producing *Escherichia coli* (SLTEC) isolated from calves with and without diarrhea. Zentralbl. Bakteriol. *276*, 243–253.

40. WIELER, L.H. *et al.* 1998. Virulence properties of Shiga toxin-producing *Escherichia coli* (STEC) strains of serogroup O118, a major group of STEC pathogens in calves. J. Clin. Microbiol. *36*, 1604–1607.
41. WILSON, M.W. and BETTELHEIM, K.A. 1980. Cytotoxic *Escherichia coli* serotypes. Lancet *1*(8161), 201.
42. YANO, T., GARCIA, M., LEITE, D.S., DE CAMARGO, I.J.B. and DE CASTRO, A.F.P. 1988. Fimbria-like adhesive factor (EAF44) from verotoxigenic *Escherichia coli* of bovine origin. Res. Vet. Sci. *45*, 418–419.
43. ZADIK, P.M., CHAPMAN, P.A. and SIDDONS, C.A. 1993. Use of tellurite for the selection of enterohaemorrhagic *Escherichia coli* O157. J. Med. Microbiol. *39*, 155–158.

CHAPTER 12

PATHOGENIC ASPECTS OF VTEC INFECTION IN NON-RUMINANT ANIMALS

CARLTON L. GYLES

Dept. of Pathobiology
Ontario Veterinary College
University of Guelph
Guelph, Ontario, Canada

Verotoxigenic *Escherichia coli* (VTEC) have been recovered from the feces or intestine of non-ruminant animals, including chickens, horses, cats, dogs, and pigs, but a role in disease has been demonstrated only in dogs and pigs.

VTEC in Avian Species

VTEC have only rarely been isolated from chickens. Researchers in several countries failed to detect VTEC in over 1400 samples of chicken feces or carcasses (8,10,32,46,48), but an O103:H- VTEC isolated from the intestinal contents of a chicken was shown to induce attaching and effacing (AE) lesions in experimentally infected chickens (54). Furthermore, the intestine of chickens may be colonized for prolonged periods following experimental infection with some strains of O157:H7 VTEC (41).

Wild birds may also carry O157 VTEC. Wallace *et al.* (60) isolated O157 VTEC from 12 of 691 *E. coli* recovered from fecal samples of wild birds (mainly gulls) at a landfill site and a bay.

Recently, Blanco *et al.* (10) reported that none of 458 *E. coli* strains from septicemia in broiler chickens nor 167 from healthy controls produced verotoxin. Parreira and Yano (44) identified a verotoxin, called VT2y, in 36 of 50 *E. coli* from swollen head syndrome (SHS) in chickens. The toxin was produced in low concentration and was neutralized by antibodies to VT2. The researchers speculated that VT2y may play a role in SHS as the target organs and lesions are similar to those seen in the VT-mediated edema disease of pigs.

VTEC in Horses

Horses have occasionally been implicated as a source of VTEC. There is one report of bloody diarrhea in a farmer from whom an O157 VTEC was isolated and shown to be identical to an O157 VTEC recovered from a horse (16). The horse had diarrhea attributable to *Salmonella typhimurium* and had been under the farmer's care. Trevena *et al.* (55) reported the isolation of the same strain of O157 VTEC from a boy with diarrhea and from a dog and a pony on a farm visited by the boy. Hancock *et al.* (26) found O157:H7 VTEC in the feces of 1 of 90 healthy horses.

VTEC in Cats

Verotoxigenic *E. coli* are carried by healthy cats (2,8,49). In a study of VTEC in healthy domestic animals, Beutin *et al.* (8) found no VTEC in 26 cats from private homes but in 9 of 39 cats from animal shelters. Eight of the nine VTEC belonged to the same serotype (O21:K5:H14). The overall frequency (9/65) was similar to that reported in healthy cats (11/66) by Smith *et al.* (49). In one study in which the researchers looked specifically for O157 *E. coli*, none were found in feces from 33 cats (26).

Epidemiological studies have been used to assess whether fecal excretion of VTEC by cats is associated with diarrheal disease. Abaas *et al.* (2) showed cytotoxicity for Vero cells in culture supernatants from 40% of 25 *E. coli* isolates from healthy cats and 95% of 22 *E. coli* from diarrheic cats. Forty-one of the 47 *E. coli* isolates were hemolytic. The verotoxin activity was neutralized with antiserum against Shiga toxin in all but one of the positive isolates. Among the 32 feline VTEC that were identified, 13 belonged to serogroup O6. There were significant differences between the healthy and diarrheic cats with respect to both frequency of excretion of VTEC and level of activity of VT. Beutin *et al.* (8) found that 8 of 9 feline VTEC, all of serotype O21:K5:H14, produced alpha hemolysin. In contrast to the report by Abaas *et al.* (2), a case-control study involving 179 cats (113 with diarrhea and 66 with normal feces) found that there was no significant association between excretion of VTEC and diarrheal illness, and that the majority of the VTEC produced a VT that differed from VT1 and VT2 (49). In each group there were 11 cats in which VTEC were detected. The detection of an atypical VT from feline isolates was also reported by other researchers (8).

VTEC in Dogs

VTEC are carried by dogs and have been implicated in disease. Beutin *et al.* (8,9) found that 3 of 63 healthy dogs were carriers of VTEC, two of which produced VT1 and VT2 and one of which produced VT2 only. Similarly, Gallien *et al.* (23) found that 1 of 25 healthy dogs excreted VTEC in its feces. A number of researchers have investigated the association of VTEC with diarrhea in dogs. In a study of 57 normal and 45 diarrheic dogs, VTEC with the *vt1* gene were found in similar percentages (8.9% and 12.3%) in isolates from diarrheic and normal dogs (25). However, VTEC with the *vt2* gene were isolated from 22.2% of diarrheal dogs but 0% of normal dogs. Dogs may transmit VTEC infections to humans through fecal carriage or contamination of their coats.

Sporadic cases of hemolytic uremic syndrome (HUS) have been reported in young dogs of several breeds, which showed a prodrome of bloody diarrhea, followed by thrombocytopenia, microangiopathic hemolytic anemia, and anuric acute renal failure (29). Necrosis and hemorrhage of the renal glomeruli and proximal tubules were noted, as were fibrin thrombi and fibrinoid necrosis of blood vessels.

A disease called cutaneous and renal glomerular vasculopathy (CRVG) has been associated with the feeding of raw ground beef to greyhounds in the US (15,17,21,27). VT and VTEC, including O157:H7, were found in samples of the ground beef and in the feces of affected dogs (21). VTEC, including O157:H7, were also recovered, albeit at a lower frequency, from healthy greyhounds. Typically, dogs are non-febrile and present with cutaneous ulcers and edema of the extremities. About 5% of the affected dogs develop HUS, which may be preceded by diarrhea. Renal lesions are most prominent in the glomeruli, in which necrosis and loss of capillary endothelial cells, and adherence of aggregated platelets to the basement membranes are observed in the early stages (27). Later, fibrin thrombi and hypertrophic endothelial cells are evident. In the skin, vasculitis of arterioles and epidermal necrosis are observed. Recently, Fenwick and Cowan (21) reported that they reproduced hemorrhagic colitis in dogs following injection of low concentrations of VT1 or VT2, but they were not successful in reproducing the kidney lesions.

VTEC and Edema Disease in Pigs

Edema disease (ED) is caused by certain types of VTEC and derives its name from the massive edema that develops in specific sites in affected pigs (6,53). The disease occurs almost exclusively in pigs but there has been one reported case of edema disease in a sea lion (20). A hemolytic *E. coli* of serotype O147:K52 was recovered from the animal but the isolate was not tested for production of VT.

Clinical Signs and Symptoms. ED of pigs is characterized by inappetence, swelling of the eyelids, a peculiar squeal, and loss of coordination, with respiratory distress as a common terminal development (6,53). There is no diarrhea nor elevation of body temperature, and some pigs die suddenly without showing any signs of illness. Cases may be sporadic or may occur in an outbreak. The major bacterial and host factors that are implicated in disease are shown in Fig. 1 and discussed below.

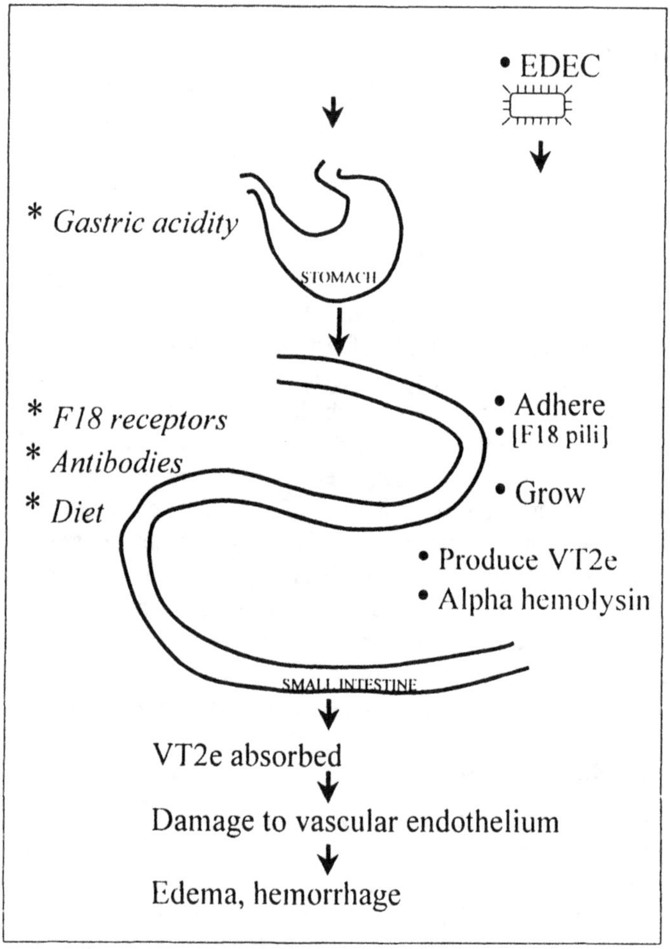

FIG. 1. BACTERIAL AND HOST FACTORS IN EDEMA DISEASE OF PIGS
EDEC = Edema disease producing *E. coli*.
* Host factors. • = Bacterial factors.

Bacterial Factors.

Edema disease toxin (VT2e). The signs, symptoms, and pathological lesions of ED are attributable to an *E. coli* toxin (37), variously called edema disease toxin, edema disease principle, verotoxin 2e, 2v or 2vp1 (VT2e, VT2v, or VT2vp1), Shiga-like toxin IIe or IIv (SLT-IIe or IIv), and Shiga toxin 2e

(Stx2e). The toxin is highly potent, and pigs injected intravenously with as little as 3 nanograms of toxin per kg body weight develop characteristic ED (37). Like other verotoxins, VT2e binds to a specific receptor in epithelial or vascular endothelial cells. VT2e can bind the glycolipid receptor globotriaosylceramide (Gb3), which is used by VT1 and VT2, but it binds preferentially to globotetraosylceramide (Gb4) (13,18). The amount of toxin produced *in vitro* by edema disease-producing *E. coli* (EDEC) is less than that produced by strains that produce VT1 and/or VT2 (24,38).

The toxin is detectable in the intestinal contents, and toxin-mediated damage to vascular endothelium can be identified in the target tissues of pigs that die from the disease (6,53). Many EDEC are non-enterotoxigenic. This is often the case for O139:K82 strains (1,30,35). However many VT2-positive *E. coli*, including O139:K82 strains, also carry genes for enterotoxins (24,30,51,61). Some researchers have observed that VT2e-positive strains that are implicated in diarrheal cases possess enterotoxin genes, while those that are involved in ED lack enterotoxin genes (11,35). However, in experimental and natural infections, strains which possess genes for both VT2e and enterotoxins can induce diarrhea alone, or diarrhea and ED (30,51).

VT2e is produced almost exclusively by EDEC, but VTEC which produce VT2e have occasionally been implicated in diarrhea and in hemolytic uremic syndrome in humans (45). The VT2e-positive strains recovered from humans belonged to O serogroup 101, a serogroup of VT2e-producing *E. coli* which has been isolated from the feces of slaughtered pigs (14).

Alpha Hemolysin. EDEC usually produce alpha hemolysin (11,23,24,52), which can damage membranes of a wide range of host cells. The hemolysin is produced during the course of disease but elimination of the genes for alpha hemolysin does not prevent a strain of VTEC from causing experimental ED (52). The pigs used in the experimental infections were unusually susceptible to infection, and it is probable that alpha hemolysin contributes to intestinal colonization, as hemolytic *E. coli* appear to have a selective advantage compared with nonhemolytic strains in the intestinal environment of recently weaned pigs (19).

F18 pili (fimbriae). F18 pili (previously F107), which are not readily expressed *in vitro*, are the pili which EDEC use to attach to intestinal epithelial cells (28,42). There are two types of F18 pili — F18ab and F18ac- and EDEC, especially those of O139, usually produce the F18ab pili (42,61). The genes for F18 pili are carried by large plasmids, which also encode alpha hemolysin synthesis, except in the case of O139:K82 isolates in which the F18 genes are chromosomal (62). It has been suggested that a capsule layer on O139:K12 strains of *E. coli* could play a role in adhesion to the intestine (39).

O antigens. The strains of *E. coli* that are most frequently isolated from pigs with edema disease belong to serotypes O138:K81:H14 or H-; O139:K12(82):H1 or H- and O141:K85:H4 (1,4,11,24,35,61). The O139 organisms tend to be the most common but the dominant serotype in a region may vary over time (1,35). Other serogroups of porcine *E. coli* which have been shown to possess the genes for VT2e and/or produce VT2e include O2:H-, O2:H?, O9:H-, O101, O107:H29, O107:H-, O120:H-, O121:H10, O125ac:H27, O154:H-, O157:KV17, and O rough:H11 (4,24). The O antigen does not appear to be required for disease as rough isolates which lack this antigen have been recovered from animals with ED.

Typically, most strains of O139:K82 *E. coli* that are tested are positive for VT2e (1,4,11,24,61), but in one report from Australia, none of 10 O139:F18ab *E. coli* from weaned pigs were positive with a probe for *vt2* (28). In one study, VT2e was produced by 12 of 12 O139:K82, 15 of 25 O138:K81, and 9 of 14 O141:K85 strains, but by only 2 of 101 O149:K91:K88ac and 5 of 66 O157:KV17 strains (24). The presence of a low percentage of VT2e producers among strains of these latter serotypes, which are porcine ETEC, suggests that there may be some horizontal transfer of the *vt2e* genes, even though they are chromosomal and have not been associated with bacteriophages. Interestingly, Smith *et al.* (50) transferred the genes for VT2e to *E. coli* K12 following coincubation in broth of the K12 organism with a wild VT2e-positive *E. coli*. VT1 and/or VT2 are only rarely produced by porcine *E. coli* (8,24).

Host and Environmental Factors in Disease. Pigs may resist ED because they lack the intestinal receptors for the F18 pili which mediate colonization by VTEC that cause edema disease. The presence or absence of the receptor is determined by a single locus with a dominant susceptibility allele and a recessive resistant allele (6). Until recently, experimental infections were needed to identify resistant and susceptible pigs (6), but *in vitro* tests can now identify the genes associated with resistance/susceptibility (56). Vogeli and coworkers (56) mapped the gene for the F18 receptor locus in pigs to the halothane linkage group on chromosome 6. They then showed that the gene for alpha (1,2) fucosyltransferase (FUT1) in pigs is closely linked to this locus and that pigs which lack the receptor have a specific mutation in this gene. They suggested that FUT1 may be required for synthesis of a glyconjugate receptor for the F18 pili. On the basis of the occurrence of the specific mutation in the gene for FUT1, resistance was present in only 1% of Swiss Landrace and 11% of Large White pigs (56).

Age influences susceptibility to edema disease through age-related expression of receptors for F18 pili. *E. coli* with F18 pili bind to intestinal epithelial cells from older pigs but not to intestinal epithelial cells from younger pigs (43).

Antibodies against EDEC or VT2e can inhibit colonization of the intestine and the toxic effects of the VT. Weaning removes pigs from sow's milk with its antibodies and other protective components, leading to enhanced susceptibility of pigs in the postweaning period, and a peak in prevalence of ED during the first 1-2 weeks after pigs are weaned (4,6,53). Sow's milk often contains antibodies to surface antigens of pathogenic *E. coli* and its addition to the diet of weaned pigs has been shown to reduce the excretion of hemolytic *E. coli* (19,47), including EDEC. Antibodies to VT2e in the serum of pigs may also affect their susceptibility to edema disease. In one study, anti-VT2e IgG antibodies were detected in 3.6% of 28 affected pigs, compared with 21.5% of unaffected pigs in the same herd during an outbreak of edema disease (22). In another study, no toxin neutralizing antibodies were detected in the sera of 45 pigs, which included pigs from herds with a history of edema disease (24).

Certain physiologic changes which occur in the intestine of the pig at the time of weaning lead to a period of temporary malabsorption that may enhance susceptibility to colonization of the intestine by *E. coli*. Another host factor which influences susceptibility to disease is rapid growth. The fastest growing pigs in a litter are most commonly affected in an outbreak of edema disease (53), but the reason for this is not known.

Diet influences the colonization of the intestine of weaned pigs by pathogenic *E. coli*. High protein diets, which are desirable for rapid weight gain, favour colonization and disease (6,51), and reduction of feed intake or use of a low nutrition diet can prevent disease.

Stresses associated with weaning predispose to development of edema disease. These are associated with separation of pigs from their dams, transportation, and mixing of pigs from a variety of sources, and are believed to facilitate intestinal proliferation of *E. coli*, including strains which cause ED.

Pathogenesis. Major factors that have been shown or suggested to be involved in pathogenesis of ED are shown in Figure 1. Certain aspects of pathogenesis are not understood and it is difficult to reproduce the disease, using virulent *E. coli* strains and apparently susceptible pigs. It is assumed that low gastric pH inhibits passage of EDEC into the intestine but the resistance of EDEC to gastric acidity has not been investigated. Among VTEC diseases, ED is unique with respect to the ETEC-type colonization of the intestine and the fact that all clinical cases appear to be the result of toxin absorption and systemic effects of VT.

Colonization of the pig's intestine by EDEC appears to develop slowly. In experimental studies, maximal colonization by EDEC required 3 to 6 days (7,51), in contrast to 2-3 days for diarrheal strains (51). This may be related to the production of F18ab pili which are less effective in mediating adherence to intestinal epithelial cells compared with F18ac pili (42). The pattern of association of the bacteria with the intestinal epithelium is similar to that described for enterotoxigenic *E. coli* and neither intimate adherence nor effacement of microvilli occurs (7,40).

VT2e, a large protein molecule, needs to be absorbed in a functional form from the intestine and transported to target tissues. However, the toxin does not appear to be absorbed from the intestine under normal conditions (58) and it is likely that pigs that are affected have an unusual ability to absorb the toxin. Receptors that bind VT2e have been detected on enterocytes in the crypt portion of villi in the small intestine of pigs (57,59) but their role in absorption of VT2e is unknown. Other bacterial products and/or host factors may contribute to the absorption of toxin. Addition of deoxycholate to the intestine allowed absorption of VT2e to occur (58) and it is possible that diet could influence secretion of bile, which could affect absorption.

Absorbed toxin binds to vascular endothelial cells in target tissues and damages them, resulting in edema and hemorrhage. The presence or absence of receptors for VT2e is believed to determine the tissue distribution of lesions. However, receptors for VT2e have been shown to be present in non-target tissues such as liver and spleen (57). These findings support the view that other factors are important in determining which organs are affected (45). Receptors for VT2e were also found in vascular smooth muscle (57), a tissue which undergoes necrosis in target organs in pigs with ED.

Pathology. Postmortem examination of pigs that have died of ED shows edema and hemorrhage in the subcutaneous tissues of the eyelids and forehead, the greater curvature of the stomach, the mesenteric lymph nodes, the colonic mesentery and the brain, especially the cerebellum (4,7,40). Light and electron microscopic examination of naturally- and experimentally-infected pigs show that intestinal colonization does not result in structural changes to enterocytes, and that there are degenerative changes in vascular endothelial cells, thrombosis in some blood vessels, and perivascular edema and necrosis of vascular smooth muscle (4,7,40).

Prevention of Disease. The likelihood of disease in a herd may be reduced by changes in management to minimize contamination of the environment with EDEC, reduce stresses associated with weaning, and supply protective antibodies. Vaccination with a toxoided crude extract of an EDEC (5), or with VT2e which was inactivated by genetic manipulation (12), protected pigs against challenge with an O139 EDEC. Vaccination with a glutaraldehyde-inactivated purified VT2e induced protection in pigs challenged with toxin (36) or by natural exposure to EDEC in herds with a history of edema disease (33). Pigs may also be protected by subcutaneous administration of anti-VT2e serum prepared in horses (34) or pigs (3). Alternatively, pigs may be protected by antibacterial antibodies which prevent intestinal colonization by EDEC. Such antibodies may be induced by oral vaccination of pigs with a live streptomycin-dependent EDEC (6) or by addition of chicken egg-yolk anti-F18ab antibodies in the feed (31).

REFERENCES

1. AARESTRUP, F.M., JORSAL, S.E., AHRENS, P., JENSEN, N.E. and MEYLING, A. 1997. Molecular characterization of *Escherichia coli* strains isolated from pigs with edema disease. J. Clin. Microbiol. *35*, 20–24.
2. ABAAS, S., FRANKLIN, A., KUHN, I., ORSKOV, F. and ORSKOV, I. 1989. Cytotoxin activity on Vero cells among *Escherichia coli* strains associated with diarrhea in cats. Am. J. Vet. Res. *50*, 1294–1296.
3. ALEXA, P., HAMOIK, J. and SALAJA, E. 1998. Prevention of edema disease of weaners by administration of immune serum to verotoxin VT2e. Proc. 15th IPVS Congress, Birmingham, UK.
4. APPEL, G. *et al.* 1990. Morphologic changes in the intestine of swine after infection with verotoxin-producing *Escherichia coli*. DTW Dtsch. Tierarztl. Wochenschr. *97*, 357–360.
5. AWAD-MASALMEH, M., SCHUH, M., KOFER, J. and QUAKYI, E. 1989. Verification of the protective effect of a toxoid vaccine against edema disease of weaned pigs in an infection model. Dtsch. Tierarztl. Wochenschr. *96*, 419–421.
6. BERTSCHINGER, H.U. and GYLES, C.L. 1994. Oedema Disease, *In Escherichia coli* in Domestic Animals and Humans, (C.L. Gyles, ed) pp. 193–219, CAB International. Oxon, UK.
7. BERTSCHINGER, H.U. and POHLENZ, J. 1983. Bacterial colonization and morphology of the intestine in porcine *Escherichia coli* enterotoxemia (edema disease). Vet. Pathol. *20*, 99–110.
8. BEUTIN, L., GEIER, D., STEINRUCK, H., ZIMMERMANN, S. and SCHEUTZ, F. 1993. Prevalence and some properties of verotoxin (Shiga-like toxin)-producing *Escherichia coli* in seven different species of healthy domestic animals. J. Clin. Microbiol. *31*, 2483-2488.
9. BEUTIN, L., GEIER, D., ZIMMERMANN, S. and KARCH, H. 1995. Virulence markers of Shiga-like toxin-producing *Escherichia coli* strains originating from healthy domestic animals of different species. J. Clin. Microbiol. *33*, 631–635.
10. BLANCO, J.E., BLANCO, M., MORA, A. and BLANCO, J. 1997. Production of toxins (enterotoxins, verotoxins, and necrotoxins) and colicins by *Escherichia coli* strains isolated from septicemic and healthy chickens: relationship with in vivo pathogenicity. J. Clin. Microbiol. *35*, 2953–2957.
11. BOSS, P., MONCKTON, R.P., NICOLET, J. and BURNENS, A.P. 1992. Detection with nonradioactively labeled probes of the toxin genes of different *E. coli* pathotypes from swine. Schweiz. Arch. Tierheilkd. *134*, 31–37.

12. BOSWORTH, B.T., SAMUEL, J.E., MOON, H.W., O'BRIEN, A.D., GORDON, V.M. and WHIPP, S.C. 1996. Vaccination with genetically modified Shiga-like toxin IIe prevents edema disease in swine. Infect. Immun. *64*, 55–60.
13. BOYD, B., TYRRELL, G., MALONEY, M., GYLES, C., BRUNTON, J. and LINGWOOD, C. 1993. Alteration of the glycolipid binding specificity of the pig edema toxin from globotetraosyl to globotriaosyl ceramide alters in vivo tissue targetting and results in a verotoxin 1-like disease in pigs. J. Exp. Med. *177*, 1745–1753.
14. CAPRIOLI, A. *et al.* 1993. Characterisation of verocytotoxin-producing *Escherichia coli* isolated from pigs and cattle in northern Italy. Vet. Rec. *133*, 323–324.
15. CARPENTER, J.L., ANDELMAN, N.C., MOORE, F.M. and KING, N.W. 1988. Idiopathic cutaneous and renal glomerular vasculopathy of greyhounds. Vet. Pathol. *25*, 401–407.
16. CHALMERS, R.M. *et al.* 1997. Vero-cytotoxin-producing *Escherichia coli* O157 in a farmer handling horses. Lancet *349*, 1816.
17. COWAN, L.A., HERTZKE, D.M., FENWICK, B.W. and ANDERSON, C.B. 1997. Clinical and clinicopathologic abnormalities in greyhounds with cutaneous and renal glomerular vasculopathy: 18 cases. J. Am. Vet. Med. Assoc. *210*, 789–793.
18. DEGRANDIS, S., LAW, H., BRUNTON, J., GYLES, C. and LINGWOOD, C.A. 1989. Globotetraosylceramide is recognized by the pig edema disease toxin. J. Biol. Chem. *264*, 12520–12525.
19. DEPREZ, P., VAN DEN HENDE, C., MUYLLE, E. and OYAERT, W. 1986. The influence of the administration of sow's milk on the postweaning excretion of hemolytic *E. coli* in the pig. Vet. Res. Commun. *10*, 469–478.
20. DIAMOND, S.S., RAFLO, C.P. and BELEAU, M.H. 1980. Edema disease in a sea lion. J. Am. Vet. Med. Assoc. *177*, 808–810.
21. FENWICK, B.W. and COWAN, L.A. 1998. Canine model of hemolytic-uremic syndrome. *In Escherichia coli* O157:H7 and other Shiga Toxin-Producing *E. coli* Strains, (J.B. Kaper and A.D. O'Brien, eds.) pp. 268–277, ASM Press, Washington, D.C.
22. FRANKE, S., GUNZER, F., WIELER, L.H., BALJER, G. and KARCH, H. 1995. Construction of recombinant Shiga-like toxin-IIv (SLT-IIv) and its use in monitoring the SLT-IIv antibody status of pigs. Vet. Microbiol. *43*, 41-52.
23. GALLIEN, P. *et al.* 1994. Detection of verotoxin-producing *E. coli* in field isolates from domestic and agricultural animals in Sachsen-Anhalt. Berl. Munch. Tierarztl. Wochenschr. *107*, 331–334.
24. GANNON, V.P., GYLES, C.L. and FRIENDSHIP, R.W. 1988. Characteristics of verotoxigenic *Escherichia coli* from pigs. Can. J. Vet. Res. *52*, 331–337.
25. HAMMERMUELLER, J., KRUTH, S., PRESCOTT, J. and GYLES, C. 1995. Detection of toxin genes in *Escherichia coli* isolated from normal dogs and dogs with diarrhea. Can. J. Vet. Res. *59*, 265–270.
26. HANCOCK, D.D., BESSER, T.E., RICE, D.H., EBEL, E.D., HERRIOTT, D.E. and CARPENTER, L.V. 1998. Multiple sources of *Escherichia coli* O157 in feedlots and dairy farms in the northwestern USA. Prev. Vet. Med. *35*, 11–19.

27. HERTZKE, D.M., COWAN, L.A., SCHONING, P. and FENWICK, B.W. 1995. Glomerular ultrastructural lesions of idiopathic cutaneous and renal glomerular vasculopathy of greyhounds. Vet. Pathol. *32*, 451–459.
28. HIDE, E.J., CONNAUGHTON, I.D., DRIESEN, S.J., HASSE, D., MONCKTON, R.P. and SAMMONS, N.G. 1995. The prevalence of F107 fimbriae and their association with Shiga-like toxin II in *Escherichia coli* strains from weaned Australian pigs. Vet. Microbiol. *47*, 235–243.
29. HOLLOWAY, S., SENIOR, D., ROITH, L. and TISHER, C.C. 1993. Hemolyic uremic syndrome in dogs. J. Vet. Int. Med. *7*, 220–227.
30. IMBERECHTS, H. *et al.* 1994. Prevalence of F107 fimbriae on *Escherichia coli* isolated from pigs with oedema disease or postweaning diarrhoea. Vet. Microbiol. *40*, 219–230.
31. IMBERECHTS, H., DEPREZ, P., VAN DRIESSCHE, E. and POHL, P. 1997. Chicken egg yolk antibodies against F18ab fimbriae of *Escherichia coli* inhibit shedding of F18 positive *E. coli* by experimentally infected pigs. Vet. Microbiol. *54*, 328–341.
32. IRWIN, R.J., MCEWEN, S.A., CLARKE, R.C. and MEEK, A.H. 1989. The prevalence of verocytotoxin-producing *Escherichia coli* and antimicrobial resistance patterns of nonverocytotoxin-producing *Escherichia coli* and *Salmonella* in Ontario broiler chickens. Can. J. Vet. Res. *53*, 411–418.
33. JOHANSEN, M., ANDRESEN, L.O., JORSAL, S.E., THOMSEN, L.K., WADDELL, T.E. and GYLES, C.L. 1997. Prevention of edema disease in pigs by vaccination with verotoxin 2e toxoid. Can. J. Vet. Res. *61*, 280–285.
34. JOHANSEN, M., ANDRESEN, L.O., THOMSEN, L.K., BUSCH, E., JORSAL, S.E. and GYLES, C.L. 2000. Prevention of edema disease in pigs by passive immunization. Can. J. Vet. Res. *64*, 914.
35. LINGGOOD, M.A. and THOMPSON, J.M. 1987. Verotoxin production among porcine strains of *Escherichia coli* and its association with oedema disease. J. Med. Microbiol. *24*, 359–362.
36. MACLEOD, D.L. and GYLES, C.L. 1991. Immunization of pigs with a purified Shiga-like toxin II variant toxoid. Vet. Microbiol. *29*, 309–318.
37. MACLEOD D.L., GYLES, C.L. and WILCOCK, B.P. 1991. Reproduction of edema disease of swine with purified Shiga-like toxin-II variant. Vet. Pathol. *28*, 66–73.
38. MARQUES, L.R., MOORE, M.A., WELLS, J.G., WACHSMUTH, I.K. and O'BRIEN, A.D. 1986. Production of Shiga-like toxin by *Escherichia coli*. J. Infect. Dis. *154*, 338–341.
39. MENO, Y., FUJIMOTO, S. and AMAKO, K. 1996. Fine structure of enterotoxemic *Escherichia coli* O139:K12 strains associated with swine edema disease. Ach. Microbiol. *166*, 357–360.
40. METHIYAPUN, S., POHLENZ, J.F. and BERTSCHINGER, H.U. 1984. Ultrastructure of the intestinal mucosa in pigs experimentally inoculated with an edema disease-producing strain of *Escherichia coli* (O139:K12:H1). Vet. Pathol. *21*, 516–520.

41. MOXLEY, R.A. and FRANCIS, D.H. 1997. Overview of Animal Models, *In Escherichia coli* O157:H7 and other Shiga Toxin-Producing *E. coli* Strains, (J.B. Kaper and A.D. O'Brien, eds) pp. 249–260. ASM Press, Washington, D.C.
42. NAGY, B. *et al.* 1997. Biological relationship between F18ab and F18ac fimbriae of enterotoxigenic and verotoxigenic *Escherichia coli* from weaned pigs with oedema disease or diarrhoea. Microb. Pathog. *22*, 1–11.
43. NAGY, B., CASEY, T.A., WHIPP, S.C. and MOON, H.W. 1992. Susceptibility of porcine intestine to pilus-mediated adhesion by some isolates of piliated enterotoxigenic *Escherichia coli* increases with age. Infect. Immun. *60*, 1285–1294.
44. PARREIRA, V.R. and YANO, T. 1998. Cytotoxin produced by *Escherichia coli* isolated from chickens with swollen head syndrome (SHS). Vet. Microbiol. *62*, 111–119.
45. PATON, J.C. and PATON, A.W. 1998. Pathogenesis and diagnosis of Shiga toxin-producing *Escherichia coli* infections. Clin. Microbiol. Rev. *11*, 450–479.
46. READ, S.C., GYLES, C.L., CLARKE, R.C., LIOR, H. and MCEWEN, S. 1990. Prevalence of verocytotoxigenic *Escherichia coli* in ground beef, pork, and chicken in southwestern Ontario. Epidemiol. Infect. *105*, 11–20.
47. SALAJKA, E., CERNOHOUS, J. and SARMANOVA, Z. 1975. Association of the colonization of the intestine by pathogenic strains of haemolytic *E. coli* in weaned piglets with withdrawal of antibody contained in the dams' milk. Documenta Veterinaria Brno *8*, 43–55.
48. SMITH, H.R., CHEASTY, T., ROBERTS, D., THOMAS, A. and ROWE, B. 1991. Examination of retail chickens and sausages in Britain for vero cytotoxin-producing *Escherichia coli*. Appl. Environ. Microbiol. *57*, 2091–2093.
49. SMITH, K.A., KRUTH, S., HAMMERMUELLER, J., GYLES, C. and WILSON, J.B. 1998. A case-control study of verocytotoxigenic *Escherichia coli* infection in cats with diarrhea. Can. J. Vet. Res. *62*, 87–92.
50. SMITH, H.W., GREEN, P. and PARSELL, Z. 1983. Vero cell toxins in *Escherichia coli* and related bacteria: transfer by phage and conjugation and toxic action in laboratory animals, chickens and pigs. J. Gen. Microbiol. *129*, 3121-3137.
51. SMITH, H.W. and HALLS, S. 1968. The production of oedema disease and diarrhoea in weaned pigs by the oral administration of *Escherichia coli*: factors that influence the course of the experimental disease. J. Med. Microbiol. *1*, 45–59.
52. SMITH, H.W. and LINGGOOD, M.A. 1971. Observations on the pathogenic properties of the K88, Hly and Ent plasmids of *Escherichia coli* with particular reference to porcine diarrhoea. J. Med. Microbiol. *4*, 467–485.
53. SOJKA, W.J. 1965. *Escherichia coli* in Domestic Animals and Poultry. Commonwealth Agricultural Bureaux, Farnham Royal, Bucks. Pp. 104–125.
54. SUEYOSHI, M. and NAKAZAWA, M. 1994. Experimental infection of young chicks with attaching and effacing *Escherichia coli*. Infect. Immun. *62*, 4066–4071.
55. TREVENA, W.B., HOOPER, R.S., WRAY, C., WILLSHAW, G.A., CHEASTY, T. and DOMINGUE, G. 1996. Vero cytotoxin-producing *Escherichia coli* O157 associated with companion animals. Vet. Rec. *138*, 400.

56. VOGELI, P. *et al.* 1997. Ein molekular test fur den nechweis des *E. coli*-F18-rezeptos: ein durchbruch im kampf gegen oedemkrankheit und absetzdurchfall beim schwein. Schw. Arch. Tierheilk. *139*, 479–484.
57. WADDELL, T.E., COOMBER, B.L. and GYLES, C.L. 1998. Localization of potential binding sites for the edema disease verotoxin (VT2e) in pigs. Can. J. Vet. Res. *62*, 81–86.
58. WADDELL, T.E. and GYLES, C.L. 1995. Sodium deoxycholate facilitates systemic absorption of verotoxin 2e from pig intestine. Infect. Immun. *63*, 4953–4956.
59. WADDELL, T.E., LINGWOOD, C.A. and GYLES, C.L. 1996. Interaction of verotoxin 2e with pig intestine. Infect. Immun. *64*, 1714–1719.
60. WALLACE, J.S., CHEASTY, T. and JONES, K. 1997. Isolation of Vero-cytotoxin-producing *Escherichia coli* O157 from wild birds. J. Appl. Microbiol. *82*, 399–404.
61. WITTIG, W., KLIE, H., GALLIEN, P., LEHMANN, S., TIMM, M. and TSCHAPE, H. 1995. Prevalence of the fimbrial antigens F18 and K88 and of enterotoxins and verotoxins among *Escherichia coli* isolated from weaned pigs. Zentralbl. Bakteriol. *283*, 95–104.
62. WITTIG, W., PRAGER, R., STAMM, M., STRECKEL, W. and TSCHAPE, H. 1994. Expression and plasmid transfer of genes coding for the fimbrial antigen in porcine *Escherichia coli* strains. Zentralbl. Bakteriol. *281*, 130–139.

CHAPTER 13

PATHOGENIC ASPECTS OF STEC INFECTIONS IN HUMANS

HERBERT SCHMIDT and HELGE KARCH

Institut für Hygiene und Mikrobiologie der Universität Würzburg
Josef-Schneider-Str. 2
D-97080 Würzburg, Germany

AND

MARTIN BITZAN

Department of Pediatrics
Wake Forest University School of Medicine and Baptist Medical Center
Medical Center Boulevard
Winston-Salem, NC

CLINICAL ILLNESS CAUSED BY HUMAN STEC INFECTION

In 1982, the Centers for Disease Control and Prevention (CDC), Atlanta, reported an outbreak of hemorrhagic colitis (HC) in Oregon and Michigan by a rare *Escherichia coli* serotype, O157:H7 (24). At the time, no one suspected that this pathogen had begun its triumphant progress around the world. Less than two decades later, *E. coli* O157:H7 has emerged as an important cause of gastrointestinal disease that has been identified throughout the world. This organism is remarkable for its abilities to produce Shiga (or Vero) toxins, a family of unique, bipartite protein toxins (85), and to cause attaching and effacing lesions on epithelial cells (30).

The close association of *E. coli* O157:H7 with a distinctive clinical entity, hemorrhagic colitis, led to the epithet "enterohemorrhagic" *E. coli* (EHEC). However, it is its potential to cause severe extra-intestinal complications, primarily the classic hemolytic uremic syndrome (HUS), and widespread outbreaks, which has earned this organism its notoriety.

At present, Shiga toxin-producing *E. coli* (STEC) comprise more than 100 serotypes, which have been isolated from patients, healthy human carriers, and from animals, food, and water (49,124).

In humans, STEC can cause a broad spectrum of diseases, ranging from non-bloody and bloody diarrhea to severe HC, and death. In a large outbreak of *E. coli* O157:H7 in Washington State involving 501 bacterial culture-positive individuals, 30% of the affected persons were hospitalized, 9% developed HUS and 0.6% died (8). However, the percentage of complications varies considerably among other reported outbreaks (51).

E. coli O157:H7 associated illness typically begins with non-bloody diarrhea, abdominal pain, low-grade fever and emesis. After one to two days, the diarrhea often becomes bloody with frequent discharges of small amounts of stool accompanied by painful abdominal cramps (106). This stage lasts four to 10 days. Severe HC may appear as "all blood and no stool". In most cases, the bloody diarrhea resolves without further complications. However, in approximately 5-10% of the patients <10 years, the colitis will progress to HUS. Classical HUS is defined by the triad of microangiopathic hemolytic anemia, thrombocytopenia and acute renal parenchymal disease. Peripheral leukocytosis is common. While increased peripheral neutrophil counts (>13/nl) during the colitic phase is associated with a several-fold increased risk of developing HUS (2,9), the association with poor renal outcome (147) has not yet been confirmed. Renal failure can develop rapidly and coincides with the onset of anemia. The degree of renal insufficiency in HUS varies widely. In most patients urine output falls precipitously, and extracellular fluid overload with edema and hypertension may ensue. Blood urea nitrogen, serum creatinine, uric acid and potassium, worsened by profound intravascular hemolysis, may rise rapidly in oliguric patients. Hypertension is occasionally severe and may be exacerbated by red blood cell transfusion. Fever is noted in one-third of the patients. Rectal prolapse has been reported in 8% and central nervous system manifestations in nearly 30% of the children with HUS (19,68). Seizures may occur secondary to electrolyte imbalance and hypertension, or herald more severe complications, especially stroke, coma, decerebration and death. Minor neurologic manifestations including lethargy, irritability and tremor are common. HUS and HC are vascular disorders, primarily affecting small vessels and the surrounding tissue. Renal glomerular capillary thrombotic microangiopathy with or without arteriolar involvement and, occasionally, extension into interlobar arteries, is considered the hallmark of diarrhea-associated HUS (103,45). HC is characterized by the involvement of small mucosal and submucosal blood vessels, primarily of the colon, which leads to intravascular thrombosis, profound submucosal edema and hemorrhage (103), and to vasoconstriction ("ischemic colitis") (10,107).

STEC disease can mimic thrombotic thrombocytopenic purpura (TTP). Historically, HUS and TTP have been portrayed as variants of the same disease entity (100). It is now apparent, however, that the latter is due to the inherited or acquired deficiency of von Willebrand factor-cleaving metalloprotease (33,133). With the increased diagnosis of STEC infections and HUS, mild or "incomplete" forms of HUS, i.e., isolated hemolytic anemia with or without thrombocytopenia, are being recognized (42,97).

HC and HUS are not limited to infections by *E. coli* O157:H7. Outside the United States, non-O157:H7 STEC serotypes are identified with increasing frequency, both in sporadic cases and in outbreaks (17,23,36,44,49,76,94). The clinical manifestations, especially systemic complications, are similar to those observed with infection by prototypic STEC O157:H7 strains. For example, in 1997, 259 STEC strains were isolated in Germany in the laboratories of Bockemühl, Karch and Tschäpe (17). One hundred six strains were of serotype O157:H7 and O157:H$^-$, whereas 153 were of non-O157 serotypes, including, in frequency order, O26:H11/H$^-$, O103:H2,H18,H$^-$, O145:H28,H$^-$ and O111:H$^-$.

MOBILE GENETIC ELEMENTS OF STEC AND THEIR CONTRIBUTION TO PATHOGENESIS AND VIRULENCE

Pathogenic STEC contain several mobile genetic elements. In *E. coli* O157:H7, a large plasmid, designated pO157, the locus of enterocyte effacement (LEE) and Stx-encoding bacteriophages have been identified. The extent to which these elements contribute to pathogenesis is at the center of ongoing research. The structure and function of LEE are described elsewhere in this book. This chapter will focus on the large plasmid pO157 and Stx-converting phages.

Virtually all *E. coli* O157:H7 isolates carry a large plasmid of approximately 90-kb in size, which has been designated pO157 (132). Although pO157 has attracted considerable research interest (20-22,30,52,113,116,132,135,145), its precise role in the pathogenicity of STEC still needs to be defined. Animal models are naturally limited to a few aspects of *E. coli* O157-associated disease. The lack of an appropriate animal model also impedes investigations into the direct influence of plasmid pO157 on the pathogenesis of the enteropathic HUS. At this stage, we will therefore provide an overview of recent advances in our understanding of the molecular organization of plasmid pO157 including results from our own laboratory.

The ability of *E. coli* O157 and other EHEC strains to cause a so-called enterohemolytic phenotype was first described in 1989 (11). This phenotype is characterized by small turbid zones of hemolysis occurring after incubation at 37°C for 18-24 h, that can be detected on blood agar plates containing defibrinated, washed sheep red blood cells. This contrasts with the extracellular α-hemolysin of *E. coli*, which produces large, clear zones of hemolysis, detectable after 4-6 h of incubation (11,115). Beutin *et al.* (11) reported a close association between the production of enterohemolysin and Shiga toxin in STEC strains of *E. coli* O157:H7 and other serotypes. Eighty-nine percent of these strains expressed a hemolysin, which was unrelated to α-hemolysin. In an attempt to clarify the genetic basis of this hemolytic activity, we cloned a 12-kb fragment from *E. coli* O157:H7 strain 933 derived pO157 which mediated strong hemolytic activity when expressed in non-hemolytic *E. coli* laboratory strains (22). A clone carrying this 12-kb fragment produced a clear zone of hemolysis of the type seen with *E. coli* laboratory strain WAF100, which harbors the recombinant α-hemolysin plasmid pSF4000 (115). Nucleotide sequence analysis of this clone revealed four open reading frames (ORF) that were highly related to sequences of the *E. coli* α-hemolysin and other so-called RTX (**R**epeats in **T**o**X**in) operons (113). The identified sequences were termed e-*hlyC*, *-A*, *-B* and *-D*, in analogy to the respective genes of the *E. coli* α-hemolysin operon. The detection of an immune response against EHEC-hemolysin in sera from patients recovering from HUS demonstrated that the protein is expressed during the course of the infection (113). This observation also suggested that a plasmid pO157-encoded protein may be involved in the pathogenesis of *E. coli* O157-induced HUS.

Analysis of DNA extending upstream of the EHEC-*hly* operon revealed the presence of a Type II secretion pathway system including 13 open reading frames. These sequences showed homology to genes of type II secretion pathway systems of other gram-negative bacteria (114). These type II systems are frequently involved in the transport of pathogenicity factors outside the bacterial cell (108). Greatest similarities were found to the pullulanase secretion pathway of *Klebsiella oxytoca* and *K. pneumoniae*, a system involving at least 14 genes and proteins (99). We named the ORFs *etpC* to *etpO* in accordance with the EHEC Type II Secretion Pathway (114). The function and specificity of this putative transport system are not yet known. On the same plasmid, we identified an ORF of 2208 bp, which we termed *kat*P because of its homology to bacterial catalase-peroxidase genes (20). KatP possesses an aminoterminal signal peptide, suggesting that it is transported through the cytoplasmic membrane. We demonstrated that the catalase-peroxidase KatP was present in wildtype *E. coli* O157 strains, mainly in the periplasm, and that it possesses both catalase and peroxidase activity (20).

Analysis of TCA-precipitable culture supernatants of *E. coli* DH5α harboring cloned DNA-fragments derived from pO157 revealed a protein of 104 kDa apparent molecular weight (*E. coli* strain DH5α/pB9). The DNA fragment responsible for the production of this protein was localized on a subclone designated pB9-5 (21). Nucleotide sequencing of plasmid pB9-5 revealed a 3900 bp ORF encoding a 1300-residue protein with a calculated molecular mass of 141.8 kDa. We termed the protein EspP. A SwissProt and NBRF-PIR protein sequence library search revealed sequence similarities between distinct regions of EspP and various secreted and surface exposed proteins. Further studies indicated that EspP is a serine protease capable of cleaving human coagulation factor V *in vitro*. Interestingly, convalescence phase sera from 5 of 6 children suffering from EHEC infections specifically reacted with purified EspP (21). A protease, termed PssA, which is similar in structure and function to EspP was isolated and characterized from a bovine *E. coli* O26 strain by Chakraborty's group (29).

In 1999, the complete DNA sequences of pO157 from *E. coli* O157:H7 strain EDL933 and from *E. coli* O157:H7 strain RIMD 0509952, derived from the outbreak in 1996 in Sakai city, Japan, were published (22,73). Several new genetic determinants were identified, some of which are candidates for virulence genes (Table 1). A formidable candidate is L7095, a gene putatively encoding 3169 amino acids (22). The first 700 residues showed strong sequence homology to the N-terminal domain of the large clostridial toxin family "LCT" (142). This family includes *Clostridium difficile* ToxA and ToxB, the *C. novyi* alpha toxin and the *C. sordelii* cytotoxin L. These AB toxins share a C-terminal domain that mediates cell entry. The N-terminus exhibits glucosyltransferase activity that modifies the G-proteins Rho, Rac and Ras, among others (22,142). Inactivation of the small G-proteins results in the breakdown of the actin filament network, depolymerization of F-actin to G-actin and destruction of intestinal tight junctions (79,84). This process may contribute to the diarrhea and results in altered regulation of the cytoskeleton. It is not yet known whether the *toxB* gene is expressed in *E. coli* O157.

The cardinal virulence trait of STEC is the ability to produce one or more Shiga toxins (Stx1, Stx2, and variants of Stx2) (37,55,87). Stx are toxic to cultured human colonic and ileal epithelial cells (80) and to endothelial cells (88). Even in the absence of cytotoxicity, Stx may stimulate endothelial cells to produce vasoactive and, possibly, pro-thrombotic mediators (13). Thus, the ability to produce Stx is quite plausibly related to the intestinal and extraintestinal manifestations of human STEC infections. This will be discussed in detail below.

Stx1 and Stx2 are encoded in the genome of lambdoid prophages (27,40,86,123). Stx-encoding phages (synonyms: Stx-converting bacteriophages, Stxϕ) are heterogeneous in their size and genetic organization (105,122). The Stx1-converting bacteriophages H-19J and H-19B, originally isolated from Stx-producing O26:H11 strain H19, and the Stx2-converting bacteriophage 933W, derived from *E. coli* O157:H7 strain EDL933, have been extensively characterized (41,53,82,83,96). Both, stx_1 and stx_2 genes in STEC O157:H7, O26 and O111 are closely associated with a *p*-like gene which is highly homologous to the *p* gene of bacteriophage λ (27). Thus, *stx* genes of different Stx-converting bacteriophages must have integrated at the same site close to the origin of replication in the bacteriophage genome. This was subsequently confirmed in a study which, in addition, demonstrated the presence of an *ileX* tRNA gene upstream of the stx_2 gene in STEC O157:H7, O26:H11, O103:H2 and O111:H2 strains (117). Neely and Friedman (82,83) characterized a 17-kb DNA segment of the Stx1-encoding phage H-19B which included an analogue of the λ Q transcription activator, the stx_1 gene and an open reading frame homologous to the λ holin lysis gene. They suggested a role for the λ-like Q protein and the holin lysis genes in the regulation of Stx1 expression and its release from the bacterial cell (83). The linkage of late phase regulation and expression of Stx was also demonstrated for Stx2-phage 933W (53,96). It was described that particular antibiotics used in human therapy enhanced the production of Stx (35).

TABLE 1.
SELECTED PUTATIVE VIRULENCE-ASSOCIATED GENES LOCATED ON pO157

Orf/Gene	Size (bp)	Predicted/Detected* function of gene product	Related to	Reference
e-*hlyC*	507	Activation of HlyA	*hlyC* of *E. coli*	(113)
e-*hlyA*	2997	EHEC-hemolysin*	*hlyA* of *E. coli*	(113)
e-*hlyB*	2121	Transport of HlyA	*hlyB* of *E. coli*	(113)
e-*hlyD*	1440	Transport of HlyA	*hlyD* of *E. coli*	(113)
etpC	876	Type II secretion system	*pulC* of *K. oxytoca*	(114)
etpD	1758	Type II secretion system	*pulD* of *K. oxytoca*	(114)
etpE	1506	Type II secretion system	*pulE* of *K. oxytoca*	(114)
etpF	1224	Type II secretion system	*pulF* of *K. oxytoca*	(114)
etpG	435	Type II secretion system	*pulG* of *K. oxytoca*	(114)
etpH	555	Type II secretion system	*pulH* of *K. oxytoca*	(114)
etpI	348	Type II secretion system	*pulI* of *K. oxytoca*	(114)
etpJ	600	Type II secretion system	*pulJ* of *K. oxytoca*	(114)
etpK	978	Type II secretion system	*pulK* of *K. oxytoca*	(114)
etpL	849	Type II secretion system	*pulL* of *K. oxytoca*	(114)
etpM	513	Type II secretion system	*pulM* of *K. oxytoca*	(114)
etpN	717	Type II secretion system	*pulN* of *K. oxytoca*	(114)
etpO	401	Type II secretion system	*pulO* of *K. oxytoca*	(114)
katP	2208	Catalase-peroxidase*	*katG* of *E. coli*	(20)
espP	3900	Serine protease*	*sepA* of *S. flexneri* *espC* of EPEC	(21)
L7095	9510	Glucosyltransferase	*toxB* of *C. difficile*	(22)
L7004	210	Hemolysin regulator	*hha* of *E. coli*	(22)
L7029	1032	Acyltransferase (LPS synthesis)	*msbB* of *E. coli*	(22)
L7031	2697	TagA (Lipoprotein)	*tagA* of *V. cholerae*	(22)

* Detected function (the others are predicted)

Furthermore, enhanced toxin production is coupled with enhanced production of Stx-phage particles (146). Antibiotics which are used as antibacterial growth promoters in animal husbandry also have the potential to strongly induce the simultaneous production of Stx and Stx-encoding phages (62). The location of *stx* genes in the phage genome may contribute to the spread of *stx* genes and, via regulation of *stx* expression, to intestinal Stx-production and human pathology.

EFFECTS OF STX ON MAMMALIAN CELLS AND IMPLICATIONS FOR THE PATHOGENESIS OF HEMORRHAGIC COLITIS AND HUS

Despite recent advances in the epidemiology and molecular characterization of *E. coli* O157:H7 and other Stx-producing *E. coli*, the pathogenesis of STEC induced, systemic disease is still poorly understood. A causal role for Stx in HUS and hemorrhagic colitis has been inferred from the following findings: (1) Stx production is common to almost all HUS-associated *E. coli* isolates, regardless of serotype (54,55,127). (2) *In vivo* produced Stx can be detected in the feces of HUS patients and experimentally infected animals (14,55,60,93). (3) HUS patients develop antibodies to Stx (34,55,101,102). The specificity of these antibodies closely follows the toxin type expressed by the patients' isolates (14,34). (4) Intravenous injection of purified Stx causes severe systemic disease in laboratory animals with histopathological features of human HC and HUS, which is prevented in the presence of Stx neutralizing antibodies (12,104). (5) In non-immune animals, circulating Stx binds rapidly to the vascular endothelium of target organs (15,104). (6) "Natural" HUS-like disease has been identified in racing greyhounds that were fed *E. coli* 157:H7-contaminated lots of ground meat (26,39) and in baboons following injection of purified Stx (128).

Although an attractive hypothesis, data to support a pathogenic role for the EHEC hemolysin (113), e.g., as a cause of the acute, non-immune hemolytic anemia that is so characteristic for human HUS, are not yet available. Similarly, no pathogenic role has been identified for any of the other recently identified pO157 plasmid-encoded, secreted STEC proteins, such as KatP, EspP or the putative ToxB-like protein.

Stx-mediated Cytotoxicity Towards Mammalian (Host) Cells

Konowalchuk noticed, in 1979, that Stx (verotoxin) is a potent cytotoxin for a limited range of primate cell lines (e.g., HeLa cells; African Green monkey kidney and Vero cells) (54,55,61). This observation directed attention to the potent cytotoxic activity of this family of bipartite bacterial protein toxins for mammalian cells as a possible mode of tissue damage (89,91). The cytotoxic effect was shown to depend on the expression of a digalactosyl residue-carrying membrane glycosphingolipid, globotriaosylceramide (Gb3), the specific receptor for Stx1, Stx2 and Stx2c (48,64-66,144) or globotetraosylceramide (Gb4), for Stx2e (28). However, the susceptibility of mammalian cells to the cytotoxic activity of Stx does not correlate strictly with receptor densities on the outer cell membrane (4,25,31). Gb3 also plays a decisive role in the intracellular (retrograde trans-Golgi) trafficking of Stx to the endoplasmic reticulum (109,111). The StxA subunit possesses intrinsic *N*-glycosidase activity, which catalyzes the hydrolytic cleavage of a specific adenine residue of the 28s rRNA of the large mammalian 60S ribosomal subunit (32,112). This posttranscriptional modification leaves the phosphodiester bond intact, but effectively interferes with the binding and coordinated function of elongation factors 1a and 2 and disrupts peptide synthesis, eventually causing the cells demise (32,87,92). The pentameric Gb3-binding Stx B subunit mediates the attachment of the holotoxin to the cell surface, which triggers internalization and targeting of the holotoxin to the Golgi and endoplasmic reticulum (3,110,111). Preferred binding to the nuclear envelope rather than the Golgi, e.g., in B cell lymphoma and ovarian carcinoma cells, has been found to correlate with increased sensitivity to Stx, possibly due to variations in the length of Gb3 fatty acid chains (3).

Stx-induced Apoptosis

Pathologic studies on biopsy or autopsy material from patients with STEC-induced HC and HUS (103), and from animals with experimental shigatoxemia, demonstrated necrotic lesions. However, when investigated, morphological features of programmed cell death were also identified. Evidence that apoptosis may indeed be an important, perhaps major pathway of Stx-mediated cytotoxicity, came from studies on the fate of Burkitt lymphoma and maturing germinal center B cells expressing the CD77/Gb3 antigen at high density. It was subsequently shown that Stx can induce programmed cell death in these and other cell lines, including Vero cells commonly employed for *in vitro* testing of Stx toxicity (46,75,125,148). Mangeney *et al.* (75) showed that cross-linking of membrane CD77/Gb3 invokes cAMP-dependent protein kinase activation and C2-ceramide synthesis in select cell types (74), and that ceramides may play a role as secondary messengers during apoptosis through activation of the stress-activated protein kinase/cJun N-terminal kinases (SAPK/JNK) and through suppression of the extracellular regulated kinase (ERK) pathway (125). These and other studies, using stx_1B gene transfection and expression in fibroblasts, established the critical role of the Stx B subunit in the initiation of the apoptotic pathway (81). While cross-linking of CD77/Gb3 on the outer leaflet of the membrane bilayer was necessary and sufficient to induce the apoptotic signal in Burkitt lymphoma cells (125), physical targeting of the B-subunit to the nuclear envelope (rather than to the Golgi apparatus) was associated with apoptosis induction in astrocytoma and ovarian carcinoma cells (4). Nakagawa *et al.* (81) provided evidence for the direct activation of the caspase effector pathway by B-subunit. In contrast, isolated transfection of the stx_1A gene induced necrotic cell damage with no DNA fragmentation (81). It remained unclear, whether protein synthesis was required to induce apoptosis. However, it is quite possible that Stx can induce necrosis or apoptosis, and that this dichotomy is cell or growth state-dependent. Interestingly, Stx has now been shown to exert anti-apoptotic effects in neutrophils, in a protein kinase C-dependent manner (67).

This observation may indeed explain the leukocytosis seen in patients with HUS.

Effect of Stx on Vascular Endothelial Cells

Early pioneering studies by Obrig *et al.* (88), that were subsequently confirmed by other groups (130,137), established that cultured human umbilical vein endothelial cells (HUVEC) die when exposed to Stx, and that this effect is cell cycle dependent, i.e., significantly more pronounced in proliferating than quiescent (endothelial) cells (98,141). It is now appreciated that HUVEC are relatively resistant to the cytotoxic effect of Stx, compared to Vero cells (88). However, endothelial cells derived from different vascular beds vary widely in their sensitivity to Stx. For example, microvascular cortical renal (71,90), glomerular capillary and dermal microvascular endothelial cells (141), are several orders of magnitude more sensitive than HUVEC or saphenal vein endothelial cells (130). The intestinal capillary network is the first epithelial cell target to be encountered as small quantities of Stx translocate across the intestinal epithelial layer. It has been demonstrated that human intestinal microvascular endothelial cells are sensitive to Stx1 and Stx2 *in vitro*. Using human microvascular endothelial cells from renal cortex, Obrig's group reported that Stx2 was about 100-fold more potent than Stx1 (71). This was not confirmed for selectively purified glomerular endothelial cells (141). Of note, Yoshida *et al.* (149) recently reported that Stx indeed induces apoptosis in primary cultures of human umbilical vein and other macrovascular endothelial cells and that apoptosis is not readily seen in later passages. The *in vivo* relevance of these findings for the pathogenesis of HC and HUS, although intriguing, awaits confirmation by *in vivo* studies. While the induction of apoptosis by Stx in susceptible microvascular endothelia would explain, in part, the renal and colonic pathology of HUS and HC, the precise mechanism(s) and signaling cascades involved have yet to be elucidated.

Stx Effects Towards Intestinal Epithelial and Renal Tubular Epithelial Cells

Translocation of Stx across the intestinal barrier to the submucosa and blood vessels is a fascinating, yet incompletely understood, process. Important topics to be addressed include: What is the mechanism of the initial diarrhea and subsequent bloody colitis that precede the HUS? What is the principal, transcellular or paracellular route of the toxin? Which receptors and intracellular structures are involved in the transport? Are there accessory, constitutive or inducible bacterial and host factors that facilitate toxin absorption? What is the biochemical basis of the remarkable age dependency of the HUS? Are there differences between Stx1 and Stx2? Unraveling these questions will aid in developing rational preventive and therapeutic strategies.

Infection of experimental animals with STEC mutants by Tzipori *et al.*

(134) indicated that the *eae* gene product, intimin, influences the colonization site, where absorption of Stx is believed to occur. Interestingly, when feeding gnotobiotic piglets with Stx2 producing *E. coli* O157:H7 strains, intimate bacterial attachment and mucosal damage were not a prerequisite for toxin translocation from the gut lumen into the circulation (134). When loops of rabbit small bowel were exposed to Stx, an augmented secretory response was obtained (77). Mobassaleh *et al.* (78) reported that microvillus membrane preparations from rabbit intestinal mucosa bound Stx via hydroxylated fatty acid-containing Gb3. Studies using rat colon loops revealed increased net absorption of NaCl and water, associated with inhibition of protein synthesis in secretory crypt cells, but not in surface cells. This corresponded to detection of Stx binding to the membrane of crypt cells (63). Acheson *et al.* (1) reported that Stx1 is translocated across intact monolayers of polarized CaCo2 and T84 cells in an energy-requiring and saturable manner, while their electrical resistance and inulin movement remained unchanged. Acheson *et al.* (1) showed that induction of Gb3 expression by butyrate treatment reduced the translocation rate of Stx1, while Izumikawa *et al.* (47) showed that Stx1 and Stx2 traversed CaCo2 cell monolayers with differing efficiencies. Philpott *et al.* (95) confirmed that Stx1 had no effect on the barrier function of T84 monolayers and provided evidence for the transcellular mode of Stx transport using immunoelectron microscopy. The toxin was observed within endosomes associated with the Golgi network and endoplasmic reticulum (95). More recently, Thorpe *et al.* (131) provided *in vitro* evidence that Stx treatment resulted in the augmented production and secretion of the chemokine IL-8, a potent neutrophil activator. Using (adult) rabbit intestinal epithelial cells, Wada *et al.* (143) demonstrated Stx-induced apoptosis in rabbit intestinal epithelial cells. Additionally, Pai *et al.* (93) and Keenan *et al.* (58), feeding rabbits Stx-producing *E. coli*, found evidence of intestinal cell apoptosis *in vivo*. Taken together, these data suggest a mode of luminal Stx uptake by toxin-insensitive intestinal epithelial cells lacking the Gb3 receptor. Conversely, increased expression of Gb3 may result in toxin binding and, possibly, cell death via apoptosis and/or the induction and release of cytokines and chemokines.

Renal tubular epithelial cells have been reported to be highly sensitive to Stx cytotoxicity *in vitro* (43,126). In mouse models of intestinal STEC infection, proximal renal tubular damage was indeed the major histological finding (57,145). Renal adenocarcinoma-derived cell lines and primary human proximal renal tubulus cell cultures, when exposed to Stx in picomolar concentrations, displayed features of apoptotic cell death (59,126). Renal tubular epithelial apoptosis has recently also been to shown to occur in human STEC infection, as demonstrated in renal biopsy samples of children with HUS (57,126). These findings suggested a role for Stx in the genesis of acute renal failure in HUS, which had not been previously appreciated (50).

THE ROLE OF INFLAMMATORY CYTOKINES AND STX-INDUCED CELL ACTIVATION

During the past few years, attention of researchers in the field has focussed on the role of inflammatory cytokines in STEC disease. Early studies by Barrett *et al.* (5-7) in rodents, using concomitant or sequential administration of Stx and LPS, suggested a disease-modifying role for these agents. Obrig's group (69,70), and van de Kar *et al.* (137) first showed that moderately "resistant" HUVEC can become more sensitive to the cytotoxic effect of Stx by "presensitization" of the cell monolayer with low concentrations of LPS, TNF-α or IL-1β, that had no apparent effect on cell viability, or on protein synthesis. Cytokine stimulation results in a moderate increase of Gb3 surface expression (69,131, 136,137), possibly via protein kinase C activation (136) and/or galactosyl transferase enzyme gene induction (72). However, whether augmented Gb3 expression sufficiently explains these observations, and whether this mechanism is of relevance *in vivo*, remains to be resolved.

A paradoxical effect of Stx on a human colonic epithelial cell line (Hct-8 cells) has been described by Thorpe *et al.* (131). The authors showed that, while blocking general mRNA translation in these cells, Stx treatment led to increased mRNA levels and secretion of the chemokine IL-8. IL-8 induction was ascribed to mitogen activated protein kinase p38 pathway-mediated transcriptional activation (131). Thus, Stx-induced increases in intestinal epithelial cell chemokine production may lead to the amplification of the host mucosal inflammatory response during STEC infection (131). Similarly, other groups showed that monocyte-like human leukemia-derived THP cells (129) and freshly isolated human monocytes (139), and also tubular epithelial and glomerular mesangial cells (121,140), could be driven by Stx to produce various cytokines. A paradigm is now emerging that Stx, at concentrations that fail to affect cell viability, can stimulate "resistant" cells to express and secrete "professional" mediators (cytokines, chemokines, vasomediators, etc).

This effect has been attributed, at least in some cell types, to increased mRNA transcript stabilization, exemplified for endothelin-1, a potent vasoconstrictor and intercellular and vascular adhesion molecule (15,16). Elegant studies reported by Harel *et al.* (38) suggested that Stx-mediated cell activation occurs *in vivo*, possibly in an organ-specific manner. These authors injected purified Stx into transgenic mice harboring a human TNF-α promoter-chloramphenicol acetyltransferase (CAT) gene construct and observed CAT mRNA and protein expression in, and surprisingly restricted to, the kidney (38). Indeed, several independent groups reported the detection of elevated plasma and/or urine levels of TNF-α, IL-1a and other cytokines and chemokines (56,138) as well as endothelin-1 (120) in children with HUS, and more recently, in the baboon model of shigatoxinemia (128). Plasma and urine measurements may not accurately reflect the production of paracrine and autocrine mediators, especially in the setting of impaired renal function or dialysis, and thus may explain discrepant results. Using molecular technology, one can now begin to address the age-dependent susceptibility to systemic complications of STEC infections. Several hypotheses have been advanced to explain why young children and seniors are more prone to developing the systemic complications of STEC infection, especially HUS lack of protective anti-toxin antibodies (101); developmentally regulated expression of Gb3 receptors in target tissues (18,77); genetic aspects (118,119); and the more vigorous inflammatory responses in children.

In summary, the significance of Stx-mediated induction of inflammatory or vasoactive mediators in cell cultures for the development of HUS and HC, although suggestive, is not known at present. Whether Stx or endotoxin- induced inflammatory cytokines play a necessary or contributory role in the evolution of systemic disease, especially HUS, has to be elucidated in additional, precise studies, preferably using a suitable animal model. At the cellular and molecular level, mechanisms of gene regulation and induction of apoptosis by Stx need to be resolved. The relationship of these events to the modification of ribosomal RNA, protein synthesis inhibition and cell cycle blockade needs to be addressed. A better understanding of these mechanisms may help to identify relevant risk factors for the devastating complications of STEC infection and to develop preventive and therapeutic strategies.

CONCLUSIONS

Based on animal experimentation (5,6,7,128), translocation of Stx from the intestinal cavity or toxin delivery by STEC that adhere tightly to the mucosal epithelium is believed to be a critical prerequisite to the development of systemic disease, i.e., HUS and HC. Although exogenous Stx was sufficient to induce characteristic pathological lesions in these animals, endogenous (host) factors may play a contributory or necessary role. Unresolved research questions are: (1) The role of additional bacterial products that may be required for, or augment, the observed effects of Stx (e.g., LPS), or that account for some facet of the hemolytic uremic syndrome (e.g., EHEC hemolysin) (113); (2) the role of host factors, especially cytokines, that are elicited in response to Stx or other bacterial mediators; (3) the modifying role of neutrophils and of oxidant stress; and (4) the reason why persons at the extremes of age (young children, seniors) are particularly prone to systemic disease following STEC infection.

REFERENCES

1. ACHESON, D.W. *et al.* 1996. Translocation of Shiga toxin across polarized intestinal cells in tissue culture. Infect. Immun. *64*, 3294–3300.
2. AKASHI, S. *et al.* 1994. A severe outbreak of haemorrhagic colitis and haemolytic uraemic syndrome associated with *Escherichia coli* O157:H7 in Japan. Eur. J. Pediatr. *153*, 650–655.
3. ARAB, S. and LINGWOOD, C.A. 1998. Intracellular targeting of the endoplasmic reticulum/nuclear envelope by retrograde transport may determine cell hypersensitivity to verotoxin via globotriaosyl ceramide fatty acid isoform traffic. J. Cell Physiol. *177*, 646–660.
4. ARAB, S. *et al.* 1998. Verotoxins inhibit the growth of and induce apoptosis in human astrocytoma cells. J. Neurooncol. *40*, 137–150.
5. BARRETT, T.J., POTTER, M.E. and STROCKBINE, N.A. 1990. Evidence for participation of the macrophage in Shiga-like toxin II-induced lethality in mice. Microb. Pathog. *9*, 95–103.
6. BARRETT, T.J., POTTER, M.E. and WACHSMUTH, I.K. 1989. Bacterial endotoxin both enhances and inhibits the toxicity of Shiga-like toxin II in rabbits and mice. Infect. Immun *57*, 3434–3437.
7. BARRETT, T.J., POTTER, M.E. and WACHSMUTH, I.K. 1989. Continuous peritoneal infusion of Shiga-like toxin II (SLT II) as a model for SLT II-induced diseases. J. Infect. Dis. *159*, 774–777.
8. BELL, B.P. *et al.* 1994. A multistate outbreak of *Escherichia coli* O157:H7-associated bloody diarrhea and hemolytic uremic syndrome from hamburgers. The Washington experience. JAMA *272*, 1349–1353.
9. BELL, B.P., GRIFFIN, P.M., LOZANO, P., CHRISTIE, D.L., KOBAYASHI, J.M. and TARR, P.I. 1997. Predictors of hemolytic uremic syndrome in children during a large outbreak of *Escherichia coli* O157:H7 infections. Pediatrics *100*, E12 (Abstract).
10. BELLAICHE, G. *et al.* 1996. *Escherichia coli* O157:H7 ischemic colitis with hemolytic-uremic syndrome. Gastroenterol. Clin. Biol. *20*, 614–615.
11. BEUTIN, L. *et al.* 1989. Close association of verotoxin (Shiga-like toxin) production with enterohemolysin production in strains of *Escherichia coli*. J. Clin. Microbiol. *27*, 2559–2564.
12. BIELASZEWSKA, M., CLARKE, I., KARMALI, M.A. and PETRIC, M. 1997. Localization of intravenously administered verocytotoxins (Shiga-like toxins) 1 and 2 in rabbits immunized with homologous and heterologous toxoids and toxin subunits. Infect. Immun. *65*, 2509–2516.
13. BITZAN, M., CHAN, M. and MARSDEN, P.A. 1998. Verotoxin induces intracellular adhesion molecule-1 (ICAM-1) expression in vascular endothelium by a novel posttranscriptional mechanism. J. Am. Soc. Nephrol. *9*, 472A. A2415 (Abstract).
14. BITZAN, M., LUDWIG, K., KLEMT, M., KÖNIG, H., BUREN, J. and MÜLLER-WIEFEL, D. 1993. The role of *Escherichia coli* O 157 infections in the classical (enteropathic) haemolytic uraemic syndrome: results of a Central European, multicentre study. Epidemiol. Infect. *110*, 183–196.

15. BITZAN, M. *et al.* 1992. Endothelial cell localization of verotoxins (VT) by immunofluorescence in the rabbit model of verotoxemia. Abstr. Annu. Meet. Am. Soc. Microbiol. *76*, B–300 (Abstract).
16. BITZAN, M.M., WANG, Y., LIN, J. and MARSDEN, P.A. 1998. Verotoxin and ricin have novel effects on preproendothelin-1 expression but fail to modify nitric oxide synthase (ecNOS) expression and NO production in vascular endothelium. J. Clin. Invest. *101*, 372–382.
17. BOCKEMÜHL, J., KARCH, H. and TSCHÄPE, H. 1998. Zur Situation der Infektionen des Menschen durch enterohämorrhagische *Escherichia coli* (EHEC) in Deutschland 1997. Bundesgesundheitsblatt *41*, 2–5.
18. BOYD, B. and LINGWOOD, C. 1989. Verotoxin receptor glycolipid in human renal tissue. Nephron *51*, 207–210.
19. BRANDT, J.R. *et al.* 1994. *Escherichia coli* O157:H7-associated hemolytic-uremic syndrome after ingestion of contaminated hamburgers. J. Pediatr. *125*, 519–526.
20. BRUNDER, W., SCHMIDT, H. and KARCH, H. 1996. KatP, a novel catalase-peroxidase encoded by the large plasmid of enterohaemorrhagic *Escherichia coli* O157:H7. Microbiology. *142*, 3305–3315.
21. BRUNDER, W., SCHMIDT, H. and KARCH, H. 1997. EspP, a novel extracellular serine protease of enterohaemorrhagic *Escherichia coli* O157:H7 cleaves human coagulation factor V. Mol. Microbiol. *24*, 767–778.
22. BURLAND, V., SHAO, Y., PERNA, N.T., PLUNKETT, G., SOFIA, H.J. and BLATTNER, F.R. 1998. The complete DNA sequence and analysis of the large virulence plasmid of *Escherichia coli* O157:H7. Nucleic. Acids. Res. *26*, 4196–4204.
23. CAPRIOLI, A. and TOZZI, A.E. 1998. Epidemiology of Shiga toxin-producing *Escherichia coli* infections in continental Europe. *In Escherichia coli* O157:H7 and other Shiga toxin-producing *E. coli* strains, (J.B. Kaper and A.D. O'Brien, eds.) pp. 38–48, American Society for Microbiology, Washington, D.C.
24. CDC. 1982. Isolation of *E. coli* O157:H7 from sporadic cases of hemorrhagic colitis — United States. MMWR. Morb. Mortal. Wkly. Rep. *31*, 580–585.
25. COHEN, A., MADRID, M.V., ESTROV, Z., FREEDMAN, M.H., LINGWOOD, C.A. and DOSCH, H.M. 1990. Expression of glycolipid receptors to Shiga-like toxin on human B lymphocytes: a mechanism for the failure of long-lived antibody response to dysenteric disease. Int. Immunol. 2, 1–8.
26. COWAN, L.A., HERTZKE, D.M., FENWICK, B.W. and ANDREASEN, C.B. 1997. Clinical and clinicopathologic abnormalities in greyhounds with cutaneous and renal glomerular vasculopathy: 18 cases. J. Am. Vet. Med. Assoc. *210*, 789–793.
27. DATZ, M., JANETZKI-MITTMANN, C., FRANKE, S., GUNZER, F., SCHMIDT, H. and KARCH, H. 1996. Analysis of the enterohemorrhagic *Escherichia coli* O157 DNA region containing lambdoid phage gene p and Shiga-like toxin structural genes. Appl. Environ. Microbiol. *62*, 791–797.
28. DEGRANDIS, S., LAW, H., BRUNTON, J., GYLES, C. and LINGWOOD, C.A. 1989. Globotetraosylceramide is recognized by the pig edema disease toxin. J. Biol. Chem. *264*, 12520–12525.

29. DJAFARI, S., EBEL, F., DEIBEL, C., KRAMER, S., HUDEL, M. and CHAKRABORTY, T. 1997. Characterization of an exported protease from Shiga toxin-producing *Escherichia coli*. Mol. Microbiol. *25*, 771–784.
30. DYTOC, M. *et al.* 1993. Multiple determinants of verotoxin-producing *Escherichia coli* O157:H7 attachment-effacement. Infect. Immun. *61*, 3382–3391.
31. EIKLID, K. and OLSNES, S. 1980. Interaction of *Shigella shigae* cytotoxin with receptors on sensitive and insensitive cells. J. Recept. Res. *1*, 199–213.
32. ENDO, Y., TSURUGI, K., YUTSUDO, T., TAKEDA, Y., OGASAWARA, T. and IGARASHI, K. 1988. Site of action of a Vero toxin (VT2) from *Escherichia coli* O157:H7 and of Shiga toxin on eukaryotic ribosomes. RNA N-glycosidase activity of the toxins. Eur. J. Biochem. *171*, 45–50.
33. FURLAN, M., ROBLES, R., SOLENTHALER, M. and LÄMMLE, B. 1998. Acquired deficiency of von Willebrand factor-cleaving protease in a patient with thrombotic thrombocytopenic purpura. Blood *91*, 2839–2846.
34. GREATOREX, J.S. and THORNE, G.M. 1994. Humoral immune responses to Shiga-like toxins and *Escherichia coli* O157 lipopolysaccharide in hemolytic-uremic syndrome patients and healthy subjects. J. Clin. Microbiol. *32*, 1172–1178.
35. GRIF, K., DIERICH, M.P., KARCH, H. and ALLERBERGER, F. 1998. Strain-specific differences in the amount of Shiga toxin released from enterohemorrhagic *Escherichia coli* O157 following exposure to subinhibitory concentrations of antimicrobial agents. Eur. J. Clin. Microbiol. Infect. Dis. *17*, 761–766.
36. GRIFFIN, P.M. 1998. Epidemiology of Shiga toxin-producing *Escherichia coli* infections in humans in the United States. *In Escherichia coli* O157:H7 and other Shiga toxin-producing *E. coli* strains, (J.B. Kaper and A.D. O'Brien, eds.) pp. 15–22. American Society for Microbiology, Washington, D.C.
37. GRIFFIN, P.M. and TAUXE, R.V. 1991. The epidemiology of infections caused by *Escherichia coli* O157:H7, other enterohemorrhagic *E. coli*, and the associated hemolytic uremic syndrome. Epidemiol. Rev. *13*, 60–98.
38. HAREL, Y., SILVA, M., GIROIR, B., WEINBERG, A., CLEARY, T.B. and BEUTLER, B. 1993. A reporter transgene indicates renal-specific induction of tumor necrosis factor (TNF) by shiga-like toxin. Possible involvement of TNF in hemolytic uremic syndrome. J. Clin. Invest. *92*, 2110–2116.
39. HERTZKE, D.M., COWAN, L.A., SCHONING, P. and FENWICK, B.W. 1995. Glomerular ultrastructural lesions of idiopathic cutaneous and renal glomerular vasculopathy of greyhounds. Vet. Pathol. *32*, 451–459.
40. HUANG, A. *et al.* 1986. Cloning and expression of the genes specifying Shiga-like toxin production in *Escherichia coli* H19. J. Bacteriol. *166*, 375–379.
41. HUANG, A., FRIESEN, J. and BRUNTON, J.L. 1987. Characterization of a bacteriophage that carries the genes for production of Shiga-like toxin 1 in *Escherichia coli*. J. Bacteriol. *169*, 4308–4312.
42. HUBER, H.C., KUGLER, R. and LIEBL, B. 1998. Infektionen mit enterohämorrhagischen *Escherichia coli* (EHEC)-Ergebnisse einer epidemiologischen Erhebung in Bayern für den Zeitraum April 1996 bis März 1997. Gesundheitswesen *60*, 159–165.
43. HUGHES, A.K., STRICKLETT, P.K. and KOHAN, D.E. 1998. Cytotoxic effect of Shiga toxin-1 on human proximal tubule cells. Kidney Int. *54*, 426–437.

44. HUPPERTZ, H.I., BUSCH, D., SCHMIDT, H., ALEKSIC, S. and KARCH, H. 1996. Diarrhea in young children associated with *Escherichia coli* non-O157 organisms that produce Shiga-like toxin. J. Pediatr. *128*, 341–346.
45. INWARD, C.D., HOWIE, A.J., FITZPATRICK, M.M., RAFAAT, F., MILFORD, D.V. and TAYLOR, C.M. 1997. Renal histopathology in fatal cases of diarrhoea-associated haemolytic uraemic syndrome. British Association for Paediatric Nephrology. Pediatr. Nephrol. *11*, 556–559.
46. INWARD, C.D. *et al.* 1995. Verocytotoxin-1 induces apoptosis in vero cells. J. Infect. *30*, 213–218.
47. IZUMIKAWA, K. *et al.* 1998. *Escherichia coli* O157 interactions with human intestinal Caco-2 cells and the influence of fosfomycin. J. Antimicrob. Chemother. *42*, 341–347.
48. JACEWICZ, M.S., ACHESON, D.W., MOBASSALEH, M., DONOHUE ROLFE, A., BALASUBRAMANIAN, K.A. and KEUSCH, G.T. 1995. Maturational regulation of globotriaosylceramide, the Shiga-like toxin 1 receptor, in cultured human gut epithelial cells. J. Clin. Invest. *96*, 1328–1335.
49. JOHNSON, R.P. *et al.* 1996. Growing concerns and recent outbreaks involving non-O157:H7 serotypes of verotoxigenic *Escherichia coli*. J. Food Prot. *59*, 1112–1122.
50. KAPLAN, B.S. 1998. Shiga toxin-induced tubular injury in hemolytic uremic syndrome. Kidney Int. *54*, 648–649.
51. KARCH, H., BIELASZEWSKA, M., BITZAN, M. and SCHMIDT, H. 1999. Epidemiology and diagnosis of Shiga toxin-producing *Escherichia coli* infections. Diagn. Microbiol Infect. Dis. *34*, 229–243.
52. KARCH, H., HEESEMANN, J., LAUFS, R., O'BRIEN, A.D., TACKET, C.O. and LEVINE, M.M. 1987. A plasmid of enterohemorrhagic *Escherichia coli* O157:H7 is required for expression of a new fimbrial antigen and for adhesion to epithelial cells. Infect. Immun. *55*, 455–461.
53. KARCH, H., SCHMIDT, H., JANETZKI-MITTMANN, C., SCHEEF, J. and KRÖGER, M. 1999. Shiga toxins even when different are encoded at identical positions in the genomes of related temperate bacteriophages. Mol. Gen. Genet. *262*, 600–607.
54. KARMALI, M.A. 1989. Infection by verocytotoxin-producing *Escherichia coli*. Clin. Microbiol. Rev. *2*, 15–38.
55. KARMALI, M.A., PETRIC, M., LIM, C., FLEMING, P.C., ARBUS, G.S. and LIOR, H. 1985. The association between idiopathic hemolytic uremic syndrome and infection by verotoxin-producing *Escherichia coli*. J. Infect. Dis. *151*, 775–782.
56. KARPMAN, D., ANDREASSON, A., THYSELL, H., KAPLAN, B.S. and SVANBORG, C. 1995. Cytokines in childhood hemolytic uremic syndrome and thrombotic thrombocytopenic purpura. Pediatr. Nephrol. *9*, 694–699.
57. KARPMAN, D. *et al.* 1998. Apoptosis of renal cortical cells in the hemolytic-uremic syndrome: in vivo and in vitro studies. Infect. Immun. *66*, 636–644.
58. KEENAN, K.P., SHARPNACK, D.D., COLLINS, H., FORMAL, S.B. and O'BRIEN, A.D. 1986. Morphologic evaluation of the effects of Shiga toxin and *E. coli* Shiga-like toxin on the rabbit intestine. Am. J. Pathol. *125*, 69–80.

59. KIYOKAWA, N. *et al.* 1998. Induction of apoptosis in normal human renal tubular epithelial cells by *Escherichia coli* Shiga toxins 1 and 2. J. Infect. Dis. *178*, 178–184.
60. KLEANTHOUS, H. *et al.* 1990. Haemolytic uraemic syndromes in the British Isles, 1985-8: association with verocytotoxin producing *Escherichia coli*. Part 2: Microbiological aspects. Arch. Dis. Child *65*, 722–727.
61. KONOWALCHUK, J., SPEIRS, J.I. and STAVRIC, S. 1977. Vero response to a cytotoxin of *Escherichia coli*. Infect. Immun. *18*, 775–779.
62. KÖHLER, B., KARCH, H. and SCHMIDT, H. 2000. Antibacterials that are used as growth promoters in animal husbandry can affect the release of Shiga toxin 2-converting bacteriophages and Shiga toxin 2 from *Escherichia coli* strains. Microbiology. *146*(5),1085–1090.
63. KRISHNAN, S., RAMAKRISHNA, B.S., KEUSCH, G.T., ACHESON, D.W., PULIMOD, A.B. and MATHAN, M. 1999. Effect of Shiga Toxin on NaCl transport in rat distal colon. Scand. J. Gastroenterol *34*, 777–783.
64. LINGWOOD, C.A. 1993. Verotoxins and their glycolipid receptors. Adv. Lipid Res. *25*, 189–211.
65. LINGWOOD, C.A. 1999. Glycolipid receptors for verotoxin and *Helicobacter pylori*: role in pathology. Biochim. Biophys. Acta *1455*, 375–386.
66. LINGWOOD, C.A. *et al.* 1987. Glycolipid binding of purified and recombinant *Escherichia coli* produced verotoxin in vitro. J. Biol. Chem. *262*, 8834–8839.
67. LIU, J. *et al.* 1999. Inhibition of neutrophil apoptosis by verotoxin 2 derived from *Escherichia coli* O157:H7. Infect. Immun. *67*, 6203–6205.
68. LOIRAT, C., BAUDOUIN, V., SONSINO, E., MARIANI-KURKDJIAN, P. and ELION, J. 1993. Hemolytic-uremic syndrome in the child. Adv. Nephrol. Necker. Hosp. *22*, 141–168.
69. LOUISE, C.B. and OBRIG, T.G. 1991. Shiga toxin–associated hemolytic-uremic syndrome: combined cytotoxic effects of Shiga toxin, interleukin-1 beta, and tumor necrosis factor alpha on human vascular endothelial cells in vitro. Infect. Immun. *59*, 4173–4179.
70. LOUISE, C.B. and OBRIG, T.G. 1992. Shiga toxin-associated hemolytic uremic syndrome: combined cytotoxic effects of shiga toxin and lipopolysaccharide (endotoxin) on human vascular endothelial cells in vitro. Infect. Immun. *60*, 1536–1543.
71. LOUISE, C.B. and OBRIG, T.G. 1995. Specific interaction of *Escherichia coli* O157:H7-derived Shiga-like toxin II with human renal endothelial cells. J. Infect. Dis. *172*, 1397–1401.
72. LOUISE, C.B., TRAN, M.C. and OBRIG, T.G. 1997. Sensitization of human umbilical vein endothelial cells to Shiga toxin: involvement of protein kinase C and NF-kappaB. Infect. Immun. *65*, 3337–3344.
73. MAKINO, K. *et al.* 1998. Complete nucleotide sequences of 93-kb and 3.3-kb plasmids of an enterohemorrhagic *Escherichia coli* O157:H7 derived from Sakai outbreak. DNA Res. *5*, 1–9.
74. MANGENEY, M., LINGWOOD, C.A., TAGA, S., CAILLOU, B., TURSZ, T. and WIELS, J. 1993. Apoptosis induced in Burkitt's lymphoma cells via Gb3/CD77, a glycolipid antigen. Cancer Res. *53*, 5314–5319.

75. MANGENEY, M., RICHARD, Y., COULAUD, D., TURSZ, T. and WIELS, J. 1991. CD77: an antigen of germinal center B cells entering apoptosis. Eur. J. Immunol. *21*, 1131–1140.
76. MARIANI-KURKDJIAN, P. *et al.* 1993. Identification of a clone of *Escherichia coli* O103:H2 as a potential agent of hemolytic-uremic syndrome in France. J. Clin. Microbiol. *31*, 296–301.
77. MOBASSALEH, M., DONOHUE-ROLFE, A., JACEWICZ, M., GRAND, R.J. and KEUSCH, G.T. 1988. Pathogenesis of shigella diarrhea: evidence for a developmentally regulated glycolipid receptor for shigella toxin involved in the fluid secretory response of rabbit small intestine. J. Infect. Dis. *157*, 1023–1031.
78. MOBASSALEH, M., GROSS, S.K., MCCLUER, R.H., DONOHUE-ROLFE, A. and KEUSCH, G.T. 1989. Quantitation of the rabbit intestinal glycolipid receptor for Shiga toxin. Further evidence for the developmental regulation of globotriaosylceramide in microvillus membranes. Gastroenterology *97*, 384–391.
79. MOORE, R., POTHOULAKIS, C., LAMONT, J.T., CARLSON, S. and MADARA, J.L. 1990. *C. difficile* toxin A increases intestinal permeability and induces Cl^- secretion. Am. J. Physiol. *259*, G165–G172.
80. MOYER, M.P., DIXON, P.S., ROTHMAN, S.W. and BROWN, J.E. 1987. Cytotoxicity of Shiga toxin for primary cultures of human colonic and ileal epithelial cells. Infect. Immun. *55*, 1533–1535.
81. NAKAGAWA, I., NAKATA, M., KAWABATA, S. and HAMADA, S. 1999. Regulated expression of the Shiga toxin B gene induces apoptosis in mammalian fibroblastic cells. Mol. Microbiol. *33*, 1190–1199.
82. NEELY, M.N. and FRIEDMAN, D.I. 1998. Arrangement and functional identification of genes in the regulatory region of lambdoid phage H-19B, a carrier of a Shiga-like toxin. Gene *223*, 105–113.
83. NEELY, M.N. and FRIEDMAN, D.I. 1998. Functional and genetic analysis of regulatory regions of coliphage H-19B: location of shiga-like toxin and lysis genes suggest a role for phage functions in toxin release. Mol. Microbiol. *28*, 1255–1267.
84. NUSRAT, A. *et al.* 1995. Rho protein regulates tight junctions and perijunctional actin organization in polarized epithelia. Proc. Natl. Acad. Sci. U.S.A. *92*, 10629–10633.
85. O'BRIEN, A.D. and HOLMES, R.K. 1987. Shiga and Shiga-like toxins. Microbiol. Rev. *51*, 206–220.
86. O'BRIEN, A.D., MARQUES, L.R., KERRY, C.F., NEWLAND, J.W. and HOLMES, R.K. 1989. Shiga-like toxin converting phage of enterohemorrhagic *Escherichia coli* strain 933. Microb. Pathog. *6*, 381–390.
87. O'BRIEN, A.D. *et al.* 1992. Shiga toxin: biochemistry, genetics, mode of action, and role in pathogenesis. Curr. Top. Microbiol Immunol. *180*, 65–94.
88. OBRIG, T.G. *et al.* 1988. Direct cytotoxic action of Shiga toxin on human vascular endothelial cells. Infect. Immun. *56*, 2373–2378.
89. OBRIG, T.G., DEL VECCHIO, P.J., KARMALI, M.A., PETRIC, M., MORAN, T.P. and JUDGE, T.K. 1987. Pathogenesis of haemolytic uraemic syndrome. Lancet *2*, 687.
90. OBRIG, T.G., LOUISE, C.B., LINGWOOD, C.A., BOYD, B., BARLEY, M.L. and DANIEL, T.O. 1993. Endothelial heterogeneity in Shiga toxin receptors and

responses. J. Biol. Chem. *268*, 15484–15488.

91. OBRIG, T.G., MORAN, T.P. and BROWN, J.E. 1987. The mode of action of Shiga toxin on peptide elongation of eukaryotic protein synthesis. Biochem. J. *244*, 287–294.
92. OGASAWARA, T., ITO, K., IGARASHI, K., YUTSUDO, T., NAKABAYASHI, N. and TAKEDA, Y. 1988. Inhibition of protein synthesis by a Vero toxin (VT2 or Shiga-like toxin II) produced by *Escherichia coli* O157:H7 at the level of elongation factor 1-dependent aminoacyl-tRNA binding to ribosomes. Microb. Pathog. *4*, 127–135.
93. PAI, C.H., KELLY, J.K. and MEYERS, G.L. 1986. Experimental infection of infant rabbits with verotoxin-producing *Escherichia coli*. Infect. Immun. *51*, 16–23.
94. PATON, A.W. *et al.* 1996. Molecular microbiological investigation of an outbreak of hemolytic-uremic syndrome caused by dry fermented sausage contaminated with Shiga-like toxin-producing *Escherichia coli*. J. Clin. Microbiol. *34*, 1622–1627.
95. PHILPOTT, D.J., ACKERLEY, C.A., KILIAAN, A.J., KARMALI, M.A., PERDUE, M.H. and SHERMAN, P.M. 1997. Translocation of verotoxin-1 across T84 monolayers: mechanism of bacterial toxin penetration of epithelium. Am. J. Physiol. *273*, G1349–G1358.
96. PLUNKETT, G., ROSE, D.J., DURFEE, T.J. and BLATTNER, F.R. 1999. Sequence of Shiga toxin 2 phage 933W from *Escherichia coli* O157:H7: Shiga toxin as a phage late-gene product. J. Bacteriol. *181*, 1767–1778.
97. PROULX, F., TURGEON, J.P., DELAGE, G., LAFLEUR, L. and CHICOINE, L. 1992. Randomized, controlled trial of antibiotic therapy for *Escherichia coli* O157:H7 enteritis. J. Pediatr. *121*, 299–303.
98. PUDYMAITIS, A. and LINGWOOD, C.A. 1992. Susceptibility to verotoxin as a function of the cell cycle. J. Cell Physiol. *150*, 632–639.
99. PUGSLEY, A.P. 1993. The complete general secretory pathway in gram-negative bacteria. Microbiol. Rev. *57*, 50–108.
100. REMUZZI, G. 1987. HUS and TTP: variable expression of a single entity. Kidney Int. *32*, 292–308.
101. REYMOND, D. *et al.* 1996. Neutralizing antibodies to *Escherichia coli* Vero cytotoxin 1 and antibodies to O157 lipopolysaccharide in healthy farm family members and urban residents. J. Clin. Microbiol. *34*, 2053–2057.
102. REYMOND, D., KARMALI, M.A., CLARKE, I., WINKLER, M. and PETRIC, M. 1997. Comparison of the western blot assay with the neutralizing-antibody and enzyme-linked immunosorbent assays for measuring antibody to verocytotoxin 1. J. Clin. Microbiol. *35*, 609–613.
103. RICHARDSON, S.E., KARMALI, M.A., BECKER, L.E. and SMITH, C.R. 1988. The histopathology of the hemolytic uremic syndrome associated with verocytotoxin-producing *Escherichia coli* infections. Hum. Pathol. *19*, 1102–1108.
104. RICHARDSON, S.E. *et al.* 1992. Experimental verocytotoxemia in rabbits. Infect. Immun. *60*, 4154–4167.
105. RIETRA, P.J., WILLSHAW, G.A., SMITH, H.R., FIELD, A.M., SCOTLAND, S.M. and ROWE, B. 1989. Comparison of Vero-cytotoxin-encoding phages from *Escherichia coli* of human and bovine origin. J. Gen. Microbiol. *135*, 2307–2318.

106. RILEY, L.W. 1987. The epidemiologic, clinical, and microbiologic features of hemorrhagic colitis. Annu. Rev. Microbiol. *41*, 383–407.
107. RILEY, L.W. *et al.* 1983. Hemorrhagic colitis associated with a rare *Escherichia coli* serotype. N. Engl. J. Med. *308*, 681–685.
108. SALMOND, G.P. and REEVES, P.J. 1993. Membrane traffic wardens and protein secretion in gram-negative bacteria. Trends. Biochem. Sci. *18*, 7–12.
109. SANDVIG, K. *et al.* 1993. Entry of Shiga toxin into cells. Int. J. Med. Microbiol Virol. Parasitol. Infect. Dis. *278*, 296–305.
110. SANDVIG, K., GARRED, O., PRYDZ, K., KOZLOV, J.V., HANSEN, S.H. and VAN DEURS, B. 1992. Retrograde transport of endocytosed Shiga toxin to the endoplasmic reticulum. Nature *358*, 510–512.
111. SANDVIG, K. and VAN DEURS, B. 1996. Endocytosis, intracellular transport, and cytotoxic action of Shiga toxin and ricin. Physiol. Rev. *76*, 949–966.
112. SAXENA, S.K., O'BRIEN, A.D. and ACKERMAN, E.J. 1989. Shiga toxin, Shiga-like toxin II variant, and ricin are all single-site RNA N-glycosidases of 28 S RNA when microinjected into Xenopus oocytes. J. Biol. Chem. *264*, 596–601.
113. SCHMIDT, H., BEUTIN, L. and KARCH, H. 1995. Molecular analysis of the plasmid-encoded hemolysin of *Escherichia coli* O157:H7 strain EDL 933. Infect. Immun. *63*, 1055–1061.
114. SCHMIDT, H., HENKEL, B. and KARCH, H. 1997. A gene cluster closely related to type II secretion pathway operons of gram-negative bacteria is located on the large plasmid of enterohemorrhagic *Escherichia coli* O157 strains. FEMS Microbiol. Lett. *148*, 265–272.
115. SCHMIDT, H., KARCH, H. and BEUTIN, L. 1994. The large-sized plasmids of enterohemorrhagic *Escherichia coli* O157 strains encode hemolysins which are presumably members of the *E. coli* alpha-hemolysin family. FEMS Microbiol. Lett. *117*, 189–196.
116. SCHMIDT, H., MAIER, E., KARCH, H. and BENZ, R. 1996. Pore-forming properties of the plasmid-encoded hemolysin of enterohemorrhagic *Escherichia coli* O157:H7. Eur. J. Biochem. *241*, 594–601.
117. SCHMIDT, H., SCHEEF, J., JANETZKI-MITTMANN, C., DATZ, M. and KARCH, H. 1997. An ileX tRNA gene is located close to the Shiga toxin II operon in enterohemorrhagic *Escherichia coli* O157 and non-O157 strains. FEMS Microbiol. Lett. *149*, 39–44.
118. SHETH, K.J., GILL, J.C., HAVENS, P.L. and LEICHTER, H.E. 1995. Racial incidence of hemolytic-uremic syndrome. Pediatr. Nephrol. *9*, 401.
119. SHETH, K.J., GILL, J.C., LEICHTER, H.E., HAVENS, P.L. and HUNTER, J.B. 1994. Increased incidence of HLA-B40 group antigens in children with hemolytic-uremic syndrome. Nephron *68*, 433–436.
120. SIEGLER, R.L., EDWIN, S.S., CHRISTOFFERSON, R.D. and MITCHELL, M.E. 1991. Endothelin in the urine of children with the hemolytic uremic syndrome. Pediatrics *88*, 1063–1066.
121. SIMON, M., CLEARY, T.G., HERNANDEZ, J.D. and ABBOUD, H.E. 1998. Shiga toxin 1 elicits diverse biologic responses in mesangial cells. Kidney Int. *54*, 1117–1127.

122. SMITH, H.W., GREEN, P. and PARSELL, Z. 1983. Vero cell toxins in *Escherichia coli* and related bacteria: transfer by phage and conjugation and toxic action in laboratory animals, chickens and pigs. J. Gen. Microbiol. *129*, 3121–3137.
123. STROCKBINE, N.A., MARQUES, L.R., NEWLAND, J.W., SMITH, H.W., HOLMES, R.K. and O'BRIEN, A.D. 1986. Two toxin-converting phages from *Escherichia coli* O157:H7 strain 933 encode antigenically distinct toxins with similar biologic activities. Infect. Immun. *53*, 135–140.
124. STROCKBINE, N.A., WELLS, J.G., BOPP, C.A. and BARRETT, T.J. 1998. Overview of detection and subtyping methods. *In Escherichia coli* O157:H7 and other Shiga toxin-producing *E. coli* strains, (J.B. Kaper and A.D. O'Brien, eds.) pp. 331–356. American Society for Microbiology, Washington, D.C.
125. TAGA, S. *et al.* 1997. Intracellular signaling events in CD77-mediated apoptosis of Burkitt's lymphoma cells. Blood *90*, 2757–2767.
126. TAGUCHI, T. *et al.* 1998. Verotoxins induce apoptosis in human renal tubular epithelium derived cells. Kidney Int. *53*, 1681–1688.
127. TARR, P.I. 1995. *Escherichia coli* O157:H7: clinical, diagnostic, and epidemiological aspects of human infection. Clin. Infect. Dis. *20*, 1–8.
128. TAYLOR, F. *et al.* 1999. Characterization of the baboon responses to Shiga-like toxin: descriptive study of a new primate model of toxic responses to Stx-1. Am. J. Pathol. *154*, 1285–1299.
129. TESH, V.L., RAMEGOWDA, B. and SAMUEL, J.E. 1994. Purified Shiga-like toxins induce expression of proinflammatory cytokines from murine peritoneal macrophages. Infect. Immun. *62*, 5085–5094.
130. TESH, V.L., SAMUEL, J.E., PERERA, L.P., SHAREFKIN, J.B. and O'BRIEN, A.D. 1991. Evaluation of the role of Shiga and Shiga-like toxins in mediating direct damage to human vascular endothelial cells. J. Infect. Dis. *164*, 344–352.
131. THORPE, C.M., HURLEY, B.P., LINCICOME, L.L., JACEWICZ, M.S., KEUSCH, G.T. and ACHESON, D.W. 1999. Shiga toxins stimulate secretion of interleukin-8 from intestinal epithelial cells. Infect. Immun. *67*, 5985–5993.
132. TOTH, I. *et al.* 1990. Influence of the 60-megadalton plasmid on adherence of *Escherichia coli* O157:H7 and genetic derivatives. Infect. Immun. *58*, 1223–1231.
133. TSAI, H.M., NAGEL, R.L., HATCHER, V.B. and SUSSMAN, I.I. 1989. Endothelial cell-derived high molecular weight von Willebrand factor is converted into the plasma multimer pattern by granulocyte proteases. Biochem. Biophys. Res. Commun. *158*, 980–985.
134. TZIPORI, S., GUNZER, F., DONNENBERG, M.S., DE MONTIGNY, L., KAPER, J.B. and DONOHUE-ROLFE, A. 1995. The role of the *eaeA* gene in diarrhea and neurological complications in a gnotobiotic piglet model of enterohemorrhagic *Escherichia coli* infection. Infect. Immun. *63*, 3621–3627.
135. TZIPORI, S. *et al.* 1987. Role of a 60-megadalton plasmid and Shiga-like toxins in the pathogenesis of infection caused by enterohemorrhagic *Escherichia coli* O157:H7 in gnotobiotic piglets. Infect. Immun. *55*, 3117–3125.
136. VAN DE KAR, N.C., KOOISTRA, T., VERMEER, M., LESSLAUER, W., MONNENS, L.A. and VAN HINSBERGH, V.W. 1995. Tumor necrosis factor alpha induces endothelial galactosyl transferase activity and verocytotoxin

receptors. Role of specific tumor necrosis factor receptors and protein kinase C. Blood *85*, 734–743.

137. VAN DE KAR, N.C., MONNENS, L.A., KARMALI, M.A. and VAN HINSBERGH, V.W. 1992. Tumor necrosis factor and interleukin-1 induce expression of the verocytotoxin receptor globotriaosylceramide on human endothelial cells: implications for the pathogenesis of the hemolytic uremic syndrome. Blood *80*, 2755–2764.
138. VAN DE KAR, N.C., SAUERWEIN, R.W., DEMACKER, P.N., GRAU, G.E., VAN HINSBERGH, V.W. and MONNENS, L.A. 1995. Plasma cytokine levels in hemolytic uremic syndrome. Nephron *71*, 309–313.
139. VAN SETTEN, P.A., MONNENS, L.A., VERSTRATEN, R.G., VAN DEN HEUVEL, L.P. and VAN HINSBERGH, V.W. 1996. Effects of verocytotoxin-1 on nonadherent human monocytes: binding characteristics, protein synthesis, and induction of cytokine release. Blood *88*, 174–183.
140. VAN SETTEN, P.A. *et al.* 1997. Verocytotoxin inhibits mitogenesis and protein synthesis in purified human glomerular mesangial cells without affecting cell viability: evidence for two distinct mechanisms. J. Am. Soc. Nephrol. *8*, 1877–1888.
141. VAN SETTEN, P.A. *et al.* 1997. Effects of TNF alpha on verocytotoxin cytotoxicity in purified human glomerular microvascular endothelial cells. Kidney Int. *51*, 1245–1256.
142. VON EICHEL-STREIBER, C., BOQUET, P., SAUERBORN, M. and THELESTAM, M. 1996. Large clostridial cytotoxins — a family of glycosyltransferases modifying small GTP-binding proteins. Trends. Microbiol *4*, 375–382.
143. WADA, Y., MORI, K. and IWANAGA, T. 1997. Apoptosis of enterocytes induced by inoculation of a strain of attaching and effacing *Escherichia coli* and verotoxin. J. Vet. Med. Sci. *59*, 815–818.
144. WADDELL, T., HEAD, S., PETRIC, M., COHEN, A. and LINGWOOD, C. 1988. Globotriosyl ceramide is specifically recognized by the *Escherichia coli* verocytotoxin 2. Biochem. Biophys. Res. Commun. *152*, 674–679.
145. WADOLKOWSKI, E.A., SUNG, L.M., BURRIS, J.A., SAMUEL, J.E. and O'BRIEN, A.D. 1990. Acute renal tubular necrosis and death of mice orally infected with *Escherichia coli* strains that produce Shiga-like toxin type II. Infect. Immun. *58*, 3959–3965.
146. WAGNER, P.L., ACHESON, D.W. and WALDOR, M.K. 1999. Isogenic lysogens of diverse Shig toxin 2-encoding bacteriophages produce markedly different amounts of Shiga toxin. Infect. Immun. *67*, 6710–6714.
147. WALTERS, M.D., MATTHEI, I.U., KAY, R., DILLON, M.J. and BARRATT, T.M. 1989. The polymorphonuclear leucocyte count in childhood haemolytic uraemic syndrome. Pediatr. Nephrol. *3*, 130–134.
148. WILLIAMS, J.M., LEA, N., LORD, J.M., ROBERTS, L.M., MILFORD, D.V. and TAYLOR, C.M. 1997. Comparison of ribosome-inactivating proteins in the induction of apoptosis. Toxicol. Lett. *91*, 121–127.
149. YOSHIDA, T. *et al.* 2000. Primary cultures of human endothelial cells are susceptible to low doses of Shiga toxins and undergo apoptosis. J. Infect. Dis. *180*, 2048–1052.

CHAPTER 14

HEALTHY ANIMALS AS CARRIERS OF STEC

CATHY A. BROWN[1]
and
BARRY G. HARMON[2]

[1]*Athens Diagnostic Laboratory*
[2]*Department of Pathology*
College of Veterinary Medicine
University of Georgia
Athens, Georgia

AND

TONG ZHAO
and
MICHAEL P. DOYLE

Center for Food Safety
Department of Food Science and Technology
Georgia Experiment Station
University of Georgia
Griffin, Georgia

INTRODUCTION

During the past fifteen years, numerous outbreaks of diarrhea and hemolytic uremic syndrome in humans have been associated with the consumption of food contaminated with Shiga-toxin producing strains of *Escherichia coli*, particularly O157:H7. Cattle were initially implicated as a source of these bacteria because most STEC-related illnesses in people were linked to the ingestion of contaminated undercooked ground beef and unpasteurized milk (3,6,58). Microbiological surveys and experimental inoculation studies have substantiated the importance of cattle as a reservoir for STEC, and have begun to elucidate the ecology of STEC in the bovine host.

CARRIAGE OF *E. COLI* O157:H7 IN CATTLE

A number of studies have investigated the natural occurrence of fecal shedding of *E. coli* O157:H7 in cattle (4,8,10,17,20,25,30,34,35,48,53,65,68). These studies have provided important information concerning *E. coli* O157:H7 carriage in cattle, including clinical signs associated with *E. coli* O157:H7 carriage, the magnitude and duration of fecal shedding, the age-incidence of fecal shedding, and possible factors influencing fecal shedding in some populations of cattle.

While there is some evidence that calves may develop attaching and effacing intestinal lesions and diarrhea associated with some *eae* positive STEC and non-STEC strains, the great majority of cattle shedding *E. coli* O157:H7 and other STEC are clinically healthy (34,38,40,57,65,66). In fact, there is a negative association between gastrointestinal disease and *E. coli* O157:H7 fecal shedding, such that shedding was detected in 3.7% of clinically normal calves compared to 0.75% of calves with diarrhea (53). Shiga toxin has not been shown to have a role in causing disease in cattle, and vascular lesions attributed to the action of Shiga toxin have not been described.

E. coli O157:H7 is worldwide in distribution, and has been isolated from 752 of 4800 cattle in the UK, 59 of 450 cattle in Italy, and from 56 of 1091 and 10 of 3570 cattle in the US (5,12,17,30). Reported individual animal and herd prevalences of *E. coli* O157:H7 fecal shedding range from 0-16.6% and from 0-70%, respectively (5,12,17,20,30,33-35). While this wide range in prevalence may reflect true population differences, differences in the sensitivity of detection methods used are likely to make significant contributions to interstudy differences, limiting comparisons between studies (1). However, in all positive surveys, the herd prevalence is much greater than the individual animal prevalence, indicating that *E. coli* O157:H7 is carried by small numbers of individual cattle, and that these individuals are widely dispersed among different cattle populations.

Calves normally acquire *E. coli* from their dams at birth, from other animals, and from their environment. Calves initially carry a very heterogeneous *E. coli* population and there is a continual turnover of O-typeable *E. coli* strains (36,37). This diversity decreases as cattle age and predominant *E. coli* serotypes become established (36). Similarly, the prevalence of STEC, and *E. coli* O157:H7 in particular, is higher in weaned calves compared to adult cattle (10,30,34,65).

The magnitude of fecal shedding is typically low. *E. coli* O157:H7 was detected by direct plating, at populations ranging from 10^3-10^5 CFU/gram of feces, in 52% of positive animals. The level of fecal shedding in the remaining cattle was $< 10^2$ CFU/gram of feces and detectable only via enrichment procedures (68).

Shedding of *E. coli* O157:H7 appears to be transient or intermittent within individual animals. Most animals that test positive for O157:H7 will be fecal culture negative within 1-3 months (5,48). While shedding among individual animals appears transient, herd infection appears to be maintained (65,68). However, the DNA profile, toxin type, and plasmid profiles of the strains of *E.*

coli O157:H7 isolated from within positive herds or animals change over time (5,65,68). This excretion of different PFGE types of *E. coli* O157:H7 at different times is consistent with clearance of *E. coli* O157:H7 followed by acquisition of different strains.

Several studies have shown an increased incidence of *E. coli* O157:H7 and non-O157 STEC fecal shedding in the summer (12,17,30,34). Cattle in Italy shed STEC at a prevalence of 17.5% in the summer and only 2.9% in the winter (12). This seasonal effect could be due to seasonal dietary practices, such as increased pasture grazing in the summer. Fecal shedding of *Escherichia coli* O157:H7 increased when sheep fed alfalfa pellets were turned out onto pasture (43). In addition, there may be increased proliferation of *E. coli* O157:H7 in the environment during warmer weather. Increased ambient temperatures may result in increased *E. coli* O157:H7 proliferation in feed or water, both of which are commonly contaminated with bovine feces on the farm (45).

Attempts have been made to identify management factors which might influence the magnitude or duration of fecal shedding of *E. coli* O157:H7. In feedlot studies, cattle on feed for the shortest period of time were more likely to be shedding *E. coli* O157:H7 (20,31). Similarly, cattle examined at the abattoir were found to shed larger populations of STEC in their feces than they had at their farm of origin (61). Proposed factors which might influence STEC carriage in these cattle include short-term food deprivation during transport, exposure of naïve cattle to shedders, dietary changes, and transport "stress" (20). In experimental studies, food deprivation does not increase the magnitude of fecal shedding of *E. coli* O157:H7 in inoculated calves (19,32). However, there is evidence that fasting cattle are more susceptible to colonization by *E. coli* O157:H7 (19). Therefore, cattle deprived of food during transport and exposed to *E. coli* O157:H7 shedders as they enter the feedlot may be more likely to be colonized by, and subsequently shed, *E. coli* O157:H7. Experimental evidence has shown that prior infection with *E. coli* O157:H7 does not influence shedding upon reexposure (13,18,44). Therefore, a difference between naïve cattle and previously exposed cattle in regards to susceptibility to infection is considered less likely. Abrupt dietary changes affect the magnitude of fecal shedding of nonpathogenic *E. coli* in cattle and *E. coli* O157:H7 fecal shedding in experimentally inoculated sheep (22,44). The effects of transport "stress" on fecal shedding of STEC have not been reported.

Dietary constituents may influence the carriage and fecal shedding of *E. coli* O157:H7. A positive association has been made between the prevalence of *E. coli* O157:H7 fecal shedding and [a] corn silage consumption in dairy herds or [b] the feeding of barley to feedlot cattle (20,33). Ionophore feed additives, which were introduced in the late 1970s as bovine growth promotors, favor the proliferation of gram-negative bacteria in the gastrointestinal tract and were weakly associated with an increased incidence of *E. coli* O157:H7 fecal shedding in one study (33). The effects of feeding a high concentrate grain diet versus a high roughage hay diet on *E. coli* O157:H7 carriage have been examined under experimental conditions (see below).

EXPERIMENTAL *E. COLI* O157:H7 CARRIAGE IN CALVES

Neonatal calves (< 36 h of age) inoculated with large populations of *E. coli* O157:H7 (10^{10} CFUs) can develop diarrhea, enterocolitis, and attaching and effacing intestinal lesions (21). The importance of natural infection with *E. coli* O157:H7 in neonates is questionable, as *E. coli* O157:H7 has not been isolated from nursing calves in the field (30) and has not been reported as a cause of diarrhea in neonates. In contrast, experimental infection studies and clinical observation of naturally infected animals indicate that *E. coli* O157:H7 is not pathogenic in adult cattle and calves older than 3 weeks of age. Except for mild transient nonhemorrhagic diarrhea and mild pyrexia in some animals, calves and adult cattle orally inoculated with 10^{10} CFUs of *E. coli* O157:H7 remain clinically healthy (13,18,39,54).

After inoculation, *E. coli* O157:H7 is shed for a longer period of time in weaned calves than in adult cows. Weaned calves shed *E. coli* O157:H7 for up to 20 weeks post inoculation, while adult cows shed for a maximum of 9 weeks post inoculation (18). Calves shed large populations of *E. coli* O157:H7 for the first two weeks after inoculation, followed by intermittent bacterial shedding in concentrations ranging from values only detectable using enrichment methods (< 50 CFU/g) to 5×10^4 CFU/g of feces (13,18, Fig. 1). The magnitude of fecal shedding is highly variable, both between calves and from day to day in individual animals (13,18,32, Fig. 1). Factors responsible for this variability in fecal shedding have not yet been identified.

E. coli O157:H7 remains confined to the gastrointestinal tract (13,18) of experimentally infected calves, where it localizes within the lumen of the rumen and colon. Histologic intestinal abnormalities, such as attaching and effacing lesions, have not been observed in experimentally infected weaned calves and adult cows. Under experimental conditions, *E. coli* O157:H7 does not appear to colonize the intestinal mucosal surface of weaned calves and adult cattle (13,18).

The importance of the rumen in the ecology of *E. coli* O157:H7 is supported by the observation that *E. coli* O157:H7 carriage is observed predominantly in ruminant species, such as cattle, sheep, goats, and deer. In calves, *E. coli* O157:H7 carriage is infrequent in calves prior to weaning and rumen development. Similarly, older veal calves maintained on a milk diet have underdeveloped rumens and *E. coli* O157:H7 is rarely isolated from their feces (12,35). After weaning and following introduction of fiber into the diet, the rumen undergoes structural and functional development, eventually comprising 65-70% of the gut volume in cattle (24). The rumen contains predominantly anaerobic bacteria, protozoa, and fungi. Facultative anaerobic enterobacteria, such as nontoxigenic *E. coli*, normally represent 1% or less of the rumen microbial population and are thought to be a nonproliferating, transient population (28,50). Likewise, under experimental conditions *E. coli* O157:H7 does not demonstrate significant proliferation in the rumen. In inoculated calves, *E. coli* O157:H7 is present in the rumen in populations that parallel, but are usually smaller than, populations in the colon (13, Fig. 1). Although present in small populations, *E. coli* O157:H7 persists in the rumen and in some calves may only be recoverable at necropsy from this site (13). *E. coli* O157:H7 persisting in the rumen may serve as a source for continued reintroduction into the colon where more substantial proliferation may occur.

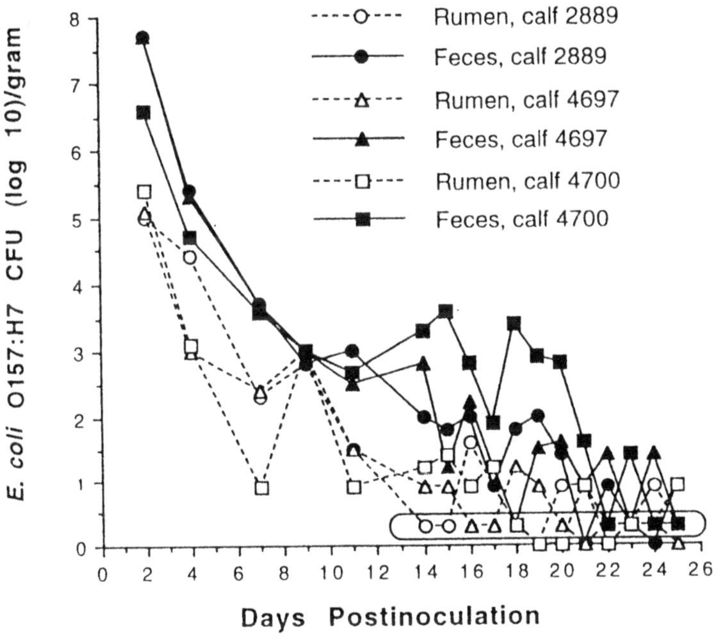

FIG. 1. DETECTION OF *E. COLI* O157:H7 IN THE RUMEN AND FECES OF CALVES FOLLOWING INOCULATION WITH 1 × 10^{10} CFU OF A FIVE-STRAIN MIXTURE OF NALIDIXIC ACID-RESISTANT *E. COLI* O157:H7

Points within the oval represent isolation of *E. coli* O157:H7 by enrichment procedures only (< 50 CFU/gram).

Factors affecting the survival, proliferation, and the development of acid resistance of *E. coli* O157:H7 in the rumen may affect the duration and magnitude of *E. coli* O157:H7 fecal shedding. Dietary changes affect the fermentative microenvironment of both the rumen and colon. In the rumen, normal bacterial fermentation of ingested substrates results in the production of volatile fatty acids, which are absorbed across the rumen mucosa and serve as an energy source for the cow. The combination of low rumen pH and high volatile fatty acid (VFA) concentration, as found in the rumen of well-fed calves, inhibits the proliferation of *E. coli* O157:H7 *in vitro* (49). Withholding feed increases rumen pH values and decreases VFA concentrations, and results in increases in ruminal *Salmonella* sp. and *E. coli* populations in cattle (14). After inoculation with *E. coli* O157:H7, fed calves also have higher rumen concentrations of VFAs and lower pH values but do not exhibit a difference in fecal shedding of *E. coli* O157:H7 compared to periodically fasted calves (19,32).

Rumen pH and VFA concentrations are also affected by dietary composition. Calves fed a grain diet, high in readily fermentable substrates, have lower rumen pH values and higher volatile fatty acid concentrations than calves fed a hay diet (63). Despite these differences, inoculated calves fed grain or hay had similar numbers of *E. coli* O157:H7 in their rumen contents and shed similar populations of *E. coli* O157:H7 in their feces throughout the 30-day study (63). In contrast, in another study inoculated grain-fed 1-2 year-old steers shed O157:H7 for a shorter duration (mean, 4 days) than did hay-fed steers (mean 39-42 days) (39). Similarly, inoculated sheep fed grass hay shed *E. coli* O157:H7 in their feces for longer, and in larger numbers, than sheep fed a corn/pelleted alfalfa diet (44).

Larger populations of nonpathogenic *E. coli* in grain-fed cattle are acid resistant compared to the colonic *E. coli* populations in hay-fed cattle (22,39). Furthermore, cattle abruptly changed from a grain to a hay diet shed smaller populations of nonpathogenic *E. coli* and a smaller percentage of these were acid resistant (22). In contrast, *E. coli* O157:H7 isolates from inoculated steers maintained on hay or grain diets exhibit a wide range of resistance to acid shock, and an effect of diet on *E. coli* O157:H7 acid resistance was not apparent (39). While the acid resistance of *E. coli* O157:H7 was unaffected by diet in this study, an effect of diet on the development of acid resistance has been demonstrated *in vitro*. *E. coli* O157:H7 preincubated in an *in vitro* fermentation system supplemented with rumen fluid from grain-fed animals exhibited increased survival in acid conditions (pH 2.0 for 2-4 h) compared to *E. coli* O157:H7 preincubated in rumen fluid from hay-fed animals (64). It has been suggested that the lower pH and higher VFA concentrations in the rumen fluid from grain-fed animals may activate pH dependent acid resistance mechanisms, enabling *E. coli* O157:H7 to survive passage through the acidic abomasum in greater numbers. These acid resistant bacteria may then enjoy a competitive advantage in the cecum and colon, where the conditions of anaerobic fermentation are similar to those in the rumen. The impact of diet on the development of acid resistant EHEC, and the significance of any observed effect, requires further investigation.

The rumen has also been suggested as a potential site of generation of new serotypes of EHEC (6). *E. coli* can transfer antibiotic resistance plasmids under conditions present in the rumen, and it is conceivable that plasmids containing virulence factors could also be transmitted (6).

OTHER STEC IN CATTLE

While O157:H7 is the most commonly identified serotype of EHEC in foodborne disease outbreaks in most countries, microbiological surveys of cattle and other animals have expanded to include identification of a variety of STEC, as more than 160 *E. coli* serotypes have been associated with human disease and 57 have been associated with HC or HUS (8,29,41). More than 200 serotypes and 82 serogroups of STEC have been isolated from healthy cattle (29).

Bovine STEC represent a heterogeneous population of bacteria, of which only a small subset are highly pathogenic to humans. Those virulence factors, in addition to Shiga toxins, which must be present to confer pathogenicity in humans are poorly defined (41,46,62). STEC types associated with severe diarrhea and HUS in humans often carry other virulence markers; 92% and 89% of these STEC types were positive for intimin (*eae* gene) and for enterohemolysin production (*ehxA* gene), respectively (9,29). Serotypes lacking the *eae* gene are more frequently found in asymptomatic carriers or in uncomplicated cases of diarrhea in adults (29).

STEC are commonly isolated from the feces of healthy cattle. STEC were identified in 63.2% of 19 cattle in Germany, 21% of 142 Swiss cattle, and 35% of 268 cows in Spain (8,11,15). While EHEC serotypes O157:H7, O26:H11, and O111:NM are associated with the majority of human outbreaks of disease, these STEC serotypes only constitute a small proportion of the total bovine STEC population. In three surveys, these EHEC serotypes represented 1.7% to 16.3% of the total typeable STEC isolated from healthy cattle (10,11,41, Table 1B). All EHEC isolates were *eae* gene positive, and 92.4% were positive for stx 1 or both stx 1 and 2 (Table 1B). In contrast, the majority of non-enterohemorrhagic STEC isolated from healthy cattle are negative for the *eae* gene (9-11, 55) and in 3 surveys only 59.1% were positive for stx 1 or both stx 1 and 2 (10,11,41, Table 1A). There is a strong association between certain O-groups and the *eae* gene, and stx 1 is more frequently associated with *eae*-positive than *eae*-negative serogroups (55, Table 1A and B). Surveys in which all STEC isolates are reported, without determining serotypes or virulence factor attributes, are probably identifying both pathogenic and nonpathogenic bacteria.

TABLE 1A.
CHARACTERISTICS OF TYPEABLE STEC ISOLATED FROM HEALTHY CATTLE

Serogroup[1]	Prevalence[2]	Positive isolates (%)				References
		Stx1	*Stx2*	*Stx1/Stx2*	*EaeA*	
O2	1/60, 7/115	0.0, 0.0	100.0, 100.0	0.0, 0.0	0.0, 0.0	Blanco 1996, 1997
O4	1/60, 3/115	0.0, 0.0	100.0, 100.0	0.0, 0.0	0.0, 0.0	Blanco 1996, 1997
O5[3]	26/487	76.9	3.9	19.2	92.3	Johnson 1997
O7	1/60	100.0	0.0	0.0	0.0	Blanco 1996
O8	4/60, 7/115	0.0, 14.3	100.0, 28.6	0.0, 57.1	0.0, 0.0	Blanco 1996, 1997
O9	1/60, 1/115	0.0, 100.0	100.0, 0.0	0.0, 0.0	0.0, 0.0	Blanco 1996, 1997
O15	2/60, 12/487	0.0, 0.0	50.0, 41.7	50.0, 58.3	0.0, 16.7	Blanco 1996, Johnson 1996
O20	5/60, 2/115	0.0, 0.0	20.0, 0.0	80.0, 100.0	0.0, 0.0	Blanco 1997, Johnson 1996
O22	6/60, 17/115, 57/487	0.0, 5.8, 7.0	20.0, 23.5, 5.3	80.0, 70.7, 87.7	0.0, 0.0, 1.8	Blanco 1996, 1997, Johnson 1996
O41	2/115	100.0	0.0	0.0	0.0	Blanco, 1997
O77	8/60, 10/115	13.5, 0.0	25.0, 90.0	62.5, 10.0	13.5, 0.0	Blanco 1996, 1997
O78	1/115	100.0	0.0	0.0	0.0	Blanco 1997
O81	1/60	0.0	0.0	100.0	0.0	Blanco 1996
O82	7/115	0.0	0.0	100.0	0.0	Blanco 1997
O88	1/60	0.0	100.0	0.0	0.0	Blanco 1996
O90	1/115	100.0	0	0.0	0.0	Blanco 1997
O91	4/115	0.0	75.0	25.0	0.0	Blanco 1997
O92	1/115	100.0	0.0	0.0	0.0	Blanco 1997
O96	1/60	0.0	0.0	100.0	0.0	Blanco 1996
O103[3]	2/115, 33/487	100.0, 100.0	0.0, 0.0	0.0, 0.0	100.0, 97.0	Blanco 1997, Johnson 1996
O105	9/115	11.1	0.0	88.9	0.0	Blanco 1997
O110	1/60	0.0	100.0	0.0	100.0	Blanco 1996
O113[3]	8/60, 20/115, 61/487	12.5, 0.0, 16.4	87.5, 95.0, 54.1	0.0, 5.0, 29.5	0.0, 0.0, 1.6	Blanco 1996, 1997, Johnson 1996
O116	2/60, 1/115, 66/487	0.0, 0.0, 3.0	100.0, 100.0, 13.6	0.0, 0.0, 83.3	0.0, 0.0, 0.0	Blanco 1996, 1997, Johnson 1996
O126	4/60	25.0	0.0	75.0	0.0	Blanco 1996

O132	1/115, 12/487	100.0, 16.7	0.0, 83.3	0.0, 0.0	0.0, 0.0	Blanco 1997, Johnson 1997
O136	2/60, 3/115	100, 33.3	0.0, 66.7	0.0, 0.0	0.0, 33.3	Blanco 1996, 1997
O138	2/60	100.0	0.0	0.0	50.0	Blanco 1996
O140	1/60	0.0	100.0	0.0	0.0	Blanco 1996
O145	12/487	33.3	66.7	0.0	91.7	Johnson 1996
O146	1/115	0.0	100.0	0.0	0.0	Blanco 1997
O149	1/60	100.0	0.0	0.0	0.0	Blanoc 1996
O153[3]	40/487	22.5	50.0	27.5	5.0	Johnson 1997
O156	52/487	7.7	82.7	9.6	7.7	Johnson 1997
O162	3/60	0.0	66.7	33.3	0.0	Blanco 1996
O163	3/115	33.3	66.7	0.0	33.3	Blanco 1997
O165	2/115	0.0	100.0	0.0	100.0	Blanco 1997
O168	1/60	100.0	0.0	0.0	0.0	Blanco 1996
O171	1/60, 8/115	0.0, 0.0	100.0, 100.0	0.0, 0.0	0.0, 0.0	Blanco 1996, 1997
Totals	542 isolates	20.7%	41.5%	38.4%	15.7%	

[1] excluding serogroups O157, O26, and O111 (Table 1B).

[2] number isolated/total typeable STEC isolates

[3] serogroups frequently associated with human disease (HC or HUS, Gyles 1998)

TABLE 1B.
CHARACTERISTICS OF COMMON ENTEROHEMORRHAGIC STEC ISOLATED FROM HEALTHY CATTLE

Serotype	Prevalence	Positive isolates (%)				References
		Stx1	*Stx2*	*Stx1/Stx2*	*EaeA*	
O157:H7	1/60, 0/115, 52/487	0.0, 5.8	100.0, 15.4	0.0, 78.8	100.0 100.0	Blanco 1996,1997 Johnson 1996
O26:H11	1/60, 2/115, 50/487	100.0, 100.0, 96.0	0.0, 0.0 2.0	0.0, 0.0 2.0	100.0 100.0	Blanco 1996, 1997 Johnson 1996
O111:NM	0/60, 0/115, 14/487	85.7	0	14.3	100.0	Blanco 1996, 1997; Johnson 1996
Totals	118 isolates	55.1%	7.6%	37.3%	100%	

STEC CARRIAGE IN OTHER ANIMAL SPECIES

Sheep and Goats

Sheep inoculated with *E. coli* O157:H7 remain clinically healthy and have been used as an experimental model of *E. coli* O157:H7 carriage in ruminants (43,44). In some microbiological surveys, STEC are more frequently isolated from healthy sheep than from cattle (8). STEC were isolated from 88% of 25 sheep, with serotypes O125, O128:H2, and O146:H4 predominating (8). STEC were recovered from 66.6% of 120 sheep and 56.1% of 66 goats in Germany, and none of these were typical enterohemorrhagic strains (7). More recently, *E. coli* O157:H7 was isolated from 15.7% of 4800 cattle and 2.2% of 1000 sheep (17). The O157:H7 isolates from sheep in this study were, like the bovine isolates, shigatoxigenic and *eaeA* gene positive. STEC and *E. coli* O157:H7 have also been isolated from 15% and 0.8% of 882 market lambs, respectively (47). The high prevalence of STEC shedding among healthy sheep and goats indicate they are also an important reservoir of STEC. However, although lambs commonly carry STEC, no human cases of STEC disease have been associated with the ingestion of contaminated lamb (47).

Chickens

Young chickens experimentally inoculated with *E. coli* O157:H7 and other STEC remain clinically healthy (2,60). Following inoculation of 1-day-old chicks, *E. coli* O157:H7 colonizes the ceca and colon and is shed in the feces at moderate levels (10^3 CFU/gram of feces) for 90 days post inoculation (2).

Despite the apparent potential for chickens to serve as carriers of STEC, microbiological surveys have not isolated STEC from chicken feces, and poultry have not been identified as a source of human STEC infections. *E. coli* O157:H7 was not isolated from 1000 fecal samples collected from chickens at slaughter in the UK (17), and STEC were not isolated from fecal swabs from 144 chickens in Germany (7). Likewise, several studies have failed to detect VTEC in retail poultry meat (51,59). While some sampled poultry meat in the USA and Thailand has tested positive for STEC (23,61), these STEC may have originated from nonpoultry sources.

Deer

Based on experimental inoculation studies and microbiological surveys, deer may be carriers of *E. coli* O157:H7. Experimentally, the carriage of *E. coli* O157:H7 by deer is similar to carriage in domestic ruminants (13,26). Inoculated deer remain clinically healthy, fecal shedding is variable in magnitude and duration, and the bacteria are present in the rumen and colon at necropsy (26).

Similar to other ruminants, deer shedding *E. coli* O157 appear to be clinically healthy (56). *E. coli* O157:H7 was isolated from 5 of 212 (2.4%) deer fecal samples collected from cattle pastures in Kansas (56). Shiga-toxin producing *E. coli* O157 were isolated from 2 of 10 farmed deer in the UK and from 2 of 108 (1.9%) deer fecal samples and 5 of 191 (2.6%) cattle fecal samples collected from a ranch in Texas (16,52). In the Texas study, the isolates from the deer and cattle had identical PFGE patterns of genomic DNA and all produced only Stx 2, suggesting a common source of infection or cross-species infection. In a study conducted in the Southeastern US, *E. coli* O157:H7 was isolated from only 3 of 919 (0.33%) deer fecal samples obtained from the ground or from hunter-killed animals (27). The 3 positive deer samples were obtained from 77 (3.8%) deer at one geographic site; *E. coli* O157:H7 was also isolated from 13 of 305 (4.3%) of cattle in the same area sampled 1 year later. There was no apparent relationship between the deer and cattle isolates in this study, as deer and cattle isolates had different PFGE patterns (27). While deer can harbor *E. coli* O157:H7, the prevalence of carriage in deer is low and the role of deer in the ecology of bovine STEC appears minimal.

Deer are a potential source of human infection, with human infection following ingestion of contaminated deer products or produce contaminated in the field by deer pellets. Homemade venison jerky prepared from a hunter-killed black-tailed deer was the source of an outbreak of *Escherichia coli* O157:H7 infection, and *E. coli* O157:H7 was recovered from 3/32 (9%) of deer pellets collected from nearby forest land (42). Contamination of downed apples with deer droppings may have been the source of an apple cider associated disease outbreak (4).

HOST SPECIFICITY OF STEC

Microbiological surveys and experimental inoculation studies have suggested that there is some degree of STEC host specificity. Examination of STEC from sheep and goats show great serologic diversity, with certain serotypes predominating in each species (9). While 42 different O-typeable serogroup of STEC were isolated from cattle in 3 surveys, 75% of these isolates belonged to 8 serogroups (O5, O22, O103, O113, O116, O153, O156, and O157; Table 1) and these strains may be well-adapted to the bovine host. In contrast, serotypes O125, O128, and O146 appear to predominate in sheep (8). These resident species-adapted populations of STEC may also have a role in the generation of transient STEC by providing *stx*-encoding bacteriophages to *stx*-negative *E. coli* (8). Pigs carry some STEC that produce a unique variant of Shiga toxin, stx2v, that produces edema disease in pigs but is unassociated with human disease.

Additionally, some degree of species specificity for different genetic strains of *E. coli* O157:H7 has been suggested by experimental inoculation studies. Five of six isolates obtained from 2 calves 17 to 28 days after inoculation with a 5 strain mixture (2 from human feces, 1 from ground beef, 2 from cattle feces) of *E. coli* O157:H7 were a cattle fecal strain; the remaining isolate was a strain isolated from human feces during a ground beef-associated outbreak of diarrhea (13). Similarly, deer inoculated with the same 5-strain *E. coli* O157:H7 mixture, in which one of the human fecal isolates was replaced with a venison isolate, carried the deer-associated strain at the termination of the study (26).

REFERENCES

1. ARMSTRONG, G.L., HOLLINGSWORTH, J. and MORRIS, J.G. 1996. Emerging foodborne pathogens: *Escherichia coli* O157:H7 as a model of entry of a new pathogen into the food supply of the developed world. Epidemiol. Rev. *18*, 29–51.
2. BEERY, J.T, DOYLE, M.P. and SCHOENI, J.L. 1985. Colonization of chicken cecae by *Escherichia coli* associated with hemorrhagic colitis. Appl. Environ. Microbiol. *49*, 310–315.
3. BELL, B.P. *et al.* 1994. A multistate outbreak of *Escherichia coli* O157:H7-associated bloody diarrhea from hamburgers, the Washington experience. JAMA *272*, 1349–1353.
4. BESSER, R.E., LETT, S.M., DOYLE, M.P., BARRETT, T.J., WELLS, J.G. and GRIFFIN, P.M. 1993. An outbreak of diarrhea and hemolytic uremic syndrome from *Escherichia coli* O157:H7 in fresh-pressed apple cider. JAMA *269*, 2217–2220.
5. BESSER, T.E., HANCOCK, D.D., PRITCHETT, L.C., MCRAE, E.M., RICE, D.H and TARR, P.I. 1997. Duration of detection of fecal excretion of *Escherichia coli* O157:H7 in cattle. J. Infect. Dis. *175*, 726–729.
6. BETTELHEIM, K.A. 1996. Enterohaemorrhagic *Escherichia coli*: A new problem, an old group of organisms. Aust. Vet. J. *73*, 20–25.
7. BEUTIN, L., GRIER, D., STEINRUCK, H., ZIMMERMANN, S. and SCHEUTZ, F. 1993. Prevalence and some properties of verotoxin (Shiga-like toxin)-producing *Escherichia coli* in seven different species of healthy domestic animals. J. Clin. Microbiol. *31*, 2483–2488.
8. BEUTIN, L., GEIER, D., ZIMMERMANN, S., ALEKSIC, S., GILLESPIE, H.A. and WHITTAM, T.S. 1997. Epidemiological relatedness and clonal types of natural populations of *Escherichia coli* strains producing shiga toxins in separate populations of cattle and sheep. Appl. Environ. Microbiol. *63*, 2175–2180.
9. BEUTIN, L. 1999. *Escherichia coli* O157:H7 and other types of verocyotoxigenic *E. coli* (VTEC) isolated from humans, animals and food in Germany. *In Escherichia coli* O157:H7 in Farm Animals (C.S. Steward and H.J. Flint, eds.) pp. 121–140, CAB International.

10. BLANCO, M. *et al.* 1996. Prevalence and characteristics of *Escherichia coli* serotype O157:H7 and other verotoxin-producing *E. coli* in healthy cattle. Epidemiol. Infect. *117*, 251–257.
11. BLANCO, M. *et al.* 1997. Distribution and characterization of faecal verotoxin-producing *Escherichia coli* (VTEC) isolated from healthy cattle. Vet. Microbiol. *54*, 309–319.
12. BONARDI, S. *et al.* 1999. Isolation of verocytotoxin-producing *Escherichia coli* O157:H7 from cattle at slaughter in Italy. Vet. Microbiol. *67*, 203–211.
13. BROWN, C.A., HARMON, B.G., ZHAO, T. and DOYLE, M.P. 1997. Experimental *Escherichia coli* O157:H7 carriage in calves. Appl. Environ. Microbiol. *63*, 27–32.
14. BROWNLIE, L.E. and GRAU, F.H. 1967. Effect of food intake on growth and survival of salmonellas and *Escherichia coli* in the bovine rumen. J. Gen. Microbiol. *46*, 125–134.
15. BURNENS, A.P., FREY, A., LIOR, H. and NICOLET, J. 1995. Prevalence and clinical significance of vero-cytotoxin-producing *Escherichia coli* (VTEC) isolated from cattle in herds with and without calf diarrhoea. J Vet. Med. Assoc *42*, 311–318.
16. CHAPMAN, P.A. and ACKROYD, H.J. 1997. Farmed deer as a source of verocytotoxin-producing *Escherichia coli* O157. Vet. Record *141*, 314–315.
17. CHAPMAN, P.A, SIDDONS, C.A., CERDAN MALO, A.T. and HARKIN, M.A. 1997. A 1-year study of *Escherichia coli* O157 in cattle, sheep, pigs, and poultry. Epidemiol. Infect. *119*, 245–250.
18. CRAY, W.C. and MOON, H.W. 1995. Experimental infection of calves and adult cattle with *Escherichia coli* O157:H7. Appl. Environ. Microbiol. *61*, 1586–1590.
19. CRAY, W.C, CASEY, T.A., BOSWORTH, B.T. and RASMUSSEN, M.A. 1998. Effect of dietary stress on fecal shedding of *Escherichia coli* O157:H7 in calves. Appl. Environ. Microbiol. *64*, 1975–1979.
20. DARGATZ, D.A, WELLS, S.J., THOMAS, L.A., HANCOCK, D.D. and GARBER, L.P. 1997. Factors associated with the presence of *Escherichia coli* O157:H7 in feces of feedlot cattle. J. Food Protection *60*, 466–470.
21. DEAN-NYSTROM, E.A., BOSWORTH, B.T., CRAY, W.C. and MOON, H.W. 1997. Pathogenicity of *Escherichia coli* O157:H7 in the intestines of neonatal calves. Infect. Immun. *65*, 1842–1848.
22. DIEZ-GONZALEZ, F., CALLAWAY, T.R., KIZOULIS, M.J. and RUSSELL, J.B. 1998. Grain feeding and the dissemination of acid-resistant *Escherichia coli* from cattle. Science *281*, 1666–1668.
23. DOYLE, M.P. and SCHOENI, J.L. 1987. Isolation of *Escherichia coli* O157:H7 from retail fresh meat and poultry. Appl. Environ. Microbiol. *53*, 2394–2396.
24. DUNCAN, S.H., SCOTT, K.P., FLINT, H.J. and STEWARD, C.S. 1999. Commenal-pathogen interactions involving *Escherichia coli* O157:H7 and the prospects for control. *In Escherichia coli* O157:H7 in Farm Animals (C.S. Steward and H.J. Flint, eds.) pp. 71–89, CAB International.
25. FAITH, N.G. *et al.* 1996. Prevalence and clonal nature of *Escherichia coli* O157:H7 on dairy farms in Wisconsin. Appl. Environ. Microbiol. *62*, 1519–1525.

26. FISCHER, J.R, DOYLE, M.P., ZHAO, T., BROWN, C.A. and SEWELL, C.T. 1997. Inoculation of deer with *E. coli* O157:H7. Proceedings Third International Symposium and Workshop on "shiga-toxin (verocytotoxin)-producing *Escherichia coli* infections". Baltimore, Maryland, June 22–26.
27. FISCHER, J.R., ZHAO, T., DOYLE, M.P. and STALLKNECHT D.E. 1999. *Escherichia coli* O157:H7 in sympatric populations of domestic cattle and free-ranging white-tailed deer. Infect. Dis. Rev. *1*, 225.
28. GRAU, F.H. and BROWNLIE, L.E. 1965. Occurrence of salmonellas in the bovine rumen. Aust. Vet. J. *41*, 321.
29. GYLES, C. *et al.* 1998. Association of enterohemorrhagic *Escherichia coli* hemolysin with serotypes of Shiga-like-toxin-producing *Escherichia coli* of human and bovine origins. Appl. Environ. Microbiol. *64*, 4134–4141.
30. HANCOCK, D.D., BESSER, T.E., KINSEL, M.L., TARR, P.I, RICE, D.H. and PAROS, M.G. 1994. The prevalence of *Escherichia coli* O157:H7 in dairy and beef cattle in Washington State. Epidemiol. Infect. *113*, 199–207.
31. HANCOCK, D.D., RICE, D.H., THOMAS, L.A., DARGATZ, D.A. and BESSER, T.E. 1997. Epidemiology of *Escherichia coli* in feedlot cattle. J. Food Protect *60*, 462–465.
32. HARMON, B.G. *et al.* 1999. Fecal shedding and rumen growth of *Escherichia coli* O157:H7 in fasted calves. J. Food Protect. *62*, 574—579.
33. HERRIOTT, D.E., HANCOCK, D.D., EBEL, E.D., CARPENTER, L.V., RICE, D.H. and BESSER, T.E. 1998. Association of herd management factors with colonization of dairy cattle by Shiga toxin–positive *Escherichia coli* O157:H7. J. Food Protect. *61*, 802–807.
34. HEUVELINK, A.E. *et al.* 1998. Occurrence of verocytotoxin-producing *Escherichia coli* O157 on Dutch dairy farms. J. Clin. Microbiol. *36*, 3480–3487.
35. HEUVELINK, A.E. *et al.* 1998. Isolation and characterization of verocytotoxin-producing *Escherichia coli* O157 strains from Dutch cattle and sheep. J. Clin. Microbiol. *36*, 878–882.
36. HINTON M. 1985. The sub-species differentiation of *Escherichia coli* with particular reference to ecological studies in young animals including man. J. Hyg (Camb) *95*, 595–609.
37. HINTON, M., LINTON, A.H. and HEDGES, A.J. 1985. The ecology of *Escherichia coli* in calves reared as dairy-cow replacements. J. Appl. Bacteriol. *58*, 131–138.
38. HOLLAND, R.E., WILSON, R.A., HOLLAND, M.S., YUZBASIYAN-GURKAN, V., MULLANEY, T.P. and WHITE, D.G. 1999. Characterization of *eae* + *Escherichia coli* isolated from healthy and diarrheic calves. Vet. Microbiol. *66*, 251–263.
39. HOVDE, C.J., AUSTIN, P.R., CLOUD, K.A., WILLIAMS, C.J. and HUNT, C.W. 1999. Effect of cattle diet on *Escherichia coli* O157:H7 acid resistance. Appl. Environ. Microbiol. *65*, 3233–3235.
40. JAMLE, B.H. *et al.* 1990. Attaching and effacing *Escherichia coli* infection as a cause of diarrhea in young calves. JAVMA *196*, 897–901.

41. JOHNSON, R.P. *et al.* 1996. Growing concerns and recent outbreaks involving non-O157:H7 serotypes of verotoxigenic *Escherichia coli*. J. Food Protect. *59*, 1112–1122.
42. KEENE, W.E. *et al.* 1997. An outbreak of *Escherichia coli* O157:H7 infections traced to jerky made from deer meat. JAMA *277*, 1229–1231.
43. KUDVA, I.T., HATFIELD, P.G. and HOYDE, C.J. 1995. Effect of diet on the shedding of *Escherichia coli* O157:H7 in a sheep model. Appl. Environ. Microbiol. *61*, 1363–1370.
44. KUDVA, I.T., HUNT, C.W., WILLIAMS, C.J., NANCE, U.M. and HOVDE, C.J. 1997. Evaluation of dietary influences on *Escherichia coli* O157:H7 shedding by sheep. Appl. Environ. Microbiol. *63*, 3878–3886.
45. LYNN, T.V. *et al.* 1998. The occurrence and replication of *Escherichia coli* in cattle feeds. J. Dairy Sci *1*, 1102–1108.
46. MAINIL, J. 1999. Shiga/verocytotoxins and shiga/verotoxigenic *Escherichia coli* in animals. Vet. Res. *30*, 235–257.
47. MCCLUSKEY, B.J. *et al.* 1999. Prevalence of *Escherichia coli* O157 and other Shiga-toxin-producing *E. coli* in lambs at slaughter. J. Vet. Diagn. Invest. *11*, 563–565.
48. RAHN, K. *et al.* 1997. Persistence of *Escherichia coli* O157:H7 in dairy and the dairy farm environment. Epidemiol. Infect. *119*, 251–259.
49. RASMUSSEN, M.A, CRAY, W.C., CASEY, T.A. and WHIPP, S.C. 1993. Rumen contents as a reservoir of enterohemorrhagic *Escherichia coli*. FEMS Microbiol. Lett. *114*, 79–84.
50. RASMUSSEN, M.A, WICKMAN, T.L., CRAY, W.C. and CASEY, T.A. 1999. *Escherichia coli* O157:H7 and the rumen environment. *In Escherichia coli* O157:H7 in Farm Animals (C.S. Steward and H.J. Flint, eds.) pp. 39–50, CAB International.
51. READ, S.C., GYLES, C.L., CLARKE, R.C., LIOR, H. and MCEWEN S. 1990. Prevalence of verocytotoxigenic *Escherichia coli* in ground beef, pork and chicken in southwestern Ontario. Epidemiol. Infect. *105*, 11–20.
52. RICE, D.H., HANCOCK, D.D. and BESSER, T.E. 1995. Verotoxigenic *E. coli* O157 colonisation of wild deer and range cattle. Vet. Record *137*, 524.
53. RICHARDS, M.S, CORKISH, A.R., SAYERS, A.R., MCLAREN, I.M., EVANS, S.J. and WRAY, C. 1997. Studies of the presence of verocytotoxic *Escherichia coli* O157 in bovine faeces submitted for diagnostic purposes in England and Wales and on beef carcasses in abattoirs in the United Kingdom. Epidemiol. Infect. *120*, 187–192.
54. SANDERSON, M.W., BESSER, T.E., GAY, J.M., GAY, C.C. and HANCOCK, D.D. 1999. Fecal *Escherichia coli* O157:H7 shedding patterns of orally inoculated calves. Vet. Microbiol. *69*, 199–205.
55. SANDHU, K.S. *et al.* 1996. Prevalence of the *eaeA* gene in verotoxigenic *Escherichia coli* strains from dairy cattle in Southwest Ontario. Epidemiol. Infect. *116*, 1–7.
56. SARGEANT, J.M., HAFER, D.J., GILLESPIE, J.R., OBERST, R.D. and FLOOD, S.J.A. 1999. Prevalence of *Escherichia coli* O157:H7 in white-tailed deer sharing rangeland with cattle. JAVMA *215*, 792–794.

57. SCHOONDERWOERD, M., CLARKE, R.C., VAN DREUMEL, A.A. and RAWLUK, S.A. 1988. Colitis in calves: natural and experimental infection with a verotoxin-producing strain of *Escherichia coli* O111:NM. Can. J. Vet. Res. *52*, 484–487.
58. SLUTSKER, L., RIES, A.A., MALONEY, K., WELLS, J.G., GREENE, K.D. and GRIFFIN, P.M. 1998. A nationwide case-control study of *Escherichia coli* O157:H7 infection in the United States. J. Infect. Dis. *177*, 962–966.
59. SMITH, H.R., CHEASTY, T., ROBERTS, D., THOMAS, A. and ROWE, B. 1991. Examination of retail chickens and sausage in Britain for vero cytotoxin-producing *Escherichia coli*. Appl. Environ. Microbiol. *57*, 2091–2093.
60. STAVRIC, S., BUCHANAN, B. and GLEESON, T.M. 1993. Intestinal colonization of young chicks with *Escherichia coli* O157:H7 and other verotoxin-producing serotypes. J. Appl. Bacteriol. *74*, 557–563.
61. SUTHIENKUL, O., BROWN, J.E., SERIWATANA, J., TIENTHONGDEE, S., SASTRAVAHA, S. and ECHEVERRIA, P. 1990. Shiga-like-toxin-producing *Escherichia coli* in retail meats and cattle in Thailand. Appl. Environ. Microbiol. *56*, 1135–1139.
62. TARR, P.I. and NEILL, M.A. 1996. Perspective: the problem of non-O157:H7 shiga toxin (verocytotoxin) producing *Escherichia coli*. J. Infect. Dis. *174*, 1136–1139.
63. TKALCIC, S. *et al.* 1998. Effect of diet of rumen proliferation and fecal shedding of *Escherichia coli* O157:H7 in calves. Proceedings International Conference on Emerging Infectious Diseases, p 82.
64. TKALCIC, S. *et al.* 1999. Effects of diet on the proliferation and acid tolerance of *E. coli* O157:H7 in rumen fluid. Conference on Emerging Diseases: Veterinary Medicine, Agriculture, and Human Health, Athens, GA, August 12–14.
65. WELLS, J.G. *et al.* 1991. Isolation of *Escherichia coli* serotype O157:H7 and other Shiga-like toxin-producing *E. coli* from dairy cattle. J. Clin. Microbiol. *29*, 985–989.
66. WIELER, L.H. *et al.* 1996. Shiga toxin-producing *Escherichia coli* strains from bovines: association of adhesion with carriage of *eae* and other genes. J. Clin. Microbiol. *34*, 2980–2984.
67. WILSON, J.B. *et al.* 1996. Vero cytotoxigenic *Escherichia coli* infection in dairy farm families. J. Infect. Dis. *174*, 1021–1027.
68. ZHAO, T., DOYLE, M.P., SHERE, J. and GARBER, L. 1995. Prevalence of enterohemorrhagic *Escherichia coli* O157:H7 in a survey of dairy herds. Appl. Environ. Microbiol. *61*, 1290–1293.

CHAPTER 15

SURVIVAL AND GROWTH OF VTEC IN THE ENVIRONMENT

D.A. MCDOWELL

Food Microbiology Research Unit
School of Applied Medical Sciences And Sport Studies
Faculty of Science
University of Ulster, Jordanstown
Newtownabbey BT37OQB, Northern Ireland

AND

J.J.SHERIDAN

Department of Food Safety
The National Food Centre
Dunsinea, Castleknock, Dublin 15, Ireland

INTRODUCTION

Escherichia coli has long been recognised as being capable of survival and growth in a range of conditions, including many natural environments. However, until relatively recently, the frequent observation of the persistence of these organisms in water and soil, and in other farm environments such as grass, silage, etc., was considered to be of limited clinical importance. Thus these organisms were generally ignored, or firmly classified as non pathogens. While enteric strains were frequently employed as indicators of general faecal contamination, signalling the possible presence of more dangerous enteric pathogens, most *E. coli* were considered to have little or no clinical significance to most humans. Some strains were recognised as important pathogenic forms, causing conditions such as "travellers diarrhoea" in humans, and a range of infections in animals (64). However, much of the scientific literature on *Escherichia coli* has focused on its exploitation as a safe, easily cultivated and genetically malleable test bed for the elucidation of procaryotic genetics and metabolism. Such emphases mean that a great deal is known about certain aspects of the activities of some strains of *E. coli*, including their impressive plasticity in the acquisition, exploitation, modulation and dissemination of useful genetic information. However, much of this knowledge has proven to be of limited value when a particular course of genetic transfers transformed one or more strains of this normally commensal organism into significant pathogens, capable of establishing life threatening and/or permanently debilitating infections in humans.

The mechanisms by which this group, the verocytotoxigenic *E. coli* (VTEC) and, more specifically, the currently most important serotype, *E. coli* O157:H7, acquired their formidable armoury of pathogenic characteristics, the mechanisms by which extremely small infectious doses can establish human infections, their clinical sequelae, and other elements of their interactions with man and animals are dealt with in other chapters in this text. In terms of more general synopses, readers are directed to a number of recent reviews (31,102,126).

The present chapter focuses on the ability of this species to persist and/or grow in farm or related environments, from whence it may gain direct or indirect access to humans and the human food chain. However, before proceeding to review the prevalence, persistence and implications of VTEC in such environments, it may be cautionary to note that data relating to the organism itself, while important, may not give an absolute index of the risks that may be subsequently presented to humans.

While most studies have focused on the dangers posed by the persistence, transmission and metabolism of *E. coli* O157:H7, disease potential is not solely the prerogative of this serotype, or of a particular set of strains or serotypes. Since the initial recognition of the importance of the above serotype, it has become increasingly clear that the pathogenic characteristics it expresses continue to be transferred more widely among other serotypes and strains of *E. coli*, conferring significant disease inducing abilities on recipient clones. Thus, from humans alone, more than 100 *E. coli* serotypes carrying *vt* genes have been identified (157), and a pattern of increasing clonal diversity among O157 carrying isolates is being observed (144,189). One implication of such diversity is that strains of *E. coli* adapted to persistence and/or growth in most farm environments will continue to acquire and maintain these undesirable pathogenic characteristics. It is therefore hardly surprising that as these genetic materials, and the abilities they impart, transfer among the *E. coli* strains which occur widely in most farm and related environments, the pathogenic characteristics for which they code, and the severe clinical hazards they entail, become more and more widely observed. This scenario demands urgent reassessment of the risks posed by such enterohaemorrhagic strains, and redefinition of the nature and scale of the steps to be taken to protect humans from the risks which may be posed by previously innocuous strains.

FACTORS INFLUENCING SURVIVAL

Any estimation of risk should recognise the considerable differences in stress resistance among strains of VTEC, and other potential hosts capable of acquiring, expressing and disseminating the "genetic package" which confers pathogenicity. Such estimates must reflect reported differences in resistance to such stresses as hydrostatic pressure (6), mild heat, oxidative and osmotic challenge (23), and reports on significant increases in resistance under stationary phase/starvation conditions (13,151). Worst case estimates of persistence of pathogenic strains of VTEC may therefore have to recognise the wide spectrum of resistances of currently non-pathogenic *E. coli*, and/or other potential recipient strains/species, as this dangerous set of pathogenic characteristics becomes more widely disseminated.

A further reason for caution in basing estimations of risk too narrowly on the characteristics of currently detected VT clones/strains of *E. coli*, is related to the roles of *vt* encoding phages. Phages are considered to play a major role in the observed transmission of *vt* and related genes within the genus and beyond (123). As yet, little is known about the actual incidence of such active gene transfers among *E. coli* under natural conditions, although transduction of genes among other species has been shown to occur at significant rates in simulated aqueous environments (148). There is some evidence that rates of transfer under natural conditions are even higher (117). As discussed later, evidence is also mounting that the widespread application of antibiotics as growth promoters and in the treatment of infectious diseases in animals and humans, may stimulate the expression, and wider dissemination of *vt* encoding phages (91,94,172,180,186). However, it is currently recognised that, as well as transmitting *vt* genes, such phages may have crucial roles as reservoirs of these genes in nature. VT encoding phages persist longer than their host bacteria in aqueous environments (123,124). They are more resistant than their host bacteria to a number of naturally occurring and deliberately applied inactivation processes, as shown in Table 1.

TABLE 1.
SURVIVAL OF VT ENCODING PHAGES

VT encoding phages persist longer than VTEC in/during	Reference
Freshwater	(21)
Seawater	(92)
Chlorinated water	(125)
Heat treatments	(125)
Disinfection treatments	(153)
Water supply treatments	(83)

SURVIVAL IN UNFAVOURABLE CONDITIONS

Uncertainties about the scale, nature and implications of injury suffered by VTEC exposed to on-farm conditions, including loss of culturability on currently used selective media (140,147), and/or entry to the viable non-culturable state (121,173), are likely to lead to failures to detect these organisms. Conclusions in relation to the presence or absence of VTEC should, but frequently do not, carry the caveat that the organism may not have been recovered, because it was present in a damaged or stressed state, or was present in numbers below the detection limit of the assay used. Thus environments which are not grossly contaminated with unstressed VTEC may provide negative results, if an insufficiently sensitive assay is used. It should, however, be borne in mind that such loci of persistence and/or routes of transmission may be significant if they enable the pathogen to access subsequent, more favourable opportunities for growth.

For the above reasons, much of the available data on the occurrence, persistence, and transmission of VTEC within and from on-farm activities and locations should be viewed as potentially underestimating the risks to humans posed by the broader movement of *vt* genes within on-farm and other environments.

PERSISTENCE OF VTEC

While initial concerns in relation to the implications of the presence of VTEC in farm environments tended to focus on transmission from animals, and most specifically cattle, fairly directly into the human food chain, it is now clear that this organism occurs widely in many if not most farming and food production activities, and their contiguous environments. This pattern of increasingly frequent detection is likely to continue, particularly as more sensitive detection methods become available. Specifically, methods combining selective enrichment/recovery with immunomagnetic separation (IMS) will facilitate detection of this organism from diverse on-farm premises and products (40,42,43,76,135).

Cattle were the first well-recognised reservoir for VTEC (42), and have consistently been confirmed as the most significant source of such organisms (62,67,114,115,188). From this primary source, VTEC can be directly transmitted to farmers working with cattle (165). There have been a number of well publicised incidents in which other humans have acquired VTEC infections during more transient contact with farms and farm animals, i.e., visits to "open" or petting farms (10,120,123,136), fairs (9,106) and festivals (47).

Despite the particular importance of cattle as sources of this organism, little is known about the ecology of VTEC on cattle farms (135). However, qualitative and quantitative aspects of the persistence, carriage, individual and herd dynamics, and the wider interactions of this pathogen in cattle and other ruminants, remain important and active areas of study (137). It is becoming clear that in cattle, as in many other animate and inanimate reservoirs of VTEC, the patterns of results obtained by the application of more sensitive detection methods mean that it may be necessary to continuously revise (upwards) estimates of the prevalence of this pathogen (100).

Recognising the early and particular significance of cattle in the ecology of VTEC, much of the earliest work on the occurrence and dissemination of this pathogen focussed on food products from this source, principally beef and beef products, and (to a lesser extent) milk and milk products. Bearing in mind that cattle are the principal source of VTEC, and that the organism is excreted in faeces, considerable attention has also been paid to the occurrence and persistence of this human pathogen in cattle faeces in the form of manure and slurry, and the surfaces on which they are deposited during farm activities (30,139,141).

As discussed in other chapters in this book, and recently reviewed (156), farm management practices, such as stocking rates, housing practices, and manure and slurry management (63) are important in the control of VTEC. Any aspect of the generation and handling of animal waste, is likely to have significant effects on the persistence of VTEC on farms, and the spread of this organism in the wider environment.

It may therefore be constructive to examine the persistence and dissemination of the pathogen within those elements of the farm environment which come directly, consistently and substantively in contact with ruminants, their products and by-products, before examining areas when contamination is more likely to be indirect, sporadic and generally of lower orders of magnitude.

Survival in the Farm Environment

The factors affecting the uptake and persistence of VTEC in cattle and other animals are still under investigation. Differences in reported carriage rates of 0.3-6.1% in cattle, and extended shedding intervals of 30 days — 1 year (98,99,188) demonstrate differences between and within herds (25). There are reports of the impact of factors such as stress (49) diet (50,87,187), differences in virulence among bacterial strains (132), and the presence of multiple sources of VTEC on farms (69). This makes it difficult to derive definitive statements about the frequency of acquisition of VTEC by, and its carriage within, host animals. Nevertheless, it is generally agreed that VTEC is ubiquitous in the population of feedlot/preslaughter cattle (68), and that cattle are a major reservoir of these organisms through such faecal shedding (115,150,156,177).

Concerns in relation to the significance of VTEC on farms, within the wider environment, and the human food chain, begin with the shedding of this organism in animal faeces. As noted by Jordan (88) the concentration of VTEC in the faeces of naturally infected cattle has been poorly described in the literature. However, recent estimates suggest that cattle, particularly calves and heifers colonised with this organism, may shed it at levels ranging from 10^2 to 10^5 cfu g^{-1} (74). There has been considerable interest in potential means for reducing such concentrations, including short term changes in the nature and composition of feedstuffs (50,87), the provision of diet additives to suppress the proliferation of VTEC in the bovine gut (187), and immunisation (104). While some or all of these approaches may prove valuable, it appears that, for the foreseeable future, cattle faeces will continue to represent a major source of VTEC and verotoxin encoding genes. Deposited within on-farm environments, these spread into the wider environment.

Survival on Animal Hide

Before considering the significance of VTEC in faeces deposited on byre floors/pasture land, it is important to reflect on the scope and significance of faeces which remains on animal hides. During normal farming procedures VTEC and other gut organisms pass in the faeces to the hide (33), and are routinely transferred to beef carcasses at slaughter (34,37,114). It is generally accepted that the extent to which faecal material adheres to animals delivered to abattoirs significantly influences the levels of microbial contamination on derived carcasses (113,143). Recognising how important it is for cattle to be clean prior to slaughter, many countries have already implemented, or are developing, clean cattle policies (as discussed more fully in other chapters of this book), although it remains virtually impossible to exclude VTEC from carcasses and beef processing plants (88), and subsequent raw meat processing and preparation (12). While "Clean Cattle" policies clearly reduce gross and/or visible contamination of carcasses, more direct, complex and potentially invasive approaches ranging from preslaughter washing (37) and carcass pasteurisation (53) to shaving (169) and chemical dehairing (149), may be necessary to directly reduce the risks posed by VTEC (14,15,37).

Survival in Faeces

VTEC appear to be well adapted to survival in cattle and other animal faeces, remaining recoverable (i.e., culturable) from such materials for extended periods (30,111,112,135) (Table 2). Studies have demonstrated survival of detectable numbers of VTEC in animal manures and slurry for significant periods ranging from several weeks to many months (31,62,74,97,174), and such material therefore remains a very important potential vehicle for transmission within herds, farms (39,65), the retail fresh food chain (145) and the wider environment (12).

TABLE 2.
THE SURVIVAL OF *E. COLI* O157:H7 IN BOVINE FAECES AND SLURRY UNDER LABORATORY CONDITIONS, AND IN THE NATURAL ENVIRONMENT

Mileau	Count (log_{10}cfu/g)		Storage		Survival (days)	Reference
	Initial	Final	Conditions	Temp (°C)		
Faeces	8.18	3.18	Closed plastic containers in field	-6.5 to 9.6	99.0	(30)
	8.00	6.50	Glass container in laboratory	18.0	54.0	(112)
	7.15	1.50	In the laboratory	23.0	80.0	(97)
	5.00	<1.0	Closed plastic bags	25.0 15.0 5.0	56.0 126.0 63.0	(62)
	3.00	<1.0		15.0	96.0	
	1.00	<1.0		15.0	96.0	
Soil	8.00	7.00	Plastic containers in laboratory	18.0	130.0	(112)
Grass	7.47	3.02	On grass surface in field	-6.5 to 19.6	99.0	(30)
Slurry	6.01 Type	(1) 0.75 (2) 2.31	Closed plastic containers in laboratory	10.0	84.0	(116)
	6.78	0.00 2.20	In laboratory	23.0 10.0	<2.0 30.0	(97)
	8.00	2.00	Glass container	18.0	9.0	(112)
	Inoculated cattle	Positive	On concrete floors	Ambient conditions	47.0	(97)

Overly prescriptive estimates of the persistence of VTEC in manure/slurry should, however, be avoided. Examination of the literature suggests that estimates of the survival of VTEC in faeces may be dictated by the duration of the assay period and the sensitivity of the detection method used. However, the general implications are clear. Appropriate handling of bovine faeces is important to control spread of this pathogen (62,174), and to limit the significant risks of direct infection of humans coming into direct contact with manure or slurry (84). Recognition of such risks may lend greater emphasis to the development of methods for the destruction of VTEC in manures involving the application of heat and or "sterilising" gases (75). However, the wider implementation of such processes poses major operational and financial challenges, and may only be applicable to intensive farming systems.

VTEC survives after direct deposit on pasture land in faeces (30,112,135), or in slurry (58,116) posing a number of direct and indirect risks. Thus the organism may be directly recycled within cattle herds (128), during grass consumption, or within silage (29) or other forage. It is clear that effective silage fermentation reduces the numbers of VTEC in ensiled contaminated or slurry treated grass. However, recent reports on the ability of VTEC to multiply rapidly in poorly fermented silage (59) suggest that inadequate silage management may represent a significant means for persistence of VTEC among ruminants fed such silage.

The ability to survive for extended periods on grass and other vegetation (30,111) presents opportunities for VTEC to gain access to and from other farmed animals, and wild herbivores, consuming contaminated forage. Although the pattern and directions of such transmission is not yet clear, a wide range of animals may be involved, with various qualitative and quantitative significance in terms of the movement of VTEC in the human food chain. Sheep, probably the second most extensively examined group of farmed herbivores, have been frequently reported to carry VTEC (26,39,41,155,162). Some studies have identified the strains isolated from these animals as human infection related strains (26,70). As noted in cattle, changes in diet (98), and other factors significantly modulate the rate of VTEC shedding. Some workers report significant seasonal differences in rates of shedding of VTEC (98).

VTEC have been detected in a wide range of other animals and their faeces, including domesticated, feral and wild species (Table 3). Interestingly, there are reports of higher frequencies of detection among primates, perhaps suggesting that humans are among the most susceptible to VTEC infections. Overall, such reports reconfirm the increasing prevalence of VTEC, and its ability to colonise many animal hosts.

Recognition of the range and scope of such frequency of occurrence of VTEC in animal faeces increases the frequently stated but infrequently achieved need to exclude faeces from on-farm food production and processing activities.

Contaminated faeces poses risks in terms of direct transmission to livestock, and cross contamination of food products, particularly in light of the increasing amount of further processing and farm shop activities that now occur on or adjacent to farm premises. As well as such direct transmission, it is also important to note the ability of VTEC to survive for extended periods, i.e., well in excess of 6 months in the faeces of such incidental species as rats and pigeons (45). Thus the concerns previously expressed in relation to persistence of VTEC in the admittedly much larger volumes of food animal (mainly bovine) faeces, apply equally to VTEC persistence in the faeces of potentially more mobile rats, pigeons, and other animals coming into contact with farmed ruminants.

TABLE 3.
REPORTS OF VTEC IN ANIMALS/FAECES

Detection of VTEC in animals or their faeces	
Goats	(39, 56, 136, 189)
Pigs	(39, 41, 56, 71, 130, 144)
Farmed deer	(42)
Horses/ponies	(20, 67)
Chicken/eggs	(41)
Cats and Dogs	(166)
Rats	(45)
Pigeons	(45)
Other vertebrates	(88)
Zoo animals	(20)
Primates	(20)

As VTEC excreted in animal faeces (exotic or otherwise) can persist for considerable periods, they can pose risks if contaminated manure and/or slurry is applied as a fertiliser to vegetable or fruit crops, especially as many of such crops are consumed in a raw or minimally prepared state. Thus risks have been identified, or outbreaks associated with, a wide range of vegetables, fruit, and derived products (Table 4). Such outbreaks have implicated citrus fruit (131) and apples, contaminated when fallen fruit come into contact with manure spread below trees in orchards and persisting on gathered fruit and derived products.

Survival in Water

As well as undesirable transfers from slurry, manure, etc., directly into other food production processes, VTEC are more generally flushed into surface and other water sources, gaining access to farm water supplies used on farms. In water, they can move from farms into the general surface and ground water environment. In immediate terms, they can re-enter animals which consume contaminated surface or well water (57,135,150,160). VTEC can also be directly transferred from positive animals during feeding/drinking (67,139,142). Humans consuming well water or other water supplies contaminated with slurry, manure and surface run off water are also at risk (3,11,74,80,109,118,134,158, 178).

TABLE 4.
POTENTIAL VEGETABLE/FRUIT VECTORS OF VTEC

Potential/established vectors of VTEC infection	
Vegetables	(1, 168)
Lettuce	(27, 170)
Tomatoes	(28, 54)
Peppers	(66)
Radish	(175)
carrot & cucumber	(1)
soft fruit (strawberries)	(185)
Citrus fruit	(131)
Fresh apples	(35, 51, 81, 82, 146, 179)
Orange juice	(107)
Apple juice	(24, 36, 46, 61, 154)
Cider	(52, 72, 167, 182)

After accessing surface water on or adjacent to farms, VTEC can disseminate more widely through the environment (85) as it persists for extended periods (days or weeks) in a range of different types of water (90,111,135). As discussed by Lisle, (108), VTEC flushed into lakes and streams from farm animals, their waste products, and other natural sources, exhibit the characteristics of a robust commensal pathogen. Thus they are capable of adapting to, and surviving, environmental stresses in an aquatic system, and of contaminating/infecting other farms, wild animals, etc., as well as posing a direct threat to human water supplies. In some aqueous environments, such survival, along with enhanced virulence and resistance to biocides and antibiotics, is enhanced by intraprotozoal growth (17). As protozoa are widely distributed in most naturally occurring bodies of water (152), the ability of *E. coli* O157:H7 to survive ingestion during protozoal "grazing", may contribute to the overall persistence of this pathogen in water, or high water content environments. Intraprotozoal persistence may be of particular significance in relation to the recovery of *E. coli* O157:H7 from on-farm water supplies such as cattle troughs, ponds, etc., (67,85,139,142,150), and should be recognised within wider strategies which aim to limit the survival and dissemination of this pathogen within farm and related environmental management practices.

The dangers of VTEC contamination of unchlorinated or insufficiently chlorinated water have been graphically demonstrated by an outbreak involving almost 250 cases of human VTEC, and 4 fatalities, after the organisms gained access to an unchlorinated municipal water supply (158), or the even starker juxtaposition of circumstances in Japan, where contamination of well water led to a number of fatalities, including kindergarten children (3).

The presence of VTEC in water has been linked to outbreaks of infection. Of particular note are outbreaks associated with the hydroponic production of foods (Table 5) which have been responsible for some of the largest outbreaks of foodborne illness (38,60,119,159,175). Attempts to develop mechanisms to protect consumers by disinfecting hydroponic products failed to eradicate VTEC introduced to the growing environment in contaminated water or seeds (79, 101).

TABLE 5.
WATERBORNE OUTBREAKS OF VTEC

Hydroponic food Production systems	Alfalfa	(110)
	Mung beans	(7)
	Radish sprouts	(79)
Recreational waters	Swimming pools/lakes	(2, 8, 18, 48, 77, 89, 105, 122, 133, 163)
	Hot tubs	(96)
	Paddling pools	(32, 73)

The persistence of VTEC, and its low infective dose, means that water borne infections can result from contact with (i.e., not necessarily purposive consumption of) contaminated water. Thus outbreaks have been traced to water based recreational activities (Table 5). The wider recognition of the occurrence of these organisms in many if not most surface, recreational and unchlorinated water supplies (93), and in sewage (123,124), has serious consequences for the human population.

However, it is also important to recognise that such increasing prevalence may increase the risks of transmission of VTEC and *vt* genes (21,83,153) from humans, the apparently preferred primate host (20), back into farm water supplies. For obvious reasons, transmission of VTEC and/or VT encoding bacteriophages from humans to animals is less likely to seize the attention of the public. However, transmission of VTEC within human faeces has been clearly demonstrated in circumstances where human personal hygiene systems are inadequate, i.e., nursery (4,5,22,32,73) or geriatric care (95,176). Thus the potential movement of these organisms by such routes, especially when sewage disposal and general standards of public health are inadequate, or disrupted, e.g., by flooding or civil unrest, may contribute to the dissemination and recycling of VTEC from human populations back into farm environments.

The capacity for wild birds to recycle pathogenic bacteria considerable distances from sewage out-falls, slurry treated pasture land and other high risk sites, leading to (re)contamination of on-farm and water supply catchment areas has been established in relation to *Salmonella* (86), *Vibrio* (103) *Campylobacter* and *Listeria* (138). As VTEC has been isolated from wild birds (171), e.g. pigeon (45), they may transfer VTEC to farm water supplies.

In narrower terms, the detection of this organism in incidental on-farm species, such as dogs (67,78) cats and rabbits (16), rats (44), other rodents (67) and flies (69,139) provide potent evidence of the increasing ubiquity of VTEC, and the many potential routes of possible recontamination of farm water supplies.

DISSEMINATION OF VTEC

As noted by Batt (19) the probability of detecting all on-farm reservoirs and dissemination routes of such pathogens is low. More significantly, it may be very difficult to limit the numbers of organisms moving from such reservoirs to humans and the human food chain. There is also good reason to expect *vt* genes to continue to spread among other *E. coli* populations, and related species and genera, through the action of VT encoding phages.

Intensive efforts in the development and application of more sensitive methods for detecting and differentiating strains of *vt*-carrying *E. coli* continue to increase our understanding of the incidence of this group of organisms. Greater knowledge confirms the complexity and diversity of their occurrence within farm environments. Thus, in 1987, five years after the first identification of VTEC as a human pathogen, the model for its transmission involved a small number of routes, i.e., cow/cow, cow to human in beef and milk, and human/human (161). Ten years on, such models have become much more complex, involving some 15 routes of transmission (161), including many forms of on-farm activity, and perhaps more worrying, significant interactions with many animate and inanimate elements of on-farm and other environments.

This chapter began by noting that *E. coli* presents a paradox, in that no other organism has been so intensively investigated in laboratory terms, as a test bed in microbial genetics, while being quite comprehensively ignored in terms of its microbial ecology within the wider environment.

The emergence of VTEC calls for more commitment to investigation of the means of its transmission. VTEC, or other *E. coli* fully capable of acting as "hosts" to *vt* genes, are firmly, if not almost universally, established in on-farm and related environments. As such, they pose increasing threats to human populations coming into direct or indirect contact with these organisms. The "normal" reaction to the recognition of a new or [re]emergent pathogen, i.e., rapid application of antibiotic therapy, has had a very disappointing record in this particular case, to the extent that antibiotic therapy has been widely considered to exacerbate the clinical course of VTEC infection in humans (65,127,181,183, 184) increasing the occurrence of systemic sequelea such as haemolytic uraemic syndrome (HUS) (129) and/or thrombotic thrombocytopenic purpura (TTP) (55).

The prophylactic application of antibiotics within farm practices induces lysogenic verocytotoxin coding bacteriophages (91), increasing the overall rates of horizontal transfer within the animal intestine (186) or the excretion of these bacteriophages into the wider environment (94). Thus, antibiotic based interventions may increase the rates of transmission of *vt* genes among host bacteria, and appear unlikely to be valuable in terms of reducing the impact of such undesirable organisms in humans or food animals.

Alternative strategies to reduce the prevalence of VTEC in farm environments could involve immunisation of cattle, leading to overall reductions of the cycling of VTEC through animals and on-farm environments. As discussed by Tauxe (161), such a novel strategy could have substantial impacts in reducing human illness. However, *vt* genes are now so widespread that immunisation programmes are unlikely to adequately suppress VTEC across all significant on-farm and contiguous environments.

CONCLUSIONS

In the long term, more effective progress in the control, suppression or exclusion of VTEC may be gained by further study of this already intensely investigated species, with greater emphasis on its ecology. Better knowledge of the interactions between VTEC, other organisms, and their sensing and control of their environments, could facilitate more effective interventions.

It is salutary to note that, in its frequently successful attempts to invade and persist in human hosts, VTEC displays an intimate knowledge of our cellular metabolism (164), for example in the "cross talk" between bacteria and intestinal cells, as part of the development of the attaching and effacing (A/E) histopathology. Perhaps when as much is known about the cellular communication, pathogenic and metabolic systems of VTEC, as this pathogen "knows" about such systems in humans, ruminants and other hosts, real progress will be made in its control in on-farm and other environments.

REFERENCES

1. ABDUL-RAOUF, U.M., BEUCHAT, L.R. and AMMAR, M.S. 1993. Survival and growth of *Escherichia coli* O157:H7 on salad vegetables. Appl. Environ. Microbiol. *59*, 1999–2006.
2. ACKMAN, D., MARKS, S., MACK, P., CALDWELL, M., ROOT, T. and BIRKHEAD, G. 1997. Swimming-associated haemorrhagic colitis due to *Escherichia coli* O157:H7 infection: evidence of prolonged contamination of a fresh water lake. Epidemiol. Infect. *119*, 1–8.
3. AKASHI, S. *et al.* 1994. A severe outbreak of haemorrhagic colitis and haemolytic uraemic syndrome associated with *Escherichia coli* O157:H7 in Japan. Eur. J. Pediatr. *153*, 650–5.
4. AL-JADER, L., SALMON, R.L., WALKER, A.M., WILLIAMS, H.M., WILLSHAW, G.A. and CHEASTY, T. 1999. Outbreak of *Escherichia coli* O157 in a nursery: lessons for prevention. Arch. Dis. Child. *81*, 60–3.
5. ALLABY, M. and MAYON-WHITE, R. 1995. *E. coli* O157 outbreak in a day nursery. Commun. Dis. Rep. CDR Wkly. *5*, R4–6.
6. ALPAS, H., KALCHAYANAND, N., BOZOGLU, F., SIKES, A., DUNNE, C.P. and RAY, B. 1999. Variation in resistance to hydrostatic pressure among strains of food-borne pathogens. Appl. Environ. Microbiol. *65*, 4248–51.
7. ANDREWS, W.H. *et al.* 1982. Microbiological hazards associated with bean sprouting. Assoc. Off Anal Chem. *65*, 241–248.
8. ANON 1996. Lake-associated outbreak of *Escherichia coli* O157:H7 — Illinois, 1995. MMWR. Morb. Mortal. Wkly. Rep. *45*, 437–9.
9. ANON 1999. Outbreak of *Escherichia coli* O157:H7 and *Campylobacter* among attendees of the Washington County Fair-New York, 1999. MMWR. Morb. Mortal. Wkly. Rep. *48*, 803–5.
10. ANON 1999. VTEC O157 outbreak associated with a farm visitor centre in North Wales. Commun. Dis. Rep. CDR Wkly. *9*, 227–230.
11. ANON 1999. VTEC O157 outbreak linked to beach holidays. Commun. Dis. Rep. CDR Wkly. *9*, 327, 330.
12. ARMSTRONG, G.L., HOLLINGSWORTH, J. and MORRIS, J.G. 1996. Emerging foodborne pathogens: *Escherichia coli* O157:H7 as a model of a new pathogen into the food supply of the developed world. Epidemiol. Rev. *18*, 29–51.
13. ARNOLD, K.W. and KASPAR, C.W. 1995. Starvation- and stationary-phase-induced acid tolerance in *Escherichia coli* O157:H7. Appl. Environ. Microbiol. *61*, 2037–9.

14. ATTENBOROUGH, M. and MATTHEWS, K.R. 2000. Food safety through the meat supply chain. J. Appl. Microbiol. *88*, 144S–148S.
15. BACON, R.T., BELK, K.E., SOFOS, J.N., CLAYTON, R.P., REAGAN, J.O. and SMITH, G.C. 2000. Microbial populations on animal hides and beef carcasses at different stages of slaughter in plants employing multiple - sequential interventions for decontamination. J. Fd. Prot. *63*, 1080–1086.
16. BALJER, G. 1997. Animal reservoirs. Fortschr Med. *115*, 84.
17. BARKER, J., HUMPHREY, T.J. and BROWN, M.W.R. 1999. Survival of *Escherichia coli* O157 in a soil protozoan: implications for disease. FEMS Microbiol. Lett. *173*, 291–295.
18. BARWICK, R.S., LEVY, D.A., CRAUN, G.F., BEACH, M.J. and CALDERON, R.L. 2000. Surveillance for waterborne-disease outbreaks — United States, 1997-1998. MMWR CDC Surveill. Summ. *49*, 1–21.
19. BATT, C. 1997. Down on the farm. Food Microbiol. *14*, 199.
20. BAUWENS, L., DE MEURICHY, W. and VERCAMMEN, F. 2000. Isolation of *Escherichia coli* O157 from zoo animals. Vlaams Diergen.skund. Tijds. *69*, 76–79.
21. BELL, R.G. 1976. The limitation of the ratio of fecal coliforms to total coliphage as a faecal pollution index. Water Res. *30*, 229.
22. BELONGIA, E.A., OSTERHOLM, M.T., SOLER, J.T., AMMEND, D.A., BRAUN, J.E. and MACDONALD, K.L. 1993. Transmission of *Escherichia coli* O157:H7 infection in Minnesota child day-care facilities. J.A.M.A. *269*, 883–888.
23. BENITO, A., VENTOURA, G., CASADEI, M., ROBINSON, T. and MACKEY, B. 1999. Variation in resistance of natural isolates of *Escherichia coli* O157 to high hydrostatic pressure, mild heat, and other stresses. Appl. Environ. Microbiol. *65*, 1564–9.
24. BESSER, R.E., LETT, S.M. and WEBER, J.T. 1993. An outbreak of diarrhoea and haemolytic uremic syndrome from *Escherichia coli* O157:H7 in freshly-pressed apple cider J.A.M.A. *269*, 2217–2220.
25. BESSER, T.E., HANCOCK, D.D., PRICHETT, L.C., MCRAE, E.M., RICE, D.H. and TARR, P.I. 1997. Duration of detection of fecal excretion of *Escherichia coli* O157:H7 in cattle. J. Infect. Dis. *175*, 726–729.
26. BETTELHEIM, K.A., BENSINK, J.C. and TAMBUNAN, H.S. 2000. Serotypes of verotoxin-producing (Shiga toxin-producing) *Escherichia coli* isolated from healthy sheep. Comp. Immunol. Microbiol. Infect. Dis. *23*, 1–7.
27. BEUCHAT, L.R. 1999. Survival of enterohemorrhagic *Escherichia coli* O157:H7 in bovine feces applied to lettuce and the effectiveness of chlorinated water as a disinfectant. J. Fd. Prot. *62*, 845–9.
28. BEUCHAT, L.R., NAIL, B.V., ADLER, B.B. and CLAVERO, M.R. 1998. Efficacy of spray application of chlorinated water in killing pathogenic bacteria on raw apples, tomatoes, and lettuce. J. Fd. Prot. *61*, 1305–1311.
29. BOLTON, D.J., BYRNE, C.M., MCGEE, P.J. and SHERIDAN, J.J. 1999. Survival and growth of Verocytoxigenic *E. coli* in the farming environment, p. 28–36. *In* Verocytotoxigenic *E. coli* in Europe. Survival and growth of verocytotoxigenic *E. coli*. The National Food Centre, Dublin, Ireland, Athens, Greece.

30. BOLTON, D.J., BYRNE, C.M., SHERIDAN, J.J., MCDOWELL, D.A. and BLAIR, I.S. 1999. The survival characteristics of an non-toxigenic strain of *Escherichia coli* O157:H7. J. Appl. Microbiol. *86*, 407–411.
31. BOLTON, F.J. and AIRD, H. 1998. Verocytotoxin-producing *Escherichia coli* O157: public health and microbiological significance. Brit. J. Biomed. Sci. *55*, 127–35.
32. BREWSTER, D.H., BROWN, M.I., ROBERTSON, D., HOUGHTON, G.L., BIMSON, J. and SHARP, J.C.M. 1994. An outbreak of *Escherichia coli* O157 associated with a children's paddling pool. Epidemiol. Infect. *112*, 441–447.
33. BROWN, C.A., HARMON, B.G., ZHOA, T. and DOYLE, M.P. 1997. Experimental *Escherichia coli* O157:H7 carriage in calves. Appl. Environ. Micro. *63*, 27–32.
34. BROWN, M. H. (ed.) 1982. Meat Microbiology. Applied Science Publishers, London.
35. BUCHANAN, R.L., EDELSON, S.G., MILLER, R.L. and SAPERS, G.M. 1999. Contamination of intact apples after immersion in an aqueous environment containing *Escherichia coli* O157:H7 J. Fd. Prot. *62*, 444–50.
36. BUCHANAN, R.L., EDELSON, S.G., SNIPES, K. and BOYD, G. 1998. Inactivation of *Escherichia coli* O157:H7 in apple juice by irradiation. Appl. Environ. Microbiol. *64*, 4533–4535.
37. BYRNE, C.M., BOLTON, D.J., SHERIDAN, J.J., MCDOWELL, D.A. and BLAIR, I.S. 2000. The effects of preslaughter washing on the reduction of *Escherichia coli* O157:H7 transfer from cattle hides to carcasses during slaughter. Letts. Appl. Microbiol. *30*, 142–5.
38. CASTRO-ROSAS, J. and ESCARTIN, E.F. 2000. Survival and growth of *Vibrio cholerae* 01, *Salmonella typhi,* and *Escherichia coli* O157:H7 in alfalfa sprouts. J. Fd. Sci. *65*, 162–165.
39. CHAPMAN, P.A. 2000. Sources of *Escherichia coli* O157 and experiences over the past 15 years in Sheffield, UK. J. Appl. Microbiol. *88*, 51S–60S.
40. CHAPMAN, P.A. and SIDDONS, C.A. 1996. A comparison of immunomagnetic separation and direct culture for the isolation of verocytotoxin-producing *Escherichia coli* O157 from cases of bloody diarrhoea, non-bloody diarrhoea and asymptomatic contacts. J. Med. Microbiol. *44*, 267–71.
41. CHAPMAN, P.A. SIDDONS, C.A., GERDAN MALO, A.T. and HARKIN, M.A. 1997. A 1-year study of *Escherichia coli* O157 in cattle, sheep, pigs and poultry. Epidemiol. Infect. *119*, 245–50.
42. CHAPMAN, P.A., SIDDONS, C.A., WRIGHT, D.J., NORMAN, P., FOX, J. and CRICK, E. 1993. Cattle as a possible source of verocytotoxin-producing *Escherichia coli* O157 infections in man. Epidemiol. Infect. *111*, 439–447.
43. CHAPMAN, P.A., WRIGHT, D.J. and SIDDONS, C.A. 1994. A comparison of immunomagnetic separation and direct culture for the isolation of verocytotoxin-producing *Escherichia coli* O157 from bovine faeces. J. Med. Microbiol. *40*, 424–7.
44. CIZEK, A., ALEXA, P., LITERAK, I., HAMRIK, J., NOVAK, P. and SMOLA, J. 1999. Shiga toxin-producing *Escherichia coli* O157 in feedlot cattle and Norwegian rats from a large-scale farm. Letts. Appl. Microbiol. *28*, 435–9.

45. CIZEK, A., LITERAK, I. and SCHEER, P. 2000. Survival of *Escherichia coli* O157 in faeces of experimentally infected rats and domestic pigeons. Letts. Appl. Microbiol. *31*, 349-352.
46. CODY, S.H., GLYNN, M.K., FARRAR, J.A., CAIRNS, K.L. and GRIFFIN, P.M. 1999. An outbreak of *Escherichia coli* O157:H7 infection from unpasteurized commercial apple juice. Ann. Intern. Med. *130*, 202–9.
47. CRAMPIN, M. *et al.* 1999. Outbreak of *Escherichia coli* O157 infection associated with a music festival. Eur. J. Clin. Microbiol. Infect. Dis. *18*, 286–8.
48. CRANSBERG, K. *et al.* 1996. Four cases of hemolytic uremic syndrome — source contaminated swimming water? Clin. Nephrol. *46*, 45–9.
49. CRAY JR., W.C., CASEY, T.A., BOSWORTH, B.T. and RASMUSSEN, M.A. 1998. Effect of Dietary Stress on Fecal Shedding of *Escherichia coli* O157:H7 in Calves. Appl. Environ. Microbiol. *64*, 1975–1979.
50. DIEZ-GONZALEZ, F., CALLAWAY, T.R., KIZOULIS, M.G. and RUSSELL, J.B. 1998. Grain feeding and the dissemination of acid resistant *Escherichia coli* from cattle. Science *281*, 1666–1668.
51. DINGMAN, D.W. 2000. Growth of *Escherichia coli* O157:H7 in Bruised Apple (Malus domestica) tissue as influenced by cultivar, date of harvest, and source. Appl. Environ. Microbiol. *66*, 1077–1083.
52. DOCK, L.L., FLOROS, J.D. and LINTON, R.H. 2000. Heat inactivation of *Escherichia coli* O157:H7 in apple cider containing malic acid, sodium benzoate, and potassium sorbate. J. Fd. Prot. *63*, 1026–1031.
53. DORSA, W.J., CUTTER, C.N. and SIRAGUSA, G.R. 1996. Effectiveness of a steam-vacuum sanitizer for reducing *Escherichia coli* O157:H7 inoculated to beef carcass surface tissue. Letts. Appl. Microbiol. *23*, 61–3.
54. DROSINOS, E.H., TASSOU, C., KAKIOMENOU, K. and NYCHAS, G.J.E. 2000. Microbiological, physico-chemical and organoleptic attributes of a country tomato salad and fate of *Salmonella enteritidis* during storage under aerobic or modified atmosphere packaging conditions at 4°C and 10°C. Fd. Control. *11*, 131–135.
55. DUNDAS, S. and TODD, W.T. 2000. Clinical presentation, complications and treatment of infection with verocytotoxin-producing *Escherichia coli*. Challenges for the clinician. J. Appl. Microbiol. *88*, 24S-30S.
56. FAGAN, P.K., HORNITZKY, M.A., BETTELHEIM, K.A. and DJORDJEVIC, S.P. 1999. Detection of shiga-like toxin (stx1 and stx2), intimin (eaeA), and enterohemorrhagic *Escherichia coli* (EHEC) hemolysin (EHEC hlyA) genes in animal feces by multiplex. PCR Appl. Environ Microbiol. *65*, 868–72.
57. FAITH, N.G. *et al.* 1996. Prevalence and clonal nature of *Escherichia coli* O157:H7 on dairy farms in Wisconsin. Appl. Environ. Microbiol. *62*, 1519–25.
58. FENLON, D.R., OGDEN, I.D., VINTEN, A. and SVOBODA, I. 2000. The fate of *Escherichia coli* and *E-coli* O157 in cattle slurry after application to land. J. Appl. Microbiol. *88*, 149S–156S.
59. FENLON, D.R. and WILSON, J. 2000. Growth of *Escherichia coli* O157 in poorly fermented laboratory silage: a possible environmental dimension in the epidemiology of *E. coli* O157. Letts. Appl. Microbiol. *30*, 118–121.

60. FETT, W.F. 2000. Naturally occurring biofilms on alfalfa and other types of sprouts. J. Fd. Prot. *63*, 625–632.

61. FOLSOM, J.P. and FRANK, J.F. 2000. Heat inactivation of *Escherichia coli* 0157:H7 in apple juice exposed to chlorine. J. Fd. Prot. *63*, 1021–1025.

62. FUKUSHIMA, H.H.K. and GOMYODA, M. 1999. Long-term survival of Shiga toxin-producing *Escherichia coli* O26, O111, and O157 in bovine feces. Appl. Environ. Microbiol. *65*, 5177–5181.

63. GARBER, L., WELLS, S., SCHROEDER-TUCKER, L. and FERRIS, K. 1999. Factors associated with fecal shedding of verotoxin-producing *Escherichia coli* O157 on dairy farms. J. Fd. Prot. *62*, 307–12.

64. GYLES, C.L. (ed.) 1994. *Escherichia coli* in Domestic Animals and Humans. CAB International, Oxford.

65. GYLES, C.L. 1998. Vaccines and Shiga Toxin producing *Escherichia coli* in Animals, *In Escherichia coli* O157:H7 and Other Shiga Toxin-Producing *E. coli* Strains. (J.B. Kaper and A.D. O'Brien, eds.) p. 434–444. ASM Press, Washington.

66. HAN, Y., SHERMAN, D.M., LINTON, R.H., NIELSEN, S.S. and NELSON, P.E. 2000. The effects of washing and chlorine dioxide gas on survival and attachment of *Escherichia coli* O157:H7 to green pepper surfaces. Fd. Microbiol. *17*, 521–533.

67. HANCOCK, D.D., BESSER, T.E., RICE, D.H., EBEL, E.D., HERRIOTT, D.E. and CARPENTER, L.V. 1998. Multiple sources of *Escherichia coli* O157 in feedlots and dairy farms in the northwestern USA. Prev. Vet. Med. *35*, 11–9.

68. HANCOCK, D.D., RICE, D.H., THOMAS, L.A., DARGATZ, D.A. and BESSER, T.E. 1997. Epidemiology of *Escherichia coli* O157 in feedlot cattle. J. Fd. Prot. *60*, 462–465.

69. HEUVELINK, A.E. *et al.* 1998. Occurrence of verocytotoxin-producing *Escherichia coli* O157 on Dutch dairy farms. J. Clin. Microbiol. *36*, 3480–7.

70. HEUVELINK, A.E. *et al.* 1998. Isolation and characterization of verocytotoxin-producing *Escherichia coli* O157 Strains from Dutch cattle and sheep. J. Clin. Microbiol. *36*, 878–882.

71. HEUVELINK, A.E., ZWARTKRUIS-NAHUIS, J.T.M., VAN DEN BIGGELAAR, F., VAN LEEUWEN, W.J. and DE BOER, E. 1999. Isolation and characterization of verocytotoxin-producing *Escherichia coli* O157 from slaughter pigs and poultry. Int. J. Fd. Microbiol. *52*, 67–75.

72. HILBORN, E.D. *et al.* 2000. An outbreak of *Escherichia coli* O157:H7 infections and haemolytic uraemic syndrome associated with consumption of unpasteurized apple cider. Epidemiol. Infect. *124*, 31–36.

73. HILDEBRAND, J.M., MAGUIRE, H.C., HOLLIMAN, R.E. and KANGESU, E. 1996. An outbreak of *Escherichia coli* O157 infection linked to paddling pools. Commun. Dis. Rep. CDR Rev. *6*, R33–6.

74. HIMATHONGKHAM, S., BAHARI, S., RIEMANN, H. and CLIVER, D. 1999. Survival of *Escherichia coli* O157:H7 and *Salmonella typhimurium* in cow manure and cow manure slurry. FEMS Microbiol. Lett. *178*, 251–257.

75. HIMATHONGKHAM, S. and RIEMANN, H. 1999. Destruction of *Salmonella typhimurium*, *Escherichia coli* O157:H7 and *Listeria monocytogenes* in chicken manure by drying and/or gassing with ammonia. FEMS Microbiol Lett. *171*, 179–182.
76. HINDLE, M.A., BOLTON, E.J., WRIGHT, P.A. and DURBAND, C.A. 1995. Improved detection of verocytotoxigenic-producing *E. coli* O157 in fecal samples by enrichment culture. PHLS Microbiol. Dig. *12*, 71–73.
77. HORNSBY-LEWIS, L. 1995. A newly recognized source of *Escherichia coli* O157:H7 infections. Gastroenterology *108*, 1597–9.
78. HORZER, J. and AWAD-MASALMEH, M. 2000. Prevalence of shigatoxinogenic *Escherichia coli* (STEC) in several species of carnivores and poultry as well as some other species. Wien. Tierarztl. Monatsschr. *87*, 154–158.
79. ITOH, Y. *et al.* 1998. Enterohemorrhagic *Escherichia coli* O157:H7 present in radish sprouts. Appl. Environ. Microbiol. *64*, 1532–1535.
80. JACKSON, S.G. *et al.* 1998. *Escherichia coli* O157:H7 diarrhoea associated with well water and infected cattle on an Ontario farm. Epidemiol Infect. *120*, 17–20.
81. JANISIEWICZ, W.J., CONWAY, W.S., BROWN, M.W., SAPERS, G.M., FRATAMICO, P. and BUCHANAN, R.L. 1999. Fate of *Escherichia coli* O157:H7 on fresh-cut apple tissue and its potential for transmission by fruit flies. Appl. Environ. Microbiol. *65*, 1–5.
82. JANISIEWICZ, W.J., CONWAY, W.S. and LEVERENTZ, B. 1999. Biological control of postharvest decays of apple can prevent growth of *Escherichia coli* O157:H7 in apple wounds. J Fd. Prot. *62*, 1372–5.
83. JOFRE, J., OLLER, E., RIBAS, F., VIDAL, A. and LUCENA, F. 1995. Potential usefulness of bacteriophages that infect *Bacteriodes fragilis* as model organisms for monitoring virus removal in drinking water treatment plants. Appl. Environ. Microbiol. *61*, 3227–3231.
84. JOHNSON, R.P. *et al.* 1999. Human infections associated with exposure to farm animals and rural environments. *In Escherichia coli* O157 in Farm Animals (C.S. Stewart and H.J. Flint, eds.), CABI Publishing, Oxon, U.K.
85. JONES, D.L. 1999. Potential health risks associated with the persistence of *Escherichia coli* O157 in agricultural environments. Soil Use and Mgt. *15*, 76–83.
86. JONES, F., SMITH, P. and WATSON, D.C. 1978. Pollution of a water supply catchment area by breeding gulls and the potential environmental health implications. J. Inst. Water Engin. Scient. *32*, 469–482.
87. JORDAN, D. and MCEWEN, S.A. 1998. Effect of duration of fasting and a short-term high roughage ration on the concentration of *Escherichia coli* biotype 1 in cattle feces. J. Fd. Prot. *61*, 531–534.
88. JORDAN, D., MCEWEN, S.A., LAMMERDING, A.M., MCNAB, W.B. and WILSON, J.B. 1999. Pre-slaughter control of *Escherichia coli* O157 in beef cattle: a simulation study. Prev. Vet. Med. *41*, 55–74.
89. KEENE, W.E. *et al.* 1994. A swimming-associated outbreak of hemorrhagic colitis caused by *Escherichia coli* O157:H7 and *Shigella sonnei*. N. Engl. J. Med. *331*, 579–84.

90. KERR, M., FITZGERALD, M., SHERIDAN, J.J., MCDOWELL, D.A. and BLAIR, I.S. 1999. Survival of *Escherichia coli* O157:H7 in bottled natural mineral water. J. Appl. Microbiol. *87*, 833–41.
91. KIMMITT, P.T., HARWOOD, C.R. and BARER, M.R. 2000. Toxin gene expression by shiga toxin-producing *Escherichia coli*: the role of antibiotics and the bacterial SOS response. Emerg. Infect. Dis. *6*, 458–465.
92. KNOTT, Y. and BEN, A.H. 1968. Bacteriophages as marine pollution indicators. Rev. Int. Oceanogr. Med. *9*, 207–217.
93. KOGURE, K. and IKEMOTO, E. 1997. Wide occurrence of enterohemorrhagic *Escherichia coli* O157 in natural freshwater environment. [Japanese] Jpn. J. Microbiol. *52*, 601–607.
94. KOHLER, B., KARCH, H. and SCHMIDT, H. 2000. Antibacterials that are used as growth promoters in animal husbandry can affect the release of Shiga-toxin-2-converting bacteriophages and Shiga toxin 2 from *Escherichia coli* strains. Microbiolgy-(UK). *146*, 1085–1090.
95. KOHLI, H.S., CHAUDHURI, A.K., TODD, W.T., MITCHELL, A.A. and LIDDELL, K.G. 1994. A severe outbreak of *E. coli* O157 in two psychogeriatric wards. J. Public Health Med. *16*, 11–15.
96. KRAMER, M.H., HERWALDT, B.L., CRAUN, G.F., CALDERON, R.L. and JURANEK, D.D. 1996. Surveillance for waterborne-disease outbreaks — United States, 1993-1994. MMWR CDC Surveill. Summ. *45*, 1–33.
97. KUDVA, I.T., BLANCH, K. and HOVDE, C.J. 1998. Analysis of *Escherichia coli* O157:H7 survival in ovine or bovine manure and manure slurry. Appl. Environ. Microbiol. *64*, 3166–74.
98. KUDVA, I.T., HATFIELD, P.G., and HOVDE, C.J. 1996. *Escherichia coli* O157:H7 in microbial flora of sheep. J. Clin. Microbiol. *34*, 431–433.
99. KUDVA, I.T., JELACIC, S., TARR, P.I., YOUDERIAN, P. and HOVDE, C.J. 1999. Biocontrol of *Escherichia coli* O157 with O157-specific bacteriophages. Appl. Environ. Microbiol. *65*, 3767–73.
100. LAEGREID, W.W., ELDER, R.O. and KEEN, J.E. 1999. Prevalence of *Escherichia coli* O157 in range beef calves at weaning. Epidemiol Infect. *123*, 291–298.
101. LANG, M.M., INGHAM, B.H. and INGHAM, S.C. 2000. Efficacy of novel organic acid and hypochlorite treatments for eliminating *Escherichia coli* O157:H7 from alfalfa seeds prior to sprouting. Int. J. Food Microbiol. *58*, 73–82.
102. LAW, D. 2000. Virulence factors of *Escherichia coli* O157 and other Shiga toxin producing *E. coli*. J. Appl. Microbiol. *88*, 729–745.
103. LEE, J.V., BASHFORD, D., DONOVAN, T., FURNISS, A. and WEST, D. 1982. The incidence of *Vibrio cholerae* in water, animals and birds in Kent, England. J. Appl. Bacteriol. *52*, 281–291.
104. LEVINE, M.M. 1998. Immunoprophylaxis of Shiga toxin producing *Escherichia coli* infections and disease: strengths and weaknesses of various strategies, *In Escherichia coli* O157:H7 and Other Shiga Toxin-Producing *E. coli* Strains (J.B. Kaper and A.D. O'Brien, eds.) p. 405–408, ASM Press, Washington.

105. LEVY, D.A., BENS, M.S., CRAUN, G.F., CALDERON, R.L. and HERWALDT, B.L. 1998. Surveillance for waterborne-disease outbreaks — United States, 1995-1996. MMWR CDC Surveill. Summ. *47*, 1–34.
106. LEWIS, R. 1999. On the trail of *E. coli* O157:H7. Scientist *13*, 10.
107. LINTON, M., MCCLEMENTS, J.M. and PATTERSON, M.F. 1999. Survival of *Escherichia coli* O157:H7 during storage in pressure-treated orange juice. J. Fd. Prot. *62*, 1038–1040.
108. LISLE, J.T., BROADAWAY, S.C., PRESCOTT, A.M., PYLE, B.H., FRICKER, C. and MCFETERS, G.A. 1998. Effects of starvation on physiological activity and chlorine disinfection resistance in *Escherichia coli* O157:H7. Appl. Environ. Microbiol. *64*, 4658–4662.
109. MACLER, B.A. and MERKLE, J.C. 2000. Current knowledge on groundwater microbial pathogens and their control. Hydrogeol. J. *8*, 29–40.
110. MAHON, B.E. *et al.* 1997. An international outbreak of *Salmonella* infections caused by alfalfa sprouts grown from contaminated seeds. J. Infect. Dis. *175*, 875–882.
111. MAULE, A. 1997. Survival of the verotoxigenic strain *E. coli* O157:H7 in laboratory scale microcosms. *In* Coliforms and *E. coli*: problem or solution? (D. Kay and C. Fricker, eds.) p. 61–65. The Royal Society of Chemistry, Cambridge, United Kingdom.
112. MAULE, A. 2000. Survival of verocytotoxigenic *Escherichia coli* O157 in soil, water and on surfaces. J. Appl. Microbiol. *88*, 71S–78S.
113. MCEVOY, J.M. *et al.* 2000. The relationship between hide cleanliness and bacterial numbers on beef carcasses at a commercial abattoir. Letts. Appl. Microbiol. *30*, 390–395.
114. MCEVOY, J.M., DOHERTY, A.M., SHERIDAN, J.J., MCGUIRE, L., MCDOWELL, D.A. and BLAIR, I.S. 1999. The incidence of *Escherichia coli* O157:H7 in faeces, rumen contents and on carcasses in a commercial Irish beef abattoir, *In* Verocytotoxigenic *E. coli* in Europe. Pathogenicity and virulence of verocytotoxigenic *E. coli*. p. 197. TEGASC, The National Food Centre, Liege, Belgium.
115. MCEVOY, J. M. *et al.* 2001. The incidence and spread of *Escherichia coli* O157:H7 at a commercial beef abattoir. *In* Verocytotoxigenic *E. coli* in Europe. Epidemology of Verocytotoxigenic *E. coli*. The National Food Centre, Dublin, Ireland.
116. MCGEE, P., BOLTON, D.J., SHERIDAN, J.J., EARLEY, B. and LEONARD, M. 2001. The survival of *Escherichia coli* O157:H7 in slurry from cattle fed different diets. Letts. Appl. Microbiol. *32*(3), 152–155.
117. MEL, S. and MEKALANOS, J.J. 1996. Modulation of horizontal gene transfer in pathogenic bacteria by in vivo signals. Cell. *87*, 795–798.
118. MICHEL, P., WILSON, J.B., MARTIN, S.W., CLARKE, R.C., MCEWEN, S.A. and GYLES, C.L. 1999. Temporal and geographical distributions of reported cases of *Escherichia coli* O157:H7 infection in Ontario. Epidemiol. Infect. *122*, 193–200.

119. MICHINO, H. *et al.* 1999. Massive outbreak of *Escherichia coli* O157:H7 infection in schoolchildren in Sakai City, Japan, associated with consumption of white radish sprouts. Am. J. Epidemiol. *150*, 787–796.
120. MILNE, L.M., PLOM, A., STRUDLEY, I., PRITCHARD, G.C. and CROOKS, R. 1999. *Escherichia coli* O157 incident associated with a farm open to members of the public. Commun. Dis. Pub. Health. *2*, 22–26.
121. MIZUNOE, Y., WAI, S.N., TAKADE, A. and YOSHIDA, S. 1999. Restoration of culturability of starvation-stressed and low-temperature-stressed *Escherichia coli* O157 cells by using H2O2-degrading compounds. Arch. Microbiol. *172*, 63–7.
122. MOORE, A.C., HERWALDT, B.L., CRAUN, G.F., CALDERON, R.L., HIGHSMITH, A.K. and JURANEK, D.D. 1993. Surveillance for waterborne disease outbreaks — United States, 1991-1992. MMWR CDC Surveill. Summ. *42*, 1–22.
123. MUNIESA, M. and JOFRE, J. 1998. Abundance in sewage of bacteriophages that infect *Escherichia coli* O157:H7 and that carry the Shiga toxin 2 gene. Appl. Environ. Microbiol. *64*, 2443–8.
124. MUNIESA, M. and JOFRE, J. 2000. Occurrence of phages infecting *Escherichia coli* O157:H7 carrying the Stx2 gene in sewage from different countries. FEMS Microbiol. Lett. *183*, 197–200.
125. MUNIESA, M., LUCENA, F. and JOFRE, J. 1999. Comparative survival of free Shiga Toxin 2-Encoding phages and *Escherichia coli* strains outside the gut. Appl. Environ. Microbiol. *65*, 5615–5618.
126. NATARO, J.P. and KAPER, J.B. 1998. Diarrheagenic *Escherichia coli*. Clin. Microbiol. Rev. *11*, 142–201.
127. NEILL, M.A. 1998. Treatment of disease due to shiga toxin producing *Escherichia coli*: Infectious disease management. *In Escherichia coli* O157:H7 and Other Shiga Toxin-Producing *E. coli* Strains. (J.B. Kaper and A.D. O'Brien, eds.) p. 357–363. ASM Press, Washington.
128. OHYA, T. and ITO, H. 1999. Experimental infection of calves with *Escherichia coli* O157:H7. J. Vet. Med. Sci. *61*, 1187–1189.
129. O'RYAN, M. and PRADO, V. 2000. Risk of the hemolytic-uremic syndrome after antibiotic treatment of *Escherichia coli* O157:H7 infections. N. Engl. J. Med. *343*, 1271.
130. OSEK, J. 1999. Prevalence of virulence factors of *Escherichia coli* strains isolated from diarrheic and healthy piglets after weaning. Vet. Microbiol. *68*, 209–17.
131. PAO, S., DAVIS, C.L. and KELSEY, D.F. 2000. Efficacy of alkaline washing for the decontamination of orange fruit surfaces inoculated with *Escherichia coli*. J. Fd. Prot. *63*, 961–964.
132. PATON, A.W. *et al.* 1996. Molecular microbiological investigation of an outbreak of hemolytic uremic syndrome caused by dry fermented sausage contaminated with Shiga-like toxin producing *Escherichia coli*. J. Clin. Microbiol. *34*, 1622–1627.
133. PAUNIO, M. *et al.* 1999. Swimming-associated outbreak of *Escherichia coli* O157:H7. Epidemiol. Infect. *122*, 1–5.

134. PEBODY, R.G. *et al.* 1999. An international outbreak of Vero cytotoxin-producing *Escherichia coli* O157 infection amongst tourists; a challenge for the European infectious disease surveillance network. Epidemiol. Infect. *123*, 217–23.
135. PORTER, J., MOBBS, K., HART, C.A., SAUNDERS, J.R., PICKUP, R.W. and EDWARDS, C. 1997. Detection, distribution and probable fate of *Escherichia coli* O157 from asymptomatic cattle on a dairy farm. J. Appl. Microbiol. *83*, 297–306.
136. PRITCHARD, G.C., WILLSHAW, G.A., BAILEY, J.R., CARSON, T. and CHEASTY, T. 2000. Verocytotoxin-producing *Escherichia coli* O157 on a farm open to the public: outbreak investigation and longitudinal bacteriological study. Vet. Rec. *147*, 259–264.
137. PRUIMBROOM-BREES, I.M. *et al.* 2000. Cattle lack vascular receptors for *Escherichia coli* O157:H7 Shiga toxins. Proc. Natl. Acad. Sci. USA *97*, 10325–10329.
138. QUESSY, S. and MEISSIER, S. 1992. Prevalence of *Salmonella* spp, *Campylobacter* spp. and *Listeria* spp. in ring-billed gulls (*Larus delawarensis*). J. Wild. Dis. *28*, 526–531.
139. RAHAN, K. *et al.* 1997. Persistence of *Escherichia coli* in dairy cattle and the dairy farm environment. Epidemiol. Infect. *119*, 251–259.
140. RAJKOWSKI, K.T. and RICE, E.W. 1999. Recovery and survival of *Escherichia coli* O157:H7 in reconditioned pork-processing wastewater. J. Fd. Prot. *62*, 731–734.
141. REINDERS, R.D., BIJKER, P.G.H. and OUDE-ELFERINK, S.J.W.H. 1999. Growth and survival of verotoxigenic *Escherichia coli* O157 in selected farm environments. Survival and Growth Of Verocytotoxigenic *E. coli*. EU Concerted Action on VTEC (CT98-3935), Athens.
142. RICE, E.W. and JOHNSON, C.H. 2000. Survival of *Escherichia coli* O157:H7 in dairy cattle drinking water. J. Dairy Sci. *83*, 2021–2023.
143. RIDELL, J. and KORKEALA, H. 1993. Special treatment during slaughter in Finland of cattle carrying an excessive load of dung: meat hygiene aspects. Meat Sci. *35*.
144. RIOS, M. *et al.* 1999. Clonal diversity of Chilean isolates of enterohemorrhagic *Escherichia coli* from patients with hemolytic-uremic syndrome, asymptomatic subjects, animal reservoirs, and food products. J. Clin Microbiol. *37*, 778–781.
145. SAMADPOUR, M. *et al.* 1994. Occurrence of Shiga-like toxin producing *Escherichia coli* in retail fresh seafood, beef, lamb, pork and poultry from grocery stores in Seattle, Washington. Appl. Environ. Micro. *60*, 1038–1040.
146. SAPERS, G.M., MILLER, R.L., JANTSCHKE, M. and MATTRAZZO, A.M. 2000. Factors limiting the efficacy of hydrogen peroxide washes for decontamination of apples containing *Escherichia coli*. J. Fd. Sci. *65*, 529–532.
147. SATA, S., OSAWA, R., ASAI, Y. and YAMAI, S. 1999. Growth of starved *Escherichia coli* O157 cells in selective and non-selective media. Microbiol. Immunol. *43*, 217–227.

148. SAYE, D.J., OGUNSEITAN, O., SAYLER, G.S. and MILLER, R.V. 1990. Transduction of linked chromosomal genes between *Pseudomonas aeruginosa* during incubation in a freshwater environment. Appl. Environ. Micro. *56*, 140–145.
149. SCHNELL, T.D., SOFOS, J.N. and LITTLEFIELD, V. 1995. Effects of postexsanguination dehairing on the microbial load and visual cleanliness of beef carcasses. J. Fd. Prot. *58*, 1297–1302.
150. SHERE, J.A., BARTLETT, K.J. and KASPAR, C.W. 1998. Longitudinal study of *Escherichia coli* O157:H7 dissemination on four dairy farms in Wisconsin. Appl. Environ. Microbiol. *64*, 1390–9.
151. SHERIDAN, J.J. and MCDOWELL, D.A. 1998. Factors affecting emergence of pathogens on food. Meat Sci. *49*, 151–167.
152. SIBILLE, I., SIME-NGANDO, T., MATHIEU, L. and BLOCK, J.C. 1998. Protozoan bacterivory and *Escherichia coli* survival in drinking water distribution systems. Appl. Environ. Micro. *64*, 197–202.
153. SOBSY, M.P. 1989. Inactivation of health related microorganisms in water disinfection processes. Water Sci. Technol. *21*, 179–195.
154. STEELE, B.T., MURPHY, N. and RANCE, C.P. 1982. An outbreak of hemolytic uremic syndrome associated with ingestion of fresh apple juice. J. Pediat. *101*, 963–965.
155. STEPHAN, R., RAGETTLI, S. and UNTERMANN, F. 2000. Prevalence of verotoxin-producing *Escherichia coli* (VTEC) in feces of healthy beef cattle and sheep in Switzerland. Schweiz. Archiv. Tierheilkd. *142*, 110–114.
156. STEWART, C.S. and FLINT, H.J. (eds.) 1999. *Escherichia coli* O157 in Farm Animals. CABI Publishing, Oxford, U.K.
157. STROCKBINE, N.A., WELLS, J.G., BOPP, C.A. and BARRETT, T.J. 1998. Overview of detection and subtyping methods, *In Escherichia coli* O157:H7 and Other Shiga Toxin-Producing *E. coli* Strains. (J.B. Kaper and A.D. O'Brien, eds.) p. 331–356. ASM Press, Washington.
158. SWERDLOW, D.L. *et al.* 1992. A waterborne outbreak in Missouri of *Escherichia coli* O157:H7 associated with bloody diarrhea and death. Ann. Intern. Med. *117*, 812–9.
159. TAORMINA, P.J., BEUCHAT, L.R. and SLUTSKER, L. 1999. Infections associated with eating seed sprouts: an international concern. Emerg. Infect. Dis. *5*, 626–634.
160. TAUXE, R.V. 1997. Emerging foodborne diseases: an evolving public health challenge. Emerg. Infect. Dis. *3*, 425–434.
161. TAUXE, R.V. 1998. Public health perspectives on immunoprophylactic strategies for *Escherichia coli* O157:H7: who or what should we immunize? *In Escherichia coli* O157:H7 and Other Shiga Toxin-Producing *E. coli* Strains (J.B. Kaper and A.D. O'Brien, eds.) p. 445–452. ASM Press, Washington.
162. THORNS, C.J. 2000. Bacterial food-borne zoonoses. Rev. Sci. Tech. Off. Int. Epizoot. *19*, 226–239.
163. TRACHTMAN, H. and CHRISTEN, E. 1999. Pathogenesis, treatment, and therapeutic trials in hemolytic uremic syndrome. Curr. Opin. Pediatr. *11*, 162–8.

164. TRAN VAN NHIEU, G. and SANSONETTI, P.E. 2000. Cell Adhesion Molecules and Bacterial Pathogens, *In* Cellular Microbiology (P. Cossart, P. Boquet, S. Normark and R. Rappuoli, eds.) p. 97–111. ASM Press, Washington, DC.
165. TREVANA, W.B., WILLSHAW, G.A., CHEASTY, T., DOMIGUE, G. and WRAY, C. 1999. Transmission of Verocytotoxin producing *Escherichia coli* O157 infection from farm animals to humans in Cornwall and West Devon. Commun. Dis. Pub. Health. *2*, 263–268.
166. TREVENA, W.B., HOOPER, R.S., WRAY, C., WILLSHAW, G.A., CHEASTY, T. and DOMINGUE, G. 1996. Verocytotoxin-producing *Escherichia coli* O157 associated with companion animals. Vet. Rec. *138*, 400.
167. ULJAS, H.E. and INGHAM, S.C. 2000. Survey of apple growing, harvesting, and cider manufacturing practices in Wisconsin: Implications for safety. J. Fd. Saf. *20*, 85–100.
168. USDA 1997. An update: *Escherichia coli* O157:H7 in humans and cattle. Animal and Plant Health Inspection Service, Centers for Epidemiology and Animal Health.
169. VAN DONKERGOED, J., JERICO, K.W.F., GROGAN, H. and THORLAKSON, B. 1997. Pre-slaughter hide status of cattle and the microbiology of carcasses. J. Fd. Prot. *60*, 1502–1508.
170. VOETSCH, A., WOJTKUNSKI, C., SWARTZ, M., MSHAR, R. and LAMBERT-FAIR, M.A. 1999. A multistate outbreak of *Escherichia coli* O157:H7 infections associated with consumption of mesclun lettuce. Arch Intern Med. *159*, 1758–1764.
171. WALLACE, G., CHEASTY, T. and JONES, K. 1997. Isolation of verocytotoxigenic producing *Escherichia coli* O157:H7 from wild birds. Appl. Environ. Micro. *82*, 399–404.
172. WALTERSPIEL, J., ASHKENAZI, S., MORROW, A. and CLEARY, T. 1992. Effects of subinhibitory concentrations of antibiotics on extracellular Shiga-like toxin I. Infection. *20*, 25–29.
173. WANG, G. and DOYLE, M.P. 1998. Survival of enterohemorrhagic *Escherichia coli* O157:H7 in water. J. Fd. Prot. *61*, 662–7.
174. WANG, G., ZHAO, T. and DOYLE, M.P. 1996. Fate of enterohemorrhagic *Escherichia coli* O157:H7 in bovine feces. Appl. Environ. Microbiol. *62*, 2567–2570.
175. WATANABE, Y. *et al.* 1999. Factory outbreak of *Escherichia coli* O157:H7 infection in Japan. Emerg. Infect. Dis. *5*, 424–8.
176. WEIGHTMAN, N.C. and KIRBY, P.J.G. 2000. Nosocomial *Escherichia coli* O157 infection. J. Hos. Infect. *44*, 107–111.
177. WELLS, J.G. *et al.* 1991. Isolation of *Escherichia coli* serotype O157:H7 and other shiga-like-toxin-producing *E. coli* from dairy cattle. J. Clin. Microbiol. *29*, 985–989.
178. WIREMAN, M. and JOB, C. 1998. Determining the risk to public water supplies wells from infective microorganisms. Wat. Well J., 63–67.
179. WISNIEWSKY, M.A., GLATZ, B.A., GLEASON, M.L. and REITMEIER, C.A. 2000. Reduction of *Escherichia coli* O157:H7 counts on whole fresh apples by treatment with sanitizers. J. Fd. Prot. *63*, 703–8.

180. WITTE, W. 1998. Medical consequences of antibiotic use in agriculture. Science *279*, 996–997.

181. WONG, C.S., JELACIC, S., HABEEB, R.L., WATKINS, S.L. and TARR, P.I. 2000. The risk of the hemolytic-uremic syndrome after antibiotic treatment of *Escherichia coli* O157:H7 infections. N. Engl. J. Med. *342*, 1930–1936.

182. WRIGHT, J.R., SUMNER, S.S., HACKNEY, C.R., PIERSON, M.D. and ZOECKLEIN, B.W. 2000. Efficacy of ultraviolet light for reducing *Escherichia coli* O157:H7 in unpasteurized apple cider. J. Fd. Prot. *63*, 563–567.

183. YOH, M., FRIMPONG, E.K. and HONDA, T. 1997. Effect of antimicrobial agents, especially fosfomycin, on the production and release of Vero toxin by enterohaemorrhagic *Escherichia coli* O157:H7. FEMS Immunol. Med. Microbiol. *19*, 57–64.

184. YOH, M. and HONDA, T. 1997. The stimulating effect of fosfomycin, an antibiotic in common use in Japan, on the production/release of verotoxin-1 from enterohaemorrhagic *Escherichia coli* O157:H7 in vitro. Epidemiol. Infect. *119*, 101–103.

185. YU, K.S., HAMILTON-KEMP, T.R., ARCHBOLD, D.D., COLLINS, R.W. and NEWMAN, M.C. 2000. Volatile compounds from *Escherichia coli* O157:H7 and their absorption by strawberry fruit. J. Agric. Food Chemist. *48*, 413–417.

186. ZHANG, X., MCDANIELS, A.D., WOLF, L.E., KEUSCH, G.T., WALDOR, M.K. and ACHESON, D.W.K. 2000. Quinolone antibiotics induce Shiga toxin encoding bacteriophages, toxin production and death in mice. J. Infect. Dis. *181*, 664–670.

187. ZHAO, T., DOYLE, M.P., HARMON, B.G., BROWN, C.A., MUELLER, P.O. and PARKS, A.H. 1998. Reduction of carriage of enterohemorrhagic *Escherichia coli* O157:H7 in cattle by inoculation with probiotic bacteria. J. Clin. Microbiol. *36*, 641–647.

188. ZHAO, T., DOYLE, M.P., SHERE, J. and GARBER, L. 1995. Prevalence of enterohemorrhagic *Escherichia coli* O157:H7 in a survey of dairy herds. Appl. Environ. Microbiol. *61*, 1290–1293.

189. ZSCHOCK, M., HAMANN, H.P., KLOPPERT, B. and WOLTER, W. 2000. Shiga-toxin-producing *Escherichia coli* in faeces of healthy dairy cows, sheep and goats: prevalence and virulence properties. Letts. Appl. Microbiol. *31*, 203–208.

CHAPTER 16

SURVIVAL AND GROWTH OF VEROCYTOTOXIGENIC *E. COLI* IN FOODS

GERALDINE DUFFY
and
PATRICIA GARVEY

Teagasc
The National Food Centre
Dunsinea, Castleknock, Dublin 15, Ireland

INTRODUCTION

Verocytotoxigenic *Escherichia coli,* in particular serotype O157:H7, have emerged in the last fifteen years as important foodborne pathogens. They are of considerable concern, not only because of their increasing incidence worldwide, but also because of the severity of the illness they can cause and the low infectious dose at which they can establish such illness. Haemorrhagic colitis, haemolytic uraemic syndrome and thrombocytopenic purpura are common complications, and this pathogen has caused severe public health and economic problems worldwide with numerous high-profile outbreaks (3,5,6,55,56,69).

In comparison with other food borne pathogens, the infectious dose for *E. coli* O157 is very low, with levels of contamination as low as 2 per 25 g reported in food implicated in infection (79). In consequence, any survival or potential for growth of *E. coli* O157:H7 can have serious food safety implications. Information on the survival of VTEC in the food chain during food processing and storage is essential for the effective development of measures to limit the survival potential of VTEC, and reduce the risks it posed. As the food industry continues to employ combinations of mild forms of food preservation, this information is of particular importance in assessing any food safety risks associated with the application of multiple hurdle strategies. Reducing the survival potential of the pathogen also limits the risks of cross contamination, and transmission of VTEC infection.

Undercooked beef burgers have been involved in many food poisoning incidents associated with this pathogen (5) but other foods including salami (6), apple cider (12), milk (71), raw milk cheeses (Reid, Chapter 10), water (55, 68) and vegetables (56,57,60) have also been implicated. Initial outbreaks of VTEC infection were traced to ground beef products (5) and subsequent investigations have established that this bacterium is present in faeces and the intestines of healthy bovines, and can contaminate meat during the slaughtering process (20). VTEC may also enter the food chain, when contaminated faeces come in contact with raw milk, fruit or vegetables, water, etc., or by cross contamination from other products or surfaces during processing, handling, storage or food preparation.

The survival or potential for growth of VTEC in foods is influenced by a range of parameters including temperature, pH, water activity, anti-microbial agents, the atmosphere in which the food is maintained and the competitive microflora. While the growth characteristics of VTEC appear broadly similar to other *E. coli*, it has been reported that the O157:H7 serotype has an atypical high tolerance to acid. This is consistent with the reported association of VTEC with low pH foods traditionally considered to be microbiologically safe, such as fermented meats (6) and fruit juices (12). Acid tolerance has been identified as a major determinant of virulence, as it improves the ability of adapted organisms to survive in the acid pH of the stomach (35).

Epidemiological evidence has shown that serogroups O26, O103, O111 and O145 are now increasingly linked to food borne illness and are often linked to HUS in children (13,19,77). In Ireland, the first outbreak of infection linked to O26 was recorded in Donegal in November 1999 (11). Despite the clinical emergence of these serogroups of VTEC, there is still limited information on their survival characteristics in the food chain. This is probably because there is no widely applied routine method for the detection of these serogroups although the recent commercial availability of immunomagnetic particles coated with strain specific antibodies is a promising development (41). Because of the paucity of information on the non O157 serogroups, this chapter will focus almost exclusively on *E. coli* O157:H7.

TEMPERATURE

Studies on the effect of temperature on the survival of *E. coli* O157:H7 in a range of foods have focussed on its survival at commonly used storage and heat inactivation temperatures.

Chilling and Chilled Storage

Data in the literature suggests that, in general, bacterial numbers may decrease, increase or remain constant during chilling. The chilling parameters (air temperature, relative humidity and air speed) which can achieve a reduction in VTEC numbers have yet to be established. However, there is concern about the possibility of survival of even small numbers of VTEC, because its low infectious dose may pose a food safety problem.

A number of studies have been carried out on the survival of *E. coli* O157:H7 in a range of foods during storage at chill temperatures (Table 1). One of the main vehicles for the transmission of VTEC to humans is undercooked ground beef. A study carried out by Heuvelink *et al.*, (44) on the survival of *E. coli* O157:H7 on raw minced beef (tartaar) stored at refrigeration temperatures showed that the pathogen numbers remaining constant for at least 3 days on meat stored at 0°C or 5°C, and for 5 days on beef stored at 7°C or 15°C. Another study carried out at a higher temperature (12°C) has demonstrated growth of the pathogen from a level of 10^3 to 10^6 cfu g^{-1} after 10 days storage under aerobic conditions (72). It is apparent that the temperature of chilled storage is of critical importance in preventing outgrowth of the pathogen and that very low temperatures may even decrease pathogen numbers. Thus Ansay *et al.* (9) reported a decrease of $\log_{10}$ 1.9 cfu g^{-1} in the number of *E. coli* O157:H7 in ground beef patties stored at very low chill temperatures (2°C) over a four week storage period.

TABLE 1.
SURVIVAL OF *E. COLI* O157:H7 IN A RANGE OF FOOD DURING STORAGE AT CHILL TEMPERATURES

Food	Storage temp (°C)	Storage time (days)	Population		Ref
			Inoculum ($\log_{10}$cfu g^{-1}/ml^{-1})	Surviving ($\log_{10}$cfu g^{-1}/ml^{-1})	
Steak tartaar	7	5	3.00	2.5	44
Ground beef	2	28		+	9
Raw beef	12	10	3.00	6.00	72
Pasteurised milk	5	28	3.00	0.4-1.00	74
Raw milk	5	28	3.00	0.5-1.0	74
Hard cheese	6-7	158	3.00	+	61
Cucumber	5	10	5.05	3.43	1
Carrots	5	14	5.31	+	1
Lettuce	5	14	5.34	4.23	1
Mineral water	15	63	3.00	0.56	48

Note + = detected after enrichment

This has important implications, not only for refrigerated storage of retail meats but also for the chilling of carcasses in beef abattoirs. Research on the effect of carcass chilling on the survival of *E. coli* O157:H7 on beef carcasses and on the use of chilling as a critical control point in beef slaughter is urgently needed. There is also evidence that during the storage of raw meat, VTEC are in competition with other members of the complex meat microflora of this material, and that their survival is affected by the nature and activities of this diverse microflora (28). During a study in which a *Lactobacillus* spp. cocktail was added to meat inoculated with *E. coli* O157:H7 and then stored at 12°C, growth of the pathogen was completely inhibited by the competitive microflora (72), although the pathogen grew in the absence of the *Lactobacillus* spp. cocktail. While adequate cooking should eliminate the risk of transmission of VTEC to humans, meats such as steak tartaar or filet americain, which are consumed with little or no cooking, are causes for concern (44).

Human infections with *E. coli* O157:H7 have been linked to raw milk contaminated with cattle faecal material (7). Pasteurised milk has also been implicated in human infections, but only where inadequate pasteurisation, or post process contamination was reported (71). Studies carried out in milk have shown that *E. coli* O157:H7 was able to grow (by $\log_{10}$ 1.0-2.0) in pasteurised milk stored at temperatures ranging from 7°C-15°C (43,52), and to a lesser extent in raw milk (74). However, in milk stored at a slightly lower temperature (5°C), the pathogen numbers actually decreased by $\log_{10}$ 1.6-2.0 cfu ml^{-1} (74). In contrast, studies on survival of VTEC on meat reported earlier in this chapter noted that *E. coli* O157:H7 numbers did not decrease but remained constant during storage on meat at 5°C. Such differences may be related to differences between the competitive microflora of these two food types. The significantly slower growth of the pathogen in raw milk, than in pasteurised milk, has been attributed to the more significant antagonistic effects of the larger background flora of raw milk. Modifications of the fat content of pasteurised milk (3.5% fat, 2% fat, skim) were not reported to affect the growth kinetics of the pathogen, regardless of the storage temperature used (i.e., 8°C, 15°C, 22°C).

The association of *E. coli* O157:H7 with raw milk raises the possibility of VTEC survival in dairy products. In hard cheeses, the potential for survival/growth of the pathogen is significantly lower than in high moisture soft and semi-soft cheeses. The additional hurdles imposed during the hard cheese manufacturing process, including low water activity and pH which develop during the curing process, and the changes in the competing microflora, reduces the survival and growth potential of the pathogen. A study on the survival of *E. coli* O157:H7 inoculated into pasteurised milk used in the manufacture of cheddar cheese (salt 1.1-1.4%, pH 4.95-5.2, 51% fat in dry matter) indicated that the pathogen could grow during cheese manufacture, and that growth of the pathogen was faster in curd than in whey (61). It has been reported that the pathogen is able to survive for at least 60 days during cheese ripening. Thus the additional hurdles imposed during cheese manufacture may be insufficient to prevent the growth/survival of the pathogen in cheese produced from milk contaminated with the pathogen. Guraya *et al.* (36) reported growth of *E. coli* O157:H7 at 12°C in skim milk and cottage cheese. The risk which VTEC poses in farmhouse cheeses, produced from unpasteurised milk in order to retain flavours and aromas which are heat sensitive, is particularly high. Reid, in Chapter 10 of this volume, reviews several outbreaks which were attributable to raw milk farmhouse cheeses.

Fruit and vegetables have also been implicated in VTEC related illness. Products implicated include apple cider (12), vegetables (56,58) and white radish sprouts (56). The sources of contamination of fruits and vegetables have included contaminated washing water (60) and improperly composted manure (22). Fruits and vegetables tend to have hydrophobic surfaces which prevent drying of underlying tissues and afford protection against injury. Cutting and bruising of fruits has been shown to expose them to contamination with VTEC. There was a stronger attachment of the pathogen to the cut surfaces of lettuce washed in water inoculated with VTEC (70) which highlights the importance of washing vegetables prior to cutting. Another study reported growth of this pathogen on sliced cucumber, cantaloupe and watermelon cubes and to a lesser extent on shredded carrots (1) stored at 12 or 21°C. It has been speculated that the growth of the pathogen on shredded carrots could be limited by the presence of 6-methoxymellein, a carrot phytoalexein known to be inhibitory to several fungi and bacteria. At storage temperatures of 4 and 5°C, Abdul-Raouf *et al.* reported no significant growth of the pathogen on vegetables but it was noted that only a small reduction in pathogen numbers occurred during 14 days storage. The ability of *E. coli* O157:H7 to multiply on vegetable tissue has implications for the cut fruit and vegetable industry. Thus, low temperature storage of such products appears essential. *E. coli* O157:H7 presents additional risk in hydroponically grown products, such as radish sprouts. In contaminated water, the pathogen has been shown to be capable of growth on all parts of the sprouts at 25°C (39). Thus, strict sanitation measures for the seeds and the hydroponic water supply are essential.

While many cases of *E. coli* O157:H7 poisoning have been attributed to contaminated surface waters, there is increasing concern that the entry of *E. coli* O157:H7 into ground water supplies may pose risks in relation to the consumption of bottled waters. *E. coli* O157:H7 inoculated into mineral water and stored at 15°C survived for 63 days, but could not be detected after 70 days (48), while another study has reported survival >309 days in mineral water stored at 22°C (75). The type of water and the autochthonous flora have a significant effect on the survival of the pathogen with survival in mineral water>sterile mineral water > sterile distilled deionised water. The pathogen is also reported to survive for longer periods in mineral water than in spring water (48).

Recent research has indicated that food contact and preparation surfaces and utensils such as plastic, stainless steel, etc., can contribute to cross contamination with pathogens including VTEC. *E. coli* O157:H7 has been shown to survive for extended periods on stainless steel at refrigeration temperature, and can grow on plastic cutting boards in the presence of meat juices (4). Recent research indicates that *E. coli* O157:H7 dies more rapidly on copper than on stainless steel (53). Effective sanitation of work surfaces and equipment remains a primary control point in reducing the risk of pathogen cross contamination.

Frozen Storage

E. coli O157:H7 is able to survive commercial freezing temperatures (approx. -20°C) and is recoverable from frozen beef burgers (9,80), ice-cream (34,37) and from frozen strawberries and grated radishes (40). Contaminated ice cubes have been linked to an outbreak of *E. coli* O111 infection in the USA in 1999 (8).

A number of studies have investigated the survival of VTEC in frozen foods. Sage and Ingham (65) investigated the survival of *E. coli* O157:H7 after freezing and thawing of ground beef patties. Reductions in pathogen numbers of $\log_{10}$ 0.62 to 2.52 cfu g^{-1} were noted during storage at -20°C for 24 h with strain differences having more significant effects on surviving numbers, than differences in thawing method (4°C for 12 h, 23°C for 3 h or microwaving at 700W for 120 s). Byrne *et al.* (14) reported that the rate of freezing (slow/rapid) of beef burger patties had no significant effect on the extent of *E. coli* O157:H7 survival. A recent study by Bredholt *et al.* (16) on the survival of *Escherichia coli* O157:H7 in commercially produced frozen ground beef patties showed no significant decrease in the number of *E. coli* O157:H7 in patties frozen and stored at -24°C for up to 3 months. Ansay *et al.* (9) reported reductions of only 1-2 logs during storage of spiked ground beef patties for 1 year at -20°C, demonstrating prolonged survival at this temperature, and confirming that safety of the product is dependent on subsequent destruction of surviving VTEC by adequate cooking. In ice-cream containing nisin, no reduction in *E. coli* O157:H7 numbers was observed over a 20-day storage period (30).

Thermal Inactivation

E. coli O157:H7 displays heat resistance characteristics which are similar to those of other *E. coli* species. Variations in heat resistance has been observed between different strains of *E. coli* O157:H7 (29). The heating matrix can have a significant effect on the thermal resistance of pathogens and it is known that the level of fat, water, protein, carbohydrates and salt in the food, and the pH and competitive microorganisms present, can alter the thermotolerance of microorganisms including *E. coli* O157:H7 (3,26,29,50).

In salami heated to 60°C, *E. coli* O157:H7 has a D value of 1.18-2.21 min (29). In lean beef at this temperature, a D value of 0.93 min was derived (46) while in turkey a D value 0.58 min was derived (3), demonstrating the effect of the food medium on modulating the survival of the pathogen during high temperature challenge. It is clear, therefore, that during the design of systems for the thermal destruction of the *E. coli* O157:H7, time and temperature combinations employed should be validated in the target product under typical processing conditions.

As has been observed for other food pathogens, the survival of *E. coli* O157:H7 during high temperature challenge is also related to the physiological state of the organism. It is well established that adverse environmental conditions can induce cell responses aimed at enhancing cell survival, and that the prior history of the cell, i.e., exposure to non-lethal stress such as acid, cold, heat, etc., can influence the organism's resistance to heat. These cross protective effects, against unrelated or previously un-encountered stresses, are probably due to the co-expression of a range of stress proteins within a single stimulon (67).

Semanchek and Golden (66) have reported that *E. coli* O157:H7 cells grown at 10°C (refrigeration temperature) were less heat resistant than those grown at 37°C. Katsui (47) suggested that bacteria adjust the biophysical properties of their cell membranes to enable normal physiological functions at lower temperatures, by changing their unsaturated fatty acids profile, and that such adapted cell membranes are more sensitive to any subsequent heat challenge.

There is conflicting evidence on the effect of acid adaptation on the thermal resistance of *E. coli* O157:H7. Broth based studies by Duffy *et al.* (27) demonstrated that pre-adaptation to acid stress provided cross-protection against subsequent heat stress, but these results were not observed when the experiment was repeated in pepperoni product (63). Williams and Ingham (78) reported that acid adaptation did not enhance heat resistance in broth but Ryu and Beuchat (64) showed that acid-adapted *E. coli* O157:H7 cells were considerably more resistant to heat than control cells heated in broth, apple cider and orange juice. A broth based study conducted by Semanchek and Golden (66) used three different *E. coli* O157:H7 strains isolated from different foods, salami, beef and apple cider. The salami isolate had the greatest resistance to heat, followed by the apple cider isolate, with the beef isolate having the lowest heat resistance. There may be a number of reasons for this pattern, including the prior acid exposure history of the strains involved. The nature of the suspending media may have influenced the extent of induction and the contribution of cross-protection reactions. Alternatively, the specific acid adaptation procedures applied may have induced some acid damage (18).

The possibility that acid exposure may impart cross-protection against heat challenge has significant implications in the application of heating processes as elements of multiple hurdle strategies designed to eliminate *E. coli* O157:H7 from some foods, in particular, fermented meats.

pH

Survival of bacteria at low pH has been identified as a major virulence determinant in a number of enteric pathogens. It offers an ecological advantage by allowing the pathogen to survive in the gastric system, and to establish infection in the colon and other organs, effectively lowering the infectious dose (35,73).

A number of *E. coli* O157:H7 outbreaks have been associated with the consumption of low pH foods including salami (6); apple cider (12) and yoghurt (57). As might be expected, considering the nature of these foods, the pathogen has been shown to be capable of surviving in fermented meats (pH 4.8) (62); apple cider (pH 3.6-4.0) (81); orange juice (pH 3.68-3.76) (51) and mayonnaise (pH 3.86 to 3.97)(42) (Table 2). In broth based studies, *E. coli* O157:H7 has been shown to be capable of growth at pH 4.0-4.5 at room temperature (25°C) (23).

TABLE 2.
EFFECT OF pH ON THE SURVIVAL OF *E. COLI* O157:H7 DURING STORAGE IN RANGE OF FOODS

Food	**pH**	**Storage time (days)**	**Population**		**Ref**
			Inoculum (log_{10}cfu g^{-1})	**Surviving (log_{10}cfu g^{-1})**	
Pepperoni	4.8	90	5.9.± 1.1	3.7± 1.61	31
Apple cider	3.6-4.0	20	5.0	1.7	81
Orange juice	3.4	25	8.0	ND	51
Mayonnaise	3.86-3.97	17	6.23	0.48	42
Buttermilk	4.4	32	4-5	~2.8	54

note ND = not detected

Fermented meat products have traditionally been considered to be microbiologically safe due to the impact of a number of inhibitory factors including the effects of fermentation (pH reduction, end products of fermentation) and added preservatives (salt, sodium nitrite, spices, etc.). However, *E. coli* O157:H7 has been shown to survive the traditional fermentation process used in the production of fermented dry sausage (33) and pepperoni (62). In both these studies, the product was inoculated with approximately 10^6 cfu g^{-1} *E. coli* O157:H7, and pathogen numbers decreased by only $\log_{10}$ 0.47 to 0.98 during fermentation and drying. Evidence of the survival of *E. coli* O157:H7 in such products led the USDA to recommend that the processing regime should achieve a $\log_{10}$ 5.0 cfu/g decline in numbers of *E. coli* O157:H7. Manipulation of the intrinsic factors in the fermentation process are unable to achieve this target, and thus additional hurdles in the process, such as the inclusion of a heat treatment step, are therefore necessary (63).

Several studies have reported the ability of *E. coli* O157:H7 to survive and even to grow during storage of fermented dairy products. *E. coli* O157:H7 inoculated into commercial products could be recovered for up to 12 days from yogurt, 28 days from sour cream and for 32-35 days from buttermilk (25,54).

Apple cider is a traditional unpasteurised apple juice product, with a pH 3.6-4.0, commonly consumed in the US. Recent food poisoning outbreaks associated with this product, and resultant changes in regulations, have made it necessary for manufacturers to investigate additional hurdles to ensure the safety of the process. The storage temperature of the apple cider has a significant effect, with the pathogen (inoculated to initial concentrations of ~ $\log_{10}$ 5.0) present at higher numbers ($\log_{10}$ 3.57 cfu g^{-1}) after 42 days when stored at 5°C, than when stored at 25°C ($\log_{10}$ 1.26 cfu g^{-1}) for 42 days. In samples stored for 56 days, the organism was undetectable in samples stored at 5°C but was detectable (by enrichment) in samples stored at 25°C for the same period (64). The addition of preservative agents such as potassium sorbate have been reported to have minimal effects on survival of *E. coli* O157:H7, although the pathogen survives in apple cider containing sodium benzoate (1%) for 1 day at 8 and 2 days at 25°C. Many manufacturers now include a pasteurisation step to eliminate the pathogen.

The low pH (3.6-4.2) of salad dressings and edible emulsions has traditionally been considered to pose a hurdle to microbial growth/survival. Depending on the storage temperature and the acidulant (acetic, lactic, citric) used, *E. coli* O157:H7 has been observed to survive in commercial mayonnaise (full fat, reduced calorie) and salad products which incorporate low pH dressings (42). This is believed to be related to the decrease in the fluidity/permeability of the bacterial cell membrane at refrigeration temperatures which prevents penetration of the acidulant into the cell and decreases the rate of the chemical reactions which damage the cell. When vinegar (acetic acid) is used as an acidulant, a more rapid death of the pathogen was observed than when lemon juice (citric acid) was used.

Three acid survival mechanisms have been identified in *E. coli* O157:H7 which contribute to its survival at low pH: an oxidative system; an arginine-dependent acid resistance system and a glutamate-dependent system (49). The stationary phase sigma factor *rpoS* is required for the oxidative system and is also partially involved in the other two systems although natural *rpoS*-negative variants of the pathogen have been identified which exhibit a lower tolerance to acid conditions (76).

Many processed foods contain organic acids that are added as flavour enhancers and preservatives, and many fermented foods contain organic acids as metabolites of the fermentation process. Exposure of the *E. coli* O157:H7 to mildly acidic conditions has been shown to induce an acid tolerance response which allows the organism to survive subsequent low pH challenges which would inactivate the bacterium in its unadapted state (17).

The acid tolerance exhibited by bacteria is also influenced by exposure to other forms of stress. Slow growing bacterial cells, such as those which are entering the stationary phase, exhibit increased acid tolerance as part of a generalised stress reaction (10,15). It has been suggested that this response could be triggered by stress conditions, such as nutrient deprivation, experienced as the cells entered the stationary phase (10).

PRESERVATIVES

E. coli O157:H7 can survive in product formulations which have higher salt (4.8%) and sodium nitrite concentrations (300 ppm). Such additive concentrations would not normally be acceptable in commercial products, i.e., these salt levels would render products organoleptically unacceptable. Similarly, such sodium nitrite concentrations are beyond the maximum legally permissible concentrations stipulated by most national legislation (62). Such reports of the ability of this organism to survive in these concentrations of additives confirm the findings of broth based studies, which noted that *E. coli* O157:H7 could tolerate higher salt and nitrite concentrations than other enteric organisms, including *Salmonella* (45). The ability of this organism to survive such adverse conditions, and recognition of its low infectious dose, means that its presence in fermented meat products and their ingredients is particularly undesirable.

As previously stated, the survival of VTEC in ground meat products destined to be consumed without further heat processing is a matter of particular concern. Such concern is heightened by suggestions that some additives commonly used in comminuted meat products, e.g., 3% lactate, have no significant effect on the survival of VTEC in these products (14).

Components present in commercial meat product formulations, but not in laboratory media, result in changes in the environment of the organism, and could significantly affect the extent to which the environment will support or suppress growth. For example, sodium nitrite, stable in a laboratory broth, is rapidly destroyed by reaction with ascorbate, present in many commercial meat products (62). Care should therefore be exercised in the extrapolation of data from broth based studies to food situations.

Garren *et al.* (32) demonstrated that induction of acid tolerance reactions in an acid tolerant acid strain of O157:H7 confer cross protection against inhibition by sodium chloride and sodium lactate. However, such protection was not observed in a weakly acid tolerant strain. Cheville *et al.* (21) found that heat and acid challenge conferred enhanced salt resistance in *E. coli* O157:H7. Such effects may be significant in foods which have traditionally been preserved by a combination of salt and low pH.

WATER ACTIVITY

E. coli O157:H7 is reported to survive well in dry foods, tolerating low a_w values (0.35-0.73). However, the extent of survival in dried foods is correlated with storage temperature. Deng *et al.* (24) found that survival of the pathogen decreased in nine commercial dry foods and ingredients, as the a_w was lowered from 0.73 to 0.35. They also noted that survival was temperature-dependent, with greatest survival reported at storage temperatures of 5°C, less survival at 21°C and least survival at 37°C. In contrast, a study by Park and Beuchat (59) showed that the survival of *E. coli* O157:H7 in potato starch powder increased as the a_w decreased from 0.75-0.78 to 0.24-0.26, although survival was correlated to temperature in a similar pattern to that reported by Deng *et al.* (24).

STORAGE ATMOSPHERE

The limited information available on the survival of *E. coli* O157:H7 under modified atmosphere packaging (MAP) or under vacuum (VAC PAC) indicates that growth under such atmospheres is not significantly slower than under aerobic conditions at 5, 10 or 20°C. Storage under MAP does not alter the tolerance of VTEC to inhibitory agents such as 3% NaCl. In damaged cells, such as those present in fermented meat products, some decline in viable cells numbers occurs under these storage atmospheres. In one study, the extent of decline in *E. coli* O157:H7 numbers was dictated by storage temperature rather than storage atmosphere, with greater declines occurring at 21°C than at -20°C or 4°C, regardless of whether product was stored in air, vacuum or CO_2 (38).

CONCLUSION

While recent reports have implicated direct and indirect contact with animal faeces and person-to-person spread as important causes of sporadic VTEC cases and of family and institutional outbreaks, food and water remain the primary agents of infection in the major high-profile VTEC outbreaks (3,12, 55,56). Undoubtedly, some of these outbreaks were attributable to specific failures during processing or treatment (55), but others resulted from enhanced survival by *E. coli* O157:H7 in foods previously considered to be microbiologically safe (6,12). As awareness increased of the risks associated with these food types, the value of performing survival studies was affirmed by the subsequent development of control measures to be applied during the processing of these high-risk foods, e.g., heating of fermented meats and pasteurisation of apple cider.

More recently, analysis of the molecular basis of survival of *E. coli* O157:H7 during exposure to different forms of stress, has established the role of generalised stress mechanisms which can impart cross protective effects against previously unencountered stresses. As the food industry proceeds to employ combinations of mild forms of food preservation, the continued investigation of these mechanisms is of particular importance in assessing any food safety risks associated with the application of multiple hurdle strategies.

It should be noted that studies on survival of VTEC have in general been carried out using *E. coli* O157:H7 and the data in the literature is based almost solely on data generated using this serotype. Although there has been considerable progress in the development of methods for the detection of clinically significant non-O157 VTEC (O111, O45, O26, O103) using immunological and molecular techniques, there remains no routine cultural method for detection of non O157 serotypes. The advent of routine methods for detection of these emergent serotypes will allow us to ascertain whether there are differences in survival and virulence characteristics between different VTEC serotypes. In the absence of data it is assumed that survival characteristics are similar among VTEC serotypes but further research in this area is clearly needed. In the meantime, current approaches based on Risk Assessment and Hazard Analysis Critical Control Point (HACCP) programmes should provide the most effective means of controlling VTEC.

REFERENCES

1. ABDUL-RAOUF, U.M., BEUCHAT, L.R. and AMMAR, M.S. 1993. Survival and growth of *Escherichia coli* O157:H7 on salad vegetables. Appl. Environ. Microbiol. *59*, 1999–2006.

2. AHMED, N.M., CONNER, D.E. and HUFFMAN, D.L. 1995. Heat resistance of *Escherichia coli* O157:H7 in meat and poultry as affected by product composition. J. Food Sci. *60*, 606–610.
3. AHMED, S. and DONAGHY, M. 1998. An outbreak of *Escherichia coli* O157:H7 in central Scotland, *In Escherichia coli* O157:H7 and other Shiga toxin-producing *E. coli* strains, (J.B. Kaper and A.D. O'Brien, eds.) p. 59–65. American Society for Microbiology, Washington, D.C.
4. AK, N.O., CLIVER, D.O. and KASPAR, C.W. 1994. Cutting boards of plastic and wood contaminated experimentally with bacteria. J. Food Prot. *57*, 16–22.
5. ANON. 1993. Update: multistate outbreak of *Escherichia coli* O157:H7 infections from hamburgers — western United States, 1992-1993. Centres for Disease Control. Morbid. Mortal. Weekly Rep. *42*, 258–263.
6. ANON. 1995. *Escherichia coli* O157:H7 outbreak linked to commercially distributed dry-cured salami — Washington and California, Centres for Disease Control Morbid. Mortal. Weekly Rep. *44*, 1.
7. ANON. 2000a. *Escherichia coli* O111:H8 outbreak among teenage campers — Texas, 1999. Morbid. Mortal. Weekly Rep. *49*, 321–324.
8. ANON. 2000b. Outbreaks of VTEC O157 infection linked to consumption of unpasteurised milk. Commun. Dis. Rep. CDR Wkly. *10*, 203–206.
9. ANSAY, S.E., DARLING, K.A. and CASPAR, C.W. 1999. Survival of *Escherichia coli* O157:H7 in ground beef patties during storage at 2, -2, 15 and then -2°C, and -20°C. J. Food Prot. *62*, 1243–1247.
10. BENJAMIN, M.M. and DATTA, A.R. 1995. Acid tolerance of enterohemorrhagic *Escherichia coli*. Appl. Environ. Microbiol. *61*, 1669–1672.
11. BIRCHARD, K. 1999. *E. coli* O26 outbreak reported in Ireland. The Lancet, *354*, 9191.
12. BISSER, R.E. *et al.* 1993. An outbreak of diarrhoea and haemolytic uremic syndrome from *Escherichia coli* O157:H7 in fresh pressed apple cider. J. Amer. Med. Assoc. *269*, 2217–2220.
13. BLANCO, M. *et al.* 1996. Prevalence and characteristics of human and bovine verotoxigenic *Escherichia coli* strains isolated in Galicia (north-western Spain). Eur. J. Epidemiol. *12*, 13–19.
14. BYRNE, C.A. 2000. Personal Communication, The National Food Centre, Dunsinea, Castleknock, Dublin 15.
15. BOOTH, I.R., THOMSON-CARTER, F., CARTER, P., MALCOLM, L. and GLOVER, J. 1999. Acid tolerance in *E. coli* O157:H7 and related organisms. Conferences proceedings on "Survival and Growth of Verocytotoxigenic *E. coli*" organised by an EU Concerted Action on VTEC (CT98-3935). Agricultural University of Athens, May 6-8th.
16. BREDHOLT, S., NESBAKKEN, T., HOLCK, A. and NISSEN, H. 2000. Survival of *Escherichia coli* O157:H7 in commercially produced ground beef patties. Conference proceedings of 4th International symposium and workshop on "shiga toxin (verocytotoxin) producing *Escherichia coli* infections". Oct 29 to Nov 2, Kyoto, Japan. Pg. 81.

17. BRUDZINSKI, L. and HARRISON, M. 1998. Influence of incubation conditions on survival and acid tolerance response of *Escherichia coli* O157:H7 and non-O157:H7 isolates exposed to acetic acid. J. Food Prot. *61*, 542–546.
18. BUCHANAN, R.L. and EDELSON, S.G. 1996. Culturing enterohemorrhagic *Escherichia coli* in the presence and absence of glucose as a simple means of evaluating the acid tolerance of stationary-phase cells. Appl. Environ. Microbiol. *62*, 4009–4013.
19. CAPRIOLI, A., LUZZI, I. and ROSMINI, F. 1994. Community wide outbreak of hemolytic uremic syndrome associated with non-O157 verocytotoxin-producing *Escherichia coli*. J. Infect. Dis. *169*, 208–211.
20. CHAPMAN, P.A. 2000. Sources of *Escherichia coli* O157 and experiences over the past 15 years in Sheffield, UK. J. Appl. Microbiol. *88* (Suppl) 51S–60S.
21. CHEVILLE, A.M., ARNOLD, K.W., BUCHRIESER, C., CHENG, C.-M. and KASPAR, C.W. 1996. *rpoS* regulation of acid, heat and salt tolerance in *Escherichia coli* O157:H7. Appl. Environ. Microbiol. *62*, 1822–1824.
22. CIESLAK, P.R. *et al.* 1993. *Escherichia coli* O157:H7 infection from a manured garden. Lancet *342*, 367.
23. CONNER, D.E. and KOTROLA, J.S. 1995. Growth and Survival of *Escherichia coli* O157:H7 under acidic conditions. Appl. Environ. Microbiol. *61*, 382–385.
24. DENG, Y., RYU, J.H. and BEUCHAT, L.R. 1998. Influence of pH on survival of *Escherichia coli* O157:H7 in dry foods and growth in reconstituted infant rice cereal. Int. J. Food Microbiol. *45*, 173–184.
25. DINEEN S.S., TAKEUCHI, K., SOUDAH, J.E. and BOOR, K.J. 1998. Persistence of *Escherichia coli* O157:H7 in dairy fermentation systems. J. Food Prot. *61*, 1602–1608.
26. DUFFY, G., ELLISON, A., ANDERSON, W., COLE, M. and STEWART, G.S.A.B. 1995. Use of bioluminescence to model the thermal inactivation of *Salmonella typhimurium* in the presence of a competitive micro-flora. Appl. Environ. Micro. *61*(9), 3463–3465.
27. DUFFY G. *et al.* 2000. Effect of pH on survival, thermotolerance, and verotoxin production of *Escherichia coli* O157:H7 during simulated fermentation and storage. J. Food Prot. *63*, 12–18.
28. DUFFY, G., WHITING, R.C. and SHERIDAN, J.J. 1999a. The effect of competitive microflora, pH and temperature on the growth kinetics of *Escherichia coli* 0157:H7. Food Microbiol. *16*(3), 299–307.
29. DUFFY, G., *et al.* 1999b. Differences in thermotolerance of various *Escherichia coli* O157:H7 strains in a salami mix. Food Microbiol. *16*, 83–91.
30. EL-HAWARY, I.I., AMAN, I.M., SHEHAB, M.M., EL-TAHLAWY, M.R. and MAHMOUD, M.R. 1998. Survival characteristics of *Escherichia coli* serotype O157:H7 in ice cream and yoghurt. Proceedings of the Eight Scientific Congress, Faculty of Veterinary Medicine, Assiut University, 15-17th November, 75–82.
31. FAITH, N.G., PARNIERE, N., LARSON, T., LORANG, T.D. and LUCHANSKY, J.B. 1997. Viability of *Escherichia coli* O157:H7 in pepperoni during the manufacture and subsequent storage of slices at 21, 4 and -20 under air, vacuum and CO_2. Int J. Food Microbiol. *37*, 47–54.

32. GARREN, D.M., HARRISON, M.A. and RUSSELL, S.M. 1998. Acid tolerance and acid shock response of *Escherichia coli* O157:H7 and non-O157:H7 isolates provides cross protection to sodium lactate and sodium chloride. J. Food Prot. *61*, 158–161.
33. GLASS, K.A., LOEFFELHOLZ, J.M., FORD, J.P. and DOYLE, M.P. 1992. Fate of *Escherichia coli* O157:H7 as affected by pH or sodium chloride and in fermented dry sausage. Appl. Environ. Microbiol. *58*, 2513–2516.
34. GOODING, C.M. and CHOUDARY, P.M. 1997. Rapid and sensitive immunomagnetic separation-polymerase chain reaction method for the detection of *Escherichia coli* O157:H7 in raw milk and ice-cream. J. Dairy Res. *64*, 87–93.
35. GORDEN, J. and SMALL, P.L.C. 1993. Acid resistance in enteric bacteria. Inf. Imm. *61*, 364–367.
36. GURAYA, R.J., FRANK, J.F. and HASSAN, A.N. 1998. Effectiveness of salt, pH and diacetyl as inhibitors for *Escherichia coli* O157:H7 in dairy foods stored at refrigeration temperatures. J. Food Prot. *61*, 1098–1102.
37. HAMMACK, T.S.P., FENG, R.M., AMAGUANA, R.M., JUNE, G.A., SHERROD, P.S. and ANDREWS, W.H. 1997. Comparison of sorbitol MacConkey and haemorrhagic coli agars for recovery of *Escherichia coli* O157:H7 from Brie, ice cream and whole milk. J. AOAC Int. *80*, 335–340.
38. HAO, Y.Y. and BRACKETT, R.E. 1993. Growth of *E. coli* O157:H7 in modified Atmosphere. J. Food Prot. *56*, 330–332.
39. HARA-KUDO, Y. *et al.* 1997. Potential hazard of radish sprouts as a vehicle of *Escherichia coli* O157:H7. J. Food Prot. *60*, 1125–1127.
40. HARA-KUDO, Y. *et al.* 2000a. Selective enrichment with a resuscitation step for isolation of freeze-injured *Escherichia coli* O157:H7 from foods. Appl. Environ. Microbiol. *66*, 2866–2872.
41. HARA-KUDO, Y., KONUMA, H., NAKAGAWA, H. and KUMAGAI, S. 2000b. *Escherichia coli* O26 detection from foods using an enrichment procedure and an immunomagnetic separation method. Letts. Appl. Microbiol. *30*, 151–154.
42. HATHCOX, A.K., BEUCHAT, L.R. and DOYLE, M.P. 1995. Death of Enterhemorrhagic *Escherichia coli* O157:H7 in real mayonnaise and reduced calorie mayonnaise dressing as influenced by initial population and storage temperature. Appl. Environ. Microbiol. *61*, 4172–4177.
43. HEUVELINK, A.E., BLEUMINK, B., VAN DEN BIGGERAAR, F.L., TE GIFFEL, M.C., BEUMER, R.R. and DE BOER, E. 1998. Occurrence and survival of verocytotoxin-producing *Escherichia coli* O157 in raw cow's milk in The Netherlands. J. Food Prot. *61*, 1597–1601.
44. HEUVELINK, A.E., ZWARTKRUIS-NAHUIS, J.T.M., BEUMER, R.R. and DE BOER, E. 1999. Occurrence and survival of verocytotoxin-producing *Escherichia coli* O157 in meats obtained from retail outlets in The Netherlands. J. Food Prot. *62* 10, 1115–1122.
45. ICMSF. 1996. *In* Microorganisms in Food, Vol. 5, Blackie Academic and Professional Press.

46. JUNEJA, V.K., SYNDER JR., O.P. and MARMER, B.S. 1997. Thermal destruction of *Escherichia coli* in beef and chicken: determination of D and Z values. Int. J. Food Microbiol. *35*, 195–308.
47. KATSUI, N., TSUCHIDO, T., TAKANO, M. and SHIBASAKI, I. 1981. Effect of preincubation temperature on the heat resistance of *Escherichia coli* having different fatty acid compositions. J. Gen. Microbiol. *122*, 357-361.
48. KERR, M., FITZGERALD, M., SHERIDAN, J.J., MCDOWELL, D.A. and BLAIR, I.S. 1999. Survival of *Escherichia coli* O157:H7 in bottled natural mineral water. J. Appl. Micro. *87*, 833-841.
49. LIN, J., SMITH, M.P., CHAPIN, K.C., BAIK, H.S., BENNETT, G.N. and FOSTER, J.W. 1996. Mechanisms of acid resistance in enterohaemorrhagic *Escherichia coli.* Appl. Environ. Microbiol. *62*, 3094–3100.
50. LINE, J.E. *et al.* 1991. Lethality of heat to *Escherichia coli* O157:H7: D-value and Z-value determinations in ground beef. J. Food Protection *54*(10), 762–766.
51. LINTON, M., MCCLEMENTS, J.M. and PATTERSON, M.F. 1999. Survival of *Escherichia coli* O157:H7 during storage in pressure-treated orange juice. J. Food Prot. *62*, 1038–1040.
52. MASSA, S., GOFFREDO, E., ALTIERI, C. and NATOLA, K. 1999. Fate of *Escherichia coli* O157:H7 in unpasteurised milk at 8 degrees C. Lett. Appl. Microbiol. *28*, 89–92.
53. MAULE, A. 1999. Environmental survival of *Escherichia coli* O157: Implications for the spread of the disease. Conferences proceedings on "Survival and Growth of Verocytotoxigenic *E. coli*" organised by an EU Concerted Action on VTEC (CT98- 3935) at Agricultural University of Athens, May 6-8th. p. 94–104.
54. MCINGVALE, S.C., CHEN, X.Q., MCKELLIP, J.L. and DRAKE, M.A. 2000. Survival of *Escherichia coli* in buttermilk as affected by contamination point and storage temperature. J. Food Prot. *63*, 441–444.
55. MCQUIGGE, M. 2000. A large outbreak of *Escherichia coli* O157:H7 and *Campylobacter* infection due to contaminated municipal drinking water. Conference proceedings of 4th International symposium and workshop on "shiga toxin (verocytotoxin) producing *Escherichia coli* infections". Oct 29 to Nov 2nd, Kyoto, Japan. pg. 36.
56. MICHINO, H. *et al.* 1999. Massive outbreak of *Escherichia coli* O157:H7 infection in schoolchildren in Sakai City, Japan, associated with consumption of white radish sprouts. Am. J. Epidemiol. *150*, 787–796.
57. MORGAN, D., NEWMAN, C.P., HUTCHINSON, D.N., WALKER, A.M., ROWE, B. and MAJID, F. 1983. Verotoxin producing *Escherichia coli* O157 infections associated with the consumption of yoghurt. Epidemiol. Infect. *111*, 181–187.
58. MORGAN, G.M., NEWMAN, C. and PALMER, S.R. 1988. First recognised community outbreak of haemorrhagic colitis due to verotoxin-producing *Escherichia coli* O157:H7 in the UK. Epidem. Inf. *101*, 83–91.
59. PARK, C.M. and BEUCHAT, L.R. 2000. Survival of *Escherichia coli* O157:H7 in potato starch as affected by water activity, pH and temperature. Letts. Appl. Microbiol. *31*, 364–367.

60. PEABODY, R.G. *et al.* 1999. An international outbreak of Vero cytotoxin-producing *Escherichia coli* O157 infection amongst tourists; a challenge for the European infectious disease surveillance network. Epidemiol. Infect. *123*, 217–223.
61. REITSMA, C.J. and HENNING, D.R. 1996. Survival of enterohemorrhagic *Escherichia coli* O157:H7 during the manufacture and curing of cheddar cheese. J. Food Prot. *59*, 460–464.
62. RIORDAN, D.C.R. *et al.* 1998. Survival of *E. coli* O157:H7 in fermented meats. J. Food Prot. *61*, 146–151.
63. RIORDAN, D.C., DUFFY, G., SHERIDAN, J.J., WHITING, R.C., BLAIR, I.S. and MCDOWELL, D.A. 2000. Effects of acid adaptation, product pH, and heating on survival of *Escherichia coli* O157:H7 in pepperoni. Appl. Environ. Microbiol. *66*, 1726–1729.
64. RYU, J.-H. and BEUCHAT, L. R. 1998. Influence of acid tolerance responses on survival, growth and thermal cross-protection of *Escherichia coli* O157:H7 in acidified media and fruit juices. Int. J. Food Microbiol. *45*, 185–193.
65. SAGE, J.R. and INGHAM, S.C. 1998. Survival of *Escherichia coli* O157:H7 after freezing and thawing in ground beef patties. J. Food Prot. *61*, 1181–1183.
66. SEMANCHEK, J.J. and GOLDEN, D.A. 1998. Influence of growth temperature on inactivation and injury of *Escherichia coli* O157:H7 by heat, acid and freezing. J. Food Prot. *61*, 395–401.
67. SHERIDAN, J.J. and MCDOWELL, D.A. 2000. Factors affecting the emergence of pathogens on foods. Meat Science *49*, (Suppl) 1, S151–S167.
68. SWERDLOW, D.L. *et al.* 1992. A waterborne outbreak in Missouri of *Escherichia coli* O157:H7 associated with bloody diarrhea and death. Ann. Intern. Med. *117*, 812–819.
69. SWINBANKS, D. 1996. Outbreak of *E. coli* infection in Japan renews concerns. Nature. *382*(6589), 290.
70. TAKEUCHI, K. and FRANK, J. 2000. Penetration of *Escherichia coli* O157:H7 into lettuce tissues as affected by inoculum size and temperature and the effect of chlorine treatment on cell viability. J. Food Prot. *63*, 434–440.
71. UPTON, P. and COIA, J.E. 1994. Outbreak of *Escherichia coli* O157:H7 infection associated with pasteurised milk supply. Lancet *344*, 1015.
72. VOLD, L., HOLCK, A., WASTESON, Y. and NISSEN, H. 2000. High levels of background flora inhibits growth of *Escherichia coli* in ground beef. Int. J. Food Micro. *56*, 219–225.
73. WANG, G. and DOYLE, M.P. 1998. Survival of enterohemorrhagic *Escherichia coli* O157:H7 in water. J. Food Prot. *61*, 662–667.
74. WANG, G., ZHAO, T. and DOYLE, M.P. 1997. Survival and growth of *Escherichia coli* O157:H7 in unpasteurised and pasteurised milk. J. Food Prot. *60*, 610–613.
75. WARBURTON, D.W., AUSTIN, J.W., HARRISON, B.H. and SANDERS, G. 1998. Survival and recovery of *Escherichia coli* O157:H7 in inoculated bottled water. J. Food Prot. *61*, 948–952.

76. WATERMAN, S.R. and SMALL, P.L.C. 1996. Characterisation of the acid resistance phenotype and *rpoS* alleles of shiga-like toxin-producing *Escherichia coli*. Infect. Imm. *64*, 2808–2811.
77. WHO. 1999. Zoonotic non-O157 shiga toxin-producing *Escherichia coli* (STEC). Report of a WHO scientific working group meeting. Berlin, Germany. WHO/CSR/APH/98.8:1–30.
78. WILLIAMS, N.C. and INGHAM, S.C. 1998. Thermotolerance of *Escherichia coli* O157:H7 ATCC 43894, *Escherichia coli* B, and an *rpoS*-deficient mutant of *Escherichia coli* O157:H7 ATCC 43895 following exposure to 1.5% acetic acid. J. Food Prot. *61*, 1184–1186.
79. WILLSHAW, G.A., THIRWELL, J., JONES, A.P., PARRY, S., SALMON, R.L. and HICKEY, M. 1994. Verocytotoxin producing *Escherichia coli* O157 in beef burgers linked to an outbreak of diarrhoea, haemorrhagic colitis and haemolytic ureamic syndrome. Lett. Appl. Microbiol. *19*, 304–307.
80. WOODY, J.M., STEVENSON, J.A., WILSON, R.A. and KNABEL, S.J. 1998. Comparison of the Difco EZ Coli rapid detection system and Petrifilm test kit-HEC for detection of *Escherichia coli* O157:H7 in fresh and frozen ground beef. J. Food Prot. *61*, 110–112.
81. ZHAO, T., DOYLE, M.P. and BESSER, R.E. 1993. Fate of enterohemorrhagic *Escherichia coli* O157:H7 in apple cider with and without preservatives. Appl. Environ. Microbiol. *59*, 2526–2530.

CHAPTER 17

PREDICTING THE FATE OF VEROCYTOTOXIGENIC *ESCHERICHIA COLI* IN FOODS

PETER J. MCCLURE

Microbiology Unit, Unilever Research
Colworth Laboratory, Colworth House
Sharnbrook
Bedford MK44 1LQ
United Kingdom

INTRODUCTION

Mathematical models for predicting the fate of microorganisms in foods have become essential tools in a number of areas related to food safety, providing valuable data to allow informed decision making. Models, in the form of equations or formulae, have been used for more than 70 years in thermal processing to predict the inactivation of microorganisms in foods. More recently, models have been developed which take into account the combined effects of a number of different factors in the prediction of the growth of foodborne pathogens and spoilage microorganisms. In the past, although interactions between factors have been studied, there has been little opportunity to interpolate combinations of factors and to predict the effects of untested combinations of conditions.

Predictive models include factors that can be manipulated or controlled and have some influence on the fate of the microorganism(s) being studied. Typical examples include temperature, pH, sodium chloride (salt) or water activity (a_w), and nitrite, organic acids or other preservatives.

The experimental systems used to generate models are generally chosen to mimic the behaviour of microorganisms in foods, although it is widely acknowledged that real food will include other factors that impact on microbial growth or survival. For this reason, it is important to consider the foods that are being modelled, and include as many relevant factors as is reasonably possible. Predictive models may also be criticised because they rarely take into account the presence and potential impacts of other microorganisms in foods. There is little doubt that the presence of other microorganisms can influence the growth of pathogens of concern, such as *E. coli* O157:H7 (21), so models including such effects should give a better reflection of the changes that may occur in particular foods. However, the accuracy of such models depends on the numbers and type of microorganisms present in the modelled food, and such factors are not easily predicted. In addition, growth of *E. coli* O157:H7 may be unaffected by natural flora contaminating some raw foods, e.g., meat (47).

The models used in food microbiology tend to be empirical in nature, since it is difficult to mathematically represent the biochemical and physical processes that determine bacterial responses such as lag time, growth rate or inactivation rate. The models are based on linear or non-linear regression techniques to estimate thermal inactivation and growth, where the kinetics are more easily predicted than for survival or non-thermal inactivation. The emergence of *Escherichia coli* O157:H7 and other verocytotoxigenic *E. coli* (VTEC) has highlighted the importance of modelling survival, principally because of the low infectious doses associated with these organisms. Unlike other organisms, such as *Listeria monocytogenes*, *Staphylococcus aureus* and *Clostridium botulinum*, where growth is an important feature in the ability to cause disease, the mere presence of low numbers of VTEC in foods is a matter of serious concern. Therefore, models that can predict survival are critical for assessing the safety of foods or raw materials potentially contaminated with VTEC. This is not to suggest that growth of VTEC in foods is unimportant. Such growth is very important as it will ultimately increase the risk of infection and likelihood of illness in those individuals exposed to the hazard, and may result in more widespread contamination and infection. However, because of the low infectious dose of such organisms, prevention of growth is not enough, and survival of small numbers of initially contaminating VTEC is sufficient to pose undesirable risks to consumers.

This chapter will cover the different types of models that have been generated to predict the fate of *E. coli* in different situations, representative of foods. It will also cover some important considerations for the generation and validation of models. Although many recent modelling studies focus on *E. coli* O157:H7, there is little evidence to suggest that this serovar, or VTEC as a group, behave any differently to other *E. coli*. There is, however, some evidence that those strains responsible for large outbreaks of illness may be more intrinsically acid tolerant than other strains (16). For this reason, it may be important to include such strains when assessing the safety of particular foods. Generally speaking, however, strains that are 'atypical' in their behaviour should only be included for specific reasons.

MODELLING GROWTH

Modelling Approaches

Models predicting the growth of microorganisms are usually generated from growth curves reflecting viable count data, although some studies used other data, such as change in optical density (35) or conductance (54). Microbial growth, when using a log transformation of the viable count, generally takes the form of a sigmoid curve and is characterised by a lag phase, an exponential or logarithmic phase, and a stationary phase. The stationary phase is usually followed by a decline in cell numbers. Models that have been used to describe this characteristic shape include the Baranyi model (5), Gompertz function (24), and 'modified' Gompertz and logistic function (58). Sigmoid curves can also be approximated using three straight lines (15). Once biologically meaningful parameters, such as lag time and growth rate, have been derived for each growth curve, the parameters are regressed on the environmental variables used in the experiment, often using quadratic functions.

Alternative approaches used to describe the effects of temperature and water activity include the Arrhenius equation (20), non-linear Arrhenius models such as the Schoolfield model (10) and 'square-root' approach (40). A full description of these and other approaches is provided in McMeekin *et al.* (39). Screening or 'siting' experiments can provide useful information for experimental design (combinations of factors to be tested) and sampling (e.g., times when samples should be taken). Screening studies can be carried out using techniques/methods that allow many factors to be studied with relatively little effort, such as gradient plates (38) and automated turbidimetric measurements (36), rather than plate counts.

Models Predicting Growth of *E. coli*

The early models predicting growth of *E. coli* did not include *E. coli* O157:H7 or other VTEC as test strains. One early approach (44) applied Arrhenius plots to six data sets from a previous study (28), which generated data sets with *E. coli*. This early example and others (27) described the effect of temperature on growth rate. A summary of the models that have been developed for use in food microbiology and able to predict the combined effects of multiple factors on growth of *E. coli* is provided in Table 1.

Models predicting the behaviour or growth of *E. coli* have also been developed in other areas. A model predicting growth of *E. coli* in response to temperature, growth medium and oxygen tension (26) has been used to determine the most effective milking frequency to control intramammary infection in milking cows. Adaptive responses to marine environmental stresses have also been modelled, enabling testing of the effects of different stresses and the role of pre-adaptation in populations of *E. coli* (33). The stress imposed when bacterial populations undergo unfavourable increases in temperature has also been modelled in *E. coli* (9). This model describes the relationship between lag time and stress duration, and has potential application in food microbiology, in optimizing non-lethal heat treatments.

TABLE 1.
EXAMPLES OF MULTIPLE-FACTOR MODELS PREDICTING GROWTH OF *E. COLI*

Factors studied	No. of strain(s)/type	Reference
Temperature, pH, sodium chloride, aerobic/anaerobic	3, *E. coli* O157:H7	11
Temperature, pH, sodium chloride, aerobic/anaerobic, nitrite	3, *E. coli* O157:H7	12
pH, lactic acid concentration	1, non pathogenic	40
Temperature, pH, sodium chloride*	4, *E. coli* O157:H7	47
Temperature, pH, sodium chloride, carbon dioxide*	4, *E. coli* O157:H7	47

* denotes models that have been validated by comparison with growth in foods

Considerations in the Development of Growth Models

As can be seen in Table 1, models intended for use in the food industry often include more than one strain, to ensure that the model generated is 'leading edge'. Strains may differ in their growth responses under different conditions, and the inclusion of more than one strain can provide data for a 'worst-case' scenario. If a_w is being studied, the nature of the particular humectant used can be important, particularly at lower a_w levels (14). The physiological status of the inoculum can influence the lag time. It is generally acknowledged that when innocula are prepared under conditions similar to those under which they will be used, lag times may be shorter than when the inoculum is prepared under different conditions from those under which they will be used, although there are exceptions to this rule. It is difficult to accommodate such variations within models, and the overall relationship between growth environment and lag time is more complex than the corresponding relationship between growth environment and maximum specific growth rate (46). A model describing the effect of temperature history on lag time has been developed for *L. monocytogenes* (2) but models describing the effects of more than one factor have yet to be developed.

Validation of models can often be overlooked. In those cases where validation studies have been undertaken (e.g., 51), models generally predict growth rates which are equal to or faster than the actual rates observed in foods. Models rarely predict growth rates slower than actual rates observed in foods, but such underestimations have been observed in some circumstances (55). Model validation is important since it provides independent evidence that supports the use of models in the prediction of the behaviour of bacteria in particular foods. It is equally important to identify those foods or conditions in which the model does not accurately predict behaviour. Indices of the performance of predictive models used in predictive microbiology have been proposed (4).

The limits of use of predictive models are also often unrecognised. Since most models are empirical, they should only be used to predict behaviour under conditions similar to those under which data were collected. In multiple-factor models, such as those in Table 1, the interpolation region is not obvious, and can be much smaller than might be expected. Since growth models are based only on conditions where growth has been observed, predictions near the boundary of growth/no growth will often overestimate growth rate. To ensure that predictions are made only within the 'domain of validity', it is advisable to define the interpolation region with a 'minimum convex polyhedron' (6). Some studies extend models, and such extensions must be supported by sufficient data in the appropriate experimental regions (34).

MODELLING THERMAL INACTIVATION

Modelling Approaches

Thermal inactivation models are probably the oldest examples of predictive models used in the food industry. Ball (3) first recognised that formulae could be derived to calculate the effects of heat processes, taking into account heating rates of cans and destruction rates for bacterial spores. This study introduced the commonly referred to "F" (time in minutes at 250°F to kill all vegetative cells/spores of a particular organism) and "*z*" (heat required in °F to decrease the population by 1 log) concepts. The '*D* value' concept (time taken for 10-fold decrease in numbers of survivors) was derived much later (49). This traditional log-linear death model assumes equal heat resistance within a population of one type of bacteria, and has been used successfully in the food industry for many years.

Despite the extensive use of this model, there are numerous examples of data that do not fit this model, with data sets commonly displaying 'tails' and/or 'shoulders'. Consequently, modelling thermal inactivation has been the subject of much debate (41) and a number of alternative approaches have recently been proposed (18,31,42) to cope with non log-linear kinetic data. "Tailing" may be an artifact of the experimental technique used even in the cases where steps were taken to eliminate it. The above novel approaches to deal with non-log linear data assume some differences in relative heat resistance among subpopulations.

In addition, other workers, rather than trying to construct multiphasic models have elected to use existing sigmoid functions, commonly used to fit growth curves. For example, the modified Gompertz (7), logistic (29) and Baranyi model (57) have all been used to describe thermal inactivation of *L.*

monocytogenes. As with growth models, so-called 'secondary' or 'second-level' models may then be fitted to the parameter(s) describing inactivation rate at each set of conditions tested. The use of a single-equation model for all data (31) has some advantages over the more common two-stage approach.

Models Predicting Inactivation of *E. coli*

In most studies investigating thermal inactivation of *E. coli*, the data are described using first-order reaction kinetics, *i.e.* *D* values. In a review of published data describing thermal inactivation of *E. coli* O157:H7 (49), the authors concluded that a heat treatment of 70°C for 2 min (or equivalent) is sufficient to deliver a 6-$\log_{10}$ reduction in cell numbers. Despite the numerous studies describing thermal inactivation of *E. coli* O157:H7, there are relatively few studies that have derived models able to predict inactivation across a range of conditions.

An early example of a predictive model describing the effects of a_w and temperature on inactivation of *E. coli* is provided in (45). More recently, models have been generated which describe the combined effects of temperature, pH and sodium chloride concentration/a_w on thermal inactivation of *E. coli*. In one of these studies (8), two functions [log linear and log logistic] were combined when fitting data to allow the inclusion of those data sets which showed significant shoulders and/or tails. This study noted that, pH and NaCl concentration, in addition to temperature, significantly affected the heat resistance of the organism, with faster inactivation rates at extreme pH values and lower concentrations of sodium chloride. Validation of the model with inactivation of *E. coli* O157:H7 in studies carried out in foods showed good agreement, with the model providing fail-safe predictions in 93% of occasions.

The kinetics of destruction of *E. coli* O157:H7 in Lebanon sausage, in relation to such factors as fermentation pH, heating temperature and time, has been modelled (22). The process-specific regression equation was developed to identify processes that would provide >5-$\log_{10}$ reduction in *E. coli* O157:H7 numbers. A model predicting the combined effects of temperature, pH, sodium chloride and sodium pyrophosphate on *D* values of a 4-strain cocktail of *E. coli* O157:H7 is also available (29).

Considerations for Inactivation Studies/models

A number of other factors have been shown to impact on the thermal inactivation of microorganisms. These include growth conditions/phase of the inoculum, strain(s) used, presence of fat in the heating medium, heating method employed and recovery conditions (e.g., aerobic *vs.* anaerobic). The factors shown to affect thermal inactivation of *E. coli* O157:H7 are summarised in Stringer *et al.* (49).

When considering application of models predicting combined effects of multiple factors in the inactivation, rather than growth of bacteria, the 'domain of validity' usually covers the whole range of conditions tested, since data can be generated in all areas of the experimental matrix. Therefore, definition of the interpolation region for thermal inactivation models is not as complex as in the case of growth models. There are also differences between thermal inactivation models and growth models in relation to validation. Predictions from growth models will often 'fail-safe' since growth in the experimental systems used to generate the models is often faster than observed under the same conditions in foods. However, the reverse frequently applies in thermal inactivation models, where the rates of thermal inactivation in foods is often slower than the rates observed in model systems (8).

MODELLING NON-THERMAL INACTIVATION

Modelling Approaches

The approaches/models used to describe thermal inactivation are also commonly used to describe non-thermal inactivation/survival. An additional model, designed to describe non-thermal inactivation has been proposed by Whiting (55).

Non-thermal Inactivation Models

The few published models that are able to predict non-thermal inactivation of *E. coli* (37) describe survival curves that had a range of different shapes, in response to factors such as pH, nitrite, sodium chloride, temperature and pH. The curves were 'fitted' using two different functions, the log-logistic function (19) and the non-linear model of Whiting (56). Even though the kinetics of survival observed were variable, it was possible to derive a mathematical model which could provide relatively accurate predictions of treatment times necessary to achieve defined reductions in numbers. Variable kinetics were also observed in relation to the survival of *E. coli* O157:H7 in eggplant salad at various pH values, temperatures and concentrations of oregano oil (48). Many of these curves exhibited long shoulders, indicating little change in bacterial numbers, before the onset of decline.

Non-thermal inactivation kinetics can also be similar to thermal inactivation kinetics, as shown by Entani *et al.* (23). A model predicting the bactericidal effect of processed vinegar (acetic acid), temperature and sodium chloride on *E. coli* O157:H7 was derived from such data (53).

CONCLUSIONS

There are a number of models available to predict the growth of *E. coli*, in response to single factors, such as temperature, or in response to multiple factors. These models have been applied in a number of different ways, by the food industry and by those working from a regulatory perspective. Armitage (1) described how processors and regulators worked together to develop justifiable criteria for the management of refrigeration during the production of hot and warm-boned meat, and the post-slaughter handling of carcasses and offals. In this application, a temperature integration function model (25) was used to integrate temperature history with the growth characteristics of *E. coli*.

Thermal inactivation kinetics show that *E. coli* O157:H7 strains do not demonstrate any unusual or unexpected behaviour compared to other *E. coli* and many other members of the enterobacteriaceae. Thermal inactivation is an important control measure in the prevention of foodborne illness, and models predicting inactivation of *E. coli* are extremely useful for establishing safe and optimized process limits. In addition to having direct application in setting process limits, thermal inactivation models can be used, in combination with other models, in risk assessment procedures, to predict the impact of cooking regimes on the likelihood of survival of vegetative pathogens (11). Indeed, predictive models are likely to be essential tools in the 'exposure assessment' and 'risk characterisation' elements of risk assessment.

Survival modelling is, arguably, the most important aspect of predictive modelling for pathogens such as *E. coli* O157:H7, particularly in relation to the survival of these organisms in products that do not rely on conventional intervention strategies like pasteurisation. In foods such as fermented dairy and meat products, where contamination of raw materials is possible, processes have to provide some degree (e.g. 5 $\log_{10}$) of inactivation, and survival models have clear application in the development of processes to meet such targets. In some products, it may only be possible to realise reductions on this scale by using combinations of factors, such as reduced pH, low a_w and elevated temperatures — conditions that are likely to 'exhaust' (32) contaminating microorganisms and accelerate the rate of their inactivation. There are, at present, few examples of such models, but this area of predictive microbiology is likely to see rapid development in the near future in response to the consumer pressures for 'natural', fresh foods.

Predictive models, in one form or another, are now used extensively throughout the food industry. Nevertheless, care should be taken during the development of models, so that they are 'fit for purpose' and not overly complex. Care must also be taken in the application of models and the user should be aware of their limitations, which are not always obvious. For example, although validation by comparison with independent observations in foods is generally considered to be important, not all models are tested in this way. Software packages allowing easy access to suites of predictive models are available for those interested in predicting the fate of microorganisms in foods, and more information can be found at http://www.arserrc.gov/mfs/pathogen.htm and at http://www.foodmicromodel.com

In the longer term, a move towards more mechanistic models is desirable, so that models capable of extrapolation may be developed. A greater understanding of the physiology of pathogens, and their response to particular conditions is required before such progress can be achieved. Some progress has been made, in that there are rare examples where fundamental understanding at the cellular level is coupled with some mathematical representation, explaining the response of microorganisms to particular stresses. One such example has been developed in relation to the osmoprotective effects of compatible solutes in *E. coli* (17). Better understanding of control mechanisms at the genetic level will also aid development of more mechanistic models. However, until these advances are made, empirical models will continue to be developed and used to predict the microbiological safety of foods.

REFERENCES

1. ARMITAGE, N. 1997. Use of predictive microbiology in meat hygiene regulatory activity. International J. Food Microbiol. *36*, 103–109.
2. AUGUSTIN, J.C., ROSSO, L. and CARLIER, V. 2000. A model describing the effect of temperature history on lag time for *Listeria monocytogenes*. International J. Food Microbiol. *57*, 169–181.
3. BALL, C.O. 1923. Thermal process time for canned food. Bulletin of the National. Research Council (U.S.) No. *37* 7 (pt. 1), 1–76.
4. BARANYI, J., PIN, C. and ROSS, T. 1999 Validating and comparing predictive models. International J. Food Microbiol. *48*, 159–166.
5. BARANYI, J., ROBERTS, T.A. and MCCLURE, P.J. 1993. A non-autonomous differential equation to model bacterial growth. International J. Food Microbiol. *10*, 43–59.
6. BARANYI, J., ROSS, T., MCMEEKIN, T.A. and ROBERTS, T.A. 1996. Effects of parameterization on the performance of empirical models used in 'predictive microbiology'. Food Microbiology, *13*, 83–91.
7. BHADURI, S. *et al.* 1991 Thermal destruction of *L. monocytogenes* in liver sausage slurry. Food Microbiol. *8*, 75–78.
8. BLACKBURN, C.W., CURTIS, L.M., HUMPHESON, L., BILLON, C.B. and MCCLURE, P.J. 1997 Development of thermal inactivation models for *Salmonella enteritidis* and *Escherichia coli* O157:H7 with temperature, pH and NaCl as controlling factors. International J. Food Microbiol. *38*, 31–44.

9. BREAND, S., FARDEL, G., FLANDROIS, J.P., ROSSO, L. and TOMASSONE, R. 1997. A model describing the relationship between lag time and mild temperature increase duration. International J. Food Microbiol. *38*, 157–167.
10. BROUGHALL, J.M., ASLOW, P. and KILSBY, D.C. 1983. Hazard analysis applied to microbial growth in foods: development of mathematical models describing the effect of water activity. J. Applied Bacteriology *55*, 101–110.
11. BROWN, M.H., DAVIES, K.W., BILLON, C.M.P., ADAIR, C. and MCCLURE, P.J. 1998. Quantitative microbiological risk assessment: Principles applied to determining the comparative risk of salmonellosis from chicken products. J. Food Protection *61*, 1446–1453.
12. BUCHANAN, R.L., BAGI, L.K., GOINS, R.V. and PHILLIPS, J.G. 1993. Response surface models for the growth-kinetics of *Escherichia coli* O157:H7. Food Microbiol. *10*, 303–315.
13. BUCHANAN, R.L. and BAGI, L.K. 1997a. Expansion of response-surface models for the growth of *Escherichia coli* O157:H7 to include sodium nitrite as a variable. International J. Food Microbiol. *23*, 317–332.
14. BUCHANAN, R.L. and BAGI, L.K. 1997b. Effect of water activity and humectant identity on the growth kinetics of *Escherichia coli* O157:H7. Food Microbiol. *14*, 413–423.
15. BUCHANAN, R.L., WHITING, R.C. and DAMERT, W.C. 1997. When is simple good enough: A comparison of the Gompertz, Baranyi, and three-phase linear models for fitting bacterial growth curves. Food Microbiol. *14*, 313–326.
16. BUCHANAN, R.L. and EDELSON, S.G. 1996. Culturing enterohaemorrhagic *E. coli* in the presence and absence of glucose as a simple means of evaluating the acid tolerance of stationary phase cells. Applied and Environmental Microbiol. *62*, 4009–4013.
17. CAYLEY, S., LEWIS, B.A. and RECORD, M.T. 1992. Origins of the osmoprotective properties of betaine and proline in *Escherichia coli* K-12. J. Bacteriology *174*, 1586–1595.
18. CERF, O., DAVEY, K.R. and SADOUDI, A.K. 1996. Thermal inactivation of bacteria — a new predictive model for the combined effect of three environmental factors: temperature, pH and water activity. Food Research International *29*, 219–226.
19. COLE, M.B., DAVIES, K.W., MUNRO, G., HOLYOAK, C.D. and KILSBY, D.C. 1993. A vitalistic approach to describe the thermal inactivation of *Listeria monocytogenes*. J. Industrial Microbiol. *12*, 232–239.
20. DAVEY, K.R. 1989. A predictive model for combined temperature and water activity on microbial growth during growth phase. J. Applied Bacteriology, *70*, 253–257.
21. DUFFY, G., WHITING, R.C. and SHERIDAN, J.J. 1999 The effect of a competitive microflora, pH and temperature on the growth kinetics of *Escherichia coli* O157:H7. Food Microbiol. *16*, 299–307.
22. ELLAJOSYULA, K.R., DOORES, S., MILLS, E.W, WILSON, R.A., ANANTHESWARAN, R.C. and KNABEL, S.J. 1998. Destruction of *Escherichia coli* O157:H7 and *Salmonella typhimurium* in Lebanon bologna by interaction of fermentation pH, heating temperature and time. J. Food Protection *61*, 152–157.

23. ENTANI, E., ASAI, M., TSUJIHATA, S., TSUKAMOTO, Y. and OHTA, M. 1998. Antibacterial action of vinegar against food-borne pathogenic bacteria including *Escherichia coli* O157:H7. J. Food Protection *61*, 959–959.
24. GIBSON, A.M., BRATCHELL, N. and ROBERTS, T.A. 1988. Predicting microbial growth: growth responses of salmonellae in a laboratory medium as affected by pH, sodium chloride and storage temperature. International J. Food Microbiol., *6*, 155–178.
25. GILL, C.O., PHILLIPS, D.M., LOEFFEN, M.P.F. and BISHOP, C. 1985. A computer program for evaluating the hygienic efficiency of meat processing procedures from product temperature history data. 34th International Congress of Meat Science and Technology, Brisbane, pp. 531–532.
26. GOLDBERG, J.J., BRAMLEY, A.J., CHEN, H.D. and PANKEY, J.W. 1995. Use of predictive modelling to evaluate the manipulation of milking frequency, temperature, and oxygen tension on growth of *Escherichia coli* in an artificial intramammary environment. American J. Veterinary Research *56*, 440–444.
27. HEITZER, A., KOHLER, H.P.E., REICHERT, P. and HAMER, G. 1991. Utility of phenomenological models for describing temperature-dependence of bacterial growth. Applied and Environmental Microbiol. *57*, 2656–2665.
28. JOHNSON, F.H., EYRING, H. and STOVER, B.J. 1974 *The theory of rate processes in biology and medicine*. John Wiley & Sons, New York, pp. 199.
29. JUNEJA, V.K., MARMER, B.S. and EBLEN, B.S. 1999. Predictive model for the combined effect of temperature, pH, sodium chloride and sodium pyrophosphate on the heat resistance of *Escherichia coli* O157:H7. J. Food Protection *19*, 147–160.
30. KAMAU, D.N., DOORES, S. and PRUITT, K.M. 1990. Enhanced thermal destruction of *Listeria monocytogenes* and *Staphylococcus aureus* by the lactoperoxidase system. Applied and Environmental Microbiol. *56*, 2711–2716.
31. KILSBY, D.C., DAVIES, K.W., MCCLURE, P.J., ADAIR, C.A. and ANDERSON, W.A. 2000. Bacterial thermal death kinetics based on probability distributions: the heat destruction of *Clostridium botulinum* and *Salmonella* Bedford. J. Food Protection *63*, 1179–1203.
32. LEISTNER, L. 2000. Basic aspects of food preservation by hurdle technology. International J. Food Microbiol. *55*, 181–186.
33. MARTIN, Y., TROUSSELLIER, M. and BONNEFONT, J.L. 1998. Adaptive responses of *E. coli* to marine environmental stresses, a modelling approach based on viability and dormancy concepts. Oceanologica Acta *21*, 951–964.
34. MASANA, N.O. and BARANYI, J. 2000. Adding new factors to predictive models: the effect on risk of extrapolation. Food Microbiol. *17*, 367–374.
35. MCCLURE, P.J., COLE, M.B., DAVIES, K.W. and ANDERSON, W.A. 1993. The use of automated turbidimetric data for the construction of kinetic models. J. Industrial Microbiol. *12*, 277–285.
36. MCCLURE, P.J., COLE, M.B. and DAVIES, K.W. 1994. An example of the stages in the development of a predictive mathematical model for microbial growth: the effects of NaCl, pH and temperature on the growth of *Aeromonas hydrophila*. International J. Food Microbiol. *23*, 359–375.

37. MCCLURE, P.J. and HALL S. Survival of *Escherichia coli* in foods. J. Applied Microbiol. *88*, 61S–70S.
38. MCCLURE, P.J. and ROBERTS, T.A. 1987. The effect of incubation time and temperature on growth of *Escherichia coli* on gradient plates containing sodium chloride and sodium nitrite. J. Applied Bacteriology *63*, 401–407.
39. MCMEEKIN, T.A., OLLEY, J.N., ROSS, T. and RATKOWSKY, D.A. 1993. *Predictive microbiology: theory and application.* Research Studies Press, John Wiley & Sons, Brisbane.
40. MCMEEKIN, T.A. *et al.* 1987. Model for the combined effect of temperature and water activity on the growth of *Staphylococcus xylosus*. J. Applied Bacteriology, *62*, 543–550.
41. MOATS, W.A., DABBAH, R. and EDWARDS, V.M. 1971 Interpretation of nonlogarthmic survivor curves of heated bacteria. J. Food Science *36*, 523–526.
42. PELEG, M. and COLE, M.B. 1998 Reinterpretation of microbial survival curves. Critical Reviews in Food Science *38*, 353–380.
43. PRESSER, K.A., RATKOWSKY, D.A. and ROSS, T. 1997. Modelling the growth rate of *Escherichia coli* as a function of pH and lactic acid concentration. Applied and Environmental Microbiol. *63*, 2355–2360.
44. RATKOWSKY, D.A., OLLEY, J.N., MCMEEKIN, T.A. and BALL, A. 1982. Relationship between temperature and growth rate of bacterial cultures. J. Bacteriology, *149*, 1–5.
45. REICHART, O. 1994 Modeling the destruction of *Escherichia coli* on the base of reaction-kinetics. International J. Food Microbiol. *23*, 449–465.
46. ROBINSON, T.P., OCIO, M.J., KALOTI, A. and MACKEY, B.M. 1998. The effect of the growth environment on the lag phase of *Listeria monocytogenes*. International J. Food Microbiol. *44*, 83–92.
47. SAAD, S.M.I. and FRANCO, B.D.G.M. 1999. Influence of raw meat natural background flora on growth of *Escherichia coli* O157:H7 in ground beef. Revista de Microbiologia *30*, 272–277.
48. SKANDAMIS, P.N. and NYCHAS, G.J.E. 2000. Development and evaluation of a model predicting the survival of *Escherichia coli* O157:H7 NCTC 12900 in homemade eggplant salad at various temperatures, pHs, and oregano essential oil concentrations. Applied and Environmental Microbiol. *66*, 1646–1653.
49. STRINGER, S.C., GEORGE, S.M. and PECK, M.W. 2000. Thermal inactivation of *Escherichia coli* O157:H7. J. Applied Microbiol. Symposium Supplement *88*, 79S–89S.
50. STUMBO, C.R., MURPHY, J.R. and COCHRAN, J. 1950. Nature of thermal death time curves for P.A. 3679 and *Clostridium botulinum*. Food Technology *4*, 321–326.
51. SUTHERLAND, J.P., BAYLISS, A.J. and BRAXTON, D.S. 1995. Predictive modelling of growth of *Escherichia coli* O157:H7 — the effects of temperature, pH and sodium chloride. International J. Food Microbiol. *25*, 29–49.
52. SUTHERLAND, J.P., BAYLISS, A.J., BRAXTON, D.S. and BEAUMONT, A.L. 1997. Predictive modelling of *Escherichia coli* O157:H7: inclusion of carbon dioxide as a fourth factor. International J. Food Microbiol. *37*, 113–120.

53. TSUJIHATA, S., ENTANI, E., ASAI, M., TSUKAMOTO, Y. and OHTA, M. 1998 Mathematical modeling to predict the bactericidal effect of processed vinegar on *Escherichia coli* O157:H7. International J. Food Microbiol. *43*, 135–138.

54. WALLENTIN LINDBERG, C. and BORCH, E. 1994. Predicting the aerobic growth of *Y. enterocolitica* O:3 at different pH-values, temperatures and L-lactate concentrations using conductance measurements. International J. Food Microbiol. *22*, 141–153.

55. WALLS, I. and SCOTT, V.N. 1996. Validation of predictive models describing the growth of *Escherichia coli* O157:H7 in raw ground beef. J. Food Protection *59*, 1331–1335.

56. WHITING, R.C. 1993. Modeling bacterial survival in unfavourable environments. J. Industrial Microbiol., *3-5*, 240–246.

57. XIONG, R., XIE, G., EDMONSON, A.S., LINTON, R.H. and SHEARD, M.A. 1999 Comparison of the Baranyi model with the modified Gompertz equation for modelling thermal inactivation of *Listeria monocytogenes*. Food Microbiol. *16*, 269–279.

58. ZWIETERING, M.H., JONGENBURGER, I., ROMBOUTS, F.M. and VAN'T RIET, K. 1990. Modelling of the bacterial growth curve. Applied and Environmental Microbiol. *56*, 1875–1881.

CHAPTER 18

RISK ASSESSMENT IN THE CONTROL OF VTEC

R.C. WHITING[1]
M.E. COLEMAN[2]
C.A. NARROD[3]
M. POWELL[4]
and
T. ROBERTS[3]

[1]U.S. Food and Drug Administration
Center for Food Safety and Applied Nutrition
[2]Food Safety and Inspection Service
[3]Economic Research Service
[4]Office of Risk Assessment and Cost Benefit Analysis
Washington, DC 20204

and

W.D. SCHLOSSER

Food Safety and Inspection Service
College Station, Texas

INTRODUCTION

Risk analysis includes three components: risk assessment; risk management; and risk communication (29). Risk analysis typically poses three questions: what can go wrong?; how likely will that happen?; and what are the consequences? (22).

Qualitative evaluations by the industry and regulatory agencies of foodborne pathogens, including VTEC, have a long history. The risk assessor collects and organizes available data and information in a quantitative manner for use by the risk managers. The policy-making needs of risk managers or decision-makers shape the risk assessment process. The risk managers articulate the scope of the assessment, pose questions, assemble multi-disciplinary risk assessment teams and specify potential mitigations that might be considered. The risk communication process involves dialogue between the technical risk assessors, policy-making risk managers and various stakeholders in industry, academia, government, and the public. Risk communication involves transfer of technical information as well as the risk assessment's limitations and attendant uncertainties.

RISK ANALYSIS

Many of the lessons learned by chemical, environmental and engineering risk assessors (30) guide microbial risk assessors as this discipline evolves. The four component paradigm for risk analysis and definitions of concepts and terminologies were adapted by several groups concerned with microbiological risks, including WHO (32), Int. Life Sciences Inst, (18), Int. Committee Microbiol. Specifications Food (17), Nat. Advisory Committee Microbial Criteria Foods (27) and Codex Committee on Food Hygiene (10). The components of a risk assessment are hazard identification, exposure assessment, hazard characterization or dose-response assessment, and risk characterization.

Hazard Identification

Hazard identification is an evaluation of the epidemiological, medical, and food microbiological data for pathogens associated with foodborne illnesses. Based upon the hazard identification, risk managers may authorize an extensive, formal quantitative risk assessment.

Exposure Assessment

The exposure assessment determines the frequency and numbers of a pathogen that the consumer is exposed to. This is typically a quantitative modeling of the prevalence and density in raw ingredients, an estimate of the growth, survival or inactivation during the stages of processing, storage, transportation and final preparation, and the frequency and amount of the food consumed. Models for microbial changes in foods are described in Chapter 17.

Hazard Characterization or Dose-Response Assessment

Hazard characterization or dose-response assessment is a crucial link between exposure estimates for servings of food and adverse public health outcomes. Data are available from human clinical trials and from epidemiological studies that support the concept that both the frequency of illness and the severity of illness increase as dose of pathogens increases (2,14). Great uncertainty exists in selecting an appropriate surrogate dose-response model for *E. coli* O157:H7, because data are lacking from human clinical trials. The two published risk assessments (6,24) have used empirical models fit to data from human clinical trials with *Shigella dysenteriae* and *flexneri* as a surrogate, even though these pathogens present different mechanisms of pathogenesis. In addition to addressing uncertainty about selection of most appropriate surrogate dose-response models for *E. coli* O157:H7, tremendous uncertainties must also be acknowledged: parameter uncertainty, the accounting for sampling and measurement error in clinical trials with small numbers of volunteers per dose group; and model uncertainty, the accounting for our ignorance about model form, the true shape and position of the dose-response relationship. Empirical models may fit the data in the observed region for data sets from human and animal clinical trials, but differ markedly in predictions at the low-dose region. Another source of uncertainty includes ignorance about how to extrapolate the models for healthy adult volunteers to more sensitive sub-populations, including children and the immunocompromised, and the models from infant animal studies to humans. Finally, the effects of the food matrix and other environmental influences are uncertain, thus for example, high-fat foods appear to protect pathogens and increase infectivity (8).

Three empirical dose-response models having sigmoid relationships between the number of microbial pathogens consumed and the probability of illness (P) have been presented:

Beta-Poisson (15)
$P = 1 - (1 + \text{dose x/beta})^{-\text{alpha}}$

Weibull-Gamma (16)
$P = 1 - [(1 + \text{dose}^{x}/\text{beta})]^{-\text{alpha}}$

Extreme value - Gompertz (8)
$P = 1 - \exp(-\exp(\text{alpha}_i + \text{beta} * \text{dose}^{x}))$

A new strategy for developing dose-response models involves bounding the overall relationship with selected data sets (13). Based on available data for shigellosis and infant diarrhea, this approach reasons that *E. coli* O157:H7 in humans is unlikely to be more pathogenic than the invasive pathogen *Shigella dysenteriae* or less pathogenic than the EPEC strains that typically cause infant diarrhea. *Shigella dysenteriae* data forms as the upper bound, and EPEC data forms the lower bound, and a uniform distribution is given to possibilities between the upper and lower bounds for any given dose. For example, at a dose of 100 *E. coli* O157:H7 cells surviving in a cooked hamburger serving, the probability of illness might be any value between upper and lower bound estimates of 0.35 and 6×10^{-7}, respectively. Another approach also relies on the same upper and lower bounds, but links prior data derived from dose-reconstruction of outbreaks (1) and FoodNet estimates (26) to predict a most likely curve, rather than assume a uniform distribution between the upper and lower bounds. Bounding approaches appear to more fully depict our uncertainty associated with dose-response modeling for *E. coli* O157:H7 than previous assumptions of shigellosis as a surrogate dose-response model (6,24).

Risk Characterization

Risk characterization combines the information from the exposure assessment and hazard characterization to describe the set of scenarios of interest to the risk managers. The output is an estimate of risk with attendant uncertainty expressed in various formats, such as the probability of illness in a specified population after consuming a single serving, the total number of annual cases or deaths for a country, or a rate of illness (cases per million persons per year).

A lesson learned from chemical risk assessment (30) is that calculations based on deterministic analyses in which upper bound estimates of parameter values for many interacting factors or "worst case" scenarios yield implausible, though conservative, final values. Instead, a quantitative risk assessment is generally a stochastic analysis, that is, each input parameter is described by a distribution rather than just a mean, median, or an upper bound estimate. Because these factors are described by distributions, the model's predictions will also be distributions, not single point estimates. Monte Carlo Simulation is a technique to perform these calculations in an iterative manner (3,7). The model repeatedly calculates the risk with random draws for each parameter distribution. The model may be run for thousands of iterations, generating thousands of predicted risk estimates and a distribution of these estimates which can be described. Monte Carlo simulation provides a much more complete description of the risk, with its attendant uncertainty, than either the deterministic point estimate or worst case approaches.

One can presume that nearly all variables in a model have an underlying or characteristic distribution. Two components contribute to this distribution, variability and uncertainty. Variations are differences that actually exist in a population, such as differences in rates of growth/decline of different strains of a pathogen, or differences in food production/processing/preparation practices between industrial plants or cooks at home. Uncertainty refers to a lack of knowledge. This includes analytical and statistical properties of microbial methodology, such as sampling and measurement error, and representativeness of the strain tested to unknown strains. It also refers to the influence of unknown factors. Typically, wider distributions would be given to parameter values from poorer quality or limited studies. The shape of a distribution also reflects these variations and uncertainties, and it may be skewed with a long tail or may have more of the distribution in the tails compared to around the median. Separating variation and uncertainty in the risk assessment may provide important information to the decision-makers. Some variation is inherent in the population and can't be reduced by process interventions. However, operators can redesign equipment and processes, regulations can try to constrain plant operations or "educate" home cooks on how to produce safer food and shift the distributions to some extent. Model uncertainty can be reduced by additional research and data collection so that distributions used in the model more closely reflect the real world.

EXAMPLES OF *E. COLI* O157:H7 RISK ASSESSMENTS

This section provides a brief discussion of two recently published risk assessments of *E. coli* O157:H7: Marks *et al.* (24) and Cassin *et al.* (6).

Marks *et al.* (24)

The primary purpose of Marks *et al.* (24) was to examine issues in conducting microbial risk assessments and identify data gaps in evaluating the consumption by adults of hamburgers prepared away from home. The exposure assessment began with the estimation from monitoring data of the occurrence and density of *E. coli* O157:H7 in raw ground beef in which 7 of 9,821 twenty-five gram samples were confirmed positive. Using the method of moments, Marks *et al.* (24) characterized the uncertainty about this apparent prevalence as a beta distribution (Beta(6.9986, 9812)).

The density of *E. coli* O157:H7, given its presence, was estimated from 8 enumerated samples of implicated frozen ground beef recovered from outbreaks. The natural log of the *E. coli* O157:H7 concentration was modeled as a t-distribution with 7 degrees of freedom. The distribution of hamburger serving sizes was estimated using USDA Continuing Survey of Food Intake by Individuals (CSFII) data recorded for approximately 12,000 individuals. Goodness-of-fit criteria (minimal residual error) were employed to select from among a class of empirical distributions, and the distribution of serving sizes was modeled as Burr type XII function with a mean value of 74.67 g. The density of *E. coli* O157:H7 in servings, given its presence, was estimated taking into account the probability of occurrence of the pathogen in 25 g portions. Servings were treated as a collection of independent 25 g portions, with possible fractional portions.

Marks *et al.* (24) modeled pathogen growth as a function of temperature utilizing a subset of *E. coli* O157:H7 broth culture data from the ARS Pathogen Modeling Program. The study by Juneja *et al.* (21) was used to model the decline of *E. coli* O157:H7 in hamburger with cooking. The log reduction due to cooking was predicted by: $\log_{10}(N_f/N_i) = 13.93\text{-}0.12{*}T$ (deg. F). Variability in both models was modeled by using a binomial distribution.

To model the dose-response relationship for *E. coli* O157:H7, Marks *et al.* (24) considered four clinical trial data sets using *Shigella* as a surrogate pathogen. Levine *et al.* (23) employed *S. dysenteriae* (M 131), Dupont *et al.*

(11,12) used *S. flexneri* (2457T), and June (20) took the infant diarrheal strains of *E. coli*. The clinical endpoint in each of the studies was human illness. The data were fitted to Beta-Poisson curves with no bacterial threshold or a four-pathogen threshold. Marks and colleagues maintained the distinction between variability between species and experiments and uncertainty from model parameters. A distribution of the probability of illness for healthy adults for an ingested dose was generated by Monte Carlo simulation.

Marks *et al.* (24) calculated the probabilities of illness per 114 g servings that were consumed by healthy adults under nine scenarios of three cooking temperatures by three processes (Table 1).

TABLE 1.
PROBABILITY OF ILLNESS (TIMES 10^6) FOR HEALTHY ADULTS CONSUMING A HAMBURGER

	Internal cooking temperature		
	Rare (130°F)	**Medium (145°F)**	**Well-done (155°F)**
		Baseline model	
Median	51	0.88	0.07
5% probability intervals	0.40 - 2016	0.006 - 261	0.0004 - 18
		Model with temperature abuse	
Median	137	2.4	0.18
5% probability intervals	0.98 - 2288	0.03 - 760	0.0012 - 33
		Model with interventions	
Median	13	0.29	0.02
5% probability intervals	0.09 - 780	0.002 - 71	0.0001 - 3

Marks *et al.* (24).

FSIS (13) continued their efforts to quantitatively model human illnesses caused by *E. coli* O157:H7 in beef in the United States and to compare these results with national estimates of illnesses derived from observed epidemiological data. The risk assessment would identify the occurrence and levels of the pathogen at points along the farm-to-table continuum and the contribution of these points to the number of human illnesses. It will quantify the effects of various mitigation strategies in reducing the number of human illnesses, and identify future research needs.

Cassin *et al.* (6)

Cassin *et al.* (6) modeled exposure from farm through consumption of hamburgers prepared at home. The risk assessment endpoints included the per meal probability of illness (for adults and children separately), hemolytic uremic syndrome (HUS) and mortality. The exposure assessment evaluated a hypothetical slaughter plant that produced 5 kg vacuum packs of beef trimmings for a retail outlet that ground the trimmings on-site.

The occurrence of *E. coli* O157:H7 in live cattle was estimated by pooling the results from six studies conducted in the U.S. and Canada between 1986-94 in which 231 of 21,318 cattle tested were shedding detectable levels of the pathogen in their feces. Using the sample estimate of the mean and variance to estimate the distribution parameters, the uncertainty about this apparent prevalence of approximately 1% was characterized as a beta distribution, with no correction made for test sensitivity to estimate the actual prevalence. Cassin *et al.* (6) modeled the concentration of *E. coli* O157:H7 in the feces of colonized animals using the data reported by Zhao *et al.* (33) to develop an empirical cumulative distribution (with a minimum of 0.1 colony forming unit per gram (cfu/g) and a maximum of 10^5 cfu/g).

The ratio of carcasses contaminated with *E. coli* O157:H7 to clean carcasses was assumed to be two to three times the ratio of shedding animals to non-shedders [Uniform (2, 3)]. The concentration of *E. coli* O157:H7 on carcass surfaces was assumed to be proportional to the fecal concentration. Based on the results of the FSIS 1992-93 *E. coli* (Biotype 1) baseline survey for steer and heifer carcasses and Australian data on the concentration of *E. coli* (Biotype 1) in cattle feces, Cassin and colleagues represented the log dilution factor for *E. coli* O157:H7 by a normal distribution. The aggregate effect of carcass decontamination treatments (e.g., hand trimming, spray washing, steam vacuuming, etc.) was assumed to be a 1 to 2.5 log reduction in counts [Uniform (1, 2.5)]. The growth/decline of *E. coli* O157:H7 on carcasses during chilling was modeled as a triangular distribution with a minimum of −2, a mode of 0, and a maximum of 5 generations.

Cassin *et al.* (6) represented the variability in the mass of a piece of trimmings by a truncated normal distribution with a mean of 300 g, a standard deviation of 100 g, a min of 50 g, and a max of 1 kg. In simulating the formation of 5 kg packages of trim, they assumed that each piece of beef was trimmed from a different carcass. The distribution for the number of pieces of trim per package (N_{TRM}) has an expected value of 16.2, and the prevalence of contaminated trimmings (P_{CTRM}) was estimated as:

$$P_{CTRM} = \frac{F_{CC} \times P_F}{(1 - P_F + F_{CC}) \times P_F}$$

Where:
F_{CC} = Carcass cross-contamination factor
P_F = Prevalence of *E. coli* O157:H7 in cattle feces.

The resultant distribution for P_{CTRM} had an expected value of 2.6 percent. The probability of at least one piece of trim in a 5 kg package being contaminated with *E. coli* O157:H7 was therefore:

$$Pr(N_{CTRM}>0) = 1-e^{-P_F \times F_{CC} \times N_{TRM}}$$

The number of contaminated pieces of trim per package (N_{CTRM}) followed a binomial distribution:

$$N_{CTRM} \sim \text{Binomial}\ (N_{TRM}, P_{CTRM})$$

The concentration of *E. coli* O157:H7 on contaminated trim (C_{CTRM}) is estimated as:

$$C_{CTRM} = C_F - F_{dil} - R_{DEC} + \log_{10} 2^{G_{PRC}}$$

Where:
C_F = fecal concentration
F_{dil} = dilution factor
R_{TRM} = decline due to carcass treatment
G_{PRC} = growth during processing

Cassin *et al.* (6) assumed that the packages of trimmings were ground by the retailer and represented the variable mass of retail packages of fresh ground beef (M_{FGB}) as a distribution with a range of 300-1000 g and a most likely value of 500 g. The probability that a package of fresh ground beef contained one or more *E. coli* O157:H7 (P_{FGB}) was modeled as follows:

$$P_{FGB} = P_{PCT} * (1 - e^{-M_{FGB} * 10 C_{PCT}})$$

The resultant distribution had an expected value of 0.02%. The concentration of *E. coli* O157:H7 in positive packages of fresh ground beef (C_{FGB}) was modeled by a Poisson process, assuming that the retail packages were sampled from the 5 kg packages of trim.

$$C_{FGB} \sim \text{Poisson}\ (M_{FGB} * 10^{C_{PCT}})$$

The resultant distribution has an expected value of –2.8 logs per g.

Assuming storage temperature ranged from 4 to 15°C (with a mode of 10°C), storage time ranged from 4 to 96 h (with a mode of 48 h), pH values between 5.1 and 6.1, and water activity values between 0.99 and 1.00, Cassin and colleagues used the Food MicroModel (Leatherhead Food Research Assoc., UK), to estimate that the growth of *E. coli* O157:H7 in ground beef during storage averaged 0.6 $\log_{10}$ cfu/g, with a 90% confidence interval of 0 to 2.1 logs. They also based their model of thermal inactivation of *E. coli* O157:H7 on Juneja *et al.* (21). A 1994 survey of Texas consumers (25) was used to estimate the proportion of the population who consumed hamburgers at various degrees of "doneness" and a mean internal temperature was associated with each cooking preference based on Jackson *et al.* (19): rare (3.0%, 54.4°C), medium rare (16.1%, 58.6°C), medium (17.9%, 62.7°C), medium well (23.4%, 65.6°C), and well (39.6%, 68.3°C). Cassin and colleagues assumed that the amount of hamburger consumed per adult meal was log normally distributed with a mean of 83 g and a standard deviation of 48 g. For children, the mean and standard deviation were assumed to be 42 and 27 g, respectively.

Based on the Beta-Poisson model for infection, Cassin *et al.* (6) developed a modified dose-response model, called the Beta-Binomial model, which represented the probability of illness from exposure to 1 organism as a Beta distribution with parameters equal to those of a Beta-Poisson dose-response curve. Like Marks *et al.* (24), Cassin and colleagues estimated the dose-response model parameters based on data from human feeding studies conducted with *S. dysenteriae* and *S. flexneri*. The regression analysis treated variability between studies as a proxy for dose-response model parameter uncertainty, resulting in a random coefficients model that differed from a model determined by data pooling. The resultant dose-response model generated a 90 percent confidence interval for the ID_{50} of approximately 1.5 to 5 $\log_{10}$ cfu.

The overall process risk model, including the exposure and dose-response submodels, generated a per meal probability of illness with a median on the order of 10^{-12}. Cassin *et al.* (6) assumed that the probability of illness at a given dose does not vary within the population but that the most vulnerable population (defined as children under the age of 5 and the elderly) has an increased propensity for severe outcomes such as HUS.

INTEGRATE THE RISK MODEL WITH ECONOMIC INFORMATION

The cost of foodborne illness ranges from 6.5 to 34.9 billion dollars annually of which *E. coli* O157:H7 contributes from 0.3 to 2.2 billion annually (5). Medical costs and productivity losses determine these costs. To the extent that interventions reduce illness, the reduction in human illness can be considered a net benefit associated with a specific intervention. The question is — how do we tie these economic costs and benefits to a risk assessment?

The previous section discussed the probability and frequency of exposure at various levels to *E. coli* O157:H7 in ground beef. The risk manager needs to evaluate this information along with the cost of various interventions from farm to fork, and what the magnitude of risk reduction would be associated with these different interventions for different size operations. In the U.S., it was mandated by Presidential Executive Order 12286 that all federal regulations having a significant impact on society (i.e., over $100 million) are required to be supported by a cost-benefit analysis.

To understand how the output of the probabilistic risk assessment model can be combined with economic analysis, we illustrated some of this risk cost trade-off in the following example (31,28). We started with a baseline case and then looked at the impact of technology adoption on pathogen control. Improvements in technology anywhere along the process might reduce the risk to consumers.

The risk assessment needed to be run with a baseline case, i.e., no improved technologies were present, to produce the distribution function, F_0, in Fig. 1. Further model runs were performed adding one or more pathogen reduction technologies anywhere along the food continuum from farm to fork. These are illustrated by the dotted distribution, F_1. This second distribution typically was shifted to the left of the first, reflecting pathogen reduction. This shift reflects the degree to which pathogens were reduced by including the new technology or preventive action.

From a risk assessment standpoint, we are not simply interested in the expected value of hamburger contamination, but rather the frequency with which hamburgers posing some level of risk occurred. Focus for the risk assessment was on the right-hand tail of the distribution, rather than the mean value. To evaluate the effectiveness of technology adoption strategies, we selected a risk tolerance threshold. By doing so, we anticipated that the change in expected pathogen frequency above the threshold compared to the baseline level represented the effectiveness of the adoption strategy. This was expressed as:

$$\Delta P \text{ (hamburger contamination above threshold)} = (F_1 \text{ (Threshold)} - F_0 \text{ (Threshold)})$$

The difference $F_1 - F_0$ represents the change in the probability that a combo bin, for example, was above the risk threshold. The area above the threshold area F_0 is the shaded area plus the striped area. The area above the threshold for distribution F_1 is the shaded area. Note that the improved technology resulted in a reduction in the area above the threshold by the striped area.

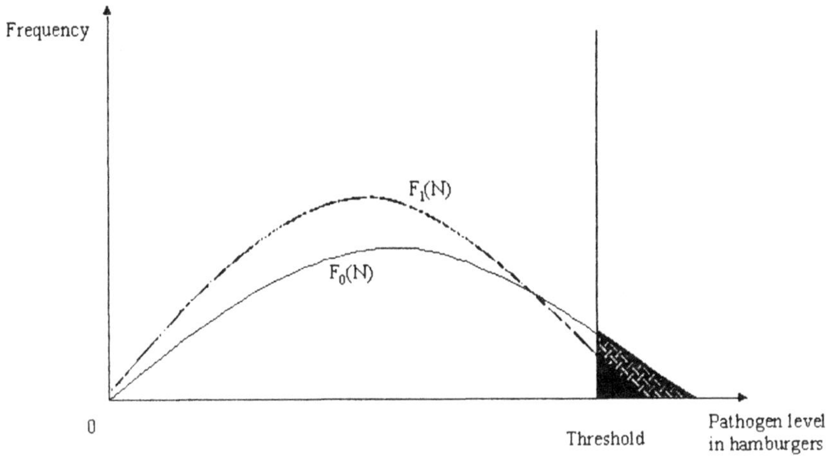

FIG. 1. DISTRIBUTION OF *E. COLI* LEVELS IN RAW HAMBURGERS

Next, the economic costs of the control strategies were combined with the predictive risk assessment model results. Figure 2 shows how four hypothetical strategies (A, B, C, and D) might be compared. The x-axis represents the marginal cost of adding one of the new strategies compared to the baseline. The y-axis represents the percentage reduction in contamination over the baseline. Strategy D can be excluded from any decision set because strategy B dominated D in the sense that B was both more effective and less costly. Choices of adoption strategies can be limited to non-dominated Strategies A, B, and C. Each firm would have a unique optimal strategy curve at a particular point in time. Which strategy a firm chose depended on the risk preferences of their customers (Does the firm get a premium price for a safer product?) and on their comparative advantage in particular risk-reduction strategies (Fig. 2).

The capability to utilize various technologies along the farm to table process resulted in different adoption costs for similar technologies. For example a plant with a stable workforce may have realized greater benefits from worker training because it had a lower likelihood of losing training value due to worker attrition than a plant with high worker turnover. Larger plants might be more likely to profit from purchasing new expensive technologies that have significant economies of scale.

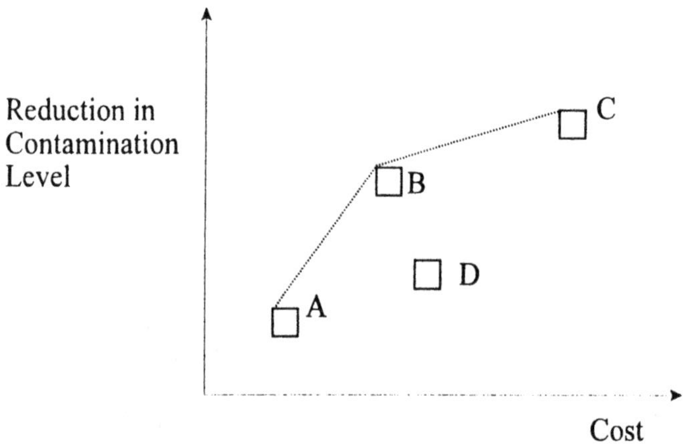

FIG. 2. COMPARISON OF ADOPTION STRATEGIES

Both small and large plants may choose not to adopt some technologies, despite their effectiveness in pathogen reduction, because there were not the proper market incentives for adoption or sufficient investment in research to develop new technologies.

RELATIONSHIP OF RISK ANALYSIS TO HACCP

The creation of a Hazard Analysis, Critical Control Point (HACCP) system involves, in part, the identification of critical processes and setting critical limits for these processes. The specification of the critical limits is typically done for each critical step without quantitative consideration of the rest of the food process. When specifying a HACCP plan for fresh juice, for example, decisions must be made about the necessity for a pasteurization step and, if it is necessary, what extent of lethality would be required. To do this, a process risk assessment is needed (4). If the estimate of the likelihood of illness from unpasteurized juice is judged to be unacceptable, the managers can request an insertion of a pasteurizing step into the process and re-estimation of the likelihood of illness.

Pasteurization needed could be specified to be a 5 $\log_{10}$ inactivation. Other mitigations that could be tested by the risk assessment are the use of less frequently/heavily contaminated ingredients, adding a washing-sanitizing step, and controlling the pH at a lower level to achieve an equivalent reduction. Through an iterative process between risk assessment and risk management, the specific process is decided upon and the critical control points specified.

Several terms have been proposed relating to controlling the risks from foodborne illnesses. First there must be a target level of safety designated as an acceptable level of risk or appropriate level of protection (9). This would typically be defined as a probability of illness in a specified individual from consuming a single serving. Setting this level is a societal value judgement, not a scientific question. It involves the level of risk individuals are willing to accept in foods they expect and enjoy. It also involves technical feasibility, costs, and retention of the expected characteristics of the food. Once the acceptable level of risk is set, the dose-response relationship estimates what frequency of exposure and numbers of pathogens constitute that level of risk. This is termed the food safety objective and sets the microbiological target for the food processor to meet or exceed. The microbiologist, process engineer and food technologist can then design the food process by testing various combinations of process steps and within-step parameters and selecting the combination that most efficiently yields the highest quality. For each critical step the objectives of that step, i.e., 5 $\log_{10}$ inactivation, is specified. This specification is termed the performance criterion. There may be several techniques to achieve a performance criteria, although conventional thermal inactivation or new technologies such as pulsed electrical fields or high pressure can achieve a 5 $\log_{10}$ inactivation. In terms of thermal pasteurization, there are many combinations of time and temperature which will yield a 5 $\log_{10}$ inactivation. The specific time-temperature combination to be used is the process criteria and becomes part of the HACCP plan.

The risk assessment, therefore, provides the underlying scientific support for the HACCP plan by the quantitative linking of all the processing steps to the incidence of illness as a consequence of consuming that food. Risk assessments provide a new frame of reference for evaluating food safety and will require effective communication to have this new paradigm accepted by the public, policymakers, industry and international trading partners.

REFERENCES

1. BELL, B.P. *et al.* 1994. A multistate outbreak of *Escherichia coli* O157:H7-associated diarrhea and hemolytic uremic syndrome from hamburgers: The Washington experience. J. American Medical Association *272*, 1349–1353.

2. BIEBER, D. *et al.* 1998. Type IV pili, transient bacterial aggregates, and virulence of enteropathogenic *Escherichia coli*. Science *280*, 2114-2118.
3. BURMASTER, D. E. and ANDERSON, P. D. 1994. Principles of good practice for the use of Monte Carlo techniques in human health and ecological risk assessments. Risk Analysis *14*, 477-481.
4. BUCHANAN, R.L. and WHITING, R.C. 1998. Risk assessment: A means for linking HACCP plans and public health. J. Food Protection *61*, 1531–1534.
5. BUZBY, J.C. and ROBERTS, T. 1996. ERS updates U.S. foodborne disease costs for seven pathogens. Food Review *19*, 20–25.
6. CASSIN, M.H., LAMMERDING, A.M., TODD, E.C.D., ROSS, W. and MCCOLL, R.S. 1998a. Quantitative risk assessment for *Escherichia coli* O157:H7 in ground beef hamburgers. Int. J. Food Microbiol. *41*, 21–44.
7. CASSIN, M.H., PAOLI, G.M. and LAMMERDING, A.M. 1998b. Simulation modeling for microbial risk assessment. J. Food Protection *61*, 1560–1566.
8. COLEMAN, M.E. and MARKS, H.M. 1998. Topics in dose-response modeling. J. Food Protection *61*, 1550–1559.
9. CODEX. 1999a. Proposed draft principles and guidelines for the conduct of microbiological risk management. Document CX/FH 99/8. Codex Alimentarius Commission, FAO/WHO, Rome.
10. CODEX. 1999b. Principles and guidelines for the conduct of microbiological risk assessments. Codex Alimentarius Commission, FAO/WHO, Rome.
11. DUPONT, H.L., HORNICK, R.B., DAWKINS, A.T., SNYDER, M.J. and FORMAL, S.B. 1969. The response of man to virulent *Shigella flexneri* 2a. J. Infectious Diseases *119*, 296-299.
12. DUPONT, H.L., HORNICK, R.B., SNYDER, M.J., LIBONATI, J.P., FORMAL, S.B. and GANGAROSA, E.J. 1972. Immunity in Shigellosis II. Protection induced by oral live vaccine or primary infection. J. Infectious Diseases *125*, 12-16.
13. FOOD SAFETY AND INSPECTION SERVICE. 1998. Preliminary pathways and data for a risk assessment of *E. coli* O157:H7 in beef. Epidemiology and Risk Assessment Division, Office of Public Health and Science, Washington, D.C. (www.fsis.usda.gov/OPHS/ecolrisk/home.htm).
14. GLYNN, J.R. and BRADLEY, D.J. 1992. The relationship between infecting dose and severity of disease in reported outbreaks of *Salmonella* infections. Epidemiol. Infec. *109*, 371-388.
15. HAAS, C.N. 1983. Estimation of risk due to low doses of microorganisms: A comparison of alternative methodologies. American J. Epidemiology *118*, 573-582.
16. HOLCOMB, D.L., SMITH, M.A., WARE, G.O., HUNG, Y.-C., BRACKETT, R.E. and DOYLE, M.P. 1999. Comparison of six dose-response models for use with food-borne pathogens. Risk Analysis *19*, 1091–1100.
17. INT. COMMISSION MICROBIAL SAFETY FOODS. 1998. Potential application of risk assessment techniques to microbiological issues related to international trade in food and food products. J. Food Protection *61*, 1075–1086.
18. INT. LIFE SCI. INST. 1996. A conceptual framework to assess the risks of human disease following exposure to pathogens. Risk Anal. *16*, 841–848.

19. JACKSON, T.C., HARDIN, M.D. and ACUFF, G.R. 1996. Heat resistance of *E. coli* O157:H7 in nutrient medium and in ground beef patties as influenced by storage and holding temperatures. J. Food Protection. *59*, 230–237.
20. JUNE, R.C., FERGUSON, W.W. and WORFEL, M.T. 1953. Experiments in feeding adult volunteers with *Escherichia coli* 55, B5, a coliform organism associated with infant diarrhea. Am. J. Hyg. *57*, 222–236.
21. JUNEJA, V.K., SNYDER, O.P., WILLIAMS, A.C. and MARMER, B.S. 1997. Thermal destruction of *Escherichia coli* O157:H7 in hamburger. J. Food Protection. *10*, 1163–1166.
22. KAPLAN, S. 1997. The words of risk analysis. Risk Analysis *17*, 407-417.
23. LEVINE, M.M. *et al.* 1973. Pathogenesis of *Shigella dysenteriae* 1 (Shiga) Dysentery. J. Infectious Disease *127*, 261-270.
24. MARKS, H.M., COLEMAN, M.E., LIN, C.-T.J. and ROBERTS, T. 1998. Topics in microbial risk assessment: Dynamic flow tree modeling. Risk Analysis *18*, 309–328.
25. MCINTOSH, W.A., CHRISTENSEN, L.B. and ACUFF, G.R. 1994. Perceptions of risks of eating undercooked meat and willingness to change cooking practices. Appetite 22, 83–96.
26. MEAD, P.S. *et al.* 1999. Food-related illness and death in the United States. Emerging Infectious Diseases *5*, 1–33.
27. NATIONAL ADVISORY COMMITTEE FOR THE MICROBIAL CRITERIA FOR FOODS. 1998. Principles of Risk Assessment for illness caused by foodborne biological agents. J. Food Protection *61*, 1071-1074.
28. NARROD, C.S., MALCOLM, S.A., OLLINGER, M. and ROBERTS, T. 2000. Pathogen reduction options in slaughterhouses and methods for evaluating their economic effectiveness. American Agricultural Economics Association Annual Meeting, Aug. 8-11, 1999. Nashville, TN. (http://agecon.lib.umn.edu/aaea99/sp99na04.pdf)
29. NATIONAL RESEARCH COUNCIL. 1983. Risk Assessment in the Federal Government: Managing the Process. Committee on Institutional Means for Assessment of Risks to Public Health, National Academy Press, Washington, DC.
30. PAUSTENBAUCH, D.J. 1995. The practice of health risk assessment in the United States (1975-1995): How the US and other countries can benefit from that experience. Human Ecological Risk Assessment *1*, 29–79.
31. ROBERTS, T., MALCOLM, S.A. and NARROD, C.A. 1999. Probabilistic risk assessment and slaughterhouse practices: Modelling contamination process control in beef destined for hamburger. PSA '99: International Topical meeting on Probabilistic Safety Assessment. American Nuclear Society, LaGrange Park, IL. pp. 809–815.
32. WORLD HEALTH ORG. 1995. Application of risk analysis to food standards issues. Report of the Joint FAO/WHO Expert Consultation. WHO, Geneva. 39 pp.
33. ZHAO, T., DOYLE, M.P., SHERE, J. and GARBER. L. 1995. Prevalence of enterohaemorrhagic *Escherichia coli* O157:H7 in a survey of dairy herds. Appl. Environ. Microbiol. *61*, 1290–1293.

CHAPTER 19

FARM MANAGEMENT PRACTICES: A SWEDISH CASE STUDY

ELISABETH BORCH

SIK Swedish Institute for Food and Biotechnology
Ideon, SE 223 70 Lund, Sweden

EVA NERBRINK

Cerealia R&D
Malmö, Sweden

IVAR VÅGSHOLM

National Veterinary Institute
Uppsala, Sweden

and

MATS TÖRNQUIST

Swedish Animal Health Service
Kävlinge, Sweden

INTRODUCTION

Escherichia coli O157 is an important food-borne human pathogen. This specific serotype, as well as other serotypes of *E. coli,* may be attributed to the group of verotoxigenic *E. coli* (VTEC). Ruminants (bovines, sheep, and goats) are considered to act as a reservoir of this bacterium (1). The long list of outbreaks occurring since the mid 1980s, reveals the transmission routes such as person-to-person, intake of food, contact with farm environment or some other environmental transmission such as swimming in contaminated water (3,11).

There are several strategies for controlling zoonoses: (a) treating the clinical cases, (b) preventing transmission between humans, (c) preventing transmission through food, drinking water, direct contact with animals or by environmental exposure, (d) preventing transmission between wildlife and domestic animals, and (e) eradicating the reservoirs of zoonotic pathogens. In order to succeed in any of the above approaches, the critical points of transmission need to be identified. Facts about the occurrence and spread of *E. coli* O157 in farm animals in the environment, and to humans, have recently been extensively reviewed (17). Farm management practices including farming intensity, housing and grouping of animals, feeding, manure/slurry management and hygienic standards influence the spread of *E. coli* O157. The possibilities of controlling *E. coli* O157 in animal production are not self evident, but some possibilities of limiting the shedding from cattle are reported, in the areas of calf management, and in feed. The grouping of calves before weaning appears to be a significant risk factor (6). The prevalence of calves shedding the bacterium increased after weaning from 1.4% to 4.8%. Some dietary factors have shown interesting results, although such results are not always consistent (15). Consumption of cotton seed and clover hay are reported to reduce the risk of shedding, while grain and corn silage are reported to increase shedding (5,10). We need a better understanding of the mechanisms behind shifts in the intestinal flora, and how these may be triggered. Gyles (7) suggested that any management changes that promote the instability of the intestinal flora, and reduce the production of volatile fatty acids in the rumen, will increase the shedding of *E. coli* O157.

The present chapter deals with experiences in Sweden, and the policy taken by competent authorities to reduce the risk of the bacterium spreading to humans. Results from studies made in an infected herd are presented. A strategy was tested for eliminating *E. coli* O157 from the farm environment. Furthermore, the movement of calves from dairy to beef herds is given special attention, since this is a considerable risk factor for the continuous re-infection of calf receiving herds.

PREVALENCE IN ANIMALS

During the period 1988-1994, the reported number of human enterohaemorrhagic *E. coli* (EHEC) cases in Sweden was less than 10 per year. During the autumn of 1995, the number of cases suddenly increased due to one or several outbreaks. In 1996-1998, between 70 and 140 cases of EHEC O157 were clinically reported annually (equivalent to 0.8-1.6/100,000 inhabitants; 2).

Over the last 4 years, a Swedish monitoring programme has indicated the presence of several animal reservoirs of VTEC (personal communication Helen Wahlström, Erik Eriksson; National Veterinary Institute; Uppsala, Sweden). During the period 1996-1999, surveys of cattle faecal specimens sampled at the slaughterhouse, showed a 0.7% prevalence of verotoxigenic *E. coli* O157 (56 positive/7,556 samples) in cattle. Monitoring of dairy herds, estimated an individual (calves and heifers) prevalence of 1.2%, and a dairy herd prevalence of around 10%. In sheep (ewes and lambs), a prevalence of 0.8% was found (5 positive/600 samples). In a survey of wild animals (Canada goose, sea gull, elk, deer, wild boar and rabbit; n = 784) only one sample (wild boar) was found to harbour verotoxigenic *E. coli* O157. In a survey of pigs, two out of 2,446 faecal samples from pigs were found to be positive for verotoxigenic *E. coli* O157 (unpublished data E. Nerbrink; Swedish Meats R&D). The positive pigs were from one outdoor herd, and from a farm where pigs and cattle rearing was combined.

SWEDISH POLICY

In Sweden a common control policy for *E. coli* O157 agreed between the public health, food and veterinary authorities, was implemented during the autumn of 1997. The objective of the policy was to harmonise the approach to control between the authorities, and in particular the means of dealing with herds in which *E. coli* O157 was found. The finding of *E. coli* O157 in animals has been notifiable since October 1996. The policy comprised non-compulsory restrictions for farms with *E. coli* O157 and for the slaughter of animals from such farms. In practice, farms were identified as positive by two means, review of results of the national monitoring programme or by laboratory detection of the bacterium in cattle faeces. In accordance with the recommendations, the cattle from infected farms were slaughtered at a reduced speed and all carcasses were analysed for *E. coli* O157. The farmer was charged with the cost of the analyses. The system was very expensive for the affected herds, and the government decided to grant financial support in order to cover expenses.

In 1999 the national policy was revised. The new policy has two aims at farm level; (1) to prevent the transmission in all herds of *E. coli* capable of causing EHEC in humans, and (2) to implement specific measures in herds connected to cases of EHEC infection in humans.

The 1999 policy gives general recommendations for ruminant herds. There should be specific bio-security procedures on all farms having regular visitors, in line with model programmes for farm visits produced with the Swedish National Farmers' Association. In addition to the required annual cleaning, barns should be disinfected during the summer. The sharing of farm equipment must be accompanied by suitable decontamination between use on each farm. Finally, the prevalence of *E. coli* O157, as well as other relevant serotypes, should be monitored, with results reported anonymously.

Specific measures are recommended for ruminant herds with a clear link to a human case of infection with enterohaemorrhagic *E. coli* (EHEC).

(a) Ruminants should not be sold to other herds without being isolated for 3 weeks, and tested twice for the presence of the EHEC, with negative results within a 14-day interval.
(b) The clean up of EHEC is verified by the absence of the organism during whole herd repeat testing.

(c) The herd should not accept visits from school or pre-school children, before the clean up has been verified.
(d) There should be specific routines for slaughtering animals from the herd, for example these animals should be slaughtered last, within the abattoir schedule. Their carcasses should be subject to microbiological analyses. Any carcass found to be positive for EHEC should be destroyed or designated for the production of heat-treated meat products.

In more general terms, animals sent for slaughter must not be dirty. Manure should be disposed of in such a way that there is no risk of contaminating vegetables, berries or drinking water. Milk or milk products should be pasteurised before being consumed. The prevalence of VTEC in food should be monitored.

MEASURES TAKEN IN A SPECIALISED BEEF HERD — RESULTS FROM A FIELD STUDY

In Sweden, there is a long-standing tradition of controlling *Salmonella* in farm animals, by identifying and slaughtering *Salmonella*-carriers, combined with the cleaning and disinfecting of barns. In the study presented below, a similar strategy was applied in a specialised beef herd, which had been found to be positive for *E. coli* O157.

DESCRIPTION OF THE HERD

The selected farm was one which had supplied cattle found at slaughter with faeces containing *E. coli* O157. This was a specialised herd which bought in groups of calves for fattening. The calves were bought at the age of two months and a live weight of 70-80 kg, and collected from dairy farms in the south-west of Sweden. Groups of about 40-60 calves were brought in four times a year. During the studied period, from August 1997 to July 1998, 283 young calves were brought into the farm. Upon arrival, calves were housed in a reception barn (I) with pens for six to eight calves, kept on straw bedding and fed milk substitute and hay. The milk was replaced with concentrate after two to four weeks. After eight to ten weeks, calves were moved 1 km to a second barn (II) where they were kept on a slatted floor in section IIA, and subsequently in section IIB. The calves were concentrate *ad-lib* and a small ration of hay or straw, for four to six months, at which time they were sent to slaughter, at a live weight of about 260 kg.

SAMPLING AND MICROBIOLOGICAL ANALYSES

Sampling was carried out on calves, environment in the barn, feed (concentrate, hay, barley and oats), feed mixer and liquid manure.

Samples were diluted 1:10-1:20 with buffered peptone water (BPW; Oxoid CM 509), and analysed using pre-enrichment, immunomagnetic separation and cultivation on SMAC and CT-SMAC agar (Handbook Dynabeads® anti-*E. coli* O157 710.03/04; Dynal AS, Oslo, Norway), and agglutination (Oxoid DR 622 latex). Biochemical confirmation was performed with Enterotube II A (Roche). Isolates of *E. coli* O157 were analysed for *vt*1, *vt*2 and *eae*A-genes by PCR and RFLP-typing was performed.

The sensitivity of the analysis for *E. coli* O157 was determined by analysing digestive tracts of calves. One-gram samples were taken from six positions in the gastro-intestinal tract; rumen, small intestine, caecum, colon and the proximal and distal part of the rectum. The sensitivity of the analysis was calculated as a/(a+c) as shown in Fig. 1.

		Digestive tract[1] +	Digestive tract[1] -
Single sample postion[2]	+	a	b
	-	c	d

FIG. 1. THE SENSITIVITY OF THE ANALYSIS FOR *E. COLI* O57 WAS DETERMINED BY TAKING OUT 1-GRAM SAMPLES FROM SIX POSITIONS IN THE DIGESTIVE TRACT OF CALVES. THE SENSITIVITY WAS CALCULATED AS a/(a+c).

[1] +, Occurrence in any of the six sampling positions; -, no occurrence.

[2] +, Occurrence in one single sample position; -, no occurrence.

E. coli O157 was quantified in samples from the rectum, by surface plating of a dilution series, on SMAC-agar and CT-SMAC and incubated at 37°C for 18-24 h. Agglutination and confirmation was performed as above.

RESULTS

Initial Status of the Herd

In August *E. coli* O157 was found in faecal material sampled from the slatted floor in two of the 14 pens which comprised the finishing section IIB. In

October all calves in section IIB were sampled, and 9 out of 40 (22%) calves were found to harbour *E. coli* O157. The positive calves were evenly distributed among the pens. During the period from August to October the bacterium had, thus, spread throughout the whole finishing section IIB. Repeated sampling of faecal material from the floor in section IIA and in the reception barn I, showed no occurrence of the bacterium during the same time period (August to October). No transmission between sections was detected, even though only a door separated sections IIA and IIB, and staff passed through this door several times per day.

Environmental sampling showed the prevalence of *E. coli* O157 in several locations in section IIB, i.e., a spade, ventilating fans, the weighing-crush, wires under the slatted floor, manure drain, stored liquid manure and poles. Several of the sampling locations were, at the time of sampling, dry and dusty, confirming good survival in uncleaned places. None of the 18 samples taken from the feed (concentrate; hay, barley and wheat-grain) or the feed mixer were positive for *E. coli* O157. All isolates from the faecal material and the environment had *vt2*- and *eaeA*-genes, and belonged to the same subtype, according to RFLP.

Strategy for the Elimination of *E. coli* O157

No new calves were moved into section IIB. Five of the positive calves had reached a normal final weight and were sent to slaughter. The four remaining positive calves were moved to a single pen row, separate from the 31 negative calves. Cleaning was performed using high pressure and subsequent disinfection using Virkon S (Ewos, Sweden). Individual sampling of all 35 calves after cleaning, revealed nine (26%) positive individuals, none of which had previously been identified as positive. These positive calves were distributed in different rows of pens, suggesting a spread of the bacterium via aerosols during cleaning. Samples from the environment revealed the presence of *E. coli* O157 in the pens, the crush and the wires below the slats.

To reduce the number of calves as much as possible, 27 were sent to slaughter, most of which had been positive on one or several occasions, or being penmates of positive calves. Thorough cleaning and disinfection was performed twice, using high pressure and Virkon S. In addition, after the floor had dried, it was powdered with lime. Nine calves were left in the section, when, in December, the calves from section IIA were moved to section IIB. Analysis of calves and pens revealed no samples positive for *E. coli* O157.

All calves in section II A were individually analysed in November, and no *E. coli* O157 was found. Three groups of new calves arrived at section I in September and December, and individually analysed for *E. coli* O157. No positive samples were found.

Status After Actions

In December, the calves in section IIA were moved to section IIB. In February and March, all calves in the three areas were individually analysed. No *E. coli* O157 was recovered. On the last occasion, two samples from the manure tank from section IIB were also negative. In May, samples of faecal material from the pen floors in section IIB still showed no *E. coli* O157. All environmental samples were also negative.

New calves were brought to the farm in January. Individual sampling two to three weeks after arrival, did not detect *E. coli* O157.

New Contamination Introduced

In April, the sampling of 60 newly arrived calves in section I demonstrated that two were positive for *E. coli* O157 one week after arrival. The calves originated from two dairy farms. In June, individual sampling demonstrated eight (13%) positive calves. Isolates (n=5) had the same RFLP-type, and harboured *vt2*- and *eae*A-genes. This RFLP-type was, however, different from the one found previously in section IIB.

All 60 calves were moved to the empty section IIB, and the 8 positive calves and their neighbours were housed in four pens in one separate row of pens. Contact, direct or indirect, was avoided between positive and negative calves. The pens harbouring the positive calves were lime-treated once a week. Six weeks after moving the calves, individual sampling showed that only four calves (all of which had been among the previously positive group of eight) were now positive. Thus, no spread of *E. coli* O157 was observed. Sampling after another four weeks confirmed that no more calves had become positive. RFLP-typing (n=6) of positive calves, determined that the organism involved was the same subtype as isolated from the calves immediately after their arrival.

Results of Individual Sampling of Calves

The occurrence of *E. coli* O157 in faecal samples was analysed for individual calves on three different occasions, starting about three weeks before slaughter (Table 1). Repeated sampling showed that the status of individual calves varied between positive and negative over the sampling period. In several of the calves, *E. coli* O157 was only detected once or twice during the course of study. Thus, analysis on a single occasion was insufficient to determine the *E. coli* O157 status of a calf.

Sampling of the Gastro-intestinal Tract

Important questions remain — how to determine the status of calves in dairy herds before being transferred to beef herds for fattening? How many samples are necessary in order to accurately determine such status? Samples were recovered at slaughter from six different positions within the gastro-intestinal tracts of calves from the infected herd. The highest number of positive samples (42-50%) was from the caecum, colon and rectum. The number of positive intestinal tracts was 71%. In 17 of the 24 gastro-intestinal tracts tested, at least one sample position provided a positive result (Table 2). The frequencies of detection in each position (Fig. 1) were 24% (rumen), 59% (caecum), 65% (colon), 70% (proximal part of the rectum) and 59% (distal part of the rectum) (Table 2). In conclusion, if a decision relies on a single negative sample taken from the rectum, about 30-40% of calves will incorrectly be considered as negative.

The sensitivity of detection is likely to depend upon the concentration of the bacterium in the faeces. In the present study, between 10 and 30 cfu/g faeces was estimated in the rectum samples (n=30). Using enrichment, prior to IMS and plating, 34% of samples were positive. On the other hand, 13% of samples contained directly countable numbers of organisms, i.e., above the threshold of 10 cfu/g by direct surface plating.

TABLE 1.
E. COLI O157 IN FAECAL SAMPLES FROM CALVES
Sampling was performed 24 days or 7 days before slaughter, and on the slaughter line

Calf	Occurrence of *E. coli* O157		
	24 days before slaughter	7 days before slaughter	At slaughter
7150	-[1]	+	+
7597	-[1]	-	+
7073	-[1]	-[1]	+
1011	-[1]	+	-
7062	+[1]	-	-
7179	+[1]	+[1]	-
1012	+	-	-
1015	-	+	-
0035	+[1]	+[1]	+
7040	+	+	+
5184	-	+	+

[1]In the same pen as a positive calf

TABLE 2.
E. COLI O157 AT SINGLE SAMPLE POSITIONS IN THE GASTROINTESTINAL TRACT OF CALVES (n = 24)
The intestinal tract was considered as positive when one or more sample positions were positive.

Calf no.	Occurrence of *E. coli* O157						
	Single sample positions						Intestinal tract
	Rumen	Small intestine	Caecum	Colon	Proximal part of rectum	Distal part of rectum	
1	-	-	-	-	+	-	+
2	-	-	-	-	-	-	-
3	-	-	-	+	+	+	+
4	-	-	-	-	-	-	-
5	+	-	+	+	+	+	+
6	-	--	-	-	-	-	-
7	-	-	-	-	-	-	-
8	-	-	+	+	-	+	+
9	+	+	-	+	+	+	+
10	-	-	+	+	+	-	+
11	-	-	+	+	+	+	+
12	+	-	+	+	+	+	+
13	-	+	+	+	+	+	+
14	-	-	-	-	+	-	+
15	-	-	+	+	+	+	+
16	-	+	-	-	-	-	+
17	+	+	+	+	+	+	+
18	-	-	+	-	-	-	+
19	-	-	-	+	-	-	+
20	-	-	-	-	-	-	-
21	-	-	-	-	-	-	-
22	-	-	-	-	+	+	+
23	-	-	-	-	-	-	-
24	-	-	+	-	-	-	+
No positive	4	4	10	11	12	10	17
% positive	17	17	42	46	50	42	71
Sensitivity[1]	24	24	59	65	70	59	

[1]See Fig. 1

Risk Assessment for the Trade in Live Calves

The risk of transmitting *E. coli* O157 by trading calves should be controlled, as the continuous inflow of calves originating from dairy herds plays an important role in the introduction of this organism into beef herds.

Based on knowledge of the prevalence of individual calves (pv) and the sensitivity (s) of the analytical method used to detect the bacterium, an estimate may be made of the status of a group of calves before being mediated. The probability of detecting false negative individuals is (12; n = number of animals tested):

$$P = (1\text{-}s)^{pv*n} \qquad \text{Eq 1}$$

Three different scenarios are considered in Table 3. Assumptions include a prevalence of 10% in a herd consisting of 100 animals, and that the herd is closed such that no new animals enter. It is also assumed that the sensitivity of the analysis is 60% (Table 2, sampling position = distal part of rectum). In the first scenario, 10 to 20 calves are tested every six months, and these should test negative before being traded. In this case, the probability that positive calves will not be detected is about 3% to 16% (Table 3).

TABLE 3.
THE PROBABILITY OF NOT DETECTING POSITIVE ANIMALS BEFORE MEDIATING CALVES TO CATTLE HERDS IN THREE DIFFERENT SCENARIOS

Scenario	Number animals tested	Probability of not detecting positive individuals (%)
A number of calves are tested every six months	2[1] x 10	16
A number of calves are tested every six months	2 x 20	3
All animals in the group to be mediated are tested[2]	10	40
All animals in the group to be mediated are tested	20	16
All animals in the group to be mediated are tested	50	1
All animals are tested, only negative are mediated	10	36
All animals are tested, only negative are mediated	20	58
All animals are tested, only negative are mediated	50	89

Animal prevalence=0.1; Sensitivity of *E. coli* O157 analyses 0.6.
[1]Tested twice a year; [2]All animals test negative before mediation

In the second scenario, all animals in the group to be traded are tested with negative results before trade. If the group consists of many animals, the probability of detecting all the positive animals is higher than in a small group of calves. If 50 calves are tested, the P-value is 0.01, which gives a 99% certainty that all calves are truly negative. Testing 20 calves gives a P-value of 0.16 (Table 3).

In a third scenario, all animals in a group are tested, but only the ones that test negative are traded. In this case, the probability of not detecting false negative individuals is calculated as (12):

$$P=1-((1-pv)/((1-pv)+pv(1-s)))^{n} \qquad \text{Eq 2}$$

For a group of 10 calves, the P-value is 36%, and for 50 animals 89%, which indicates the probability that at least one false negative calf is traded.

The third scenario is the least satisfactory alternative. The final choice, between testing all animals in a group to be traded, or testing a herd twice a year, depends on the number of calves to be traded (Table 3).

DISCUSSION

E. coli O157 was isolated between August 1997 and December 1997 from faecal material or the environment in one out of three sections at the studied farm. All isolates had *vt2*- and *eaeA*-genes, and belonged to the same subtype, according to RFLP. One possible scenario is that one strain was established in the environment, and subsequently spread to the calves, and among the calves, and via utensils. This is in agreement with Shere *et al.* (16) who reported a high similarity among isolates from the same farm. In positive herds, the shedding is intermittent as a result of an environmental source, and contaminated water is one likely vehicle.

A single cleaning treatment was not sufficient to remove *E. coli* O157 from the environment, and repeated cleaning and disinfection was necessary. This was to be expected since the dirt was firmly deposited in the environment in areas which were never empty or cleaned routinely. As a conclusion, practical cleaning should, thus, be performed on a routine basis in barns and sections housing calves and cattle, in order to prevent establishment of VTEC in the environment. Thorough cleaning and disinfection using Virkon S in combination with lime treatment of the slatted floor was able to remove the bacterium. At the same time, all VTEC-positive calves were sent to slaughter to avoid fresh contamination of the environment. Results indicated that the bacterium was spread by aerosols during cleaning, and the areas should, thus, be emptied before cleaning.

No positive samples (calves, environment) were found in the finishing sections between December and May, after the actions taken to eliminate *E. coli* O157 from the farm. Was this due to the success of the actions or was it only due to the cold season? A tendency towards seasonal shedding has been reported by Hancock *et al.* (8), with rates being highest in June, July and September. On the contrary, Mechie *et al.* (13) reported peaks in the shedding in a dairy herd in May-July and November after cattle had been housed. The movement of animals and animal to animal contact are factors that may be as important as a seasonal variation.

Repeated sampling for a period of three weeks before slaughter of individual calves showed that some animals tested positive for *E. coli* O157 on only one or two occasions during the study. The concentrations of *E. coli* O157 present were low ($\leq$ 30 cfu/g faeces), and the sensitivity of the analysis of rectal samples was only 40%. Calves and cattle are reported to shed *E. coli* O157 intermittently, ranging from 0-10^5 cfu/g faeces (18,16). In cattle herds, the presence of *E. coli* O157 in the intestinal tract is a transient event (9,14). For a naturally infected dairy herd, the period of shedding was four weeks or less (13). Infection with *E. coli* O157 is no protection against re-infection (4). In an environment where animals are continuously reinfected, herds may remain positive for a long time.

In conclusion, it was possible to control a cattle herd with *E. coli* O157 by slaughtering positive animals, and by applying thorough procedures for cleaning and disinfecting the barn environment. Furthermore, it has been demonstrated that significant advances in the control of *E. coli* O157, during trading of calves, can be achieved.

ACKNOWLEDGEMENTS

The Swedish Farmers' Research Programme supported the work performed on the farm with *E. coli* O157. The PCR and RFLP-analysis performed by Dr. S. Löfdahl, Swedish Institute for Infectious Disease Control, and the *E. coli* O157 sampling and analysis performed by Mrs. B.-M. Wahlquist, Swedish Meats R&D are gratefully acknowledged.

REFERENCES

1. ANON. 1997. Verotoxin producing *Escherichia coli* (VTEC)- report of the EU Scientific Veterinary Committee, 17th September. XXIV/B3/ScVC/0013/1997 Final.
2. ANON. 1998. Annual Report 1998. Swedish Institute for Infectious Disease Control.

3. CHAPMAN, P.A. 1999. *Escherichia coli* O157: 14 Years' Experience. *In Escherichia coli* O157 in Farm Animals (C.S. Stewart and H.J. Flint, eds.) pp. 99–119. CABI Publishing, Oxon, UK.
4. CRAY, W.C. and MOON, H.W. 1995. Experimental infection of calves and adult cattle with *E. coli* O157:H7. Appl. Environ. Microbiol. *58*, 1335–1343.
5. DIEZ-GONZALEZ, F., CALLAWAY, T.R., KIZOULIS, M.G. and RUSELL, J.B., 1998. Grain feeding and the dissemination of acid-resistant *Escherichia coli* from cattle. Science *281*, 1666–1668.
6. GARBER, L.P. *et al.* 1995. Risk factors for fecal shedding of *Escherichia coli* O157:H7 in dairy calves. J. Am. Vet. Med Assoc. *207*(1), 46–49.
7. GYLES, C.L. 1999 *E. coli* O157:H7 and other VTEC in animals. The Infectious Disease Review, *1*(2), 127–129.
8. HANCOCK, D.D., BESSER, T.E., KINSELL, M.L., TARR, P.I., RICE, D.H. and PAROS, M.G. 1994. The prevalence of *Escherichia coli* O157: H7 in dairy and beef cattle in Washington State. Epidemiol. Infection *113*, 199–207.
9. HANCOCK, D.D., BESSER, T.E., RICE, D.H., HERIOTT, D.E. and TARR. P.I., 1997. A longitudinal study of *E. coli* O157 in fourteen cattle herds. Epidemiol. Infection *118*, 193–195.
10. HERRIOTT, D.E., HANCOCK, D.D., EBEL, E.D., CARPENTER, L.V., RICE, D.H. and BESSER, T.E. 1998. Association of herd management factors with colonization of dairy cattle by shiga toxin-positive *Escherichia coli* O157. J. Food Prot. *61*(7),802–807.
11. JOHNSON, R.P. *et al.* 1999. Human Infection Associated with Exposure to Farm Animals and Rural Environments. *In Escherichia coli* O157 in Farm Animals (C.S. Stewart and H.J. Flint, eds.) CABI Publishing, Oxon, UK.
12. MACDIARMID, S.C. 1993. Risk analysis and the importation of animals and animal products. OIE Rev. Sci. Tec. *12*(4), 1093–1107.
13. MECHIE, S.C, CHAPMAN, P.A. and SIDDONS, C.A. 1997. A fifteen month study of *Escherichia coli* O157:H7 in a dairy herd. Epidemiol. and Infection *118*, 17–25.
14. RAHN, K. 1997. Persistence of *Escherichia coli* O157:H7 in dairy cattle and dairy farm environment. Epidemiol. Infection *119*, 251–259.
15. RASMUSSEN, M.A., WICKMAN, T.L., CRAY JR, W.C. and CASEY, T.A. 1999. *Escherichia coli* O157:H7 and the rumen environment. *In Escherichia coli* O157 in Farm Animals (C.S. Stewart and H.J. Flint, eds.) CABI Publishing, Oxon., UK.
16. SHERE, J.A., BARTLETT, K.J. and KASPAR, C.W. 1998. Longitudinal study of *Escherichia coli* O157: H7 dissemination on four dairy farms in Wisconsin. Appl. Environ. Microbiol. *64*, 1390–1399.
17. STEWART, C.S. and FLINT, H.J. (eds). 1999. *Escherichia coli* O157 in Farm Animals. CABI Publishing, Oxon., UK.
18. ZHAO, T., DOYLE, M.P., SHERE, J.A. and GARBER, L. 1995. Prevalence of enterohaemorrhagic *E. coli* O157:H7 in a survey of dairy herds. Appl. Environ. Microbiol. *61*, 1290–1293.

CHAPTER 20

CONTROL OF VTEC IN THE MEAT INDUSTRY

HILDE NISSEN

MATFORSK
Norwegian Food Research Institute
Osloveien 1, 1430 Ås, Norway

AND

TRULS NESBAKKEN

Norwegian Meat Research Centre
PO Box 396 Økern
0513 Oslo, Norway

VTEC occurs among the normal microflora of the gastrointestinal tract of animals and can be transferred to carcass meat during dressing. As a result, VTEC has been found on carcasses and in processed meat such as minced or ground beef. VTEC can survive freezing, and frozen products such as beefburgers may represent a hazard to the consumer if inadequately cooked. Beef meat is an important source of VTEC infection in humans while pig and poultry meat probably is not (14). However, a study from the Netherlands concluded that pigs and poultry might be a source of VTEC strains characteristic of those causing illness in man (15).

In addition to direct meat-borne contamination, illness may also arise if food handlers contaminate meat and meat products during processing. There are several points in the food chain from farm to table at which control measures can be taken to prevent the spread of pathogens from mammalian slaughter animals, via meat and meat products to man (Table 1). This table shows that it is possible to reduce or limit the spread if strict hygienic procedures are used during dressing. During the operations following dressing, i.e., chilling, cutting and deboning, further spread of VTEC can occur (4).

ABATTOIR

Cattle slaughter is an open process which presents many opportunities for the contamination of the carcass with VTEC. There are, in fact, no points in the process where such hazards can be completely eliminated. Accordingly, HACCP (Hazard Analysis Critical Control Point) and GMP (Good Manufacturing Practice) must focus on limiting the spread of such undesirable organisms. A proper hazard analysis is the basis for the identification of CCPs (Critical Control Points) on the processing line, the specification of critical limits to be used in the monitoring of the process, the corrective actions to be taken when the process is not under control and, finally, the verification of the effectiveness of the HACCP plan. Action at the CCPs related to slaughtering practices can only achieve partial control, and the opportunities to prevent the risks are limited (33). Some CCPs are in fact control points (CPs) that can be controlled by GMP.

Some additional comments on slaughtering of pigs are included under a separate heading.

TABLE 1.
THE POTENTIAL FOR INTRODUCING CONTROL MEASURES AGAINST *ESCHERICHIA COLI* O157:H7 DURING THE FOOD PRODUCTION CHAIN FROM FARM TO TABLE

Stage in production process	**Potential effect of preventive action**
Herd/Flock	+?
Meat Inspection	-
Slaughter	++
Cutting/deboning	-
Processing	+

(++ = good, + = limited, - = probably little)
Adapted from Nesbakken and Skjerve (23)

TRANSPORTATION TO SLAUGHTER AND LAIRAGE (CP)

The effects of stress on the animals may have important consequences in terms of shedding and spread of VTEC. Stress can predispose latently infected animals to shed high numbers of VTEC as a result of increased peristaltic activity. The resultant contamination of the environment (trucks, equipment, etc.) favours the spread of VTEC among slaughter animals and subsequently through the lairage and on the slaughter line. It is legally permissible to keep animals in the lairage for up to 72 hours, however, ideally, all animals should be slaughtered on the day of arrival, as such a policy reduces the risk of spreading VTEC.

Preventive actions to be taken at this CP: Gentle handling of animals. Cleaning and disinfection of vehicles and lairage.

STUNNING AND KILLING (CP)

To avoid contamination of VTEC, blood to be used in human food should be recovered in a clean environment. Knives must be cleaned and heat disinfected before the animal is bled and the trachea and oesophagus should remain intact during the bleeding process.

Preventive actions to be taken at this CP: Cleaning and disinfection of hands and tools.

SKINNING (CP)

Dirty animals have implications for the skinning process. Adjustments may be made depending on how dirty each lot of animals is judged to be. Such adjustments may include:

- rejection of dirty lots
- washing of the animals
- hide trimming or clipping
- slaughter of dirty animals at the end of the kill period
- reducing the speed of the slaughter line and/or adding extra people at certain stations
- compensation for extra time or yield loss

Removal of hides should be carried out in a manner that avoids contact between the outside of the skin and the carcass. Hands and equipment that touch the outside of the skin should not come in contact with the underlying carcass meat, in order to avoid transferring VTEC.

Preventive action to be taken at this CP: Clean animals. Cleaning and disinfection of hands and tools.

EVISCERATION (CCP)

Particularly during evisceration, VTEC may be transferred to the carcass from the intestines, stomach content, oral cavity and oesophagus. The critical operations are circumanal incision, removal of the intestinal tract and the pluck set. The oesophagus should be sealed so that rumen contents do not leak from the oesophagus at any stage. A technique termed "rodding" may be used to free the oesophagus from the trachea and diaphragm, and to carefully position a rubber ring or plastic clip on the oesophagus close to the diaphragm.

In a UK study, *E. coli* O157 were isolated from 7 (28%) of 25 carcasses of rectal-swab positive cattle and from 2 (8%) of 25 carcasses of rectal swab-negative cattle (8). The authors concluded that VTEC contamination of carcasses during slaughter and processing may form a route for contamination of beef and beef products, and thereby transmit VTEC to man. In Scandinavian countries, it is common practice to seal off the loosened rectum using plastic bags. Any such procedure, which prevents the dissemination of VTEC in faeces to the carcass and subsequently to the cut meat, is of great significance in the hygienic production of beef. Care must also be taken when the intestines are being removed, as there is a risk of puncture allowing faecal material to be spread over the carcass. Well-fed cattle appear less likely to become reservoirs for *E. coli* O157 (29), an observation which should have implications for slaughter practices.

Preventive actions to be taken at this CCP: Sealing of oesophagus. Enclosure of rectum. Disinfection of hands and tools.

SPLITTING THE CARCASS (CP)

Carcasses can be split using automatic splitting machines, which pose a risk if the splitter or saw comes in contact with the rectal incision, leading to the spread of VTEC. After splitting each carcass, the splitting machinery should always be disinfected before re-use.

Preventive actions to be taken at this CP: Cleaning and disinfection of the splitting machines/saws.

MEAT INSPECTION AND GRADING (CP)

During post-mortem meat inspection and grading, palpations and incisions can give rise to cross contamination with VTEC.

Preventive action to be taken at this CP: Disinfection of hands and tools.

DECONTAMINATION OF CARCASSES (CP)

Outbreaks of foodborne illnesses and deaths associated with the consumption of ground beef containing *E. coli* O157:H7 and other VTEC have prompted the meat industry to enhance its efforts to lower the levels of this and other pathogens in meat by more closely monitoring production processes (20). The USDA Food Safety Inspection Service has proposed that all slaughter establishments should apply at least one antimicrobial treatment or other approved intervention procedure to livestock and poultry carcasses. Various treatments have been designed to decontaminate carcasses, including the use of sanitising agents such as hot water sprays (28), organic acid sprays (31) or combinations of these treatments (7). Steam vacuuming has also shown some promise in terms of reducing pathogens on meat (10,18). These treatments and others, e.g., ionizing and non-ionizing irradiations, ultrasonics and high pressure, have been reviewed by Corry *et al.* (9).

In the US, steam pasteurisation of whole carcasses has become more and more common in the last 5 years and today about 60% of all carcasses are steam pasteurised. A study made by Nutsch *et al.* (26) indicated that steam pasteurisation is very effective in a commercial setting as a means of reducing overall bacterial populations on freshly slaughtered beef carcasses. In the EU-funded project "Decontamination of meat and meat products using combinations of sub-atmospheric pressure and organic acids" (ERB FAIR CT96-1027) (24) it was found that steam treatments at 75°C for 10 seconds reduced *E. coli* O157:H7 on beef by 2-3 logs, while the inclusion of organic acids gave larger reductions in microbial numbers. Some degree of cooking of the beef was, however, apparent. Risks still exist, in that some pathogens may survive such decontamination processes, and there is a continuing risk of re-contamination of the meat during subsequent cutting and packaging. When decontaminated and untreated beef were inoculated with *E. coli* O157:H7 and stored under conditions of mild temperature abuse (10°C) the highest rates of bacterial multiplication were noted with vacuum-packed materials (Fig. 1) (24). Even if a post-process contamination is likely to include non-pathogenic, competitive organisms and the storage temperature of raw meat is usually below 10°C, these results give cause for concern. In our view, the use of decontamination methods like steam pasteurisation may be worthwhile for "at risk" materials, e.g., carcasses from VTEC-positive herds. However, in most cases, good hygiene will be sufficient to achieve adequate levels of safety.

CHILLING (CCP)

Prompt and effective chilling of carcasses is necessary to prevent multiplication of VTEC. To achieve this, carcasses must be appropriately spaced to allow adequate circulation of cold air.

Preventive actions at this CCP: Carcasses must be chilled as fast as possible, to the lowest temperature possible, commensurate with normal ageing of the meat.

ADDITIONAL COMMENTS ON THE SLAUGHTERING OF PIGS

In order to reduce the likelihood of carcass contamination of VTEC and other intestinal bacteria during the slaughter of pigs, it is essential to withdraw feed for up to 12 h pre slaughter to empty the stomach of the animal. Scalding of pigs should be carried out using water at 60°C or more. Singeing or flaming effectively reduces VTEC at the carcass surface, but can usually only be done when the skin is dry. Cleaning and disinfection of polishing equipment, including the lashes, preferably by a cleaning-in-place (CIP) system, is particularly important. If these precautions are not properly observed, VTEC could grow overnight on the polishing equipment and contaminate carcasses processed during the next working day.

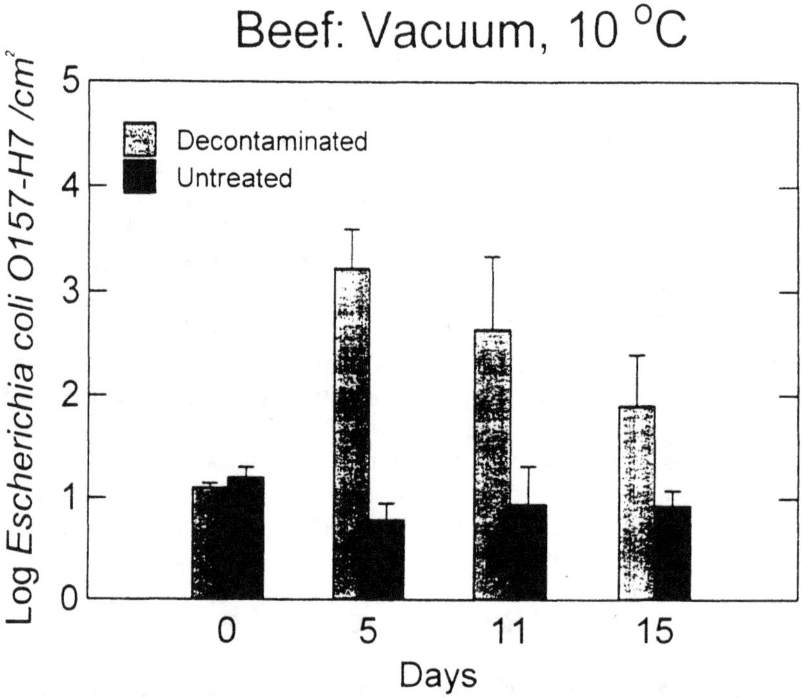

FIG. 1. BACTERIAL GROWTH (log CFU/cm²) ON DECONTAMINATED AND UNTREATED BEEF (INOCULATED WITH A LOW INOCULUM (ABOUT 12/cm²) OF *ESCHERICHIA COLI* O157:H7), STORED UNDER VACUUM AT 10°C. MEANS + SE (n=6)

SURVEILLANCE PROGRAMME FOR VTEC ON CARCASSES

Most slaughter facilities provide ample opportunities for cross-contamination from cattle contaminated with *E. coli* O157 to subsequently processed carcasses (17).

In a Swedish study, samples were collected from 650 beef carcasses. On each carcass, a total area of 1,400 cm^2 was examined. Three (0.5%) carcasses were found positive for *E. coli* O157, with two isolates positive for the *vt*2 and *eae*A genes. Two epidemiological studies in Sweden indicated *E. coli* O157 frequencies of 0.3% and 2.5% in faecal samples from cattle (22).

The Norwegian surveillance programme, monitoring *E. coli* O157 on beef carcasses, started in 1998, collecting samples from 2253 carcasses. *E. coli* O157:H7 was detected on one (0.04%) of the carcasses, of which only one was positive for the *stx*2 gene (19). An epidemiological study in Norway indicated a frequency of 0.3% in faecal samples from cattle (34).

The Danish surveillance programme, monitoring *E. coli* O157 in cattle, collected faecal samples from 2,400 animals in 60 herds. Of these, 3.8% of the animals and 17% of the herds were found to be positive. Fifty cattle from these herds were slaughtered, of which two (4%) of the carcasses were positive for *E. coli* O157 (1).

PRACTICES FOR DEALING WITH CATTLE FROM POSITIVE HERDS

In Norway and Sweden, animals from positive herds are slaughtered at the end of the day, on slaughter lines operating at reduced speed and/or with additional operatives at certain stations on the slaughter lines. In addition, these countries carry out sampling of carcasses from positive herds and heat treatment or destruction of VTEC-positive carcasses.

CUTTING AND DEBONING (CP)

If VTEC gain access to meat, during handling processes or through contamination from the slaughter environment (conveyer belts, cutting tables, tools, etc.), they may spread or even grow in the meat while it is in the cutting/deboning department.

Preventive actions at this CP: Meat should not be held for an extended period in the cutting department, and should be kept at less than 7°C. Cleaning and disinfection procedures should limit or prevent cross- or recontamination of meat.

MEAT PROCESSING

Ground Meat

Efforts to improve overall hygiene during slaughter and meat processing have already resulted in reductions of the total bacterial load of meat. However, the currently employed measures are not specific to pathogens and the harmless background flora is also reduced. Jay (16) has expressed concern that pathogen reduction strategies will make products potentially more hazardous due to parallel reductions in the levels of the background flora. Ground beef is a high-risk product because pathogens which can become mixed into the ground product may not be destroyed if the ground product does not receive an adequate heat treatment before the product is consumed. Good hygiene in all stages in the food production is therefore necessary in order to keep the number of pathogens low. Vold *et al.* (35) found that a high background flora, consisting mainly of lactic acid bacteria, inhibited growth of *E. coli* O157:H7 in ground beef stored at 12°C. Thus, appropriate lactic acid bacteria as competitive or "protective" cultures may be used.

Cooked and Chilled Meat

Most meat products are cooked before being eaten, although there are some exceptions such as "beef tartar" and most fermented sausages. "Rare" steaks may be raw inside, but should achieve surface temperatures sufficient to kill any pathogens present. In ground beef, contaminated bacteria may be distributed throughout the product, such that any VTEC present may occur both on the surface and in the centre of the product. It has been shown that during cooking, the internal temperature of many beef burgers in the US does not exceed 60°C (3). Such temperatures are insufficient to kill VTEC, and viable bacteria are most often found in large and/or thick burgers (27). The association of several outbreaks of *E. coli* O157:H7 infection with the consumption of undercooked beefburgers shows that these are high-risk products.

Other meat products which present risks include cooked cold cuts, such as ham and bologna. In most cases, these products receive adequate heat treatment during initial processing. However, as they are usually eaten without further heat treatment, they pose a risk if there is cross contamination during handling and slicing. Even though strict hygienic measures are employed, it is impossible to guarantee that no pathogen is present or that the product has not been held at "abuse" temperatures. In such products, the use of a protective culture may inhibit growth of pathogenic bacteria. Thus, Bredholt *et al.* (5) found that the application of a protective culture of lactic acid bacteria (LAB) on ham inhibited growth of *E. coli* O157:H7 at 10°C (Fig. 2. a,b). Sensory analysis did not detect any changes in the taste of the treated product. However, although the presence of LAB did inhibit further growth of *E. coli* O157:H7, it did not produce a reduction in the numbers of *E. coli* O157:H7 on the product. Because of the low infectious dose (2-2000 cells) of this pathogen (6), strict hygienic measures to prevent initial contamination are still important.

Freezing

Meat products are often frozen to extend shelf life, but it is not clear how this process influences the survival of VTEC *E. coli*. Some authors report little or no reduction of *E. coli* O157:H7 numbers in ground meat which has been frozen (11). More recently, Ansey *et al.* (2) found that the numbers of 4

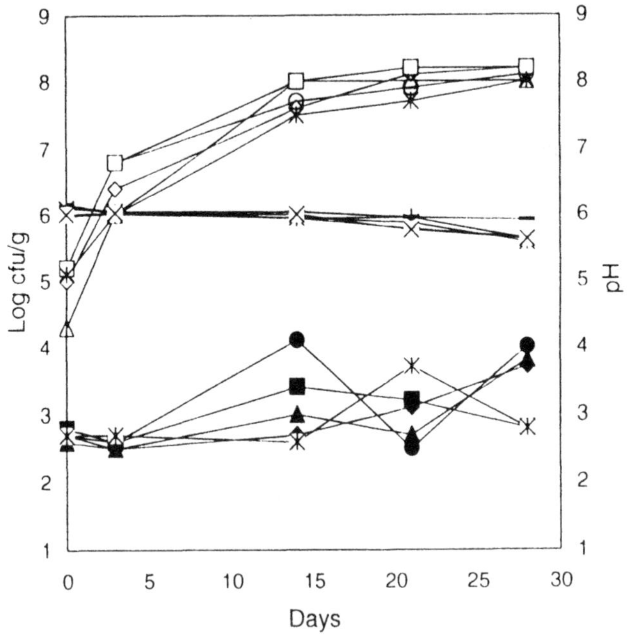

FIG. 2. (a) INHIBITION OF *E. COLI* O157:H7 ON COOKED, SLICED, VACUUM-PACKAGED HAM AT 10°C BY ADDITION OF 10^4-10^5 cfu/g OF SELECTED LAB STRAINS

Filled symbols = *E. coli* O157:H7, open symbols = LAB strains, pH profiles are shown;

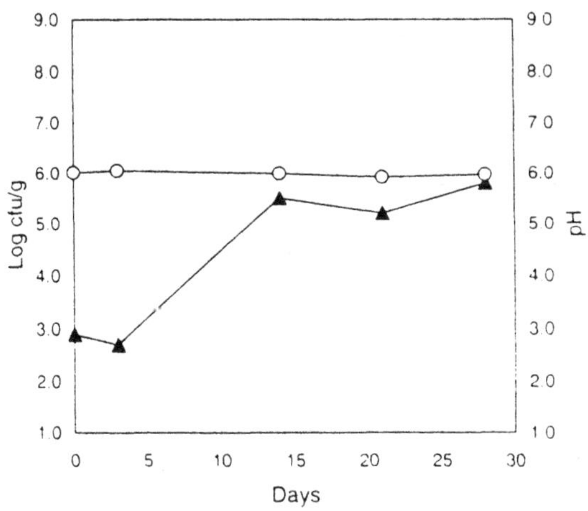

FIG. 2. (b) GROWTH OF *E. COLI:* O157:H7 ON COOKED, SLICED, VACUUM-PACKAGED HAM AT 10°C WITHOUT ADDED LAB

▲ = *coli* O157:H7, ○ = pH.

individual strains of *E. coli* O157:H7 inoculated into ground beef and stored at -20°C for 12 months declined by approximately 1.0 $\log_{10}$ CFU/g (although the numbers of a control strain of *E. coli* declined by 2.0 $\log_{10}$ CFU/g). Other workers have reported larger and varying reductions (30). Such conflicting results may be caused by differences in the fat content of products, the nature of the thawing procedure and resuscitation medium, or the variability of survival among different strains. Faith *et al.* (12) found that *E. coli* O157:H7 numbers in pepperoni declined by nearly 5 log during storage at 21°C for 60 days. When the sausage was stored at 4°C or frozen, the reduction was less than 2 log. Thus, in most cases low temperature will not be enough to achieve large reductions in the numbers of VTEC in the food.

Meat Packaging

Centrally packaged retail cuts of red meats are economical because they require less manpower, equipment and space. They also reduce the chances of cross-contamination due to the continuous process operation. Several methods are used to prolong the shelf life of packaged meat.

Irradiation

Raw packaged meat may be irradiated to obtain a longer shelf life. U.S. Food and Drug Administration (FDA) has recently approved (effective February, 2000) irradiation of red meat to reduce or inactivate possible pathogens, especially VTEC, although such products must prominently display an internal irradiation (radura) symbol.

Irradiation by medium doses (3.0-7.0 kGy) effectively destroys vegetative cells of pathogenic bacteria. However, some bacteria, including *E. coli* O157:H7, can be destroyed by significantly lower doses. Thus 1.8 kGy offers a minimum critical limit for the destruction of 10^6 CFU of *E. coli* O157:H7 per g ground beef (21). Other practical and marketing difficulties exist in relation to the wider application of irradiation. There may, for example, be problems in finding suitable polymers for packaging products to be irradiated (13). More importantly, in most European countries there is a strong sentiment against irradiated food.

Modified Atmosphere (MAP)

Vacuum packaging is a long established technique, which can extend the shelf life of meat stored at low temperature. However, it has some disadvantages, i.e., products may become a dark colour and some juice will be drawn from the meat. MAP with high CO_2 and low O_2 will also have a long microbial shelf life, but may give discoloration of the meat. Vacuum packaging and MAP with high CO_2 and low O_2 are therefore mostly used for storage and distribution. When MAP is used for retails, gas compositions of around 70-75% O_2 and 25-30% CO_2 are the most common.

In Norway, which is sparsely populated, with small stores located far apart, most meat is sold pre-packed in MA. For the past decade the Norwegian meat industry has been using a gas mixture of 60-70% CO_2, 30-40% N_2 and 0.3-0.5% CO (the CO comes ready mixed in the N_2 from the supplier). The reason for adding CO to the gas mixture is that it will produce a long-lasting cherry-red colour of the meat (32). The high CO_2 concentration in the mixture will prolong the microbiological shelf life and inhibits bacterial growth most efficiently at low temperatures. MAP meat stored at -1.5°C may thus have a shelf life of up to 6 months.

At abuse temperatures (>8°C) pathogens like *E. coli* O157:H7 and *Salmonella spp*. may grow and increase the health risks to consumers. A recent study (25) compared the growth of several pathogens in ground beef packed in a high CO_2/low CO (0.4% CO/60% CO_2/40% N_2), and a high O_2 mixture (70% O_2/30% CO_2), as commonly used in many European countries, and chub packs (vacuum). Growth of *E. coli* O157:H7 at 10°C in ground beef was nearly totally inhibited in both the high CO_2/low CO mixture and the high O_2 mixture, but there was more growth in the chub packs (Fig. 3) (25). The combination of high CO_2 and low storage temperature thus seem to inhibit growth of *E. coli*

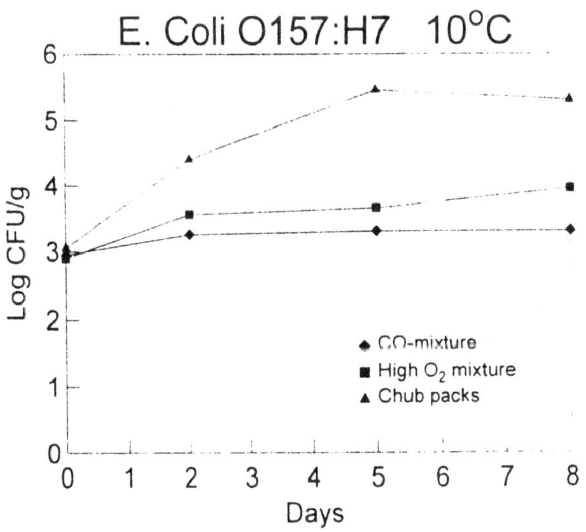

FIG. 3. GROWTH OF *E. COLI* O157:H7 INOCULATED IN GROUND BEEF PACKED IN HIGH CO_2/LOW CO MIXTURE (0.4% CO/60% CO_2/40% N_2), HIGH O_2 (70% O_2/30% CO_2) OR IN CHUB PACKS (VACUUM), STORED AT 10°C.

O157:H7, and extend shelf life. As noted above, use of a high (60-70%) CO_2 concentration requires the presence of a low concentration of CO to obtain a stable red colour of the meat. However, CO is currently not permitted as a component of modified atmospheres in the EU.

CONCLUSION

In practice it is not possible to rear animals that are free from VTEC, and in more general terms, bacterial pathogens like VTEC cannot be eliminated from meat by any of the measures presently available. However, much can be done to minimise their occurrence by ensuring high standards of hygiene in the abattoir and throughout the food chain, principally by the application of Hazard Analysis Critical Control Point (HACCP) systems or similar principles. Although the producers of meat and meat products have a responsibility to produce safe products, that responsibility has to be shared with retailers and the consumers themselves. It is particularly important that retailers take appropriate care in handling of products and temperature control during storage and display. Domestic kitchen hygiene is also important, and new products, new methods of preparation, and new eating habits make effective consumer education even more important.

REFERENCES

1. ANON. 1999. Forekomst af O157 i kvægbesætninger. Zoonose-Nyt *6*(5), 20.
2. ANSEY, S.E., DARLING, K.A. and KASPAR, C.W. 1999. Survival of *Escherichia coli* O157:H7 in ground beef patties during storage at 2, -2, 15 and then -2, and -20°C+. J. Food Protect. *62*, 1243–1247.
3. BELL, B.P. *et al.* 1994. A multistate outbreak of *Escherichia coli* O157:H7-associated bloody diarrhea and uremic syndrome from hamburgers. JAMA *272*, 1349–1353.
4. BORCH, E., NESBAKKEN, T. and CHRISTENSEN, H. 1996. Hazard identification in swine slaughter with respect to foodborne bacteria. Int. J. Food Microbiol. *30*, 9–25.
5. BREDHOLT, S., NESBAKKEN, T. and HOLCK, A. 1999. Protective cultures inhibit growth of *Listeria monocytogenes* and *Escherichia coli* O157:H7 in cooked, sliced, vacuum- and gas-packaged meat. Int. J. Food Microbiol. *53*, 43–52.
6. BUCHANAN, R.L. and DOYLE, M.P. 1997. Foodborne disease significance of *Escherichia coli* O157:H7 and other enterohemoragic *E. coli*. Food Technol. *51*, 69–76.

7. CASTILLO, A., LUCIA, L.M. GOODSON, K.J., SAVELL, J.W. and ACUFF, G.R. 1998. Comparison of water wash, trimming, and combined hot water and lactic acid treatments for reducing bacteria of fecal origin on beef carcasses. J. Food Protect. *61*, 823–828.
8. CHAPMAN, P.A., SIDDONS, C.A., WRIGHT, D.J., NORMAN, P., FOX, J. and CRICK, E. 1993. Cattle as a possible source of verocytotoxin-producing *Escherichia coli* O157 infections in man. Epidemiol. Infect. *111*, 439–447.
9. CORRY, J.E.L, JAMES, C., JAMES, S.J. and HINTON, M. 1995. *Salmonella, Campylobacter* and *Escherichia coli* O157:H7 decontamination techniques for the future. Int. J. Microbiol. *28*, 187–196.
10. DORSA, W.J., CUTTER, C.N. and SIRAGUSA, G.R. 1997. Effects of steam-vacuuming and hot water spray wash on the microflora on refrigerated beef carcass surface tissue inoculated with *Escherichia coli* O157:H7, *Listeria innocua* and *Clostridium sporogenes* J. Food Protect. *60*, 114–119.
11. DOYLE, M.P. and SCOENI, J.L. 1984. Survival and growth characteristics of *Escherichia coli* associated with hemmorrhagic colitis. Appl. Env. Microbiol. *48*, 855–856.
12. FAITH, N.G. *et al.* 1998. Survival of *Escherichia coli* O157:H7 in full and reduced-fat pepperoni and manufacture of sticks, storage of slices at 4°C or 21°C under air and vacuum and baking of slices on frozen pizza at 135, 191 and 246°C. J. Food Protect. *61*, 383–389.
13. FARKAS, J. 1998. Irradiation as a method for decontaminating food. Int. J. Food Microbiol. *44*, 189–204.
14. FRIES, R. *et al.* 1996. Microbial control in the meat industry. 7. Bacterial pathogens on raw meat and their properties (G.C. Mead and M.H. Hinton, eds.). University of Bristol Press, Bristol, UK, 34 pp.
15. HEUVELINK, A.E., ZWARTKRUIS-NAHUIS, J.T.M., VAN DEN BIGGELAAR, F.L.A.M., VAN LEEUWEN, W.J. and DE BOER, E. 1999. Isolation and characterization of verocytotoxin-producing *Escherichia coli* O157 from slaughter pigs and poultry. Int. J. Food Microbiol. *52*, 67–75.
16. JAY, J.M. 1997. Do background microorganisms play a role in the safety of foods. Trends in Food Science & Technology *8*, 421–424.
17. JORDAN, D., MCEWEN, S.A., LAMMERDING, A.M., MCNAB, W.B. and WILSON, J.B. 1999. Pre-slaughter control of *Escherichia coli* O157 in beef cattle: a simulation study. Prev. Vet. Med. *41*, 55–74.
18. KOCHEVAR, S.L., SOFOS, J.N., BOLIN, R.R., REAGAN, J.O. and SMITH, G.C. 1997. Steam vacuuming as a pre-evisceration intervention to decontaminate beef carcasses. J. Food Protect. *60*, 107–113.
19. KRUSE, H. 1999. Resultater for 1998 fra SNTs overvåkingsprogram vedrørende enterohenorrhagisk *E. coli* (EHEC) i ferskt storfekjøtt. Nor. Vet. Tidsskr. *111*, 178.
20. MACDONALD, K.L. and OSTERHOLM, M.T. 1993. The emergence of *Escherichia coli* O157:H7 infection in the United States. The changing epidemiology of foodborne disease (editorial comment) JAMA. *269*, 2264–2266.

21. MOLINS, R.A. and MOTARJEMI, Y. 1997 Irradiation: A critical point in ensuring the microbial safety of foods. World Congress on Food Hygiene, The Hague 1997. Handout Aug. 28, p. 19–25.
22. NERBRINK, E., BORCH, E. and LÖFDAHL, S. 1999. Occurrence of *Escherichia coli* O157 and verotoxinogenic bacteria in faecal samples from cattle and on beef carcasses. *In* Food Microbiology and food safety into the next millenium (A.C.J. Tuijtelaars, R.A. Samson, F.M. Rombouts and S. Notermans, eds.) pp. 75–76, Proceedings of the Seventeenth International Conference of the International Committee on Food Microbiology and Hygiene (ICFMH), 13-17 September, Veldhoven, The Netherlands.
23. NESBAKKEN, T. and SKJERVE, E. 1996. Interruption of microbial cycles in farm animals from farm to table. Meat Sci. *43*, 47–57.
24. NISSEN, H., MAUGESTEN, T. and LEA, P. 2001. Survival and growth of *Escherichia coli* O157:H7, *Yersinia entercolitica* and *Salmonella enteritidis* on decontaminated and untreated meat. Meat Science *57*, 291–298.
25. NISSEN, H., ALVSEIKE, O., BREDHOLT, S., HOLCK, A. and NESBAKKEN, T. 2000. Comparison between the growth of *Yersinia enterocolitica, Listeria monocytogenes*, *Escherichia coli* O157:H7 and *Salmonella* spp. in ground beef packed by three commercially used packaging techniques. Int. J. Food Microbiol. *59*, 211–220.
26. NUTSCH, A.L. *et al.* 1997. Evaluation of a steam pasteurization process in a commercial beef processing facility. J. Food Protect. *60*, 485–492.
27. PHILLIPS, C.A. and ROSCOE, N. 1996. Survival of *Escherichia coli* O157:H7 in ground beef during normal cooking procedures. Nutr. Food Sci. *2*, 23–26.
28. POWELL, V.H. and CAIN, B.P. 1987. A hot water decontamination system for beef sides. CSIRO, Food Research Quarterly. *47*, 79–84.
29. RASMUSSEN, M. A., CRAY JR., W.C., CASEY, T.A. and WHIPP, S.C. 1993. Rumen contents as a reservoir of enterohemmorrhagic *Escherichia coli*. FEMS Microbiol. Lett. *114*, 79–84.
30. SAGE, J.R. and INGHAM, S.C. 1998. Survival of *Escherichia coli* O157:H7 after freezing and thawing in ground beef patties. J. Food Protect. *61*, 1181–1183.
31. SMULDERS, F.J.M. 1995. Preservation by microbial decontamination of the surface treatment of meats by organic acids. Chapter 12, *In* New Methods of food preservation, (Gould, ed.) pp. 253–282. Elsevier: Amsterdam-New York-Oxford.
32. SØRHEIM, O., NISSEN, H. and NESBAKKEN, T. 1999. The storage life of beef and pork packaged in an atmosphere with low carbon monoxide and high carbon dioxide. Meat Science *52*, 157–164.
33. TOMPKIN, R.B. 1992. Corrective action procedures for deviations from the critical control point critical limits. *In* HACCP Principals and Applications (M.D. Pierson and A. Corlett Jr., eds.) pp. 72–82. Chapman and Hall, New York.
34. VOLD, L., JOHANSEN, B.K., KRUSE, H., SKJERVE, E. and WASTESON, Y. 1998. Occurrence of shigatoxigenic *Escherichia coli* O157 in Norwegian cattle herds. Epidemiol. Infect. *120*, 21–28.
35. VOLD, L., HOLCK, A., WASTESON, Y. and NISSEN, H. 2000. The background flora inhibits growth of *Escherichia coli* O157:H7 in ground beef. Int. J. Food Microbiol. *56*, 219–225.

CHAPTER 21

CONTROL OF VTEC IN NON-MEAT FOOD PRODUCTS

C.P. O'LOUGHLIN and M.E. UPTON

Department of Industrial Microbiology
University College Dublin
Belfield, Dublin 4, Ireland

Consumption of foods derived from cattle, especially undercooked ground beef and raw milk have accounted for most outbreaks of Verocytotoxigenic *Escherichia coli* (VTEC) infection (14,45). However, a number of cases have implicated several unique and seemingly innocuous foods as likely vehicles of infection, among them are unpasteurised apple cider (5,9,15), yoghurt (35), white radish sprouts (32), lettuce (26), potatoes (11,34) and mayonnaise (44). While some of these outbreaks may be due to contamination from raw meat products, faecal contamination as a result of inadequate sanitation or poor standards of food processing or personal hygiene, has been identified as being the primary cause.

The foods that pose the greatest threat of VTEC infection are those where the organism may be present, even in low numbers, in the raw material, where no process exists to reduce or eliminate it and where it may survive or even proliferate during processing or in the finished product. The aim of this chapter is to highlight the key sources of VTEC contamination of non-meat products associated with VTEC infection, such as apple cider, fresh vegetables and dairy products, and to identify measures necessary to control such potential hazards.

APPLE CIDER

The consumption of apple cider, a popular and traditional beverage of North America, has been implicated in several outbreaks of *E. coli* O157:H7 infection in recent years (5,9,15). The term "cider" in Britain and Ireland describes fermented apple products with a high alcohol content, however, apple "cider" in the US is a fresh pressed non-fermented apple juice which has not undergone pasteurisation. It should be clearly differentiated from apple juice, which is pasteurised by heating. Having a low pH, typically between 3.3 and 4.1 (31), apple beverages were not previously regarded as potentially hazardous. However, the emergence of *E. coli* O157:H7 as an acid tolerant, pathogenic contaminant of apple cider has caused concern among consumer groups, regulatory bodies and the juice manufacturing industry. The US Food and Drug Administration (FDA) have published regulations requiring that packaged fruit juices, which have not been subjected to a process designed to reduce target pathogen populations by 5 $\log_{10}$ units, carry the following warning statement "WARNING: This product has not been pasteurised and, therefore, may contain harmful bacteria which can cause serious illness in children, the elderly and persons with weakened immune systems" (20).

The production of apple cider is a fairly simple process. The fruit is harvested at the orchard, transported to the cider mill, graded, washed and milled to a pulp. The juice is extracted by subjecting the apple pulp to pressures up to about 1,500 kg in a press. Juice is then collected, possibly filtered, bottled and sold on the retail market. The shelf life of this product is usually less than 7 days.

E. coli is not a normal contaminant of apples. It is thought that apples and apple ciders become contaminated through contact with faecal material in the irrigation water or the soil, or by improper handling or contamination in the manufacturing process. Animal waste can be introduced to the orchard as manure or by allowing access of domestic or wild animals. The practice of using windfall apples, which may be contaminated with faecal material on the ground, represents a significant hazard, and a number of outbreaks of *E. coli* O157:H7 infection associated with apple cider have pinpointed the use of such fallen apples as the source of contamination (5,15). However, the practice of harvesting apples from the ground is commonplace, as demonstrated in 1991 when a survey of fresh pressed apple juice manufacturers in New England, USA revealed that all respondents (36 companies) used windfall apples in their fruit juice production (5). Even when a company adopts a policy of accepting only hand-picked fruit, in the absence of properly and consistently applied control procedures, such a policy may be difficult to enforce, as demonstrated in a case involving a large producer of apple cider in the US (15).

A number of measures can be taken to reduce the possibility that faecally contaminated fruit can enter the cider producing process. Artificial fertilisers offer the safest alternative to manure, although animal waste which has been adequately composted over several months would also be suitable. Accessibility to orchards by domestic or wild animals should be prevented. Cody *et al.* (15) isolated *E. coli* O157:H7 from fresh deer faeces collected from the locality of an orchard implicated in an outbreak of *E. coli* O157:H7 infection.

As well as such direct contacts, the water used to irrigate orchards should be free of enteric pathogens and of potable quality. Ultimately the practice of harvesting apples from the ground should be avoided. Juice manufactures should ensure that growers supply fruit that is clean and of high quality, devoid of faecal contamination.

Effective controls during processing will further minimise the potential hazard posed by *E. coli* O157:H7. Fruit used for the production of apple cider usually undergo a selection process, during which spoilt, damaged or infected fruit are removed. In addition, contaminating debris such as foliage, sand and soil are removed by washing, typically with chlorinated water. The use of chlorinated water for fruit washing prevents the water acting as a source of contamination and also prevents the spread of cross contamination throughout the batch of fruit during washing. Inadequate washing procedures represent a significant hazard as they could distribute pathogenic contaminants within a large batch of fruit, especially if bulk wash tanks are used. The level of free chlorine in the water in wash tanks needs to be properly controlled to maintain adequate biocidal activity, which would otherwise decline through contact with organic matter. In addition, the water in bulk wash tanks should be regularly changed. In terms of process parameters, apples should be submerged, agitated and held for an adequate period in the wash tanks.

Errors in such processes and parameters can have serious consequences. Cody *et al.* (15) identified the incorrect use of a phosphoric acid wash as a factor contributing to an outbreak of *E. coli* O157:H7 infection from unpasteurised commercial apple juice. Washing, even with chlorinated water, results in minimal reduction in microbial populations, usually less than 100-fold and should not be relied upon to eliminate pathogenic bacteria that may be present on produce at levels exceeding 5 to 10 cfu/cm^2 (6). However, the use of chlorinated water reduces the likelihood of a contaminated product reaching the consumer and such washing should be carried out whenever possible.

Juice production methods must minimise contamination of the final product. This requires extensive cleaning and sanitation of food contact surfaces. Small scale producers of cider typically process batches of fruit through common equipment, e.g., hammer mill, apple presses, juice extraction units, bulk storage tanks and filling lines. To prevent batch to batch cross contamination a vigorous cleaning and sanitation regime should be implemented between batches. As most small-scale cider producers are located on farms where animal manure is normally present, the product is open to contamination from the processing environment and workers if sanitation practices are inadequate.

Although apple cider, due to its acid concentration and low pH, was at one time generally regarded as an unlikely vehicle of infection with pathogenic bacteria, it is now clear that these properties, on their own, are insufficient to ensure product safety. The observation that *E. coli* O157:H7 can survive for up to several weeks in apple cider at refrigeration temperatures (33,46), has led to the supplementation of products with food preservatives such as potassium sorbate and sodium benzoate. However, these products have been shown to have minimal lethal effects on *E. coli* O157:H7. Sodium benzoate at a concentration of 0.1% can achieve a 5-$\log_{10}$ unit reduction in *E. coli* O157:H7 numbers over

2 to 10 days at 8°C (46). However, such concentrations of this material, or even at lower concentrations, i.e., as low as 0.025% have been reported to generate off-flavours (37), and are therefore unlikely to be widely used. Uljas and Ingham (41) recently reported a combination of simple nonpasteurisation treatments including short term storage at 25 or 30°C, freeze-thaw and the addition of organic acids, as yielding the 5-$\log_{10}$ unit reduction in *E. coli* O157:H7 numbers required by the FDA. This multiple hurdle approach offers a simple, inexpensive alternative to pasteurisation, and could be applicable and affordable within small apple cider mills.

FRESH FRUIT JUICES

An increasing consumer demand for healthy, "near-fresh" quality, minimally processed convenience foods has spawned the phenomenon of fresh fruit and juice "bars". From local supermarket to "trendy" high street, juice bars offer a wide variety of freshly squeezed fruit and vegetable juices. In general such juices are unpasteurised and are judged by consumers to be superior in flavour to heat processed juices. Although the vast majority of unpasteurised juices produced are consumed without reported incident of foodborne illness, consumption of unpasteurised citrus juices have been associated with food poisoning (8,39). Such fruit juices remain an obvious safety concern, bearing in mind the ability of *E. coli* O157:H7 to withstand low pH, particularly at low temperatures, and the low infective dose of the organism.

As unpasteurised fruit juice production processes do not include a treatment which is effective in the elimination of possible contaminating pathogens, it is imperative that all possible avenues through which the product may become contaminated should be identified and controlled. The use of high quality produce, extensive cleaning and sanitation of all fruit, processing equipment and food contact surfaces, coupled with proper training of personnel in the principle of food hygiene, is essential. An effective Hazard Analysis Critical Control Point (HACCP) based quality control system encompassing each stage of the juice making process, from fruit production to retail, will reduce the risk of VTEC infection and ensure a better quality of product.

Public health policy with regard to the growing fresh juice industry remains unclear. Unpasteurised fresh juices sold in restaurants, juice bars and many retail establishments in the US (an estimated 5 million gallons per year), are exempt from the aforementioned FDA labelling requirements (2). Opponents to mandatory pasteurisation claim that the unique flavour and quality of the juice is lost during heat treatment. The FDA has claimed that required pasteurisation "would eliminate the incentive to develop alternative methods than can accomplish the same purpose". There is an obvious need for discussion amongst public health regulators, consumer groups and producers/retailers, to fully determine the health risks associated with the fresh juice industry, and to establish effective means to reduce the risk to an acceptable level of risk, assuming that such a level of acceptable risk can be agreed. The best approach probably involves a combination of HACCP, public education and warning labels, to ensure the safe availability of unpasteurised fruit juice to the consumer.

SALAD VEGETABLES

Although regarded as health foods, fresh produce pose a significant public health concern and have been associated with several outbreaks of foodborne disease, including VTEC infection. In 1996 the largest series of outbreaks of *E. coli* O157:H7 infection to date occurred in Japan where 9,451 cases were reported, predominantly among children attending Sakai City elementary schools. White radish sprouts served with school lunches were implicated as the vehicle of infection (32).

Fruit and vegetables are susceptible to faecal contamination, primarily during growth and harvesting. The application of manure to fields used for the cultivation of produce represents a significant hazard as the practice may introduce VTEC to the food chain. Handling of manured vegetables, particularly potatoes, has been implicated with *E. coli* O157:H7 infection (11,13,34). The use of untreated animal waste as a fertiliser should be avoided. Safer alternatives include artificial fertilisers and properly composted manure. If manures are to be used, they should be applied to the field several months before crops are planted, to allow decline in applied pathogen numbers, and should not be applied after crops have been planted. Access to fields of fruit and vegetable crops to domestic and wild animals should be prevented. Hilborn *et al.* (26) reported an outbreak of *E. coli* O157:H7 infection due to the consumption of mesclun lettuce, and implicated cattle grazing near the lettuce fields as the probable source of faecal contamination.

The quality of the water used for irrigation of crops should be well controlled, free of contamination and of potable quality. However, in developing countries the quality of water may be compromised by inadequate or unsanitary sewage systems. Local rivers and streams used to irrigate crops may be contaminated with animal and human waste. Producers should ensure that adequate sanitation facilities are made available to field workers.

Effective handling and sanitation during processing and packaging are essential to prevent the introduction of new contaminants or cross contamination within produce. Chlorinated water (50 to 200 ppm) is widely used to sanitise fresh produce for retail sale (6). The pH and concentration of free chlorine of the wash water used in such processes should be regularly monitored, and maintained at appropriate levels. Bulk wash tanks, spray nozzles and pipelines should be systematically and effectively sanitised. Sufficient time should be allowed for chlorine spray or immersion to be effective. Care should be taken to avoid overloading bulk wash tanks. However, it should be recognised that chlorine treatments merely reduce microbial populations on produce by approximately 100 fold and should not be relied upon to eliminate pathogens.

Salad vegetables for retail sale are typically chopped or sliced and packaged in containers, sometimes under modified atmosphere. To prevent the processing environment and equipment from becoming sources of cross contamination, an effective sanitation program should be adopted. The processing facility should be of a hygienic design, staff should be trained in food sanitation, appropriate cleaning compounds and sanitisers should be used correctly, and effective cleaning procedures should be employed.

Sprouted vegetables have been implicated as vehicles in several recent outbreaks of *E. coli* O157:H7 infection (10,32). Seeds destined for sprouting are harvested from the field and transported to sprouting facilities. There they are re-hydrated by soaking in chlorinated water before germination, under controlled conditions of humidity and elevated temperatures, and completion in greenhouses. The chief source of VTEC contamination of sprouted vegetables is the harvested seed. If the seeds are contaminated, the sprouting process provides a humid and warm (20 to 25°C) environment ideal for growth of the organism, and its distribution throughout an entire batch. Hara-Kudo *et al.* (25) reported the number of *E. coli* O157:H7 exceeded 10^7 cfu/g of sprouts produced from inoculated seeds. The treatment of seeds with chlorinated water or other disinfectants can fail to eliminate *E. coli* O157:H7. The International Sprout Growers Association (27) recommendation of a chemical pre-soak of seeds in either NaOCl (2,000 ppm), $Ca(OCl)_2$ (1,800 ppm) or H_2O_2 (60,000 ppm) for a maximum of ten minutes followed by a fresh water rinse step, has also been shown to be ineffective (40). *E. coli* O157:H7 can spread to edible parts of radish sprouts from contaminated seeds (28). Once established, the organism is very difficult to eradicate from the cotyledons of radish sprouts and is capable of surviving a range of chemical treatments (28,40).

As noted above, the germination process provides the ideal opportunity for VTEC organisms, if present, to proliferate. Producers can reduce the risk of sprout associated disease by controlling the possible sources of contamination. It is critical that consignments of seed are free from enteric contaminants, including *E. coli* O157:H7. Seed should be purchased from a reputable supplier who takes the appropriate measures to avoid faecal contamination of the raw seed and who implements a HACCP based quality control system. Microbiological examination (incorporating a suitable sampling plan), for the presence of coliforms and *E. coli* O157:H7 in bulk consignments of seed and germination trials may have limited use, should nevertheless be employed. It is crucial that the quality of water and growth substrates in germination rooms and greenhouses are of a high quality and free of faecal contamination.

DAIRY PRODUCTS

E. coli O157:H7 is a pathogen of particular concern to the dairy industry. Not only has the organism been isolated from dairy herds (24) but dairy products such as raw milk (12,19,29), pasteurised milk (1,42) and yoghurt (35) have been implicated as vehicles of transmission.

It is not surprising to find raw milk associated with outbreaks of enteric infections, as it unavoidably becomes contaminated with faecal material during the milking process. The practice of good animal husbandry will ensure the health of the dairy herd and reduce the risk of pathogens being shed into the milk from an infected udder. Several hygienic precautions should be taken to prevent the transfer of pathogens during milking. Faecal material should be removed from the udder prior to milking using an effective washing procedure. A clean environment, including bedding, milking parlour equipment and floors will reduce the soiling of udders. Milking equipment such as teat cups, pipelines, filters and bulk storage vessels may provide foci for pathogens to colonise and need to be regularly subjected to effective cleaning and sanitation regimes if the spread of contamination to further batches is to be avoided. To reduce the growth of contaminant bacteria and slow the rate of spoilage, freshly collected milk should be rapid cooled to 4°C (and ideally to 2°C) within 30 minutes of milking. Wang *et al.* (43) reported the failure of *E. coli* O157:H7 to grow in raw milk stored at 5°C and noted a decrease in numbers following prolonged holding at such temperatures. Raw milk is usually transported to dairies under refrigeration in motorised bulk milk tankers. Such tankers should be cleaned between batches using adequate "cleaning-in-place" techniques.

VTEC are not unusually heat resistant and are destroyed by the high temperature, short time process (71.7°C, 15 s) which is commonly used to pasteurise milk (16). Effective control of pasteurisation is, however, vital to ensure safety of dairy products. Faulty pasteurisation has been implicated in an outbreak of *E. coli* O157 food poisoning (1). Effective cleaning and sanitation regimes, conducted at regular intervals, together with good equipment design and construction, serve to reduce the risk of contamination during dairy product manufacture.

Although *E. coli* O157:H7 is destroyed by pasteurisation, post-pasteurisation contamination in dairy products is of particular concern because of the low infectious dose (7,38) and acid tolerance of this pathogen (4,30). Contamination may arise from one of several sources. Inadequate segregation of raw and finished product areas exposes finished product to contamination with raw milk. Dairy products such as cheese curd may be handled frequently during manufacture and are open to contamination. Personnel handling raw milk should be excluded from handling finished products. In addition, dairy equipment which is handled following sanitation is at risk of being contaminated. A contaminated water supply is an important possible source of VTEC, thus water used in dairies should be of potable quality. Stored water should be chlorinated (2-5 ppm chlorine) to maintain suitable microbiological quality. The dairy house and equipment should be of a design which allows for easy and rapid drainage. The persistence of watery milk residues and puddles in equipment and the dairy house respectively, pose potential hazards, as coliforms numbers may build up in such residues, and can create a major source of contamination. Adequate ventilation should be installed to prevent a build up of condensation on surfaces as droplets may introduce VTEC onto cheese during the ripening process in maturation rooms.

Products such as cheese, cream and butter made from raw milk have a fuller flavour than products made from pasteurised milk. In view of the organisms' acid tolerance, resistance to fermentation by-products, survival at refrigeration temperatures and salt tolerance, Bell and Kyriakides (3) identified soft cheeses made from raw milk as products where the risk of VTEC infection is of particular concern. Unpasteurised cheese has been implicated with an outbreak of haemolytic uraemic syndrome in France (18). The prevalence of *E. coli* in soft and semi-soft cheeses at 40-50% (22,23) demonstrates that the microorganism can survive the cheese manufacturing process. *E. coli* O157:H7 has, for example, been reported to survive the manufacture process of Camembert and Feta cheese (36).

Studies have shown VTEC to be an intermittent contaminant of raw milk, present at low levels and at a low frequency (17,24). Nevertheless, raw milk must be considered to present a hazard, as the potential exists for it to harbor the pathogen. Consumers should be made aware that, regardless of what precautions are taken in the production and handling of raw milk, no guarantee for safety can ever be given. Ultimately the only effective way to stop raw milk associated disease is to stop people from drinking raw milk. However, as experience has shown, such an objective is next to impossible to achieve. The retail sale of raw milk for direct consumption has been made illegal in Ireland since 1997, yet a recent survey revealed that of farm families questioned, 84% consumed raw milk (21).

CONCLUSION

Processors of unpasteurised fruit juice, ready-to-eat fresh fruit and vegetables, and raw milk dairy products must realise that potential risk of VTEC infection from their products cannot be entirely eliminated, but can only be reduced. Food safety management systems incorporating HACCP should be applied to all stages of the food industry, from production, processing and distribution to consumption of the final product. A combined strategy, which ensures raw food material of a high quality grade, implementing a HACCP system which limits or excludes pathogens, and following strict adherence to good manufacturing and sanitation practices, will not only increase the quality of the product but will reduce the risk of VTEC infection. Several examples of preventative measures to control VTEC in non-meat food products are shown below.

Control of raw materials

- Animal and human waste should not be used to fertilise crops intended for human consumption.
- Crops should be irrigated with water free of faecal contamination.
- Access of domestic and wild animals to crop fields and orchards should be prevented.
- Fresh produce should be transported in a way to minimise damage and to prevent entry of VTEC into these products.
- The use of windfall apples in the production of apple beverages should be avoided.
- Faecal contamination of raw milk should be minimised during milking.
- Raw milk should be transported and stored under refrigeration prior to further processing.

Control of processing

- Raw fruit and vegetable juices should undergo pasteurisation.
- Produce should be washed with chlorinated water.
- Raw milk should be correctly pasteurised.
- Post-pasteurisation contamination of dairy products should be prevented.
- Food handlers should receive training on the principles of food hygiene and the application of HACCP.

Control at retail

- Cross contamination between raw and ready-to-eat foods should be avoided.
- Products derived from unpasteurised milk should be labelled, warning consumers of the possible risk of VTEC infection associated with their consumption.
- Food handlers should be trained in the principles of food hygiene.

REFERENCES

1. ANON. 1999. *E. coli* O157 food poisoning outbreak linked to milk. Veterinary Record, March, *274*.
2. APPLEBAUM, R. 1998. Unpasteurized juice: why is RDA taking chances with food safety? Food Technology *52*, 180.
3. BELL, C. and KYRIAKIDES, A. 1998. *E. coli: A practical approach to the organism and its control in foods*. Blackie Academic & Professional, London.
4. BENJAMIN, M.M. and DATTA, A.R. 1995. Acid tolerance of enterohemorrhagic *Escherichia coli*. Appl. Environ. Microbiol. *61*, 1669–1672.
5. BESSER, R.S. *et al.* 1993. An outbreak of diarrhea and hemolytic uremic syndrome from *Escherichia coli* O157:H7 in fresh-pressed apple cider. J. American Medical Association *269*, 2217–2220.
6. BEUCHAT, L.R., NAIL, B.V., ADLER, B.B. and CLAVERO, M.R.S. 1998. Efficacy of spray application of chlorinated water in killing pathogenic bacteria on raw apples, tomatoes, and lettuce. J. Food Protection *61*, 1305–1311.
7. BOLTON, F.J., CROZIER, L. and WILLIAMSON, J.K. 1996. Isolation of *Escherichia coli* O157 from raw meat products. Letters in Appl. Microbiol. *23*, 317–321.
8. CENTER FOR DISEASE CONTROL. 1995. Outbreak of *Salmonella* Hartford infection among travellers to Orlando, Florida. EPI–AID Trip Report, 62.
9. CENTER FOR DISEASE CONTROL AND PREVENTION. 1997. Outbreaks of *Escherichia coli* O157:H7 infection associated with drinking unpasteurized apple cider–October 1996. Annals of Emergency Medicine *29*, 645–646.
10. CENTER FOR DISEASE CONTROL AND PREVENTION. 1997. Outbreaks of *Escherichia coli* O157:H7 infection associated with eating alfalfa sprouts — Michigan and Virginia, June-July 1997. Morbidity and Mortality Weekly Report *46*, 741–744.
11. CHAPMAN, P.A. 1999. *Escherichia coli* O157: 14 years experience in Sheffield, UK. *In Escherichia coli* O157 in Farm Animals (C.S. Stewart and H.J. Flint, eds.) pp. 99–119, CABI Publishing, Oxon, UK.
12. CHAPMAN, P.A., WRIGHT, D.J. and HIGGINS, R. 1993. Untreated milk as a source of verotoxigenic *E. coli* O157:H7. Veterinary Record *133*, 171–172.
13. CIESLAK, P.R., BARRETT, T.J. and GRIFFIN, P.M. 1993. *Escherichia coli* O157:H7 infection from a manured garden. (Letter). Lancet *342*, 367.
14. CLARK, A., MORTON, S., WRIGHT, P., CORKISH, J., BOLTON, F.J. and RUSSELL, J. 1997. A community outbreak of verocytotoxin-producing *Escherichia coli* O157 infection linked to a small farm dairy. Communicable Disease Report, CDC Rev. *7*, R206–R211.
15. CODY, S.H. *et al.* 1999. An outbreak of *Escherichia coli* O157:H7 infection from unpasteurized commercial apple juice. Annals of Internal Medicine *130*, 202–209.
16. D'AOUST, J.Y., PARK, C.E., SZABO, R.A., TODD, E.C.D., EMMONS, D.B. and MCKELLAR, R.C. 1988. Thermal inactivation of *Campylobacter* species, *Yersinia enterocolitica*, and hemorrhagic *Escherichia coli* O157:H7 in fluid milk. J. Dairy Science *71*, 3230–3236.

17. DE LOUVOIS, J. and RAMPLING, A. 1998. One fifth of samples of unpasteurised milk are contaminated with bacteria. British Medical J. *316*, 625.
18. DESCHENES, G. *et al.* 1996. Cluster of cases of haemolytic uraemic syndrome due to unpasteurised cheese. Pediatric Nephrology *10*, 203–205.
19. DUNCAN, L., CARTER, A., CARLSON, A.K., BORCZYK, A. and KARMALI, A. 1987. Outbreak of gastrointestinal diseases. Ontario. Canadian Diseases Weekly Report *13*, 5–8.
20. FOOD AND DRUG ADMINISTRATION. 1998. Food labelling: warning and notice statements; labelling of juice products. Federal Register *63*, 20486–20493.
21. Food Safety Authority of Ireland. 1999. *The prevention of E. coli O157:H7 infection: a shared responsibility.*
22. FRANK, J.F. and MARTH, E.H. 1978. Survey of soft and semisoft cheese for presence of fecal coliforms and serotypes of enteropathogenic *Escherichia coli*. J. Food Protection *41*, 198–200.
23. GLATZ, B.A. and BRUDVIG, S.A. 1980. Survey of commercially available cheese for enterotoxigenic *Escherichia coli*. J. Food Protection *43*, 395–398.
24. HANCOCK, D.D., BESSER, T.E., KINSEL, M.L., TARR, P.I., RICE, D.H. and PAROS, M.G. 1994. The prevalence of *Escherichia coli* O157:H7 in dairy and beef cattle in Washington State. Epidemiology and Infection *113*, 199–207.
25. HARA-KUDO, Y. *et al.* 1997. Potential hazard of radish sprouts as a vehicle of *Escherichia coli* O157:H7. J. Food Protection *60*, 1125–1127.
26. HILBORN, E.D. *et al.* 1999. A multistate outbreak of *Escherichia coli* O157:H7 infections associated with consumption of mesclun lettuce. Archives of Internal Medicine *159*, 1758–1764.
27. INTERNATIONAL SPROUT GROWERS ASSOCIATION. 1997. Sanitary guidelines for the growing and packing for sale of fresh sprouts. International Sprout Growers Association, Sacramento, California.
28. ITOH, Y. *et al.* 1998. Enterohemorrhagic *Escherichia coli* O157:H7 present in radish sprouts. Appl. Environ. Microbiol. *64*, 1532–1535.
29. KEENE, W.E. *et al.* 1997. A prolonged outbreak of *Escherichia coli* O157:H7 infections caused by commercially distributed raw milk. J. Infectious Diseases, *176*, 815–818.
30. LEYER, G.J., WANG, L.-L. and JOHNSON, E.A. 1995. Acid adaptation of *Escherichia coli* O157:H7 increases survival in acidic foods. Appl. Environ. Microbiol. *61*, 3752–3755.
31. MATTICK, L.R. and MOYER, J.C. 1983. Composition of apple juice. J. Association of Official Analytical Chemists *66*, 1251–1255.
32. MICHINO, H. *et al.* 1999. Massive outbreak of *Escherichia coli* O157:H7 infection in schoolchildren in Sakai City, Japan, associated with consumption of white radish sprouts. American J. Epidemiology *150*, 787–796.
33. MILLER, L.G. and KASPAR, C.W. 1994. *Escherichia coli* O157:H7 acid tolerance and survival in apple cider. J. Food Protection *57*, 460–464.
34. MORGAN, G.M. *et al.* 1988. First recognized community outbreak of haemorrhagic colitis due to verotoxin-producing *Escherichia coli* O157:H7 in the UK. Epidemiology and Infection *101*, 83–91.

35. MORGAN, D., NEWMAN, C.P., HUTCHINSON, D.N., WALKER, A.M., ROWE, B. and MAJID, F. 1993. Verotoxin producing *Escherichia coli* O157 infections associated with the consumption of yoghurt. Epidemiology and Infection *111*, 181–187.
36. RAMSARAN, H., CHEN, J., BRUNKE, B., HILL, A. and GRIFFITHS, M.W. 1998. Survival of bioluminescent *Listeria monocytogenes* and *Escherichia coli* O157:H7 in Soft Cheeses. J. Dairy Science *81*, 1810–1817.
37. SALUNKHE, D.K. 1955. Sorbic acid as a preservative for apple juice. Food Technology *9*, 590.
38. SALYERS, A.A. and WHITT, D.D. 1994. *Bacterial Pathogenesis - A Molecular Approach*. Pp. 190–204, ASM Press, Washington, DC.
39. SINGH, B.R., KULSHRESHTHA, S.B. and KAPOOR, K.N. 1996. An orange juiceborne outbreak due to enterotoxigenic *Escherichia coli*. J. Food Science Technology India, *32*, 504.
40. TAORMINA, P.J. and BEUCHAT, L.R. 1999. Behaviour of enterohemorrhagic *Escherichia coli* O157:H7 on alfalfa sprouts during the sprouting process as influenced by treatments with various chemicals. J. Food Protection *62*, 850–856.
41. ULJAS, H.E. and INGHAM, S.C. 1999. Combinations of intervention treatments resulting in 5-$\log_{10}$-unit reductions in numbers of *Escherichia coli* O157:H7 and *Salmonella typhimurium* DT104 organisms in apple cider. Appl. Environ. Microbiol. *65*, 1924–1929.
42. UPTON, P. and COIA, J.E. 1994. Outbreak of *Escherichia coli* O157 infection associated with pasteurised milk supply. Lancet *344*, 1015.
43. WANG, G., ZHAO, T. and DOYLE, M.P. 1997. Survival and growth of *Escherichia coli* O157:H7 in unpasteurised and pasteurised milk. J. Food Protection, *60*, 610–613.
44. WEAGANT, S.D., BRYANT, J.L. and BARK, D.H. 1994. Survival of *Escherichia coli* O157:H7 in mayonnaise and mayonnaise-based sauces at room and refrigerated temperatures. J. Food Protection *57*, 629–631.
45. WILLSHAW, G.A. *et al.* 1994. Vero cytotoxin-producing *Escherichia coli* O157 in beefburgers linked to an outbreak of diarrhoea, haemorrhagic colitis and haemolytic uraemic syndrome in Britain. Letters in Appl. Microbiol. *19*, 304–307.
46. ZHAO, T., DOYLE, M.P. and BESSER, R.E. 1993. Fate of enterohemorrhagic *Escherichia coli* O157:H7 in apple cider with and without preservatives. Appl. Environ. Microbiol. *59*, 2526–2530.

CHAPTER 22

CLINICAL MANAGEMENT OF *E. COLI* O157 INFECTION

W.T.A. TODD
and
S. DUNDAS

Monklands Hospital
Airdrie
Lanarkshire L6 OJS, United Kingdom

AND

J. COIA

Scottish E. coli O157 Reference Laboratory
Dept. of Clinical Microbiology
Western General Hospital
Edinburgh EH4 2XU, Scotland

INTRODUCTION

VTEC infections, particularly those associated with *E. coli* O157:H7, have over the last two decades become a significant public health problem in the developed world (Table 1). Although, when compared to the totality of food poisoning notifications, they remain in the minority, these organisms are very important due to the capacity to induce the life threatening complications of Haemolytic Uraemic Syndrome (HUS) or the closely related Thrombotic Thrombocytopaenic Purpura (TTP). These complications occur in 2-7% of sporadic cases but have been reported in up to 20% of cases in outbreaks. Clinical management involves the early recognition of appropriate signals in the history, presentation and laboratory indices of the disease and the institution of relevant management to prevent further deterioration.

EPIDEMIOLOGY

The Epidemiology of VTEC especially *E. coli* O157 has become much clearer in the last five years. The reservoir of infection is clearly the gut of healthy cattle (39) but the organism has been isolated/associated with horses (21), sheep (42,77), deer (22,26), goats (61) and other animals such as geese

(91). Association or contact with farm animals has been recognised as a frequent association in the transmission of infection to humans (9,61,91) and is particularly important as the popularity of school visits to farm environments increases. Seasonal shedding of the organism by large herbivores links closely with the clear seasonality of infection in the warm summer months in both the northern and southern hemispheres (23). The nature of animal feed is also important, with increased shedding in animals fed on hay as opposed to grain (44). In early outbreaks infection was clearly transmitted by surface contamination of butcher meat, especially beef. Minced beef, or hamburgers made from it, were well-publicised vectors. *E. coli* O157 is recognised to have a very low infecting dose for humans (13), possibly as low as 50 viable organisms (89). This was not only compatible with the established route of infection, but also explained how contamination of fomites and their spread from butcher's equipment to cold, cooked meat products could cause such a massive outbreak as the Central Scotland outbreak. *E. coli* O157 has been shown to survive in manure, standing water and water run off from grazing land, water troughs, etc., and in recent months has increasingly been recognised as a waterborne infection (7,46). The use of water from grazing land for the irrigation of edible vegetable (salad) has caused major outbreaks of infection from mesculun lettuce (92) and radish shoots (60,94). Outbreaks have been associated with paddling pools, beaches and other water sources (10,69). The common housefly *Musca domestica* has been demonstrated to act as a vector of infection with the organism actively surviving in the insect's gut (51,63].

TABLE 1.
INCIDENCE* OF COMMON FOOD POISONING BACTERIA OVER THE LAST 3 YEARS

Organism	*Campylobacter*		*Salmonella*		*E. coli* O157	
Year/ source	England/ Wales	Scotland	England/ Wales	Scotland	England/ Wales	Scotland
1996	43337	5098	28983	3266	660	506
1997	50177	5523	32596	3349	1087	422
1998	58059	6375	23420	2109	890	216

*data provided by PHLS and SCIEH

Escherichia coli O157 is shed in the stools from human cases for several weeks, with younger children carrying the organism for longer than older adults. The median shedding time for the under five is 17 days after onset of symptoms (40). One case in the Central Scotland outbreak shed the organism for 75 days (Dr. K. Liddle, personal communication). Secondary attack rates of up to 22%

have been reported in young children (13). Person to person spread is well recognised and has been important in causing outbreaks in schools, long term institutions, families and day care facilities (20,70), and has been demonstrated for non O157 VTEC strains (16). It has also been recognised as a source of nosocomial infection and infection in laboratory personnel (28). Family studies show that symptom free infection occurs frequently (86). Most reported *Escherichia coli* O157 infection still occurs in outbreaks, but a recent population prevalence study showed that 39% of bloody stools that yielded an enteric pathogen were *Escherichia coli* O157 positive, suggesting that this organism remains significantly under recognised as a cause of sporadic diarrhoea (82).

In summary then, the epidemiology of this infection has been clarified over the last decade (Table 2). The clinician must now seek relevant information about farm animal contact, and must consider the possibility of waterborne infection. The relevance of the small infecting dose cannot be overemphasised and fruit and vegetable sources of infection must also be considered. The well-recognised association with minced beef products and cross contamination from uncooked meat to cooked products obviously remains. As in any potential cases of food poisoning, the association with other potential cases is very important and information must be sought with respect to possible person to person spread, particularly amongst family members or other close contacts.

TABLE 2.
EPIDEMIOLOGY OF *E.COLI* O157:H7

RESERVOIR	Healthy Cattle (gut) Seasonal shedding (summer months) Contaminated water courses
TOLERANCE	Acid, Fermentation Drying
INFECTING DOSE	< 50 organisms
MODE OF TRANSMISSION	Faecal - oral Person to person Occupational (nurses, microbiologists)
VEHICLES OF TRANSMISSION	[any article contaminated with cattle faeces, impure water or untreated human sewage] Contaminated water / manure/ run-off from grazing land Fresh (uncooked) meat products (surface) and ground meat Improperly cooked or contaminated cooked meat Contaminated vegetables. e.g.: Lettuce, radish sprouts, unpasteurised apple juice. [possible: house-fly]

CLINICAL FEATURES

Escherichia coli O157 was first associated with a distinct clinical syndrome of haemorrhagic colitis in 1982 leading to the eponym of Enterohaemorrhagic *Escherichia coli* or EHEC. Observation from large outbreaks has allowed clear documentation of the clinical features of this condition. These are summarised in Table 3.

TABLE 3.
CLINICAL FEATURES OF *E. COLI* O157:H7 INFECTION

CLINICAL FEATURE	PERCENTAGE INCIDENCE
Vomiting	30-60
Fever	<30
Abdominal Pain	70-80
Diarrhoea	95
Bloody diarrhoea	>70
Complication Rate (HUS/TTP)	2-7 in sporadic, 20 in some outbreaks
Asymptomatic infection	10-15

The average incubation period is 3 days but a range of 1-8 days has been recorded. Increased understanding of the complex epidemiology of this infection may well allow closer definition of this range over time. The clinical manifestations of the infection are diverse. They range from symptom free carriage which occurs in up to 10% of cases (86) through non-bloody diarrhoea and frank haemorrhagic colitis to the complications of HUS/TTP and death.

Typically the initial symptoms of infection are abdominal cramps with watery, non-bloody diarrhoea. In up to 70% of cases the diarrhoea will become frankly blood stained, usually within 48 hours of onset. The mean duration of diarrhoea is 7 days. Abdominal pain and tenderness can be severe, often prompting an initial diagnosis of surgical pathology. Vomiting occurs only in a significant minority (30-60%) of cases and fever is reported in less than 30% of cases (59). Systemic complications develop in approximately 10% of cases and appear after 5-7 days of clinical illness. The case fatality rate varies between 3-7% (39).

The clinical features that have been shown to distinguish *Escherichia coli* O157 or other VTEC infections from other enteric pathogens are a history of bloody diarrhoea, visibly and uniformly blood stained stools, lack of reported or observed fever, a leukocyte count greater than $10 \times 10^9/l$ and abdominal tenderness on examination. Obviously, no single feature is specific but these features in combination are highly suggestive of infection with *Escherichia coli* O157 (82).

The differential diagnosis nevertheless remains wide. In the young infant it must include intessuception or an enterocolitis from organisms such as staphylococci and clostridia. Gastrointestinal infection with *Campylobacter* or *Shigella* is part of the differential diagnosis at any age, as is inflammatory bowel disease. In the elderly, *Clostridium difficile* related pseudomembranous colitis, or colonic lesions of a surgical nature such as colonic tumours or ischaemic colitis, form the most obvious differential diagnoses. The primary care physician or the admitting officer, whether medical or surgical, must be aware of VTEC as a potential cause of diarrhoea, bloody or not, in cases across the age range. This is especially relevant when the constellation of clinical symptoms and signs described above is present.

Complications

Gastrointestinal Complications. The most common gastrointestinal complication is rectal prolapse, which occurs in approximately 10% of patients. More severe gastrointestinal complications are not infrequent and include ischaemic colitis, appendicitis, oesophageal stricture and large bowel perforation (38,64). Gangrene requiring partial colectomy occurs in 2% of cases.

Systemic Complications. The most severe systemic complications are HUS and TTP. HUS is a triad of microangiopathic haemolysis, thrombocytopenia and acute renal impairment. TTP was traditionally described as a thrombotic microangiopathy (TMA) (microangiopathic haemolysis and thrombocytopenia), with neurological symptoms and signs predominating over renal impairment. In the context of *E. coli* O157 infection HUS and TTP are a consequence of the same pathological microvascular process and can be thought of as overlapping syndromes. Complete HUS/TTP is usually established 6 days after the onset of gastrointestinal symptoms, although laboratory evidence of TMA is likely to be evident at an earlier stage. Thrombotic microangiopathy alone, without clinical complications, may be detected when monitoring blood samples. The renal presentation of established HUS/TTP is a rising urea and creatinine, culminating in oliguria. Neurological features of HUS/TTP are reported to occur in 25% of cases and usually present as drowsiness, cerebrovascular accidents, seizure and coma. Pleural and pericardial effusions occur commonly, particularly in the elderly with HUS/TTP. Pancreatitis and diabetes mellitus have also been described as complications of the systemic microvascular process. The mortality rate from HUS/TTP in children is 5%, however the mortality rate in the elderly is much higher. The two outbreaks reporting the largest number of adult cases have recorded mortality rates of 88% (20,34).

PATHOGENESIS

An elementary understanding of the pathogenic mechanisms of this illness may help to improve the appreciation of possible interventions that may influence the clinical outcome in this infection.

The most important virulence characteristic of the organism is its ability to produce one or more verocytotoxin (or Shiga toxin S1 or S2). These toxins are phage encoded and consist of 5 B subunits and one A subunit. They bind through the B subunit to globotriasylceramide (Gb_3) found in varying amounts in membranes of eukaryotic cells. The A subunit once absorbed acts to block internal cellular protein synthesis.

Toxin production alone, however, is not sufficient to cause disease, which also requires the production of haemolysin and intimin along with other factors necessary for the production of an "attachment-effacement" lesion in the gut mucosa (66). The organism, which never leaves the gut lumen, therefore causes disease by closely adhering to the mucosal brush border and causing local disruption. Such disruption may be sufficient to cause the initial non-bloody diarrhoea.

Toxin production mediates both local and systemic disease. Local effects on the intestine cause the development of bloody diarrhoea with histopathologically evident haemorrhage and oedema of the lamina propria. Post-diarrhoeal HUS/TTP results from microvascular disease when the toxins enter the bloodstream and bind to Gb_3 receptors on endothelial cells. These are especially abundant in kidneys and brain (66) and are related to the Pk Blood group antigen on the surface of red blood cells (11,31). The process of microvascular damage is thought to be mediated by pro-inflammatory cytokines such as TNF-α and interleukins (54). Endothelial damage triggers the formation of large von Willebrand factor multimers (62) which cause platelet aggregation leading to thrombocytopenia, disruption to passing erythrocytes (haemolysis) and occlusion of renal and or cerebral microvasculature leading to the serious renal impairment and neurological problems (seizures, CVA) noted in resultant disease.

Thus, the clinician has an imperative to attempt to diagnose VTEC infections using the foregoing clinical markers. Once the possible diagnosis is entertained or has been confirmed by microbiological means, certain features can then be used as prognostic markers to indicate potential development of complications. The possibility of effective therapeutic intervention, either with synthetic toxin receptor therapy to block toxin action or with therapeutic plasma exchange (34,88) (although not confirmed by a Randomised Controlled Trial), increases the pressure on clinicians to reach a speedy diagnosis of early complications, so that appropriate action can be taken.

PROGNOSTIC INDICATORS

Various risk factors have been identified as associated with the development of HUS/TTP and other systemic complications of infection (Table 4). The most consistent risk factor is extremes of age. Cases <5 yrs (32,33) and >65 yrs (20, 78) of age have been shown to be at the greatest risk of developing HUS/TTP. Retrospective studies of outbreaks, including the massive outbreak in Sakai City in 1996, have largely focused on risk factors for HUS/TTP in the paediatric population. In children fever, bloody stools (70) vomiting (12,82) and the use of antimotility agents (12,25) have been identified as other potential risk factors in the progression to HUS/TTP. Early analysis of the cases in the Central Scotland outbreak has suggested a number of interesting associations with the development of HUS/TTP in adults. Clinical features associated with HUS/TTP were a shorter incubation (less than four days) period, fever or tachycardia on admission. The severity of gastrointestinal symptoms was not correlated with HUS/TTP.

TABLE 4.
RISK FACTORS FOR THE DEVELOPMENT OF HUS
[those shown in parenthesis are doubtful]

Bloody Diarrhoea
Age <5 yrs, >65 yrs
Elevated WBC on admission
Low serum albumin on admission
[Fever]
[vomiting]
[Use of anti-motility agents]
[Female sex]

Despite evidence that this organism has a high tolerance to acid (89), adults in the Scottish cohort with low gastric acid (previous gastrectomy or taking proton pump inhibitors H2 receptor antagonists) were also at increased risk of HUS/TTP.

LABORATORY RESULTS AS PREDICTORS OF HUS/TTP AND MORTALITY

A raised white blood cell count has been consistently shown to predict both the development of HUS/TTP (70,93) and the likelihood of there being a poor outcome from this complication. Recent evidence from the 1996 Sakai City outbreak demonstrated a significant association between a rise in both leukocyte count and in C-reactive protein level and the development of HUS (48). Analysis of the Central Scotland outbreak has confirmed a white cell count greater than $20 \times 10^9/l$ and an absolute neutrophil count $>15 \times 10^9/l$, within 48 hours of symptom onset, as the earliest laboratory predictors of HUS/TTP. Hypoalbuminaemia was also an early predictor of the development of HUS/TTP in this outbreak, however, early hypoalbuminaemia may be, in part, a reflection of lower albumin levels linked with advancing age or pre-existing illness in this cohort.

Monitoring of Hospitalised Cases

All cases admitted to hospital should have the laboratory markers of Full Blood Count (FBC), Blood Film, Lactate Dehydrogenase (LDH), Urea and Electrolytes checked on admission. Any patients exhibiting abnormal values in relation to these indices, including an isolated neutrophilia, should have the bloods repeated on a daily basis until there is clear resolution. In the face of unstable clinical condition or worsening symptoms it is also appropriate to observe blood values on a daily basis. It must be recognised that these laboratory markers may be abnormal in affected individuals who do not develop HUS/TTP or TMA and when analysed singly they carry little premonitory significance.

MANAGEMENT

Gastrointestinal Infection

In general, strategies for the management of patients with acute gastrointestinal infection focus on supportive measures, with careful attention to the correction of fluid and electrolyte balance. In *E. coli* O157 infection, abdominal pain is prominent and analgesia and antispasmodic therapy are often required. There has been concern that opiate analgesia and antimotility agents could delay clearance of the organism and therefore increase toxin absorption. Retrospective assessment of outbreaks has provided evidence to support an association between the use of antimotility and antidiarrhoeal agents and progression to HUS/TTP (12,24,43). Universally, the recommendation is that antimotility and antidiarrhoeal agents should be avoided in *E. coli* O157 infection.

The role of antibiotics in the management of *E. coli* O157:H7 infections has not been clearly defined. Clinical experience with *Shigella dysenteriae* had raised concern that antibiotic therapy may increase the release of Shiga toxin by *E. coli* O157 (19). Two retrospective studies of *E. coli* O157 outbreaks found that patients who developed HUS were more likely to have been treated with antibiotics (20,70). Two studies that antibiotics had a protective effect (25,55), while others found no evidence of any effect (12,40). In the only randomised prospective study, in which treatment was started late, and small numbers were involved, cotrimoxazole had no significant effect on progression to HUS (74). Retrospective analyses of the massive 1996 Sakai City outbreak in Japan serve only to add to this conflict. One study found that children treated with fosfomycin within 2 days of illness onset had a lower risk of HUS than children who began treatment later or received no antibiotic (45). A second study concluded that, in cases admitted to hospital, 4-quinolones prevented HUS and treatment with fosfomycin was associated with HUS (81). This clinical observation is contradicted by recent in vitro evidence, which suggests that 4-quinolones increase the release of verotoxin (50). In the Central Scotland outbreak of *E. coli* O157, no relationship was demonstrated between early treatment with ciprofloxacin and HUS/TTP, however only a small number of cases, judged to be at risk of severe infection, received early antibiotic therapy according to national guidelines (35). The most contentious evidence resulting from the Central Scotland outbreak is that relating to antibiotic administration before or during the acute phase of illness. Patients coincidentally treated with antibiotics in the four weeks prior to the onset of *E. coli* O157 related symptoms were at increased risk of progression to HUS/TTP. It may be that subtherapeutic or inappropriate antibiotic therapy increases the release of verotoxin, an effect that has been demonstrated in vitro (98). The currently available evidence suggests that the role of antibiotics in HUS/TTP may be dependent on their mode of action and the timing of their administration. Therefore the safety or otherwise of the different antibiotic classes in the treatment of *E. coli* O157 remains an area of debate.

HUS/TTP

Initial treatment of HUS/TTP is supportive and aimed at managing the different complications such as fluid and electrolyte abnormalities, anaemia, thrombocytopaenia, hypertension and seizures. Of these patients, 75% will require either red blood cells or platelet transfusions. The finding that changes in leucocyte count and albumin were observed earlier than changes in the laboratory parameters defining HUS/TTP suggests that by the time the diagnosis of HUS/TTP is made, the pathological process is already well established. Progression from thrombotic microangiopathy (TMA) to the complete syndrome of HUS/TTP seems to follow a different pattern in adults than in children. Children tend to develop acute renal failure as an early manifestation of HUS/TTP with 50% becoming oliguric, requiring peritoneal or haemodialysis. In adults, however, renal disease may progress more slowly and adults often develop life threatening neurological or cardiorespiratory problems before the onset of acute renal failure. In adults, a therapeutic window for the treatment of the extended microvascular sequelae of TMA may therefore exist prior to indications for renal dialysis. In adults with idiopathic HUS, and in patients of all ages with TTP, a general consensus still advocates the use of therapeutic plasma exchange (TPE). This is controversial in *E. coli* O157 associated HUS/TTP as experience of TPE in the treatment of HUS/TTP induced by this organism is limited. In idiopathic HUS/TTP, it is still unclear whether the benefit of therapeutic plasma exchange (TPE) is in the removal of an injurious factor from the patient's plasma or the infusion a factor deficient in the patient, but contained in the donor fresh frozen plasma (FFP), or both. In the context of *E. coli* O157 infection it is postulated that TPE both removes circulating verotoxin and replaces factors depleted in plasma during the process of endothelial damage and thrombus formation. Experience from the Central Scotland outbreak has suggested that TPE may be clinically effective treatment when given to adults with *E. coli* O157 induced HUS/TTP (34). Twenty-two adults developed HUS/TTP (mean age 71 years) and 16 of these were treated with TPE. The overall mortality in adults with HUS/TTP was 45%. The only comparative "control" group available was a historical one in which the mortality rate associated with HUS/TTP in the elderly with *E. coli* O157 was 88%. The outbreak occurred in a Canadian nursing home. Cases of HUS/TTP were hospitalised and managed with full supportive therapy excluding TPE. TPE is an expensive and intensive procedure and its role in the treatment of *E. coli* O157 induced HUS/TTP still needs to be determined definitively.

Outbreak Management

In an outbreak situation, a management protocol should be agreed between key public health, primary and secondary care physicians. Information should be distributed to managing clinicians on the clinical course of *E. coli* O157 infection, monitoring strategies for HUS/TTP, criteria for admission and guidelines for treatment of patients with uncomplicated gastrointestinal infection and HUS/TTP.

Children and the elderly, who are at greatest risk of HUS/TTP, should be targeted for laboratory monitoring. Laboratory monitoring is also required in patients of all ages with hypochlorhydria, immunosuppression or clinical evidence of severe infection. FBC, Blood Film, Urea and electrolytes and LDH should be checked on initial presentation. Cases with a neutrophilia, evidence of red cell haemolysis, abnormal renal function or thrombocytopaenia should have all blood analyses repeated within 48 hours and the interval for further monitoring dictated by the trend in abnormalities. Monitoring should continue for at least seven days or until all abnormalities, including neutrophilia, are clearly reversing. Cases with normal blood results on first presentation should be screened again if clinically indicated. The need for hospital admission is likely to be dictated by the clinical condition of the patient, however, cases with progressive laboratory abnormalities should also be admitted, and the criteria for this should be defined in the management protocol. Clinicians should be advised to avoid antidiarrrhoeal agents. Given the current debate, advice regarding the use of antibiotics in gastrointestinal infection and TPE in HUS/TTP remains a dilemma for those determining management protocols.

TABLE 5.
MANAGEMENT OF *E. COLI* O157 INFECTION

1/Established management

Fluid replacement	Replace Established losses Replace ongoing losses Replace maintenance requirements
Pain Control	Adequate analgesia +/- antispasmodic
Appropriate monitoring	Standard + LDH, U/E, platelets, FBC
Renal failure (established)	Haemodialysis

2/Controversial management

Anti-motility agents	Use with extreme care
Antibiotics	Caution (see text)
Therapeutic Plasma Exchange	Use in appropriate cases

Long-term Sequelae

Complications of both gastrointestinal and systemic disease occur. The most common complication of gastrointestinal disease is post-infectious irritable bowel syndrome (IBS) which may occur in up to 30% of cases. Gastrointestinal features documented during the acute illness do not seem to predict subsequent development of IBS. Extended prospective assessment is required to assess the chronicity of gastrointestinal symptoms in patients after *E. coli* O157. Gallbladder disease and late colonic stricture have been reported as complications of the gastrointestinal disease. Late colonic strictures are thought to be caused by chronic bowel ischaemia, the pathological evidence for which has been demonstrated in biopsies taken during the acute phase (38).

E. coli O157 is responsible for more than 90% of HUS cases in children (79) and this organism now is the most common cause of chronic renal failure in children. Of these children, 9% develop end stage renal disease after HUS. Reduced glomerular filtration rate (GFR), increased total urinary protein, increased microalbuminuria and reduced renal functional reserve have been reported in up to 30% of cases (36). Neurological sequelae are reported in 4% of cases and pancreatitis and diabetes mellitus may also occur as late sequelae of HUS/TTP. Again prospective assessment is required to assess the incidence of these complications in the young and the old.

PREVENTION OF NOSOCOMIAL VTEC INFECTION

VTEC Infection as a Risk in the Health-care/Day-care Environment

The risk of secondary spread of VTEC infection in the domestic setting is well recognised, and has been discussed elsewhere. However, it is equally apparent that such transmission is by no means confined to the home. Many such outbreaks have occurred in a wide range of institutional settings (including hospitals), residential care facilities for the elderly, nursery school and day-care facilities for pre-school children (27).

There are a number of reasons why such outbreaks may occur. To some extent these reflect particular properties of the organism. On the other hand, the risks may be ascribed in part to the particular patient group affected, or to the way in which particular clinical presentations or groups of symptoms are commonly handled within the institutional or clinical setting. The following sections will address these various aspects in more detail.

Unfortunately, as with so much of our knowledge of these organisms, the majority of information is available for and refers to VTEC O157. This is not surprising as much of the data is derived from outbreak-related studies, and globally the majority of outbreaks of infection have been due to *E. coli* O157. However, it is generally believed that those measures which are considered appropriate for the control of VTEC O157 are adequate for the prevention of transmission of infection by other VTEC organisms.

Risk of Nosocomial VTEC Infection Related to Current Knowledge of the Properties of the Organisms

The key properties of the organisms which are likely to be pertinent to the propensity for secondary transmission of VTEC infection are: low infectious dose, environmental persistence, and resistance to physical and/or chemical agents and treatments. Studies that have directly addressed these questions in detail are at present limited, but this is currently an area of active research interest (56,76). Much of our present understanding of these factors relates to information derived from outbreak studies.

Waterborne transmission has been a feature of a number of outbreaks. In some instances this has been associated with contamination of potable supplies (7,30,87). In other cases the recreational use of water either via sharing of childrens paddling pools, or related to swimming (37,49), has been associated with secondary spread. These examples, coupled with the generally high rate of secondary infection in outbreaks (27,73), all suggest that the infectious dose for this organism is small. As will be discussed later in the current chapter, this conclusion is further supported by the well-documented occurrence of laboratory-acquired infection in the absence of identified lapses in good laboratory practice.

Direct human challenge studies have not been performed for obvious reasons, and there is a lack of suitable animal models that are felt to accurately model the potential risk of infection. However, some more direct measure of the possible infectious dose is provided by quantitative culture of foodstuffs epidemiologically implicated in episodes of infection. These have demonstrated levels of contamination as low as 2 organisms per 25 g of incriminated food (97). In summary, currently available data suggest that ingestion of somewhere between 10 and 100 organisms is sufficient to cause VTEC O157 infection.

It is increasingly apparent that VTEC O157 are more persistent in the environment than is commonly believed to be the case for "typical" *E. coli*. Various studies have highlighted the ability of the organism to survive in soil for relatively long periods of time (57), and carriage in migratory bird populations has been demonstrated. With regard to the potential of nosocomial transmission, it is noteworthy that recent "in vitro" studies have demonstrated that some strains remain viable on stainless steel surfaces for periods of time in excess of 35 days (58).

The majority of available data suggest that VTEC are not particularly heat tolerant in comparison with their less pathogenic counterparts. Equally they appear to be sensitive to many of the commonly employed disinfectant agents which would be expected to have activity against strains of *Escherichia coli* (67).

However, it is now well recognised that these organisms may demonstrate enhanced acid tolerance, at least in comparison with many of the commonly studied laboratory strains of *E. coli* (14). This is thought to be of most significance in relation to the potential risks posed by contamination of fermented meat products that rely upon their relative acidity for their microbiological safety. However, this may increase the susceptibility to VTEC O157 infection of patients with hypochlorhydria, either as a result of medication or underlying disease (90).

Additional Problems Posed in Specific Settings

It is clear from the foregoing discussion that there are a number of characteristics of these organisms that might be expected to enhance their propensity for secondary transmission in general. However, it is readily apparent that there are a number of additional factors which relate to the setting in which that spread occurs, which can potentiate such transmission.

With any low dose pathogen transmissible by the faecal-oral route, standards of personal hygiene, particularly in relation to toilet practices, are of paramount importance. It is well recognised that poor standards of personal hygiene may prevail in day care facilities for young children. This factor in tandem with the propensity for severe sequelae of VTEC infection in this age group yields significant difficulties in the management and control of outbreaks (3,83). One US study has demonstrated a secondary attack rate of 22% in day care centres (13). Furthermore, there are frequently social and economic pressures to keep children at nursery or day care facilities. It is not unheard of for a child excluded in one outbreak to turn up as the index case of an outbreak in another day care centre, where the child has been sent, to allow parents to return to work.

Poor standards of personal hygiene may also be a feature in other settings. Serious outbreaks of *E. coli* O157 infection have been described in institutions for mentally retarded persons (70), and other episodes have involved psychogeriatric facilities (52).

The clinical presentation of VTEC O157 may itself be an important factor in subsequent nosocomial spread. *E. coli* O157 infection may commonly be present with non-bloody diarrhoea, which may result in diagnostic delay and failure to implement appropriate infection control measures. Equally, the severe abdominal pain and frankly bloody stools that are a hallmark of this disease may initially suggest a surgical diagnosis such as intussusception in children or acute diverticulitis in the elderly. Again this may lead to delay in establishing the infective aetiology and unnecessary exposure of staff and other patients to the risk of infection (95).

As well as these specific features which increase the likelihood of secondary transmission of VTEC infection, other factors which increase the risk of nosocomial infection in general will also increase the occurrence of secondary VTEC O157 in this setting. These would include inadequate or inappropriately trained staff, which increases the likelihood of lapses in proper enteric precautions. Equally, lack of appropriate isolation facilities may further facilitate secondary spread. Even where appropriate staffing and physical resources are available, the lack of adequate documented infection control policies and procedures may frustrate attempts to minimise nosocomial transmission (29).

Review of Information on Current Knowledge of Nosocomial Outbreaks that Have Occurred

Most episodes of secondary VTEC infection in the domestic setting have occurred either in residential care facilities for the elderly, or in nursery and day care establishments (84). It is likely that this reflects the balance of risk factors outlined above.

Outbreaks in the nursery and day care setting may be associated with serious sequelae, and may present significant challenges for their adequate control (3,83). There is some evidence to suggest that faecal shedding of organisms may be prolonged in this group (86).

As has previously been discussed, many of the features which predispose to secondary infection in the day care setting may also feature in facilities for mentally and physically handicapped or psychogeriatric patients. Serious nosocomial outbreaks of *E. coli* O157, which affected residents and staff, were described by Pavia and colleagues (70) in institutions for mentally retarded persons. In 1990 a severe outbreak of VTEC O157 infection occurred in Lanarkshire, Scotland involving two wards of a large psychogeriatric hospital, which affected 8 patients (of whom 4 died) and 3 staff members (52). Detailed epidemiological investigation implicated food brought into the hospital as the potential source, but subsequent nosocomial transmission affecting both staff and patients clearly occurred. In 1992, 5 patients in a different psychiatric hospital in the west of Scotland were involved in a similar incident (80).

The morbidity and mortality that may arise from outbreaks of *E. coli* O157 infection amongst elderly patients should not be underestimated (78). In a detailed study of a large outbreak in a US nursing home, which affected 55 of 169 residents and 18 of 137 staff members, there was clear evidence of person-to-person transmission involving patients and carers and an overall mortality rate of 36% (20). In those 12 residents who developed HUS, the mortality rose to 88%. In 1996, 20 residents and staff in a residential home for the elderly in Glasgow were affected in an outbreak of *E. coli* O157 infection, initially of foodborne origin, but with subsequent nosocomial transmission (4). More recently in Arbroath, Scotland a similar nursing home episode was described, again affecting staff and patients, with subsequent secondary spread (6).

It is important to remember that outbreaks have not been confined to psychiatric hospitals, day care or chronic care facilities. There have now been a number of documented episodes in secondary and tertiary acute care facilities. In 1988, Karmali and colleagues in Toronto described VTEC O157 infection, and the subsequent development of haemorrhagic colitis (HC) in a nurse (47) who had been caring for a child with HUS. The nurse was subsequently able to identify at least one lapse of enteric precautions that had arisen while caring for the child. In 1995, the index case in a nosocomial outbreak in a residential care home for the elderly in the north of England was admitted to a district general hospital for further treatment. In the course of that admission nosocomial transmission to another patient on the same ward, who subsequently developed severe HC, is thought to have occurred (Dr. L. Lighton, personal communication). In a similar incident in the north-east of England, nosocomial transmission of VTEC O157 to two patients and three members of nursing staff occurred from the index case admitted to a district general hospital. Two patients died in this episode, although it is noteworthy that 2 of the infected staff remained asymptomatic throughout (96).

In all such episodes, it is essential that there is thorough and careful investigation to determine the exact source of infection and the nature of any subsequent transmission, which cannot be assumed to be nosocomial. An outbreak in an acute-care hospital in Toronto that involved 8 patients, 10 staff members and 3 voluntary workers was subsequently traced to the consumption of contaminated iceberg lettuce in a green salad prepared within the hospital kitchen (72). This contrasts with a district general hospital outbreak in central Scotland (8) in which 16 patients and 11 staff members were affected in a continuing care ward. Microbiological and epidemiological investigations revealed that all of the cases had consumed contaminated birthday cake prepared outside the hospital and subsequently consumed at a ward party (65).

Advice on Reducing the Risk of Nosocomial VTEC Infection, and How Ongoing Transmission Can Be Controlled

The prevention and control of secondary transmission of VTEC infection in the domestic setting is crucially dependent upon the application of good hygienic practices. The importance of formal locally-agreed written policies and protocols for the adequate documentation of these procedures cannot be overemphasised. All staff must be adequately trained in their implementation and compliance monitored. Sufficient resources must be made available to ensure that the appropriate standards can be maintained at all times. Detailed revised guidelines for the control of VTEC infections have recently been published (85), and the major features are outlined below.

All institutions must provide adequate toilet facilities that are appropriate to the needs and capabilities of the client group. Good hand hygiene is paramount in the prevention of transmission of all types of infection, and staff must not only be fully trained themselves, but must be satisfied that those in their care are aware of how and when to wash their hands. Shared towels should be avoided as far as is possible. In the nursery and day care setting, and in some adult groups where there is significant intellectual or physical impairment, this will often entail routine direct supervision of handwashing after toileting and before meals.

It is essential that episodes of contamination with blood and/or faeces are dealt with promptly according to locally agreed policies and procedures for cleaning, disinfection, treatment of infected linen and disposal of clinical waste. The potential for contamination of equipment, children's toys, sand pits, paddling pools, etc., particularly in the outbreak situation must not be overlooked.

Particular difficulties may arise in dealing with cases of VTEC infection in residential care homes for the elderly. The availability of single room accommodation with handwashing facilities for symptomatic patients should be considered a priority. If possible, sharing of toilets, commodes, towels and other equipment should be avoided while patients are symptomatic, and adequate arrangements should be in place for the subsequent cleaning and disinfection of such items, prior to their return to general use. All patients must be assessed on an ongoing basis by a medical practitioner to determine their requirement for further treatment and/or referral.

In the acute care setting the requirement for source isolation with enteric precautions should be considered for cases of undiagnosed diarrhoea. It should also be remembered that VTEC infection may not present with classical clinical manifestations such as HC or HUS, and diarrhoea may not infrequently be non-bloody. Equally, the severe abdominal pain and frankly bloody stools that are a hallmark of this disease may initially suggest a surgical diagnosis such as intussusception in children or acute diverticulitis in the elderly. This underlines the need for a high level of awareness of the potential infection hazard posed by any patient presenting with undiagnosed diarrhoeal symptoms, and the importance of sending stool specimens for culture.

Disposable bedpans, disposed of in a suitable macerator, are preferable to non-disposables. Where the latter are used, appropriate arrangements for disposal of waste, cleaning and disinfection must be implemented.

Staff should wear disposable gloves and aprons when dealing with patients with diarrhoea, handling commodes and bedpans, or dealing with episodes of soiling and contamination. Gloves and aprons should be removed prior to leaving isolation rooms, and hands must be thoroughly washed and dried. Some workers have advocated the use of full protective gowns for known or suspected cases of VTEC infection (96).

Exclusion from Work/School

All food handlers with diarrhoea must be excluded from work until they are asymptomatic and have had normal formed stools for 48 hours. Food handlers who are cases or household contacts of VTEC infection should also have two consecutive negative faecal samples taken after recovery and at least 48 hours apart.

Health care, pre-school nursery, or other staff with diarrhoea, who have direct or indirect contact via serving food with susceptible groups, should be excluded from work until they are asymptomatic and have normal formed stools. If they are cases or household contacts of VTEC infection they should also have two consecutive negative faecal samples taken after recovery and at least 48 hours apart.

In general, children under 5 with diarrhoea should not attend daycare or pre-school nursery facilities until they have recovered and have formed stools. Nappies and nappy changing present particular hazards, and adequate facilities, in a separate easily cleaned and disinfected area, must be available.

Children under 5 with VTEC infection should be excluded from nurseries and playgroups until they are asymptomatic and until they have two consecutive negative faecal samples taken after recovery and at least 48 hours apart. Household contacts under 5 years of VTEC cases should be excluded from nurseries and playgroups until the case is asymptomatic. Consideration should be given to obtaining microbiological evidence of clearance in these contacts. Excluded children should not be allowed to attend other facilities.

In older children and adults unable to implement adequate personal hygiene, cases of VTEC infection and household contacts should be excluded from school/work until they have two consecutive negative faecal samples taken after recovery and at least 48 hours apart.

Asymptomatic household contacts of cases who are not in any of the above categories need not be excluded from school/work, and need not routinely submit faecal samples for examination. However, all such individuals should be advised of the potential risk of cross infection within the household, and the importance of appropriate measures, particularly hand hygiene, in its prevention.

PREVENTION OF LABORATORY-ACQUIRED VTEC INFECTION IN THE CLINICAL LABORATORY

VTEC Infection as a Particular Risk in the Clinical Diagnostic Laboratory

The severity of illness that often occurs as a result of VTEC infection is such that medical practitioners may be more inclined to submit specimens to the diagnostic microbiology laboratory. Patients are not uncommonly hospitalised, which again increases the likelihood that specimens will be sent. Those properties of VTEC which make them particularly challenging from clinical and infection control perspectives, namely their potential for severe sequelae and low infectious dose, are equally problematic in terms of the potential for, and hazards arising from, laboratory-acquired infection. Unfortunately, this potential hazard has been realised in the occurrence of a number of laboratory-acquired infections.

Review of Current Knowledge of Episodes of Laboratory-acquired Infection Which Have Occurred

A probable laboratory-acquired case involving a hospital laboratory technician was described during the course of a surveillance study in Washington State (68). In a separate episode, faecal culture and serology confirmed a presumptive laboratory-acquired case of VTEC O157 infection (15). In this case the laboratory worker developed renal failure requiring haemodialysis after having handled cultures of the organism daily prior to the onset of her illness. In two separate incidents (18,75), microbiology laboratory workers developed severe haemorrhagic colitis after working with *E. coli* O157, despite any identified lapses in good laboratory practice. In a further episode a trainee biomedical scientist became infected with VTEC O157, and subsequently developed HUS, after handling a quality control (QC) specimen which contained the organism (5).

Advice on Safe and Efficient Procedures to Assist Clinical Laboratories to Process Samples Potentially Contaminated With VTEC Organisms

There was a clear need to minimise the risk of future occupational exposure of laboratory staff in the light of the potential hazards illustrated by these cases and other undocumented incidents that had been reported directly to regulatory bodies. In July 1996 the UK Health and Safety Executive (HSE) put forward a proposal incorporating a recommendation to the European Commission that verocytotoxigenic *E. coli*, including *E. coli* O157, should be re-classified as Hazard Group 3 organisms (but with a facility for derogation from full containment level 3). Interim guidance was issued to all interested parties and relevant agencies in December 1996. The proposal to the European Commission was formally adopted on October 7, 1997 (97/59/EC) for implementation by member states by March 31, 1998. In the UK the relevant guidance was issued by the Advisory Committee on Dangerous Pathogens (ACDP) as a supplement (2) to existing documentation on the categorisation of biological agents according to hazard and categories of containment. In recognition of the fact that these agents do not primarily infect via the airborne route, this advice recommended that VTEC organisms should be subjected to derogation from the full Hazard Group 3 precautions in a similar manner to the existing guidance for *Salmonella typhi* and *Shigella dysenteriae* Type I (1).

This guidance has a number of consequences for diagnostic and food microbiology laboratories. Most importantly, the new guidance does not significantly alter requirements for processing most routine samples. Initial examination and culture of specimens for a range of pathogens, including VTEC organisms, should continue to be performed under containment level 2 conditions, as required in the UK by the Control of Substances Hazardous to

Health 1994 (COSHH) regulations. However, if presumptive VTEC organisms are isolated, or if the laboratory is to undertake the processing of any specimen in which there is strong clinical or epidemiological likelihood that these organisms are present, further processing should be performed in containment level 3 conditions. Examples of this would include cultures of stool specimens from patients with HUS; cultures of stool specimens associated with ongoing outbreaks where VTEC infection is known or suspected; or work required to further identify or characterise putative VTEC isolates. Obviously if the laboratory is handling known isolates of such organisms, the same conditions apply. In the diagnostic laboratory setting, further work on presumptive VTEC isolates should be restricted to that which is essential for immediate clinical diagnostic and management requirements.

In food laboratories similar considerations should prevail. It would be entirely appropriate to conduct routine work under Containment Level 2 conditions unless, or until, there is any reason to suspect that VTEC organisms may be present. Once such organisms have been identified, or there are good grounds for considering they may be present, containment level 3 precautions must be employed. In many cases this may entail transfer of material to another laboratory with the appropriate facilities.

Individual laboratories have a duty to ensure that specific hazards posed by the procedures used by them for detection and isolation of these organisms are properly addressed. Although *E. coli* O157 has been derogated from the full Hazard Group 3 precautions as outlined in the preceding section, the range and extent of containment measures employed must be determined by local risk assessment under COSHH of all proposed procedures. This assessment must take particular account of the specific hazards posed by those procedures, especially if the method is likely to result in significant enrichment and/or concentration of the organisms; present a risk of aerosol formation; or involves handling large volumes of culture material which would be difficult to contain or decontaminate safely should spillage occur. For example, although the derogation from full category 3 precautions states that a microbiological safety cabinet is not essential, if the local risk assessment under COSHH determines that the proposed manipulation carries an aerosolisation risk, then a suitable Class I or Class III cabinet must be used. Similarly, if large-volume broth enrichments are proposed, even on lower risk samples, such cultures should be handled in a containment level 3 suite, which can be more readily disinfected should any spillage occur, rather than in a general laboratory area. However, it must always be remembered that containment measures are not a substitute for meticulous technique and frequent and thorough handwashing which are the cornerstone of prevention of laboratory-acquired enteric infection (71).

Laboratory-acquired infection has been associated with handling quality control (QC) materials (80). In order to minimise the risk of such incidents, laboratories should review procedures to ensure that unnecessary handling of potentially pathogenic VTEC isolates is minimised. This includes the use of non-toxigenic strains (e.g., NCTC 12900) for internal and external quality assurance work.

THE FUTURE

In terms of prevention of infection, work is already underway in vaccine development, and vaccination has been shown to prevent *E. coli* O157 infection in animals. Possible future treatments for *E. coli* O157 infection include oral verotoxin binding resins (88) and verotoxin neutralising antibodies to limit or prevent HUS/TTP (53). Greyhound dogs have recently been found to develop renal disease similar to HUS when exposed to verotoxin (41). This discovery should provide greater opportunity for investigation of the pathogenesis of verotoxin induced endothelial damage and the assessment of novel treatment modalities. The mainstay of prevention at present, however, is the education of farm workers, the enforcement of abattoir standards, the implementation of stricter licensing regulations for meat retailers and the promotion of good hygiene in consumers. There is also a need for greater epidemiological vigilance to enable the early identification, investigation and limitation of outbreaks.

CONCLUSION

Seventeen years after its recognition, outbreaks and sporadic infections attributed to *E. coli* O157 continue to increase. Acute gastrointestinal and systemic complications are frequent and severe. Risk factors for the development of HUS/TTP have been identified, of which the extremes of age and elevated leucocyte count are consistent. Recent evidence from the central Scotland outbreak suggests that individuals who are taking gastric acid lowering drugs or antibiotics coincidental with their infection with *E. coli* O157 may also be at increased risk. Clinical management, in particular the role of antibiotics in gastrointestinal infection and the role of TPE in HUS/TTP, remains controversial. These controversies in management could be resolved by a properly executed randomised controlled trial (which may prove logistically difficult), or by an international register of cases. Significant chronic sequelae of infection do occur and their frequency and severity are likely to become manifest over the next decade.

REFERENCES

1. ADVISORY COMMITTEE ON DANGEROUS PATHOGENS. 1995. Categorisation of biological agents according to hazard and categories of containment (Fourth Edition) HMSO, London.
2. ADVISORY COMMITTEE ON DANGEROUS PATHOGENS. 1999. Supplement to: Categorisation of biological agents according to hazard and categories of containment (Fourth Edition, 1995) HSE, London.
3. ALLABY, M.A.K. and MAYON-WHITE, R. 1995. *Escherichia coli* O157: outbreak in a day nursery. Communicable Diseases Report *5*, R4–6.
4. ANON. 1996. *Escherichia coli*. SCIEH Weekly Report *30*(96/23), 125.
5. ANON. 1996. *Escherichia coli* O157 infection acquired in the laboratory. CDR Weekly Report *6*, 239.
6. ANON. 1997. Outbreak of *E. coli* infection in a nursing home in Arbroath. SCIEH Weekly Report *31*(97/06), 29.
7. ANON. 1999. Outbreak of *Escherichia coli* O157:H7 and *Campylobacter* among attendees of the Washington County Fair — New York 1999. Morbidity Mortality Weekly Report *48*, 803–805.
8. ANON. 1997. Update on nosocomial outbreak of *E. coli* O157 infection at Falkirk and District Royal Infirmary (FDRI). SCIEH Weekly Report *31* (97/22), 113.
9. ANON. 1999. VTEC O157 outbreak associated with a farm visitor centre in North Wales. Commun. Dis. Rep. CDR Wkly. *9*(26), 227, 230.
10. ANON. 1999. VTEC O157 outbreak linked to beach holidays. Commun. Dis. Rep. CDR Wkly. *10*;*9*(37), 327, 330.
11. ARMSTRONG, G.D., FODOR, E. and VANMAELE, R. 1991. Investigation of Shiga-like toxin binding to chemically synthesized oligosaccharide sequences. J. Infect. Dis. *164*(6), 1160–1167.
12. BELL, B.P., GRIFFIN, P.M., LOZANO, P., CHRISTIE, D.L., KOBAYASHI, J.M. and TARR, P.I. 1997. Predictors of hemolytic uremic syndrome in children during a large outbreak of *Escherichia coli* O157:H7 infections. Pediatrics *100*, e12.
13. BELONGIA, E.A., OSTERHOLM, M.T., SOLER, J.T., AMMEND D.A., BRAUN, J.E and MACDONALD, K.L. 1993. Transmission of *Escherichia coli* O157:H7 infection in Minnesota child day-care facilities. J. American Medical Association *269*, 883–888.
14. BOOTH, I.R., THOMSON-CARTER, F., CARTER, P., MALCOLM, L. and GLOVER, J. 1999. Acid tolerance in *E. coli* O157:H7 and related organisms. *In* Survival and Growth Of Verocytotoxigenic *E. coli*. Proceedings of a meeting organised by an EU Concerted Action on VTEC (CT98-3935) in Athens, May 6–8.
15. BOOTH, L. and ROWE, B. 1993. Possible occupational acquisition of *Escherichia coli* O157 infection. Lancet *342*, 1298–1299.
16. BOUDAILLIEZ, B. *et al.* 1997. Possible person to person transmission of *Escherichia coli* O111 associated hemolytic uremic syndrome. Pediatr. Nephrol. *1*(1), 36–39.

17. BREWSTER, D.H., BROWN, M.I., ROBERTSON, D., HOUGHTON, G.L., BIMSON, J. and SHARP, J.C.M. 1994. An outbreak of *Escherichia coli* O157 associated with a children's paddling pool. Epidemiology Infection *112*, 441–447.
18. BURNENS, A.P., ZBINDEN, R., KAEMPF, L., HEINZER, I. and NICOLET, J. 1993. A case of laboratory acquired infection with *Escherichia coli* O157:H7. Zentralblatt fur Bakteriologie *279*, 512–517.
19. BUTLER, T., ISLAM, M.R., AZAD, M.A.K. and JONES, P.K. 1987. Risk factors for development of hemolytic uremic syndrome during shigellosis. J. Pediatr. *110*, 894–897.
20. CARTER, A.O. *et al.* 1987. A severe outbreak of *Escherichia coli* O157:H7-associated haemorrhagic colitis in a nursing home. New England J. Medicine *317*, 1496–1500.
21. CHALMERS, R.M. *et al.* 1997. Vero-cytotoxin-producing *Escherichia coli* O157 in a farmer handling horses. Lancet *349*, 1816.
22. CHAPMAN, P.A. and ACKROYD, H.J. 1997. Farmed deer as a potential source of verotoxin producing *Escherichia coli* O157. Vet. Rec. *141*(12), 341–345.
23. CHAPMAN, P.A., SIDDONS, C.A., MALO, A.T.C. and HARKINS, M.A. 1997. A 1-year study of *Escherichia coli* O157 in cattle, sheep, pigs, and poultry. Epidemiol. Infect. *119*, 245–250.
24. CIMOLAI, N., BASALYGA, S., MAH, D.G., MORRISON, B.J. and CARTER, J.E. 1994. A continuing assessment of risk factors for the development of *Escherichia coli* O157:H7- associated hemolytic uremic syndrome. Clin. Nephrol. *42*, 85–89.
25. CIMOLAI, N., CARTER, J.E., MORRISON, B.J. and ANDERSON, J.D. 1990. Risk factors for the progression of *Escherichia coli* O157:H7 enteritis to the hemolytic uremic syndrome. J. Pediatr. *116*, 589–592.
26. CODY, S.H., GLYNN, M.K., FARRAR, J.A., CAIRNS, K.L. and GRIFFIN, P.M. 1999. An outbreak of *Escherichia coli* O157:H7 infection from unpasteurized commercial apple juice. Ann. Intern. Med. *130*(3), 202–209.
27. COIA, J.E. 1998. Clinical, microbiological and epidemiological aspects of *Escherichia coli* O157 infection. FEMS Immunology Medical Microbiol. *20*, 1–9.
28. COIA, J.E. 1998. Nosocomial and laboratory-acquired infection with *Escherichia coli* O157. J. Hospital Infection *40*, 107–113.
29. COMPTROLLER AND AUDITOR GENERAL. 2000. The management and control of hospital acquired infection in acute NHS trusts in England. National Audit Office, London.
30. DEV, V.J., MAIN, M. and GOULD, I. 1991. Waterborne outbreak of *Escherichia coli* O157. Lancet *337*, 1412.
31. DOHI, H. *et al.* 1999. Synthesis of an artificial glycoconjugate polymer carrying Pk-antigenic trisaccharide and its potent neutralization activity against Shiga-like toxin. Bioorg. Med. Chem. *7*(9), 2053–2062.
32. DRUMMOND, K. 1985. Haemolytic uraemic syndrome — then and now. N. Engl. J. Med. *312*, 116–118.

33. DUNCAN, L., CARTER, A., CARISON, J., BORCZYK, A. and KARMALI, M. 1986. Outbreak of gastrointestinal disease in Sarnia, Ontario. Ontario Dis. Surveillance Rep. *7*, 604–610.
34. DUNDAS, S., MURPHY, J., SOUTAR, R.L., JONES, G.A., HUTHINSON, S.J. and TODD, W.T.A. 1999. Effectiveness of therapeutic plasma exchange in the 1996 Lanarkshire *Escherichia coli* O157:H7 outbreak. Lancet *354*, 1327–1330.
35. FARTHING, M. *et al.* 1996. The management of infective gastroenteritis in adults. A consensus statement by an expert panel convened by the British Society for the Study of Infection. J. Infect. *33*, 143–152.
36. FITZPATRICK, M., SHAH, V., TROMPETER, R. and DHILLON, M. 1991. Long term outcome after childhood haemolytic uraemic syndrome. BMJ. *303*, 489–492.
37. FRIEDMAN, M.S., ROELS, T., KOEHLER, J.E., FELDMAN, L., BIBB, W.F. and BLAKE, P. 1999. *Escherichia coli* O157:H7 outbreak associated with an improperly chlorinated swimming pool. Clinical Infectious Diseases *29*, 298–303.
38. GRIFFIN, P.M., OLMSTEAD, L.C. and PETRAS, R.E. 1990. *Escherichia coli* O157:H7-Associated Colitis. Gastroenterology *99*, 142–149.
39. GRIFFIN, P.M. and TAUXE, R.V. 1991. The Epidemiology of infection caused by *Escherichia coli* O157:H7, other enterohaemorrhagic *E. coli* and the associated haemolytic-uraemic syndrome. Epidemiol Rev. *13*, 60–98.
40. GRIFFIN, P.M. 1995. *Escherichia coli* O157:H7 and other enterohaemorrhagic *Escherichia coli*. *In* Infections of the gastrointestinal tract N (M.J. Blaser, P.D. Smith, J.I. Ravdin, H.B. Greenberg and R.L. Guerrant, eds.) 739–761, New York, Raven Press.
41. HERTZKE, D.M., COWAN, L.A., SCHONING, P. and FENWICK, B.W. 1995. Glomerular ultrastuctural lesions of idiopathic cutaneous and renal glomerular vasculopathy of greyhounds. Vet. Pathol. *32*(5), 451–459.
42. HEUVELINK, A.E. *et al.* 1998. Isolation and characterization of verocytotoxin-producing *Escherichia coli* O157 strains from Dutch cattle and sheep. J. Clin. Microbiol. *36*(4), 878–882.
43. HONDA, T. 1999. Factors influencing the development of hemolytic uremic syndrome caused by enterohemorrhagic *Escherichia coli* infection. From a questionnaire survey to in vitro experiment. Pediatr. Int. *41*, 209–212.
44. HOVDE, C.J., AUSTIN, P.R., CLOUD, K.A., WILLIAMS, C.J. and HUNT, C.W. 1999. Effect of cattle diet on *Escherichia coli* O157:H7 acid resistance. Appl. Environ. Microbiol. *65*(7), 3233–3235.
45. IKEDA, K., IDA, O., KIMOTO, K., TAKATORIGE, T., NAKANISHI, N. and TATARA, K. 1999. Effect of fosfomycin treatment on prevention of hemolytic uremic syndrome accompanying *Escherichia coli* O157:H7 infection. Clin. Nephrol. *52*(6), 357–362.
46. JACKSON, S.G., GOODBRAND, R.B., JOHNSON. R.P., ODORICO V.G. and ALVES, D. 1998. *Escherichia coli* O157:H7 diarrhoea associated with well water and infected cattle on an Ontario farm. Epidemiol. Infect. *120*(1), 17–20.
47. KARMALI, M.A. *et al.* 1988. Hospital-acquired *Escherichia coli* O157:H7 associated haemolytic uraemic syndrome in a nurse. Lancet (1), 526.

48. KAWAMURA, N., YAMAZAKI, T. and TAMAI, H. 1999. Risk factors for the development of *Escherichia coli* O157:H7 associated with hemolytic uremic syndrome. Pediatr. Int. *41*, 218–222.
49. KEENE, W.E. *et al.* 1994. A swimming-associated outbreak of hemorrhagic colitis caused by *Escherichia coli* O157:H7 and *Shigella sonnei*. New England J. Medicine *331*, 579–584.
50. KIMMITT, P.T., HARWOOD, C.R. and BARER, M.R. 1999. Induction of type 2 Shiga toxin synthesis in *Escherichia coli* O157 by 4-quinolones. Lancet *353*, 1588–1589.
51. KOBAYASHI, M. *et al.* 1999. Houseflies: not simple mechanical vectors of enterohemorrhagic *Escherichia coli* O157:H7. Am. J. Trop. Med. Hyg. *61*(4), 625–629.
52. KOHLI, H.S., CHAUDHURI, A.K.R., TODD, W.T.A., MITCHELL, A.A.B. and LIDDELL, K.G. 1994. A severe outbreak of *E. coli* O157 in two psychogeriatric wards. J. Public Health Medicine *16*, 11–15.
53. LISSNER, R., SCHMIDIT, H. and KARCH, H. 1996. A standard immunoglobulin preparation produced from bovine colostra shows antibody reactivity and neutralising activity against Shiga-like toxins and EHEC hemolysin of *Escherichia coli* O157:H7. Infection *24*, 378–383.
54. LOUSI, C.B. and OBRIG, T.G. 1991. Shiga toxin-associated hemolytic uremic syndrome: Combined cytotoxic effects of Shiga toxin, interleukin I'3, and tumour necrosis factor alpha on human vascular endothelial cells in vitro. Infect. Immun. *59*, 4173–4179.
55. MARTIN, D.L., MACDONALD, K.L., WHITE, K.E., SOLER, J.T. and OSTERHOLM, M.T. 1990. The epidemiology and clinical aspects of the hemolytic uremic syndrome in Minnesota. N. Engl. J. Med. *323*, 1161–1167.
56. MAULE A. 1999. Environmental survival of *Escherichia coli* O157: Implications for spread of disease. *In* Survival and Growth Of Verocytotoxigenic *E. coli*. Proceedings of a meeting organised by an EU Concerted Action on VTEC (CT98-3935) in Athens, May 6-8.
57. MAULE, A. 1997. Survival of the verotoxigenic strain *E. coli* O157:H7 in laboratory-scale microcosms. *In* Coliforms and *E. coli* — Problem or solution? (D. Kay and C. Fricker, eds.) Cambridge: The Royal Society of Chemistry 61–65.
58. MAULE, A. 1997. The survival of verocytotoxigenic *Escherichia coli* on surfaces and the effect of disinfection. PHLS 22nd annual scientific conference. Warwick.
59. MEAD, P.S. and GRIFFIN, P.M. 1998. *Escherichia coli* O157:H7. Lancet *352*, 1207–1212.
60. MICHINO, H. *et al.* 1999. Massive outbreak of *Escherichia coli* O157:H7 infection in schoolchildren in Sakai City, Japan, associated with consumption of white radish sprouts. Am. J. Epidemiol. *150*(8), 787–796.
61. MILNE, L.M., PLOM, A., STRUDLEY, I., PRITCHARD, G.C. and CROOKS, R. 1999. *Escherichia coli* O157 incident associated with a farm open to members of the public. Commun. Dis. Public Health 2(1), 22–26.

62. MOAKE, J.L. and MCPHERSON, PD. 1989. Abnormalities of von Willebrand factor multimers in thrombotic thrombocytopenic purpura and the hemolytic-uremic syndrome. Am. J. Med. *87*, 3-9N-3-15N.
63. MORIYA, K. *et al.* 1999. Verotoxin-producing *Escherichia coli* O157:H7 carried by the housefly in Japan. Med. Vet. Entomol. *13*(2), 214–216.
64. MORRIS, K.P., COULTHARD, M.G. and MATTHEWS, J.N. 1991. Predicting outcome after childhood hemolytic uraemic syndrome. Clin. Nephrol. *36*(5), 263–264.
65. MURDOCH, P.S., RILEY, A.J. and O'BRIEN, S. 1997. Report of the outbreak control team convened to investigate and control an outbreak of *E. coli* O157 at Falkirk and District Royal Infirmary in May and June.
66. NATAVO, J.P. and KAPER, J.B. 1998. Diarrhoeagenic *Escherichia coli*. Clin. Microbiol. Rev. *11*, 1–60.
67. OIE, S., KAMIYA, A., TOMITA, M., KATAYAMA, A., IWASAKI, A. and MIYAMIRA, S. 1999. Efficacy of disinfectants and heat against *Escherichia coli* O157:H7. Microbios. *98*, 7–14.
68. OSTROFF, S.M., KOBAYASHI, J.M. and LEWIS, J.H. 1989. Infections with *Escherichia coli* O157:H7 in Washington State: The first year of statewide disease surveillance. J. American Medical Association *262*, 355–359.
69. PAUNIO, M. *et al.* 1999. Swimming-associated outbreak of *Escherichia coli* O157:H7. Epidemiol. Infect. *122*(1), 1–5.
70. PAVIA, A.T. *et al.* 1990. Haemolytic-uraemic syndrome during an outbreak of *Escherichia coli* O157:H7 infections in institutions for mentally retarded persons: Clinical and epidemiological observations. J. Paediatrics *116*, 544–551.
71. PHILLIPS, G. and OLD, D.C. 1997. Laboratory-acquired VTEC infection. J. Hospital Infection *35*, 72.
72. PRESTON, M. *et al.* 1997. Hospital outbreak of *Escherichia coli* O157:H7 associated with a rare phage type. Canada Communicable Disease Report *23*, 33–36.
73. Prevention and Control of enterohaemorrhagic *Escherichia coli* (EHEC) infections: Report of a WHO consultation. Geneva, WHO, 1997.
74. PROULX, F., TURGOEN, J.P., DELAGE, G., LAFLEUR, L. and CHICOINE, L. 1992. Randomised controlled trial of antibiotic therapy for *Escherichia coli* O157:H7 enteritis. J. Paediatr. *121*, 299–303.
75. RAO, G.G., SAUNDERS, B.P. and MASTERTON, R.G. 1996. Laboratory acquired verotoxin producing *Escherichia coli* (VTEC) infection. J. Hospital Infection *33*, 228–230.
76. REINDERS, R.D., BIJKER, P.G.H. and OUDE-ELFERINK, S.J.W.H. 1999. Growth and survival of verotoxigenic *Escherichia coli* O157 in selected farm environments. *In* Survival and Growth Of Verocytotoxigenic *E. coli*. Proceedings of a meeting organised by an EU Concerted Action on VTEC (CT98-3935) in Athens, May 6–8.
77. RUBINI, S. *et al.* 1999. Verocytotoxin-producing *Escherichia coli* O157 in sheep milk. Vet. Rec. *144*(2), 56.

78. RYAN, C.A. *et al.* 1986. *Escherichia coli* O157:H7 diarrhoea in a nursing home: Clinical, epidemiological and pathological findings. J. Infectious Diseases *154*, 631–638.
79. SEIGLER, R.L. 1995. The hemolytic uremic syndrome. Pediatr. Clin. North Am. *42*, 1505–1529.
80. SHARP, J.C.M., REILLY, W.J., COIA, J.E., CURNOW, J. and SYNGE, B.A. 1995. *Escherichia coli* O157 in Scotland: an epidemiological overview 1984–1994. PHLS Microbiology Digest *12*, 134–140.
81. SHIOMI, M., TOGAWA, M., FUJITA, K. and MURAJA, R. 1999. Effects of early fluoroquinolones in hemorrhagic colitis due to *Escherichia coli* O157:H7. Pediatr. Int. *41*(2), 228–32.
82. SLUTSKER, L., RIES, A.A., MALONEY, K., WELLS, J.G, GREENE, K.D. and GRIFFIN, P.M. 1998. A nationwide case-control study of *Escherichia coli* O157:H7 infection in the United States. J. Infect. Dis. *177*, 962–966.
83. SPIKA, J.S., PARSONS, J.E., NORDENBERG, D., WELLS, J.G., GUNN, R.A. and BLAKE, P.A. 1986. Haemolytic uraemic syndrome and diarrhoea associated with *Escherichia coli* O157:H7 in a day care centre. J. Paediatrics *109*, 287–291.
84. SU, C. and BRANDT, L.J. 1995. *Escherichia coli* O157:H7 infection in humans. Annals of Internal Medicine *123*, 698–714.
85. SUBCOMMITTEE OF THE PHLS ADVISORY COMMITTEE ON GASTROINTESTINAL INFECTIONS. 2000. Guidelines for the control of infection with Verocytotoxin producing *Escherichia coli* (VTEC). Communicable Disease and Public Health *3*, 14–23.
86. SWERDLOW, D.L. and GRIFFIN, P.M. 1997. Duration of faecal shedding of *Escherichia coli* O157:H7 among children in day-care centres. Lancet *349*, 745–746.
87. SWERDLOW, D.L *et al.* 1992. A waterborne outbreak in Missouri of *Escherichia coli* O157:H7 associated with bloody diarrhoea and death. Annals of Internal Medicine *117*, 812–819.
88. TAKEDA, T., YOSHINO, K., ADACHI, E., SATO, K. and YAMAGATA, K. 1999. In vitro assessment of a chemically synthesised Shiga toxin receptor analogue attached to chromosorb P (Synsorb Pk) as a specific absorbing agent of Shiga toxin 1 and 2. Microbiol. Immunol. *43*(4), 331–337.
89. TILDEN JR., J. *et al.* 1996. A new route of transmission for *Escherichia coli* infection from dry fermented salami. Am. J. Public Health *86*(8pt1), 1142–1145.
90. TODD, W.T.A. 2000. Central Scotland *E.coli* O157 outbreak: Clinical update. In supplement to SCIEH Weekly Report *34* (2000/05), 5–6.
91. TREVANA, W.B., WILLSHAW, G.A., CHEASTY, T., DOMIGUE, G. and WRAY, C. 1999. Transmission of Verocytotoxin producing *Escherichia coli* O157 infection from farm animals to humans in Cornwall and West Devon. Commun. Dis. Pub. Health *2*, 263–268.
92. VOETSCH, A., WOJTKUNSKI, C., SWARTZ, M., MSHAR, R. and LAMBERT-FAIR, M.A. 1999. A multistate outbreak of *Escherichia coli* O157:H7 infections associated with consumption of mesclun lettuce. Arch. Intern. Med. *159*(15), 1758–1764.

93. WALTERS, M.S., MATTHEI, I.U., KAY, R. and DHILLON, M.J. 1989. The polymorphonuclear leucocyte count in childhood haemolytic uraemic syndrome. Paediatr. Nephrol. *3*, 130–134.

94. WATANABE, Y., OZASA, K., MERMIN, J.H., GRIFFIN, P.M. and MASUDA, K. 1999. Factory outbreak of *Escherichia coli* O157:H7 infection in Japan. Emerg. Infect. Dis. *5*(3), 424–428.

95. WEIGHTMAN, N.C. and BARNHAM, M. 1998. *Escherichia coli* O157 masquerading as 'rectal bleeding': a further problem for infection control. Clinical Microbiology and Infection *4*, 667.

96. WEIGHTMAN, N.C. and KIRBY, P.J.G. 2000. Nosocomial *Escherichia coli* O157 infection. J. Hospital Infection *44*, 107–112.

97. WILLSHAW, G.A., THIRWELL, J., JONES, A.P., PARRY, S., SALMON, R.L. and HICKEY, M. 1994. Verocytotoxin-producing *Escherichia coli* O157 in beefburgers linked to an outbreak of diarrhoea, haemorrhagic colitis and haemolytic uraemic syndrome in Britain. Letters in Appl. Microbiol. *19*, 304–307.

98. WOLF, L.E., ACHESON, D.W., LINCICOME, L.L. and KEUSCH, G.T. 1997. Abstracts from 3rd International Symposium and Workshop on Shiga Toxin Producing *Escherichia coli* Infections.

CHAPTER 23

VEROCYTOTOXIGENIC *E. COLI*: SAFE LABORATORY PRACTICES

HEATHER J. SHEELEY

Centre for Applied Microbiology and Research (CAMR)
Porton Down, Salisbury
Wiltshire, SP4 OJG, United Kingdom

INTRODUCTION

The recognition of verocytotoxin producing *E. coli* (42) as a serious human pathogen in 1982, led to a need for those laboratories handling uncooked meats and animal tissue to consider the risk of infection from a new group of bacteria. A wide range of laboratories including food surveillance, clinical, microbiological, immunological, veterinary, animal husbandry, nephrology, gastroenteric and environmental and ecology, all need to assess the risk arising from standard procedures. Not all of these laboratory workers will be familiar with adequate methods for control of infection. A further issue is that increasing interest in a particular organism also increases the level of work being undertaken, and raises the numbers of laboratory staff potentially exposed. This pattern has been observed in other emerging infections, where the increased activity associated with a pathogen increases the incidence of laboratory acquired infections (15). In July 1997 the European and WHO bodies classified the O157 serogroup of *E. coli* and other non-O157 VTEC as a Hazard Group 3 pathogens requiring specialist handling (19).

Not all *E. coli* are pathogenic to man, and the emergence of O serogroups producing verotoxin as foodborne pathogens has been quite dramatic. These organisms have a very low infectious dose, survive fairly rigorous environmental conditions, are readily transmitted and cause severe disease. A number of outbreaks within such serogroups since 1989 have invariably been associated with O157:H7, especially in North America and the United Kingdom.

The progress of the disease these organisms cause involves three to four days incubation during which colonisation of the bowel leads to diarrhoea and abdominal cramps, and may be associated with fever similar to that induced by *Campylobacter*. In most cases the three days of symptoms of bloody diarrhoea is self limiting. However 6% of cases may progress to haemolytic uremic syndrome (HUS) necessitating dialysis, and/or a range of neurological sequelae, chronic renal failure and death. This pattern of chronic or life-threatening complications is particularly likely in cases involving the very young and the very old (28,44).

E. coli is usually found in the gut of cattle, other ruminants and other mammalian species, and included at least 100 serotypes of verocytotoxin-producing (VTEC), not all of which are pathogenic to man (34,46). Serotypes associated with HUS include O157:H7, O111:NM, O103:H2, O48:H2. VTEC has been identified within at least six different serogroups O111, O26, O113, O121, O145, O157. Others may well join this list as pathogens of particular concern. All these strains should be treated with great caution and tissues from known reservoirs, i.e., cattle (11), dogs (2), horses (49), swine (12), deer (29, 40), cats (2), birds (21,50), sheep (31), geese (1), goats (43), and flies (22), and materials associated with them (such as milk and excreta) should be regarded as potentially infected.

The number of outbreaks and the number of individuals associated with each outbreak have enabled accurate detection of the source(s) of infection (56). Outbreaks in North America have been mostly associated with undercooked ground beef (16,18,20,42,55,56). There have, however, been outbreaks associated with a wide range of other food products including venison (29,34), salami (7,48), raw milk (3,10,13), yoghurt (36), salads (6,26,35), fruit juices (8), mayonnaise (57), vegetables (37) and potable and recreational water (27,41, 51-53). Study of the above cases have also provided precise estimates of the infective dose, i.e., between 10-100 organisms (56). This very low infectious dose means that very sensitive detection methods have had to be developed for the detection of very low concentrations of these organisms in foods.

PERSISTENCE

A number of studies have examined the persistence of these organisms in a range of foods, and in harsh environments, e.g., water and animals faeces (30,32,33,54). Such studies have shown *E. coli* O157 is not unusually heat resistant at 68.3°C compared to Salmonella, but can survive acid conditions (pH4-5) (34) and drying at 62°C for 6-8 h (17). Chemical treatment of carcasses and fruit juices with mild acids and chelating salt, including at elevated temperatures, has produced only small reductions in, or brief suspension of, growth (47).

ROUTE OF INFECTION

There is little information on infection due to inhalation of *E. coli* cells (5). The major route of infection is oral, through ingestion of contaminated food.

Therefore, within the laboratory setting, the procedures used by workers handling potentially infected material must endeavour to prevent contamination of surfaces, and the highest possible standards of personal and laboratory hygiene are essential.

PRECAUTIONARY MEASURES DURING SAMPLE RECEIPT

Sampling in the field or on receipt at the laboratory must follow good hygienic practice. Personal protective equipment such as gloves, overalls and boots used at the potentially contaminated site must be bagged for subsequent decontamination treatment. Samples require proper labelling and, if they are to be transported, they must be shipped in accordance with international regulations on the transfer of dangerous pathogens (25). Samples should be received directly into an appropriate containment laboratory. Packages must never be unpacked in general areas or unpacked by porters or secretaries. Facilities to deal with spills and damaged packages should be readily available.

LABORATORY DESIGN

The classification of *E. coli* O157 as a Hazard Group 3 pathogen means that specialised containment facilities must be available if work is to be conducted safely. Samples known or strongly suspected to contain infected material must be handled under the appropriate level 3 containment and conditions. Diagnostic and low risk samples may be dealt with under less stringent handling requirements and lower levels of containment, but only under professional advice and after sound risk assessment (Table 1).

The laboratory should be separate from any non-compatible activities and/or areas such as food preparation, or kindergartens. Access should be restricted to those who are adequately trained in dealing with, and aware of, the hazards. Arrangements should be made to safeguard non-scientific and support staff, such as cleaners, porters and office staff. There should be a barrier airlock at the entrance to the laboratory, to ensure the correct air inflow is maintained and to provide an area for the donning of suitable protective clothing. Ideally, such suitable clothing should include a back-fastening gown and gloves. A washbasin with hands free taps and disposable towels should be provided close to the exit door (Fig. 1).

The main laboratory should be free of unnecessary equipment, consumable items and paperwork. Multi-use laboratories are not suitable, nor are shared arrangements where the additional activities entail different levels of risk. Media, buffers and reagents should be made up elsewhere and brought into the laboratory in prepared form. Only materials that are made safe by autoclaving or a validated decontamination method may leave the facility. Transfer of infected material must be within a robust leak-proof container and such containers should only be opened within containment. Material must never be transferred to a laboratory with a lower containment classification if there is a risk of subsequent contamination or incomplete treatment.

TABLE 1.
CONTAINMENT LEVEL 3

	Essential	Best Practice
Smooth easily cleanable walls, floors and ceiling	✓	
All penetrations sealed	✓	
Sealed windows	✓	
All extracted air HEPA filtered	✓	
Supply and extract ventilation fans/dampers interlocked to prevent pressurisation	✓	
Alarmed ventilation failure	✓	
Air lock with hand washing basin and change of clothing	✓	
Access control	✓	
Sealable for Fumigation	✓	
All parts of working area visible from outside		✓
Sinks and drains with deep traps	✓	
Hands free operation of taps	✓	
Minimal furniture and fittings		✓
Dedicated equipment	✓	
Autoclave within or accessible from suite	✓	
Hands free telephone or intercom		✓
Class 1 or III Microbiological safety cabinets or system which provides an equal protection	✓	

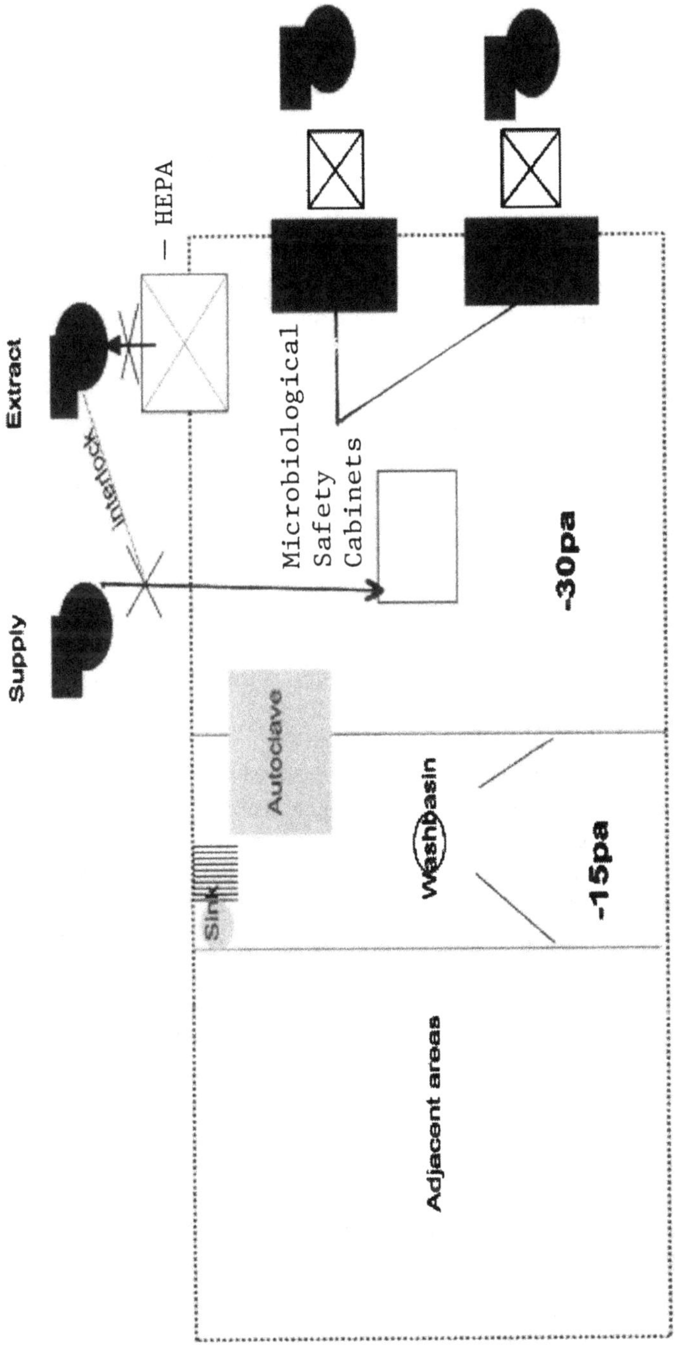

FIG. 1. TYPICAL LAYOUT OF CONTAINMENT LABORATORY

Efficient ventilation is critical to secondary containment. An inward flow of air must be maintained at all times when work is in progress, i.e., the containment laboratory must have a negative differential pressure relative to general areas. This may be achieved in a small room by use of the cabinet extraction system. Where the room is mechanically provided with supply air, the supply and extraction systems must be interlocked to prevent pressurisation. The relevant High Efficiency Particulate (HEPA) filters should be positioned as close to the room as is practicable. Any ductwork before the extract filter must be without leaks, tested before use and retested on a regular basis. Ideally, air should be supplied at high level and extracted at low-level as this draws air from the breathing zone as it sweeps the room. However, there is evidence that layouts which supply and extract air at high levels can be effective in providing good airflow patterns (24).

The laboratory must contain all its own equipment which should be selected with particular reference to its containment performance such as sealed rotors, incubators with viewing panels, etc. Care should also be taken in relation to room and furniture finishes, thus absorbent chair upholstery, cupboards with crevices, porous surfaces and inaccessible areas are unsuitable.

MICROBIOLOGICAL SAFETY CABINETS (MSC)

Class II microbiological safety cabinets (9) are unsuitable for high titre dangerous pathogens, as they can be adversely affected by environmental factors. Simple motions such as arm movements, general laboratory traffic and the use of nearby equipment will disturb air patterns in these units. In particular, care must be taken in the use of open fronted cabinets to prevent disruption of the curtain of air being drawn in through the aperture. Work patterns must be carried out with due regard to the quantities of material needed within the cabinet area, the frequencies of inward and outward movements.

Class III cabinets provide higher levels of worker and product protection. Issues of operative fatigue and dexterity remain, but these can be overcome by adequate training and good planning of work. Ideally the atmosphere within MSCs should be ducted to the environment via a single HEPA filter. Considerable care in the engineering design and operation of systems is required to maintain proper balancing of room pressures under all circumstances. The simplest design will be the most reliable and easily maintained. Any failure of the cabinet or room ventilation should trigger automatic audible and visual warnings, to operatives in the immediate vicinity, and elsewhere through building management or monitoring systems.

The laboratory must be sealable to permit fumigation. The system should have external control with remote switch, dosing and ventilation controls. The usual fumigant is vaporised formaldehyde, but there are other less noxious alternatives. The readiness of fumigation systems must be tested before beginning to work with hazardous organisms in new or altered facilities. Fumigation is a highly hazardous procedure and must be carefully planned and monitored with adequate emergency support. Leakage to adjacent areas must be monitored.

LABORATORY PROCEDURES

All laboratory procedures should be designed and operated so they reduce the spread of contamination, reduce aerosols and prevent release. Work with *E. coli* O157:H7 and similar pathogens, particularly those with such low infective doses, demands the highest levels of discipline and procedure care to prevent release and infection.

Each laboratory should have its own specific Code of Practice, detailing all procedures, maintenance, equipment logs, servicing records, disinfection policy, personal protective equipment, disposal routes and emergency procedures.

As noted above, the equipment for use within Containment Level 3 should be designed to contain including contained rotors, leak-proof test tubes and vials, operated in a manner which minimises the use of glassware within procedures which provide the best practicable containment solutions for connections and transfers.

Gloves and back-fastening gowns or equivalent must be worn at all times within the suite. All procedures with infected material must be conducted under containment within a MSC or contained equipment. All materials manipulated within cabinets must be contained or treated before removal.

Any infected materials should be stored within the suite in secure, properly labelled packaging, and kept separate from any other materials which are lower hazard groups, or which may be handled under lower levels of containment. An inventory of stock should be maintained.

WASTE TREATMENT

All used material, whether or not they have been directly contaminated, must be treated before further disposal. This can be achieved by physical or chemical treatment.

The Autoclave

The suite should include or have direct access to an autoclave validated to achieve a load treatment of 121°C for at least 15 min. Heavily contaminated, and/or dense material may require treatment at a higher temperature or for a longer period of exposure. Factors to be borne in mind in the design and selection of the autoclave include manual handling requirements, load volume and type, processing of condensate in the drain and the nature of local permissions required for release of material to the sewers. Depending on the type of load, safe transfer from the cabinet to the autoclave, can be achieved using nylon autoclavable bags or rigid semi-enclosed containers. If it is necessary to transport materials outside the containment suite, they should be sealed in a rigid container. Waste materials to be autoclaved will include gowns, reusable glass/plastic ware and equipment and all disposable materials such as gloves, paper and plastic.

Disinfection and Decontamination

While many items can be discarded or recycled after autoclaving, some equipment cannot be autoclaved. Large equipment, including electronic devices, can be successfully fumigated many times without damage. Gross surface contamination can be removed by wiping with detergent or disinfectant prior to fumigation.

Surfaces can be decontaminated using normal concentrations of disinfectants. Thus, sodium hypochlorite at a concentration of 130 ppm available chlorine reduces *E. coli* O157 numbers by 7 logs within 5 min (33). Commercial formulation of disinfectants containing both cationic and anionic surfactants are even more effective (33). It is important that an "as-used" test of the efficacy of any disinfectant is part of the planned programme of work, as the efficiency of disinfectants is highly dependent on factors such as pH, water hardness, the presence of organic material and salts, and the contact time. Routine disinfection of instruments or surfaces should use fresh disinfectants at appropriate concentrations, to ensure a complete kill. Routine topical cleaning is vital to maintain high standards of cleanliness. Unless there has been a spill outside the cabinet it should not be necessary to disinfect floors or walls, as disinfectants are hazardous and unnecessary exposure is to be avoided. Fumigation, or wiping outer surfaces or immersion of waste or samples must precede their removal from the cabinet. A system of verification and declaration of decontamination should accompany any equipment that is removed from the area or made available for service or repair.

MAINTENANCE

The Laboratory Code of Practice should contain the details of the maintenance intervals for ventilation systems, and equipment including MSC, and other in-service repairs. Access to the contained areas by non-laboratory staff must be carefully controlled and the subject of specific risk assessment.

LABORATORY MANAGEMENT AND ORGANISATION

The person responsible for the Containment Suite should be an experienced microbiologist, who will take an active part in the maintenance of practices and risk assessment of the work. Risk assessment and procedure should be subject to regular review and update as information becomes available. In the case of emerging diseases such review may need to be more frequent. Sufficient resources should be available to maintain standards in the laboratory and prevent short cuts. A Containment Level 3 laboratory is an expensive facility and, its work can not be hurried if safe practices are to be maintained. Where possible nontoxinogenic strains should be used as surrogates in research protocols.

TRAINING AND DISCIPLINE

Only staff who are competent and adequately trained, with extensive experience in microbiology and laboratory procedures, may work within a containment Level 3 facility. All individuals associated with the facility must be aware of the hazards and all procedures which affect them such as alarms, access, etc. Training and procedures should be recorded and reviewed, and it is paramount that the reasons for the containment procedures are fully understood. It is good practice to test and reinforce this understanding, by "on the job" training and evaluation of competence. Inexperienced staff must receive close supervision and high levels of support.

HEALTH SURVEILLANCE

Staff who handle dangerous pathogens should be monitored for laboratory acquired illness and for existing conditions incompatible with working with dangerous pathogens, particularly those causing severe kidney disease. Records of exposure should be maintained, and the personal physicians of those working with *E. coli* O157 and other VTEC should be kept informed.

EMERGENCIES

In the unlikely event of spillage of pathogens, adequate procedures must be in place and followed. These will always include the immediate evacuation and, if necessary decontamination at a holding area. Health risk assessment and, if necessary, treatment will involve medical and safety input by the responsible person. No exposure, however trivial, should be ignored or go unreported. Preparation before the event should include establishment of means of executing a planned facility decontamination regime.

HEALTH ISSUES FOR LABORATORY WORKERS

Current levels of reporting of occupationally acquired VTEC cases may underestimate the true frequency of infection, but there have been a number of laboratory related infections (4,14,38,39). Those who routinely handle risk-associated materials should be fully aware of the safe working practices. However, as only 18% of laboratory acquired infection can be attributed to a specific event (15), and thus the cause(s) of the majority of such illness are unknown, it is important all staff are aware of the relevant symptoms. Arrangements for active health surveillance should include monitoring of all and any illness and absence which might be linked to food poisoning, to identify any possible linkages to laboratory acquired illness infection. In principle a link should be assumed, until indications to the contrary are established. General Practitioners with laboratory workers on their panel should be sensitised to the possibilities of laboratory acquired infection through the use of an awareness card, carried by laboratory workers. Plans and notifications should encompass the possibilities and implications for familial and community transmission (23,45). The potential for transfer to other members of the family is particularly important in the case of *E. coli* O157:H7 because this organisms has such a low infectious dose. It should also be borne in mind that individuals can remain in a transient subclinical or carrier state, having no clinical symptoms, but still passing on the organism.

OTHER SEROTYPES OF CONCERN

While most concerns in relation to these organisms are currently focused on O157:H7, our understanding of the relative risks posed by the different serotypes may change over time. As discussed, other serotypes of concern have significance in some geographic areas, thus safe practices should be kept under continual review in light of advancing understanding of the particular risk factors posed by other serotypes.

SUMMARY

In wider terms, the most significant safety issues raised by the emergence of *E. coli* O157:H7 are clearly concerned with the wider public health and general consumer confidence in the safety of food. Nevertheless, the well being of laboratory, field and clinical scientists must not be overlooked. In recent years, a great deal of information has been gained about this organism and the diseases it causes. Currently established biosafety precautions need to be reinforced, developed and appraised in light of this additional knowledge.

REFERENCES

1. ALLISON, L., CARTER, P. and THOMPSON-CARTER, F. 1997. Presented at the VTEC'97: 3rd International Symposium and Workshop on Shiga toxin (verocytotoxin)-producing *Escherichia coli* infections.
2. BEUTIN, L., KNOLLMANN-SCHANBACHER, G., REITSCHEL, W. and SEEGER, H. 1996. Animal reservoirs of *Escherichia coli* O157:[H7]. Veterinary Record *139*, 70–71.
3. BORCZYK, A.A., KARMALI, M., LIOR, M. and DUNCAN, L. 1987. Bovine reservoir for verotoxin-producing *Escherichia coli* O157. Lancet *ii*, 98.
4. BURNENS, A.P., ZBINDEN, R., KAEMPF, L., HEINER, I. and NICOLET, J. 1993. A Case of Laboratory Acquired Infection with *Escherichia coli* O157. Zbl Bakt. *279*, 512–517.
5. CDC, and NIH. 1999. Biosafety in Microbiology and Biomedical Laboratories, 4 ed. US Government Printing Office.
6. CDCP. 1997. *Escherichia coli* O157:H7 infections associated with eating alfalfa sprouts — Michigan and Virginia. Mortality and Morbidity Weekly Report *46*, 741–744.
7. CDCP. 1995. *Escherichia coli* O157:H7 outbreak linked to commercially distributed dry-cured salami — Washington and California. Mortality and Morbidity Weekly Report *44*, 157–160.
8. CDCP. 1997. Outbreak of *Escherichia coli* O157:H7 infection and cryptosporidiosis associated with drinking unpasteurized apple cider — Connecticut and New York. Mortality and Morbidity Weekly Report *46*, 4–8.
9. CEN, E.C.F.S. 1999. Biotechnology — Performance Criteria for Microbiological Safety Cabinets.
10. CHAPMAN, P., WRIGHT, D. and HIGGINS, R. 1993. Untreated milk as source of verotoxgenic *E. coli* O157. Veterinary Record *133*, 171–172.
11. CHAPMAN, P., WRIGHT, D. and NORMAN, P. 1989. Verotoxin-producing *Escherichia coli* infections in Sheffield: cattle as a possible source. Epidemiology and Infection *102*, 439–445.
12. CHAPMAN, P.A., SIDDONS, C.A., CERDAN MALO, A.T. and HARKIN, M.A. 1997. A 1 yr Study of *E. coli* O157 in cattle, sheep, pigs and poultry. Epidemiology and Infection *119*, 245–250.
13. CLARKE, A., MORTON, S., WRIGHT, P., CORKISH, J., BOLTON, F. and RUSSELL, J. 1997. A community outbreak of verocytotoxin producing *Escherichia coli* O157 infection linked to a small farm dairy. Community Disease Reports CDR Review.
14. COIA, J. 1998. Nosocomial and laboratory-acquired infection with *Escherichia coli* O157. J. Hospital Infection. *40*(2), 107–113.

15. COLLINS, C. 1983. Laboratory acquired infections. Butterworths & Co, London.
16. DOYLE, M. and PADHYE, V. 1989. *Escherichia coli*, Foodborne Bacterial Pathogens. Marcel Dekker, Inc., New York.
17. DOYLE, M. and SCHOENI, J. 1989. Survival and growth characteristics of *Escherichia coli* associated with hemorrhagic colitis. Appl. Environ. Microbiol. *48*, 855–856.
18. DOYLE, M.P., ZHAO, T., MENG, J. and ZHAO, S. 1997. *Escherichia coli* O157:H7, *In* Food Microbiology: Fundamentals and Frontiers. (M. Doyle, L. Beuchat and T.J. Montville, eds.) p. 171–191. ASM, Washington.
19. EEC. 1993. Classification of Biological Agents capable of causing infection (93/88/EEC) and amendments including "Categorisation 98", Third ed. EEC.
20. GRIFFIN, P.M. *et al.* 1991. The epidemiology of infections caused by *Escherichia coli* O157:H7 and other enterohemorrhagic *E. coli* and the associated hemolytic uremic syndrome. Epidemiology Review *13*, 60–98.
21. HANCOCK, D., BESSER, T., RICE, F., EBEL, E., HERRIOTT, D. and CARPENTER, L. 1997. Sources of *Escherichia coli* O157 in feedlots and dairy farms. Previews of Veterinary Medicine.
22. HOLLINGSWORTH, J. and KAPLAN, B. 1999. Food safety in the United States, *In Escherichia coli* O157:H7 and Other Shiga Toxin-Producing *E. coli* Strains (J. Kaper and A. O'Brien, eds.), p. 111. ASM, Washington.
23. HOLTON, D. *et al.* 1992. Presented at the Program and Abstracts of the 32nd Interscience Conference on Antimicrobiol Agents and Chemotherapy.
24. HUGHES, H.C., HUGHES, V. and REYNOLDS, P.S.D. 1988. A Comparison of the Effects of State of the Art Diffusers on Animal Room Ventilation. Pharmaceutical Engineering May/June, 66–72.
25. IATA, INTERNATIONAL AIR TRANSPORT ASSOCIATION. 2000. Dangerous Goods Regulations, 41th ed., Montreal-Geneva.
26. IZUMIYA, H. *et al.* 1997. Molecular typing of enterohemorrhagic *Escherichia coli* O157:H7 isolates in Japan by using pulsed field gel electrophoresis. J. Clinical Microbiol. *35*, 1675–1680.
27. JONES, I.G. and ROWORTH, M. 1996. An Outbreak of *Escherichia coli* O157 and campylobacteriosis associated with contamination of drinking water. Public Health *110*(5), 227–282.
28. KARMALI, M. 1989. Infection by Vero cytotoxin-producing *Escherichia coli*. Clinical Microbiology Reviews *2*, 15–38.
29. KEENE, W. *et al.* 1997. An outbreak of *Escherichia coli* O157:H7 infections traced to jerky made from deer meat. JAMA *277*, 1229–1231.
30. KEEVIL, C., WALKER, J., MAULE, A. and JAMES, B. 1999. Presented at the A European study on the animal, food and biomedical aspects of Verocytotoxinogenic *E. Coli* in particular serotype O157:H7 an emerging pathogen, Ekpedeftiko Kentro Ktmatikis.
31. KUDVA, I.T., HATFIELD, P.G. and HOVDE, C.J. 1996. *Escherichia coli* O157:H7 in microbial flora of sheep. J. Clinical Microbiol. *34*, 431–433.
32. LEYER, G., WANG, L.-L. and JOHNSON, E. 1995. Acid adaptation of *Escherichia coli* increases survival in acidic foods. Appl. Environ. Microbiol. *62*, 3094–3100.

33. MAULE, A. 1999. Environmental survival of *Escherichia coli* O157: Implications for the spread of the disease. Conferences proceedings on "Survival and Growth of Verocytotoxigenic *E. coli*" organised by an EU Concerted Action on VTEC (CT98-3935) at Agricultural University of Athens, May 6-8th. P. 94–104.
34. MENG, J. and DOYLE, M.P. 1998. Microbiology of Shiga Toxin-Producing *Escherichia coli* in Foods *In Escherichia coli* O157:H7 and Other Shiga Toxin-Producing *E. coli* Strains. (J.M. Kaper and A.D. O'Brien, eds.) p. 92–108, American Society for Microbiology, Washington.
35. MERMIN, J. *et al.* 1997. Presented at the VTEC '97: 3rd International Symposium and Workshop on Shiga Toxin (Verocytotoxin)-producing *Escherichia coli* Infections.
36. MORGAN, D., NEWMAN, C., HUTCHINSON, D., WALKER, A., ROWE, B. and MAJID, F. 1993. Verotoxin producing *Escherichia coli* O157 infections associated with the consumption of Yoghurt. Epidemiology Infection *111*, 181–187.
37. MORGAN, G.M. *et al.* 1988. First recognised community outbreak of haemorrhagic colitis due to verotoxin-producing *Escherichia coli* 0157:H7 in the UK. Epidemiology Infection *101*, 83-91.
38. PHILLIPS, G. and OLD, D.C. 1997. Laboratory acquired VTEC infection. J. Hospital Infection *35*, 72.
39. PHLS. 1996. *Escherichia coli* O157 infection acquired in the laboratory. Communicable Disease Report *6*(28).
40. RICE, D., HANCOCK, D. and BESSER, T. 1995. Verotoxinogenic *E. coli* O157 colonisation of wild deer and range cattle. Veterinary Record *137*, 524.
41. RICE, E., JOHNSON, C., WILD, D. and REASONER, D. 1992. Survival of *Escherichia coli* O157:H7 in drinking water associated with a waterborne disease outbreak of hemorrhagic colitis. Letters in Appl. Microbiol. *15*, 38–40.
42. RILEY, L.W. *et al.* 1983. Haemorrhagic colitis associated with rare *Escherichia coli* serotype. New England J. Medicine *308*, 681–685.
43. ROBINS-BROWNE, R.M., ELLIOTT, E. and DESMARCHELIER, P. 1998. Shiga Toxin-Producing *Escherichia coli* in Australia, *In Escherichia coli* O157:H7 and Other Shiga Toxin-Producing *E coli* Strains (Kaper, ed.) p. 66–72. ASM Press, Washington.
44. ROWE, P., ORRIBINE, E. WELLS, G. and MCLAINE, P. 1991. The Epidemiology of hemolytic uremic syndrome in Canadian children 1986-1988. J. Pediatrics. *119*, 218–224.
45. SPIKA, J., PARSONS, J., NORDENBERG, D., WELLS, J., GUNN, R. and BLAKE, P. 1986. Hemolytic uremic syndrome and diarrhoa associated with *Escherichia coli* O157:H7 in a day care centre. J. Pediatrics *109*, 287–291.
46. STROCKBINE, N.A., WELLS, J., BOPP, C. and BARRETT, T. 1999. Overview of Detection and Sub-typing methods, *In Escherichia coli* O157:H7 and Other Shiga Toxin-Producing *E. coli* Strains (J. Kaper and A. O'Brien, eds.) p. 331–328. ASM, Washington.
47. TEO, Y., RAYNOR, T., ELLAJOSYULA, K. and KNABEL, S. 1996. Synergistic effect of high temperature and high pH on the destruction of Salmonella enteritidis and *Escherichia coli* O157:H7. J. Food Protection *59*, 1023–1030.

48. TILDEN, J. *et al.* 1996. A new route of transmission for *Escherichia coli*: infection from dry fermented salami. American J. Public Health *86*, 1146–1145.
49. TREVENA, W., HOOPER, R., WRAY, C., WILLSHAW, G., CHEASTY, T. and DOMINGUE, G. 1996. Verocytotoxin-producing *Escherichia coli* O157 associated with companion animals. Veterinary Research *138*, 400.
50. WALLACE, J., CHEASTY, T. and ROWE, B. 1997. Isolation of Vero cytotoxin-producing *Escherichia coli* O157 from wild birds. J. Appl. Microbiol. *82*, 399–404.
51. WANG, G. and DOYLE, M. 1996. Survival of *Escherichia coli* O157:H7 in drinking and recreational water. Abstracts of the Annual Meeting of the International Association of Milk, Food and Environmental Sanitarians: 78.
52. WANG, G. and DOYLE, M.P. 1998. Survival of enterohemorrhagic *Escherichia coli* O157:H7 in water. J. Food Protection *61*, 662–667.
53. WARBURTON, J., CHEASTY, T. and JONES, K. 1998. Survival and recovery of *Escherichia coli* O157:H7 in inoculated bottled water. J. Food Protection *8*, 948–952.
54. WASTESON, Y. 1999. Survival and growth of *E. coli* O157 and non-O157 VTEC in meat and meat products, and the influence of stresses in this environment on their survival: a short review and potential for future research. Conferences proceedings on "Survival and Growth of Verocytotoxigenic *E. coli*" organised by an EU Concerted Action on VTEC (CT98-3935) at Agricultural University of Athens, May 6-8th. P. 94–104.
55. WILLSHAW, G., SMITH, H., ROBERTS, D., THIRWELL, J., CHEASTY, T. and ROWE, B. 1993. Examination of raw beef products for the presence of Vero cytotoxin-producing *Escherichia coli*, particularly those of serotype O157. J. Applied Bacteriology *75*, 420–426.
56. WILLSHAW, G.A., THIRWELL, J., JONES, A.P., PARRY, S., SALMON, R.L. and HICKEY, M. 1994. Verocytotoxin-producing *Escherichia coli* 0157 in beef burgers linked to an outbreak of diarrhoea, haemorrhagic colitis and haemolytic uraemic syndrome in Britain. Letters in Appl. Microbiol. *19*, 304–307.
57. ZHAO, T. and DOYLE, M. 1994. Fate of enterohemorragic *Escherichia coli* O157:H7 in commercial mayonnaise. J. Food Protection *57*, 780–783.

CHAPTER 24

VEROCYTOTOXIGENIC *E. COLI* – LEGAL ASPECTS

RAYMOND ELLARD

Food Safety Authority of Ireland
Abbey Court
Lower Abbey St.
Dublin 1, Ireland

INTRODUCTION

Keeping on the right side of the law is a complex affair. It is often stated that ignorance is no defence, but given the seemingly ever growing and changing body of rules, the ignorant deserve some sympathy. Enforcement agencies are not always as sympathetic as one might wish.

Often, when referring to the law, little distinction is drawn between the wider legal system and the actual body of laws that form a vital part of that system. Further distinctions are possible within the body of laws as they are derived from different sources, such as common law, the constitution or legislation. The law can also be further considered in divisions such as criminal law, constitutional law or private law. Contract law is the body of law that governs voluntary relationships between two or more parties and identifies agreements that the law considers to be binding. The law of torts is concerned with private wrongs including negligence and its main purpose is to provide compensation for an injured party. Criminal law defines conduct that is prohibited and provides punishment for breach of its prohibitions.

Given the ubiquitous nature of Verocytotoxigenic *E. coli* (VTEC) and its potential to cause harm, the possibility therefore exists for it to be the cause or focus of legal action involving, for instance, breach of contract, negligence or criminal charges. Coverage of all of these areas is beyond the scope of this chapter. Rather this chapter will focus on legislation which impinges on VTEC. Legislation, also called statute law, is law (Acts) enacted by the governments, while secondary legislation (statutory instruments) is made by bodies or individuals (usually Ministers) on whom that power is conferred by statute.

It may come as no surprise that the references to VTEC in legislation are few and far between. Nevertheless, that does not imply that the legal controls on its presence are equally few. Much of national legislation in Europe contains provisions, the broad effect of which are to provide for preventive measures against VTEC, or for reactive measures where its presence denotes a threat to individual or public health.

There are many tools in use to combat the threat posed by VTEC. Legislation is one such tool. It should also be noted that legislation is generally reactive and is often a compromise. The law can provide a high level of protection but it may not be an ideal solution to a problem.

The legislation, which governs VTEC, can be considered under five broad headings

- Disease Surveillance
- Food Safety
- Water
- Product Liability and Product Safety
- Health, Safety and Welfare at Work

DISEASE SURVEILLANCE

Arrangements at European Level

Since the adoption of Article 129 of the Treaty establishing the European Community, the Member States, through the Council of Health Ministers and the Commission have taken more direct coordinated action in the field of health.

A decision (Decision 2119/98/EC) was made to establish a permanent network of epidemiological surveillance at Community level. Only specified categories of communicable diseases are included in the network, but VTEC could fall into any three of the specified categories namely

- Food-borne diseases
- Water-borne diseases
- Nosocomial infections

The network is used for two main purposes

- Epidemiological surveillance, and
- An early warning and response system for the prevention and control of these diseases.

A European Decision is binding on those to whom it is addressed; in this case the decision is addressed at Member States. Depending on the situation, the decision envisages coordinated community wide action being taken as appropriate.

Surveillance

For surveillance purposes, the network brings into direct communication the European Commission and existing structures in the Member States that collect information on communicable disease. For early warning, the system will connect the public health authorities in each Member State.

Early Warning

A further European Decision (2000/57/EC) sets down the event to be reported and the procedures for information, consultation and cooperation under the early warning and response system. The Decision provides for the following type of events to be reported:

- Outbreaks extending to more than one Member State.
- Clustering of cases if pathogenic agents are a possible cause and there is a risk of propagation between Member States.
- Clustering of cases of disease of a similar type outside the Community, if pathogenic agents are a possible cause and there is a risk of propagation to the Community.
- The appearance or resurgence of a communicable disease or an infectious agent which may require timely, coordinated Community action to contain it.

The procedures to be followed are dependent on the level of risk. Three categories of action level are described:

- Activation level 1: information exchange
- Activation level 2: potential threat
- Activation level 3: definite threat

In addition, when an event arises, Member States must without delay provide suitable information material to concerned professionals and the general public, and inform them of any control measures adopted.

The current European Communicable disease network, is a "network of networks", based on a number of disease specific networks (Table 1).

RASFF

The RASFF (European Rapid Alert System for Foodstuffs) is additional to the disease specific networks. It is operated by the European Commission (DG SANCO) to rapidly inform Member States about problems or risks concerning food which does not meet food safety requirements, or which is improperly labelled and as such poses a risk to consumers. The system is limited to food products that pose a risk to health that goes beyond the territory of a single Member State.

The legal basis for this system is found in the Council Directive 92/59/EC on general product safety in particular Article 8. Member States are obliged to notify the Commission when they have knowledge or suspicion that

- A food stuff poses a serious risk to health and safety of consumers, and
- It is probable that the foodstuff is on the market in another Member State

The Commission circulates the information and Member States must take such follow up action as is necessary.

TABLE 1.
EU SURVEILLANCE NETWORKS ON INFECTIOUS DISEASES

EU Network	Function
EWGLI	European travel related legionnaires disease network
Enter-net	International human gastrointestinal infections network
EISS	European Influenza Surveillance Scheme
Euro-GROG	European Influenza network
Euro TB	European TB network
EARSS	European Anti-microbial Resistance Surveillance System
EuroHIV	European HIV monitoring network

Zoonoses

- *Council Directive 92/117/EEC concerning measures for the protection against specified zoonoses and specified zoonotic agents in animals and products of animal origin in order to prevent outbreaks of food-borne infections and intoxications*

This Directive has a two step approach. Firstly, Member States must collect information on the epidemiology of various zoonoses and secondly, based on that information, proposals for appropriate control measures are foreseen. Zoonosis is defined as any disease and/or infection which is likely to be naturally transmitted from animals to man. Each year Member States must make a detailed submission to the Commission on the prevalence of zoonoses and zoonotic agents.

FOOD SAFETY

Food Business

The primary responsibility for food safety rests with the operator of a food business, and there are numerous rules on food safety applicable to the food industry. Most rules originate from European Directives, which now amount to over twenty separate legal texts on hygiene alone. All of the rules, in one way or another, are principally designed to achieve the same purpose, the production of safe wholesome food in clean and suitable premises. The rules generally address all the stages of production and similar aspects of safety such as design of premises, staff hygiene and training, handling, processing, storage, distribution, transport and cleaning. In recent years the hygiene rules have been supplemented with the additional stipulation that businesses adopt the HACCP (Hazard Analysis and Critical Control Point) system, and HACCP is now a legal requirement. Food safety rules are changing from the prescriptive approach of the past, to a new preventative approach of setting objectives and placing responsibility clearly on operators to develop, control and check their own food safety systems.

Therefore, if VTEC is a potential hazard in the business of any food business operator, it must by law be identified and managed, adequate controls must be put in place, and these must be monitored and verified.

Simplification of EU Legislation

In the recent White Paper on Food Safety, the European Commission announced its intention to create within Europe a coherent and transparent set of food safety rules.

Among the proposals are the developments of a General Food Law, which will embody the principles of food safety, including:

- farm to table approach
- primary responsibility for safety rests with food operators
- traceability
- risk analysis
- application of the precautionary principle to risk management
- use of science and other legitimate factors in decision making, and
- efficient controls and effective enforcement

Directives from Brussels have been developed piecemeal and augmented over a considerable period. Separate rules were developed addressing different sectors of the industry. Two separate Directorate Generals (DGs) have been involved, resulting in what many might argue were separate approaches to food hygiene and different hygiene regimes according to whether the food is of animal or plant origin. The Commission now intend to address this issue, and to introduce a new comprehensive Regulation that will recast existing legal requirements to introduce consistency and clarity throughout the food production chain.

The proposal will merge the current "horizontal" directive on Food Hygiene (93/43/EEC) with a number of "vertical" directives including those on fresh meat, fresh poultry meat, pigmeat from third countries, meat products, mince meat, milk and milk products, egg products, bivalve molluscs, fishery products, rabbit and game and wild game.

In so doing, a single approach to hygiene will be adopted across all sectors for foods of both animal and non animal origin. There will also be a single approach to HACCP with the application of all seven of its principles. So no matter what sector a food operator is in, the rules will be clear, given that the proposal is to be in the form of Regulation and not a Directive. The proposal will also contain provisions relating to on-farm hygiene. The proposed Regulation also envisages the increased development of guides to good hygiene practice.

Guides to Good Hygiene Practice

There has been a whole new approach to the way food hygiene and safety rules are written. At one time, regulations were quite prescriptive in that the "do's and don'ts" were set down in detail. Now the law lays down objectives which have to be reached, but does not prescribe, in precise detail, how those objectives are to achieved. This gap can be bridged by a guide to good practice. Directive 93/43/EEC on the hygiene of foodstuffs provides that guides should be developed within each food sector, or by a national standards body, and must be approved by the regulatory authorities as properly addressing legal requirements.

Directive 93/43/EEC also provides for the development of guides at European level, but so far none have been developed in this manner.

Powers of Authorised Officers

European directives generally provide that inspection officials should have adequate powers to carry out their duties. For instance the Directive on the official control of foodstuffs (89/397/EEC) requires Member States to ensure that inspectors have the right to carry out the controls specified in the directive. In all of the existing national regulations that transpose European Directives, enforcement officials are provided with general powers of entry, inspection and sampling. In some cases detention, seizure and destruction of unfit food may be warranted.

Official Control Laboratories

All Member States of the European Union are obliged to have an official food control system, including official food laboratories. It is a legal requirement (under Directive 93/99/EEC) that such official laboratories are accredited to European Standard EN45001. Thus official microbiological laboratories engaged in the detection of VTEC, will have to include such testing within the scope of their accreditation.

Restrictions on Employment of Infected Persons

- European Communities (Hygiene of Foodstuffs) Regulations, 1998

  These Regulations make it an offence for the proprietor of a food business to allow any person known or suspected to be suffering from, or likely to be a carrier of, a disease likely to be transmitted through food to work in any food handling area in any capacity in which there is a likelihood of directly or indirectly contaminating food with pathogenic microorganisms.

Microbiological Criteria and Standards

In many of the current European Directives applicable to food or water, microbiological criteria, i.e., guidelines or standards, have been set. Compliance with standards is obligatory. Many EU countries are now redrafting their guidelines on microbiological criteria of foods and the guidance on VTEC will be simple. In the case of ready to eat food, "not detected" in 25 g will be a satisfactory result, whereas "detected" in 25 g will be unacceptable.

Current food related European Directives that refer to VTEC include

- Council Directive 91/492/EEC (OJ L268, p1, 24/09/91) of 15 July 1991 laying down the health conditions for the production and the placing on the market of live bivalve molluscs as amended by Council Directive 95/71/EEC (OJ L332, p40, 30/12/95).
- Council Directive 91/493/EEC (OJ L268, p15, 24/09/91) of 22 July 1991 laying down the health conditions for the production and placing on the market of fishery products
- Council Directive 92/46/EEC on milk and milk products
- Council Directive 94/65/EC of 14 December 1994

Although not strictly a microbiological criteria, there is a provision within Commission Directive 91/321/EEC that infant formulae and follow-on formulae shall not contain any substance in such quantity as to endanger the health of infants and young children.

WATER SUPPLIES

European Legislation

- Council Directive 80/778/EECC of the 15 July 1980 on the quality of water intended for human consumption (OJ No. L229, 5/12/98, p1) as amended by
- 91/692/EEC (OJ No. L 377, 31/12/91, p48)
- Council Directive 98/83/EC of the 3 November 1998 on the quality of water intended for human consumption (OJ No. L 331, 5/12/98, p32)

Currently the quality of water is governed by Council Directive 80/778EEC. This has been revised by Council Directive 98/83/EC on the quality of water intended for human consumption. The latter directive revises and tightens existing *E. coli* standards, to take account of the latest scientific and medical advice and World Health Organisation recommendations. Member States are required to transpose the new Directive into law by the end of 2000 and to meet most of its standards by the end of 2003.

The objective of this Directive is to protect human health from the adverse affect of any contamination of water intended for human consumption, by ensuring that it is wholesome and clean.

Under the terms of the Directive, Member States will have to ensure that water intended for human consumption is clean and wholesome. Supplies of less than 10 m^3 a day may be exempted from the provisions of the Directive, but only provided that the population affected are informed of the exemption, and properly advised. The directive defines wholesome and clean waters as that which

(a) is free any micro-organisms and parasites which in numbers and concentrations constitute a potential danger to human health, and
(b) meets the minimum requirements set out in Annex 1, Parts A and B (microbiological and chemical parameters)

The Annex sets down the parameters and parametric values. Limits are set in the case of *Escherichia coli* (*E. coli*) as 0 per 100 ml.

The directive also sets specifications for methods of analysis. For *Escherichia coli*, the reference method is ISO9308-1. Laboratories engaged in water analysis must also have an analytical quality control system that is checked by an approved third party.

Apart from setting standards for water supplies, Member States must also monitor all water supplies, for a range of parameters, including *E. coli*. Failure to meet any parametric value, (including *E. coli*), must be investigated, remedial action taken, and where a potential danger to human health exists, the use of the supply in question must be prohibited or restricted.

Natural Mineral Water

- Directive 80/777/EEC on the approximation of the laws of the Member states relating to the exploitation and marketing of natural mineral waters, as amended by
- Directive 96/70EC of the European Parliament and of the Council on the approximation of the laws of the Member states relating to the exploitation and marketing of natural mineral waters

These directives deal with waters extracted from the ground and essentially bottled at source without treatment or with minimum treatment. There are two classes of water: "natural mineral waters" and "spring water". Both natural mineral waters and spring waters must meet the specified microbiological criteria. The waters, at source and during marketing must be free of *Escherichia coli* in any 250 ml sample.

PRODUCT LIABILITY AND PRODUCT SAFETY

Defective Products

- Council Directive 85/374/EEC on the approximation of the laws, regulations and administrative provisions of the Member States concerning liability for defective products

The main feature of this directive is that it introduced a "no fault liability". Producers are liable for damage caused by a defect in his product. Products include all "moveables". Primary agricultural products, except those that have undergone initial processing, are excluded. However given the uncertainty that surrounds this definition, it is intended to clarify it by extending the scope to include all primary agricultural products.

Consumers or purchasers of a product can seek compensation for any injury provided they can prove that

- the product was defective
- injury was caused, and
- there is a causal relationship between the defect and the injury.

Clearly there is therefore scope for legal action in the case of products contaminated with VTEC. The directive defines producers as

- The manufacturer of a finished product
- The producer of any raw material
- The manufacturer of a component part
- Any person who by putting his name, trade mark or other distinguishing feature on the product presents himself as its producer
- Any person who imports a product into the European Union for sale, hire leasing or any form of distribution, and
- The supplier, where the producers of the product cannot be identified.

A product is defective when it does not provide the safety that a person is entitled to expect, taking all circumstances into account including the presentation of the product, the use to which it could reasonably be expected that the product would be put and the time when the product was put into circulation. Certain defences are provided for producers, such as the "state of the art" defence, that is if he/she can prove

- that the state of scientific and technological knowledge at the time he put the product into circulation was not such as to enable the existence of the defect to be discovered.

Producers are also not liable if they can prove that

- they did not put the product into circulation
- the defect did not exist at the time the product was put into circulation
- the defect is due to compliance with mandatory regulations issued by public authorities.

There is a financial threshold, which means that a successful complainant has to bear the first 500 Euros. Member States can, if they wish, limit producers' liability for death and injury caused by a defective product to Euro 70 million.

Safe Products

- Council Directive 92/59/EEC on general product safety

This directive was intended to cover products the safety of which is not governed by other more specific safety requirements. Most foodstuffs are therefore covered by this directive. The purpose of this Directive is to ensure that product placed on the market are safe. "A product" means any product intended for consumers or likely to be used by consumers, supplied whether for consideration or not in the course of commercial activity. A "safe product" is defined as any product which under normal or reasonably foreseeable conditions of use does not present any risk, or only presents minimal risks, compatible with the product's use and taking the following into account:

- The characteristics of the product, including its composition, packaging, instructions for assembly and maintenance,
- The effect on other products, where it is reasonably foreseeable that it will be used with other products,
- The presentation of the product, the labelling, any instructions for use and disposal and other indication or information provided by the producer
- The categories of consumers at serious risk when using the product, in particular children.

A "dangerous product" means any product that is not a "safe product". Producers are legally obliged to only place safe products on the market. A producer means the manufacturer of the product, where is established within the European Union, the manufacturer's representative when the manufacturer is not established in the EU, or other professionals in the supply chain, insofar as their activities may affect the safety properties of a product.

Producers are obliged, within the limits of their activities, to provide consumers with relevant information to enable them to assess the risks inherent in a product where such risks are not immediately obvious without adequate warnings. Provision of warning does not however exempt a producer from the obligation to only place safe products on the market.

Member States must establish systems to monitor compliance. It should perhaps be noted that Member States are permitted to adopt appropriate measures to temporarily prohibit marketing of products which they suspect are dangerous, or to prohibit the marketing of proven dangerous products.

It is under this Directive that the Rapid Alert System for Food was established (see section 2). The Directive sets out the obligations on Member States to inform the European Commission and also provides for emergency measures at European level.

HEALTH AND SAFETY AT WORK

Protection Against Biological Agents

- Council Directive 90/679/EEC1 of 26 November, 1990 as amended by
- Council Directive 93/88/EEC,
- Commission Directive 95/30/EC,

- Commission Directive 97/59/EC,
- Commission Directive 97/65/EC,

Rules on health and safety at work have been harmonised in Europe. There is a general obligation on employers to provide adequately for the health, safety and welfare of their employees. Health and safety has to be managed and to do this, employers must develop a safety management plan. In developing the plan, the workplace hazards must be identified and a risk assessment made of each hazard. Controls must then be selected and implemented to eliminate or reduce the risks.

Workers must be protected from risks related to exposure to biological agents at work. Biological agents are classified under European rules (in ascending order of severity) as Group 2, Group 3 and Group 4 agents. Verocytotoxigenic strains of *Escherichia coli* have been classified as Group 3. This defines it as a biological agent that can cause severe human disease, presents a serious hazard to employees and/or may present a risk of spreading to the community, although effective prophylaxis or treatment is usually available.

It is therefore incumbent on employers, in particular laboratories engaged in the detection and isolation of VTEC, to carry out a risk assessment on the likely exposure of staff to those strains. Where risks of exposure are revealed by the assessment, such exposure must be prevented and, if this is not technically practicable, the appropriate level of protection must be provided. The Directives specify containment levels and containment measures linked to the use of Group 2, 3 or 4 biological agents.

APPENDIX
NOMENCLATURE OF VEROCYTOTOXINS

FLEMMING SCHEUTZ, The International *Escherichia* and *Klebsiella* Centre (WHO), Department of Gastrointestinal Infections, Statens Serum Institut, 5 Artillerivej, 2300 Copenhagen S, Denmark
LOTHAR BEUTIN, Division of Emerging Bacterial Pathogens (P13), *Escherichia coli* Reference Lab, Robert Koch-Institut, Nordufer 20, D13353 Berlin, Germany
DENIS PIÉRARD, Vrije Universiteit Brussel, Academisch Ziekenhuis, Microbiology, Laarbeeklaan 101, Brussels, B-1090, Belgium
HENRY R. SMITH, Laboratory of Enteric Pathogens, Central Public Health Laboratory, 61, Colindale Avenue London, NW9 5HT England

Nomenclature of Verocytotoxins

Many different designations have been used in the literature to describe and differentiate Verocytotoxins, which has caused much confusion in a number of areas. At present, there are two "schools of nomenclature": based on Shiga toxin terminology, or Verocytotoxin terminology. Table 1 presents known toxin genes and toxins, and an updated proposal for these two terminologies. Previous designations in the literature including the original Shiga-Like Toxin (SLT) nomenclature are listed as synonyms.

This Appendix covers all the existing nomenclature, and presents an updated version of the two currently accepted kinds of terminology as a compromise which should not offend the broad spectrum of researchers working in this field of study.

Present criteria for inclusion as one of the suggested toxin types i.e. 1, or 2 to 2f are: (1) Antigenic variability, (2) Difference in toxicity for culture cells and/or animals, (3) The capacity to be activated, (4) Difference in affinity and use of receptor (due to differences in the B subunit of the toxin), (5) Significant differences in DNA or amino acid sequences

Information on the association between the different toxins, the epidemiology and the clinical course and outcome of disease in humans, is currently limited. We suggest, in future, emerging information on epidemiological traits and/or associations with disease should form additional criteria in assigning toxins to existing types of Verocytotoxins.

The distinction between toxin types 2c and 2d in Table 1 is based on differences in the A subunit determining whether the toxin is activatable (Stx2d/VT2d) or non-activatable (Stx2c/VT2c). The other subtypes described in the literature which have not been tested for all properties have been tentatively grouped with toxin type 2c, primarily based on similarities in nucleotide sequences that places them in a phylogenetically related Stx2/VT2 cluster including all variants. DNA homology is very high (>95%) between Stx2/VT2, Stx2c/VT2c (including variants Stx2-O22/VT2-O22, Stx2-O157-TK-51/VT2-O157-TK-51, Stx2-OX3/b-O31/VT2-OX3/b-O31, Stx2-O48/VT2-O48) and Stx2d/VT2d in what has been referred to as group 1 (11). On the other hand, Stx2-OX3/a-O31/VT2-OX3/a-O31, Stx2-O111-PH/VT2-O111-PH, and Stx2-O118/VT2-O118 in what has been referred to as group 2 (11), share a high degree of DNA homology, which is again distinctly lower (<95%) with group 1 (11,21). The extent of, and differences between, toxicity for tissue culture cells and/or animals and their capacity to be activated are not fully established for all the types. Further analysis against all the criteria necessary to allow definitive placement with appropriate other toxin types is required. In the absence of a comprehensive nomenclature, suffixes are added after the O group of source organisms, and, when necessary, the original strain designation. Stx2/VT2 subtype toxins found in the same original strain are suffixed /a, /b, etc., as in Stx2-OX3/a-O31/VT2-OX3/a-O31 and Stx2-OX3/b-O31/VT2-OX3/b-O31.

TABLE 1.
DESIGNATIONS OF *stx*/*vtx* GENES AND THEIR PRODUCTS (Stx/VT) [§]

Toxin Type*	Toxin gene	Toxin	Prototype Organisms (reference)	Serotype of prototype organism	Previous designation or synonym for toxin gene	Previous designation or synonym for toxin
1	stx_1/*vtx*1	Stx1/VT1	EDL933 (1)	O157:H7	*slt*-I	SLT-I
			H19 (2)	O26:H11		
			H30 (2)	O26:H11		
	$stx_{1\text{-O103}}$/*vtx*1-O103	Stx1-O103/VT1-O103	PMK1 (3)	O103:H2		
	$stx_{1\text{-O111-PH}}$ /*vtx*1-O111-PH	Stx1-O111-PH /VT1-O111-PH	PH (4)	O111:H-		SltI/PH
	$stx_{1\text{-OX3}}$/*vtx*1-OX3	Stx1-OX3/VT1-OX3	131/3 (5)	OX3:H8		SltI/OX3
	$stx_{1\text{-O48}}$/*vtx*1-O48	Stx1-O48/VT1-O48	94C (5)	O48:H21		SltI/O48
	$stx_{1\text{-O111-CB168}}$/*vtx*1-O111-CB168	Stx1-O111-CB168/VT1-O111-CB168	CB168 (5)	O111:H-		SltI/CB168
2	stx_2/*vtx*2	Stx2/VT2	EDL933 (1)	O157:H7	*slt*-II	SLT-II
2c	stx_{2c}/*vtx*2c	Stx2c/VT2c	E32511 (6)	O157:H-		Stx2v/VT2v
					slt-IIc	SLTIIc
	$stx_{2\text{-O22}}$/*vtx*2-O22	Stx2-O22/VT2-O22	KY-O19 (7)	O22:H-		VT2v(pKTN1054)
	$stx_{2\text{-O157-TK-51}}$/*vtx*2-O157-TK-51	Stx2-O157-TK-51/VT2-O157-TK-51	TK-51 (7)	O157:H7		VT2v(pKTN1050)
	$stx_{2\text{-OX3/a-O31}}$/*vtx*2-OX3/a-O31	Stx2-OX3/a-O31/VT2-OX3/a-O31	O31 (8)	OX3:H21	$stx_{2vOX392}$	SLT-II/OX3 / VT2d-OX3 [†]
	$stx_{2\text{-OX3/b-O31}}$/*vtx*2-OX3/b-O31	Stx2-OX3/b-O31/VT2-OX3/b-O31	O31 (9)	OX3:H21	$stx_{2vOX393}$	SLT-II/OX3/2 / VT2d-OX3/2 [†]
	$stx_{2\text{-O48}}$/*vtx*2-O48	Stx2-O48/VT2-O48	94C (10)	O48:H21		SLT-II/O48
	$stx_{2\text{-O111-PH}}$ /*vtx*2-O111-PH	Stx2-O111-PH /VT2-O111-PH	CB168 (10)	O111:H-	stx_{2vO111}	SLT-II/O111 / VT2d-O111 [†]
	$stx_{2\text{-O118}}$/*vtx*2-O118	Stx2-O118/VT2-O118 **	EH250 (11)	O118:H12		Stx2d/VT2d-Ount
2d [†]	stx_{2d} /*vtx*2d	Stx2d/VT2d	B2F1 (12)	O91:H21		SLT-IIvh
	stx_{2d1}/*vtx*2d1	Stx2d1/VT2d1	B2F1 (12)	O91:H21	*stx*2vha/*vtx*2vha	Stx2vha/VT2vha
					*stx*2ha	Stx2vh-a/VT2vh-a
						VT2v-a
						SLT-IIvha
	stx_{2d2} /*vtx*2d2	Stx2d2/VT2d2	B2F1 (12)	O91:H21	*stx*2vhb/*vtx*2vhb	Stx2vhb/VT2vhb
					*stx*2hb	Stx2vh-b/VT2vh-b
						VT2v-b
						SLT-IIvhb

Toxin Type*	Toxin gene	Toxin	Prototype Organisms (reference)	Serotype of prototype organism	Previous designation or synonym for toxin gene	Previous designation or synonym for toxin
2e	*stx*$_{2e}$/*vtx*2e	Stx2e/VT2e	412 (13) S1191 (14)	O139:K12:H1 O139	*slt*-IIv *slt*-IIva *slt*-IIe	SLT-IIv SLT-IIva ‡ SLTIIe/VTe VT2vp VT2vp1
2f ¤	*stx*2f/ *vtx*2f	Stx2f/VT2f	H.I.8 (15) T4/97 (16)	O128:H2 O128:H2	*stx*$_{2ev}$/*vtx*2ev *slt*-IIvhc *slt*-IId †	Stx2ev/VT2ev Stx2vp2/VT2vp2 VTev SLTIIvhc SLT-IId/VT2d †

§ Data compiled from references (7,12,15,17-26).

* Toxin types are defined according to antigenic variability, differences in toxicity for tissue culture cells and/or animals, their capacity to be activated by mouse elastase and differences in DNA or amino acid sequences.

** Stx2-O118/VT2-O118 (formerly known as VT2d-Ount) is non-activatable (Denis Piérard and Alison O'Brien; unpublished); The original strain has been re-typed as O118:H12 (Lothar Beutin and Flemming Scheutz, unpublished)

† There are several toxins suffixed by d in the literature: The Stx2d/VT2h toxins of O91:H21 (27), the VT2d (= *stx*$_{\text{2-Ogroup/ strain designation and/or year}}$) variants by Paton, A.W. *et al.*, (8-10), Piérard, D. *et al.* (11) and the SLT-IId/VT2d (= Stx2f/VT2f) toxin produced by strain H.I.8 (serotype O128:H2) as proposed by Gyles, C.L. (26). We support that the d designation is reserved to activatable Stx2/VT2 toxins as proposed by Melton-Celsa, A.R. (27).

‡ This designation has also been referred to for the Stx2f/VT2f toxin produced by strain H.1.8 (serotype O128:H-).

¤ The nucleotide sequence of the former *stx*$_{2ev}$/*vtx*2ev of strain H.I.8 (serotype O128:H2) is nearly identical to the recently published *stx*2f/*vtx*2f found in strain T4/97 (serotype O128:H2) from feral pigeons (16). As its nucleotide sequence is distinctly different from both the other Stx2/VT2 toxins and variants and Stx1/VT1 we support the proposal of renaming *stx*$_{2ev}$/*vtx*2ev as *stx*$_{2f}$/*vtx*2f

REFERENCES

1. O'BRIEN, A.D., NEWLAND J.W., MILLER S.F., HOLMES R.K., SMITH H.W. and FORMAL S.B. 1984. Shiga-like toxin-converting phages from *Escherichia coli* strains that cause hemorrhagic colitis or infantile diarrhea. Science *226*, 694–696.
2. KONOWALCHUK, J., SPEIRS, J.I. and STAVRIC, S. 1977. Vero response to a cytotoxin of *Escherichia coli*. Infect. Immun. *18*, 775–779.
3. MARIANI-KURKDJIAN, P. *et al.* 1993. Identification of a clone of *Escherichia coli* O103:H2 as a potential agent of hemolytic-uremic syndrome in France. J. Clin. Microbiol. *31*, 296–301.
4. PATON, A.W., PATON, J.C., GOLDWATER, P.N., HEUZENROEDER, M.W. and MANNING, P.A. 1993. Sequence of a variant Shiga-like toxin type-I operon of *Escherichia coli* O111:H-. Gene *129*, 87–92.
5. PATON, A.W., BEUTIN, L. and PATON, J.C. 1995. Heterogeneity of the amino-acid sequences of *Escherichia coli* shiga-like toxin type-I operons. Gene *153*, 71–74.
6. SCHMITT, C.K., MCKEE, M.L. and O'BRIEN, A.D. 1991. Two copies of shiga-like toxin II-related genes common in enterohemorrhagic *Escherichia coli* strains are responsible for the antigenic heterogeneity of the O157:H⁻ strain E32511. Infect. Immun. *59*, 1065–1073.
7. LIN, Z. *et al.* 1993. Cloning and sequencing of 2 new verotoxin-2 variant genes of *Escherichia coli* isolated from cases of human and bovine diarrhea. Microbiol. Immunol. *37*, 451–459.
8. PATON, A.W., PATON, J.C., HEUZENROEDER, M.W., GOLDWATER, P.N. and MANNING, P.A. 1992. Cloning and nucleotide sequence of a variant Shiga-like toxin II gene from *Escherichia coli* OX3:H21 isolated from a case of sudden infant death syndrome. Microb. Pathogen. *13*, 225–236.
9. PATON, A.W., PATON, J.C. and MANNING, P.A. 1993. Polymerase chain reaction amplification, cloning and sequencing of variant *Escherichia coli* Shiga-like toxin Type-II Operons. Microb. Pathogen. *15*, 77–82.
10. PATON, A.W., BOURNE, A.J., MANNING, P.A. and PATON, J.C. 1995. Comparative toxicity and virulence of *Escherichia coli* clones expressing variant and chimeric shiga-like toxin type II operons. Infect. Immun. *63*, 2450–2458.
11. PIÉRARD, D., MUYLDERMANS, G., MORIAU, L., STEVENS, D. and LAUWERS, S. 1998. Identification of new verocytotoxin type 2 variant B-Subunit genes in human and animal *Escherichia coli* isolates. J. Clin. Microbiol. *36*, 3317–3322.
12. ITO, H., TERAI, A., KURAZONO, H., TAKEDA, Y. and NISHIBUCHI, M. 1990. Cloning and nucleotide sequencing of vero toxin 2 variant genes from *Escherichia coli* O91:H21 isolated from a patient with hemolytic uremic syndrome. Microb. Pathogen. *8*, 47–60.

13. GYLES, C.L., DE GRANDIS, S.A., MACKENZIE, C. and BRUNTON, J.L. 1988. Cloning and nucleotide sequence analysis of the genes determining verocytotoxin production in a porcine edema disease isolate of *Escherichia coli*. Microb. Pathogen. 5, 419–426.
14. WEINSTEIN, D.L., JACKSON, M.P., SAMUEL, J.E., HOLMES, R.K. and O'BRIEN, A.D. 1988. Cloning and sequencing of a Shiga-like toxin type II variant from *Escherichia coli* strain responsible for edema disease of swine. J Bacteriol. *170*, 4223–4230.
15. GANNON, V.P., TEERLING, C., MASRI, S.A. and GYLES, C.L. 1990. Molecular cloning and nucleotide sequence of another variant of the *Escherichia coli* Shiga-like toxin II family. J. Gen. Microbiol. *136*, 1125–1135.
16. SCHMIDT, H., SCHEEF, J., MORABITO, S., CAPRIOLI, A., WIELER, L.H. and KARCH, H. 2000. A new Shiga toxin 2 variant (Stx2f) from *Escherichia coli* isolated from pigeons. Appl. Environ. Microbiol. *66*, 1205–1208.
17. KOKAI-KUN, J.F., MELTON-CELSA, A.R. and O'BRIEN, A.D. 2001. Elastase in intestinal mucus enhances the cytotoxicity of Shiga toxin type 2d. J. Biol. Chem. *275*, 3713–3721.
18. TYLER, S.D., JOHNSON, W.M., LIOR, H., WANG, G. and ROZEE, K.R. 1991. Identification of verotoxin type 2 variant B subunit genes in *Escherichia coli* by the polymerase chain reaction and restriction fragment length polymorphism analysis. J. Clin. Microbiol. *29*, 1339–1343.
19. LIN, Z., KURAZONO, H., YAMASAKI, S. and TAKEDA, Y. 1993. Detection of various variant verotoxin genes in *Escherichia coli* by polymerase chain reaction. Microbiol. Immunol. *37*, 543–548.
20. TAKEDA, Y., KURAZONO, H. and YAMASAKI, S. 1993. Vero toxins (Shiga–like toxins) produced by enterohemorrhagic *Escherichia coli* (verocytotoxin-producing *E. coli*). Microbiol. Immunol. *37*, 591–599.
21. BASTIAN, S.N., CARLE, I. and GRIMONT, F. 1998. Comparison of 14 PCR systems for the detection and subtyping of stx genes in Shiga-toxin-producing *Escherichia coli*. Res. Microbiol. *149*, 457–472.
22. JOHNSON, W.M., TYLER, S.D., WANG, G. and LIOR, H. 1991. Amplification by the polymerase chain reaction of a specific target sequence in the gene coding for *Escherichia coli* verotoxin (VTe variant). FEMS Microbiol. Lett. *68*, 227–230.
23. CALDERWOOD, S.B. *et al.* 1996. Proposed new nomenclature for SLT (VT) family. ASM News *62*, 118–119.
24. KARMALI, M.A., LINGWOOD, C.A., PETRIC, M., BRUNTON, J. and GYLES, C. 1996. Maintaining the existing phenotype nomenclatures for *E. coli* cytotoxins. ASM News *62*, 167–169.
25. SCOTLAND, S.M. and SMITH, H.R. 1997. Vero cytotoxins. *In Escherichia coli*. Mechanisms of Virulence. First Ed. (M. Sussman, ed.) Cambridge: The Press Syndicate of the University of Cambridge. pp. 257–280.
26. GYLES, C.L. 1994. *Escherichia coli* verotoxins and other cytotoxins. *In Escherichia coli* in Domestic Animals and Humans. (C.L. Gyles, ed.) Wallingford: CAB International. pp. 365–398.

27. MELTON-CELSA, A.R. and O'BRIEN, A.D. 1998. Structure, biology, and relative toxicity of Shiga toxin family members for cells and animals. *In Escherichia coli* O157:H7 and Other Shiga Toxin-Producing *E. coli* Strains. 1st Ed. (J.B. Kaper and A.D. O'Brien, eds.) pp. 121–128. American Society for Microbiology Press, Washington, DC.

INDEX

Abattoir, 123
Acid adaptation, 311
Acid resistance, 267
Acid survival, 314
Actin, 216
Aerosol, 363
AFLP, 100
Alfalfa sprouts, 169
Animal hide, 284
Antibiotic treatment, 28
Antibiotics, 5, 401
Antimotility agents, 400
Anuric acute renal failure, 229
Apoptosis, 248, 250, 252
Apple cider, 169, 309, 313, 381
Argentina, 166
Australia, 167
Austria, 163
Automated PCR, 32, 77

Bacteriophage, 245
  typing, 98
Baranyi model, 325
Beef burger patties, 310
Belgium, 163
Birds, wild, 227
Bloody diarrhea, 213, 227, 229
Bottled waters, 309
Brazil, 128, 166
Buffalo, 222
Buffered peptone water, 60
Butter, 388

Caecum, 360
Calf feeders, 116
Calf-barns, 116
Calves, 213, 358
Camembert cheese, 208, 388
Cameroon, 167
Canada, 124, 164
Canary Islands, 164
Cantaloupe, 186
Captivate™ O157, 63
Carrots, 309
Cats, 153, 227-228
Cattle farms, 116
  grading, 370
  transportation, 368
  troughs, 289
Cefixime-tellurite SMAC, 65
Cheese, 201, 388
Chicken, 155, 227, 272
Chile, 152, 166
Chilling, 306, 371
Chlorine, 383, 385
Chromogenic agar, 66, 93
Ciprofloxacin, 401
Clinical features, 396
Clinical management, 393
Clinical material, 28
Coccidia, 215
Coleslaw, 169
Colon, 360
Colony DNA hybridisation assays, 37
Colony hybridisation assay, 38
Colony immunoblot, 37
Commercial kits, 40, 95
Competitor microflora 62
Containment Level 3, 429
Coronavirus, 215
Cotrimoxazole, 401
Cow feeders, 116
Cream, 388
Creche, 163
Critical Control Points, 368
Croatia, 164
Cryptosporidia, 215
Cucumber, 309
Cutaneous ulcers, 229
Cytoskeleton, 216-217
Cytotoxicity assays, 33
Czech Republic, 122, 163

Dairy products, 182, 188, 308
Deboning, 373
Deer, 222, 272
  meat, 184
Deterministic analyses, 340
Diarrhoea, 396
Dietary factors, 116
Direct immunofluorescent filter techniques, 72
Disinfection, 358
Dogs, 153, 227-229, 236
Dose-Response, 338
Dynabeads, 19, 63

E. coli broth, 59
E. hermannii, 34, 65
Ecology, 114, 266, 282
ELISA, 34, 72, 96
Enterohaemolysin, 219-220, 243
  agar, 67
Enterohemolysin test, 36
Enterohemolysin-Agar, 45
Enterotoxemia, 150
EOH medium, 67
European directives, 439
Evisceration, 369
Evolution, 2
Exposure assessment, 330, 338

F18 fimbrial factor, 151
F18 pili, 231-232
F18 receptor locus, 232
Faecal leukocytes, 28
Fairs, 170
False positives 17, 69, 76
Farm, 282
  environment, 283
  visits, 170
Farmhouse cheese, 209, 309
Feed, 116, 358
Fermented meats, 184, 313
Fimbriae, 216
Finland, 163
Fish, 156
Flies, 156
Flow cytometry, 65
Fluorocult™ *E. coli* O157:H7 agar, 68
Formaldehyde, 427
Fosfomycin, 401
Foxes, 156
France, 163
Freezing, 310
Fruit, 168, 309
Fruit juices, 186
Fumigation, 426

Gb3, 231, 247-248, 250-252
Gb4, 231, 247
Germany, 163
GN broth (Hajna formulation), 61
Goats, 135, 221, 272
  milk cheese, 206
Gompertz function, 325
G-proteins, 245
Grass, 286
Greyhounds, 229
Ground water, 288
Gulls, 155, 227

HACCP, 367, 439
Hamburgers, 182, 341
Hazard Group 3 pathogens, 421
Health and safety, 445
Healthy cattle, 269
Heat resistance, 310
Heat-stable toxin, 220
Hemorrhagic coli agar, 68
Herd prevalence, 264
High efficiency particulate filters, 426
Hong Kong, 167
Horses, 153, 227-228, 234
Hungary, 163
Hydroponic cultivation, 309

Immunisation, 291
Immunoblot assays, 19, 72
Immunomagnetic separation, 12, 19, 63, 93
Interspecies transmission, 154
Intimin, 217, 222
Iran, 167
Ireland, 123, 163
Irradiation, 376
Israel, 164
Italy, 164, 264

Japan, 166
Jerky, 184
Juice, 348

Laboratory acquired infections, 430
Laboratory design, 423
Laboratory procedures, 427
Lairage, 368
Lake-water, 170
Lambda-like phage, 4
Laser cytometry, 93
Latex agglutination, 96
  kits, 35, 37, 69
Lauryl tryptose broth, 60
LEE, 216-217, 221
Legal aspects, 435
Lettuce, 169, 185, 381
Linear regression techniques, 324
Lithuania, 164
Long-term sequelae, 403

Manure, 116, 283
Mayonnaise, 186, 381
Meat processing, 373
Mesenteric lymph nodes, 218
Mettwurst, 185
Microvilli effacement, 217
Milk, 182, 188
Mineral water, 310
Modelling, 324, 327, 329
Modified atmosphere, 376
Modified atmosphere packaging, 315
Modified eosin methylene blue agar, 67
Modified tryptone broth, 18
Monte Carlo Simulation, 340
Multilocus enzyme electrophoresis, 98
Multilocus sequence typing, 100
Multiplex PCR, 32, 39, 95
Musca domestica, 156
Music festivals, 170
Mussels, 156

Necropsy lesions, 215
Netherlands, 163
New Zealand, 167
Non-linear regression techniques, 324
Norway, 122, 163

Outbreak Management, 402
Oxyrase®, 63

Pasteurised milk, 387
Pasteurization, 349
Pathology, 234
Patient information, 29
PCR inhibitors, 31
Pepperoni, 311
Petrifilm™ HEC Test, 64-66
Phage induction, 5
Phage-typing, 98
Phagocytosis, 216
Phenol red sorbitol agar with MUG, 67
Phosphoric acid, 383
Pigeons, 155
Pigs, 150, 227, 229-234
Plasmid analysis, 98

Pneumovirus, 155
pO157, 218, 243-245, 247
Poland, 164
Post-weaning diarrhoea, 152
Potassium sorbate, 383
Potatoes, 169, 381
Poultry, 155, 272
Predictors of HUS/TTP, 400
Preservatives, 314
Product withdrawal, 173
Prognostic indicators, 399
Protozoa, 288
Pulsed-field gel electrophoresis, 99

Q transcription activator, 245

Rabbit, 156
Radish sprouts, 185, 309
Rainbow™ agar O157, 66-67
Rats, 156
Raw milk, 308, 387
RFLP, 99
Risk assessment, 330
Risk characterisation, 330, 340
Risk factors, 164
Risk groups, 28
Roast beef, 182
Rodding, 369
Rotavirus, 215
rpoS, 314

Sakai City, 166, 245
Salad dressings, 313
Salad vegetables, 385
Salami, 311
Salt water, 170
Sample receipt, 423
Sample transport, 29
Sampling, 26
Scandinavia, 163
Scotland, 162
Sea lion, 229
Seasonal variation, 120-121, 265
Seeds, 386
Semi-soft cheeses, 388
Sentinel centres, 172
Serbia, 164
Serine protease, 219, 244
Serodiagnosis, 43, 96
Sewage, 290
Shedding, 265
  and diet, 265
  and stress, 265
  effect of housing, 123
  effect of stocking, 123
  rates, 120
Sheep, 135, 220-221, 272
Shellfish, 156
Slide agglutination, 35
Slovenia, 122, 164
Slurry, 283
Sodium benzoate, 383
South Africa, 167
South Korea, 167
Spain, 164
Spring water, 310
Sprouted vegetables, 386
Steam pasteurisation, 371
Stochastic analysis, 340
Stunning, 369
Swaziland, 167
Sweden, 122, 163
Swimming, 170-171
Switzerland, 163

Thailand, 167
Thermal inactivation, 310
Thermotolerance, 311
Thrombotic thrombocytopenic purpura, 242
Thrombotic vascular lesions, 154
Tryptone soya broth, 59

TSB with cefixime, tellurite and vancomycin, 60
Turkey, 155, 164
Turkey-roll, 184
Type II secretion system, 219, 244
Type III secretion system, 217

United Kingdom, 162, 264
United States, 124, 165, 264
Unpasteurised cheese, 388
Unpasteurised milk, 308
Uruguay, 166

Vaccination, 222, 234
Vacuum, 315
Vegetables, 168, 309
Vero cell assay, 33, 94
Verotoxin kits, 71
Viable non-culturable state, 282

Washed sheep blood agar, 36, 67
Water, 287
- supplies, 168
- troughs, 116
- well, 170-171

Watermelon, 186, 309
White Paper on Food Safety, 439
White radish sprouts, 167, 169, 381
Wild bird, 155
Wild ruminants, 113

Yoghurt, 387

www.ingramcontent.com/pod-product-compliance
Lightning Source LLC
LaVergne TN
LVHW011901200926
848484LV00002B/497